Active Filters for Integrated-Circuit Applications

DISCLAIMER OF WARRANTY

The technical descriptions, procedures, and computer programs in this book have been developed with the greatest of care and they have been useful to the author in a broad range of applications; however, they are provided as is, without warranty of any kind. Artech House, Inc., and the author of the book titled *Active Filters for Integrated-Circuit Applications* make no warranties, expressed or implied, that the equations, programs, and procedures in this book or its associated software are free of error, or are consistent with any particular standard of merchantability, or will meet your requirements for any particular application. They should not be relied upon for solving a problem whose incorrect solution could result in injury to a person or loss of property. Any use of the programs or procedures in such a manner is at the user's own risk. The editors, author, and publisher disclaim all liability for direct, incidental, or consequent damages resulting from use of the programs or procedures in this book or the associated software.

For a listing of recent titles in the *Artech House Microwave Library*, turn to the back of this book.

Active Filters for Integrated-Circuit Applications

Fred H. Irons

ARTECH HOUSE
BOSTON | LONDON
artechhouse.com

Library of Congress Cataloging-in-Publication Data
Irons, Fred H.
    Active filters for integrated-circuit applications / Fred H. Irons
        p. cm.—(Artech House microwave library)
    Includes bibliographical references and index.
    ISBN 1-58053-896-7 (alk. paper)
    1. Electric filters, Active. 2. Integrated circuits—Design and construction.
    I. Title. II. Series.
    TK 7872.F5I76 2005
    621.3815'324—dc22   2005041997

British Library Cataloguing in Publication Data
Irons, Fred H.
    Active filters for integrated-circuit applications.—(Artech House microwave library)
    1. Electric filters, Active 2. Integrated circuits—Design and construction
    I. Title
    621.3'815324
    ISBN 1-58053-896-7

Cover design by Igor Valdman

© 2005 ARTECH HOUSE, INC.
685 Canton Street
Norwood, MA 02062

All rights reserved. Printed and bound in the United States of America. No part of this book may be reproduced or utilized in any form or by any means, electronic or mechanical, including photocopying, recording, or by any information storage and retrieval system, without permission in writing from the publisher. All terms mentioned in this book that are known to be trademarks or service marks have been appropriately capitalized. Artech House cannot attest to the accuracy of this information. Use of a term in this book should not be regarded as affecting the validity of any trademark or service mark.

International Standard Book Number: 1-58053-896-7

10 9 8 7 6 5 4 3 2 1

Contents

Preface		ix
Acknowledgments		xiii
1 Introduction		**1**
1.1	Filter Terminology	2
	1.1.1 Filter Component Values	8
	1.1.2 An Active Filter Definition	9
	1.1.3 Conclusion	11
1.2	Problems	12
2 Review of Circuit Analysis Concepts		**15**
2.1	Network Parameter Matrices	15
2.2	Network Scale Factors	19
2.3	Frequency Transformations of Passive Filters	24
	2.3.1 LP to LP	24
	2.3.2 LP to HP	25
	2.3.3 LP to BP	27
	2.3.4 LP to BP Network Functions	31
	2.3.5 BP Filter Element Values	34
	2.3.6 LP to BS	38
2.4	Impedance Transformations of Passive Filters	40
	2.4.1 The Internal Loop Scaling Procedure	41
	2.4.2 The Double Terminated Network and Its Dual	47
2.5	Summary	49
2.6	Problems	51
3 Frequency Effects in Feedback Circuits		**57**
3.1	The Operational Amplifier	58
	3.1.1 Some Application Procedures	64
	3.1.2 Single-Pole Open-Loop Gain	66
	3.1.3 Double-Pole Open-Loop Gain	67
	3.1.4 Triple-Pole Open-Loop Gain	70
3.2	An Operational Amplifier Dual	72

		3.2.1 Invariant Frequency Response Example	74
	3.3	Summary	75
	3.4	Problems	76
		References	80

4 Some Opamp Design Considerations — 81

- 4.1 The Current Mirror . . . 81
  - 4.1.1 The Widlar Mirror . . . 84
- 4.2 The Differential Amplifier Input Stage . . . 87
  - 4.2.1 A Single Transistor Amplifier . . . 90
  - 4.2.2 Small-Signal Frequency Response . . . 93
  - 4.2.3 Signal Representation . . . 97
  - 4.2.4 Second-Order Effect on Amplifier Model . . . 99
- 4.3 The Second Stage . . . 101
- 4.4 The Third Stage . . . 108
  - 4.4.1 Complementary Emitter Follower Output Stage . . . 113
- 4.5 All Together Now—A Three-Stage Opamp . . . 119
  - 4.5.1 Temperature Compensation of Output Offset . . . 121
  - 4.5.2 Case Study—A Current Controlled Opamp . . . 124
- 4.6 Conclusion . . . 126
- 4.7 Problems . . . 127
- References . . . 134
- Appendix 4A Matlab and Transistor Modeling . . . 135
- 4A.1 Modeling BJT Static Response . . . 135
- 4A.2 Modeling Dynamic Response . . . 143
- 4A.3 Summary . . . 149

5 Operational Design of Active Filters — 153

- 5.1 The Opamp as a Signal Processor . . . 154
  - 5.1.1 The Buffer Voltage Follower . . . 154
  - 5.1.2 The Noninverting Multiplying Buffer . . . 155
  - 5.1.3 The Noninverting Summing Multiplier . . . 155
  - 5.1.4 The Inverting Summing Multiplier . . . 156
  - 5.1.5 The Inverting Integrator . . . 157
- 5.2 Analog Operational Circuit Example . . . 158
  - 5.2.1 Adding Transmission Zeros . . . 164
- 5.3 State Variable Filters . . . 166
- 5.4 Cascade Methods . . . 169
  - 5.4.1 The Cascade Concept . . . 170
  - 5.4.2 A Single-Amplifier Quadratic Factor Circuit . . . 183
  - 5.4.3 The Twin-T Circuit . . . 186
  - 5.4.4 The Biquad Circuit . . . 191
- 5.5 Problems . . . 200
- Reference . . . 215

CONTENTS   vii

6 Network Sensitivity and Leapfrog Filters — 217
- 6.1 A Filter Sensitivity Definition 217
  - 6.1.1 A Sensitivity Property for Terminated Passive Filters 222
- 6.2 The Leapfrog Filter Architecture 223
  - 6.2.1 Leapfrog Example of Sensitivity Performance 231
  - 6.2.2 The Elliptic Filter LP Topology 236
  - 6.2.3 The All-Pole BP Filter Topology 239
  - 6.2.4 Topology for BP Filters with Finite Transmission Zeros ... 244
- 6.3 Summary 247
- 6.4 Problems 248
  - Reference 257

7 Switched Capacitor Concepts — 259
- 7.1 The CMOS Switch or Transmission Gate 259
  - 7.1.1 The Switch Clock Rate and Sampling 262
  - 7.1.2 Switch Configurations and Parasitic Capacitance 264
  - 7.1.3 The Equivalent Resistance Concept 267
  - 7.1.4 Typical Resistance and Clock Frequencies 271
- 7.2 Switched Capacitors and Analog Operations 274
  - 7.2.1 Switched Capacitor s-Plane Distortion 287
  - 7.2.2 Precompensated Network Functions 294
  - 7.2.3 Scale Factors and Bandpass Filter Considerations 298
- 7.3 Problems 304
  - References 311

8 The Approximation Problem — 313
- 8.1 Traditional Methods 313
  - 8.1.1 Ideal LP Filter Characteristics 313
  - 8.1.2 Critical Frequencies and Steady-State Response 317
  - 8.1.3 The Butterworth Polynomials 319
  - 8.1.4 The Chebyshev Polynomials 324
  - 8.1.5 Inverted Chebyshev Polynomials 330
  - 8.1.6 A General Form for the LP Transmission Function 333
- 8.2 General Methods 337
  - 8.2.1 A Fourier Series Solution 338
  - 8.2.2 Finding Polynomial Ratio Network Functions 353
  - 8.2.3 Network Function Phase Versus Loss Function Phase 362
  - 8.2.4 Obtaining Polynomials from a Phase Response 364
  - 8.2.5 Optimizing Polynomial Parameters 374
- 8.3 Problems 385
  - References 390
- Appendix 8A Approximation Details 390
- 8A.1 Derivation of the Hilbert Transform 390
- 8A.2 Program Description 393
- 8A.3 Code Listing 395

About the Author	403
Index	405

Preface

Most of the material in this text stems from material developed and taught before my retirement in 2000 at the University of Maine. The purpose of writing this text is to document and organize material into one cohesive entity to illustrate a useful approach to preserving and teaching a modern version of the art of filter design. Strictly speaking, analog design procedures are essential to the design of high-speed sampling circuits such as the switched capacitor filter. So, in order to teach design procedures for high-performance digital devices, analog design considerations can not be totally ignored. In addition, active filter circuits (e.g., switched capacitors) offer competitive choices as alternatives to digital filters in many applications. In a sense, active filters are special-purpose analog computer signal processors.

Better courses need to be developed as techniques and curricula must evolve, and that is one of the purposes for writing this text. Another reason is that the material should also be useful for the practicing integrated circuit designer who wants to learn about the theory and design of active filters. The material presented here has practical use, and the text is organized so that specific procedures can be used for reference purposes.

Some material contained in this course is available in different texts and technical papers. However, most of the material has not been organized into a single text with sufficient examples and homework problems for anyone desiring to learn the subject on his or her own. So, with the encouragement of students and teaching colleagues, my special course notes have been assembled as a text on the subject of active filter design. The resulting text contains many practical procedures that are involved in the design and testing of active filter circuits. Some examples of previously unpublished topics are as follows:

1. An emphasis upon opamp gain redistribution in multiopamp circuits to control dynamic range performance;

2. Precorrection for s-plane distortion introduced by sampled integrators in the switched capacitor circuit concept;

3. Broadband, multitone testing of switched capacitor circuits;

4. State variable implementation using switched capacitor procedures;

5. Examples showing that the state variable filter sensitivity is competitive to that obtained from the leapfrog topology;

6. LC network equivalences useful for leapfrog design procedures based upon prototype LC networks;

7. A useful algorithm that finds minimum order network functions (polynomial ratios) directly from specified loss functions.

From a teaching point of view, the material is organized as a one semester course divided into four subject areas of roughly 10 lectures (or 100 pages) each. Students are assumed to have completed junior-senior college years or to have experience working in the area of analog integrated circuit design. Required background includes basic courses in linear system theory and electronics, and it is helpful to have had experience using computer tools such as Matlab and SPICE or μCAP.

The first subject area (Chapters 1 through 3) reviews circuit analysis techniques, introduces the prototype concept together with frequency scaling and circuit transformations, and explores issues of nonideal opamp behavior. These subjects provide a background for circuit manipulation. For ease of understanding new ideas, all subsequent sections use only a simple frequency-dependent opamp model.

The second subject area (Chapters 4 and 5) introduces active circuits with emphasis upon the opamp and the state variable filter. Scaling for gain-frequency distribution and dynamic range is discussed and illustrated as a most important design consideration. The biquad circuit is only introduced as a special form of the state variable filter. Cascade synthesis procedures are illustrated, and fundamental differences between cascade and state variable filters are discussed. This provides background for subsequent development of the leapfrog structure.

The third subject area (Chapters 6 and 7) introduces sensitivity factors and methods used to estimate filter response parameter dependence upon component values. This leads to the leapfrog topology and the ability to design filters from LC prototypes in order to obtain responses that have low sensitivity to component nominal value variations. Procedures are then developed to convert the leapfrog circuits to switched capacitor implementations.

The fourth subject area (Chapter 8) introduces the student to the Approximation Problem. It provides a good closure to the topic to have some understanding of how to find a suitable network function. Topics include curve fitting to a generalized transmission function form, the use of an iterative active structure based on Fourier theory, and traditional polynomials such as the Butterworth, Chebyshev, and inverted Chebyshev polynomials, as well as a generalized polynomial form.

The goal of this course is to give readers a workable introduction to the active filter design area and to provide a foundation that is useful when and if they should pursue the subject at a more advanced level. The material is presented with sufficient detail so that readers should be able to handle a wide range of filter design problems in a practical and flexible manner. The course is not a "look up the answer" approach to filter design. It is a design course and provides that kind of experience (e.g., design examples are solved to find different candidate circuits, and discussion indicating why some are better solutions than are others is provided). Such reasons include undesired parasitic response (due to inadequate models), bad element value ratios, and possible tuning difficulties in production.

PREFACE

Experience shows that the course is popular, and clearly, filter design is a popular subject with undergraduate engineers.

Finally, it should be emphasized that this material assumes that the student has access to and has learned how to use math tools such as Matlab and circuit simulators such as SPICE or μCAP. The text examples and homework rely heavily on the use of such tools. In particular, simulation is used extensively to evaluate the correctness of candidate designs. When proper modelling is used in the design process for linear filters, simulation is a valid tool used to measure the validity of a given design. Hence, simulation is sometimes referred to as a measured response. It is a fact that actual measurements for a properly constructed filter should virtually coincide with theoretically predicted responses. Thus, simulation is a valuable tool that is used extensively, especially for complex integrated circuits, where the cost of making a design error is prohibitive. There is no need to produce fabrication masks and special tools for circuit designs that do not even pass simulation measurement testing.

This text does not dwell significantly on how to use computer programs, but it does provide extensive examples of code used to compute various unusual procedures (e.g., polynomial manipulations). A CD for all example files, including the approximation program, `apprx8`, is included to facilitate work with text problems and examples.

Acknowledgments

I am grateful to the Electrical and Computer Engineering Department for supplying both an office to work in and computer equipment to facilitate this task in my retirement years. In addition, I thank special colleagues, Professors Al Whitney, Don Hummels, and Duane Hanselmann, who provided much encouragement and assistance to me. The book probably would not have been completed without their interest and helpful discussions about how to handle different topics and software procedures. In addition, I want to acknowledge the many students who provided feedback to me over the several years that this material was taught as a technical elective using only lecture notes. They encouraged the writing of a text to produce an organized story and to publish the many techniques that were used to teach this particular subject. I hope this book will meet their expectations.

It is important to note my gratitude for the support and encouragement received from everyone at Artech House, in particular, Mark Walsh, Barbara Lovenvirth, and Rebecca Allendorf. Everyone has been very patient, helpful, and encouraging to me as I learned how to get through the camera-ready copy process for an entire book.

As far as the women in my life are concerned, recognition is certainly due to the department administrative assistant, Janice Gomm, who in her own inimitable manner cheerfully and efficiently assisted in the typing of the text for the LaTeX editor. She also kept after me to actually complete the project, and I am indebted to her for her concern to see the job completed. In addition, I also want to acknowledge the most important support of my wife, Sally, throughout this task. Her longtime support and patience with my always wanting to *go to work* has made a big difference in the accomplishment of lifetime career goals.

Chapter 1

Introduction

Why study filter theory? What are filters? Where are they used? Can you just get them on integrated circuit chips like many other circuit functions? There are many questions to answer to get started on a journey to understand the electronic filters of today.

Filters are everywhere. Many times they are key elements required to optimize the performance of a complex system. Band-limiting filters are used in front of analog-to-digital converters to prevent signal energy from overlapping in the aliased sampled signal spectrum. They are used to smooth the digital-to-analog reconstructed signals in synthesized speech or other information. Weather-mapping facsimile receiver/transmitters are an example of this digital sampling, transmission, and signal reconstruction process. Weather maps are transmitted continuously and are liberally used by the news media. Filters are used to compensate for frequency response so that closed-loop systems will be stable and provide optimum response to suddenly applied perturbations. Gated ranging radar circuits, such as those used in the Lunar Excursion Module (LEM), use active electronic filters to improve resolution as the target landing spot is approached.

Large aperture seismic arrays (LASAs) are installed underground all around the world. Each installation employs many receiver paths that have 2-Hz lowpass active filters on their input. These filters are designed and produced to exacting theoretical specifications that cannot be met (in size, space, and cost) with other technologies. The filters are guaranteed to outlast other components since it is very expensive to dig up and repair any receiver element. Whoever manufactures an element that fails has to pay for all repair and replacement costs. Thousands of active filters deployed around the world survived without failure. Now it is possible to take earth tremor measurements for granted. Properly designed active filters are reliable and dependable in their performance.

Numerous other applications could be cited. Many are familiar and use large numbers of filters in the design of a system. Many systems can be designed with commercial filters. A filter manufacturer, like any other manufacturer, likes to sell as many units of a known (reliable) product as possible. Such items cost less to the user and also make more money for the manufacturer. So it is always possible to buy standard filters from catalog listings. When such filters meet the required performance, then that is the solution that should be used in practice.

Then why study filter theory? There are at least two reasons. One is practical; the other is academic. From a practical consideration, the design engineer should not sacrifice system performance unnecessarily for the sake of using a conveniently available filter just because there is an uncertainty about where and how to design the exact or even the optimum filter required for a job. It also may be desirable to include the filter in the package with the application, or if large volume is involved, it may be cheaper to manufacture the filter in-house. These are just two examples of why it can be useful to know filter design theory. The circuit design engineer should know enough about filter theory to evaluate such trade-offs when they occur in practice.

An academic reason for studying filter design theory is that it is a good introduction to the engineering design process. In addition, most upper-level students already possess the skills they need in order to perform useful and original filter design; now they need to be exposed to some known procedures. The skills and problem solving techniques learned in the filter design problem are found to be transferrable to other design problems. Filter design is to circuit theory as mystery writing is to novel writing. The mystery is a perfect novel in that it has a well-defined form with beginning, middle, and end. Filter design involves the specifications for a requirement, the selection of a suitable design to meet all specifications, and the implementation and final testing of a suitable prototype for the job. Due to the fact that linear circuit models are adequate to describe the actual response of filter components, the circuits implemented in a given design will always provide the predicted response. This is what makes filter theory the complete design problem. For most students it is usually a rewarding experience to see how they can put together what they have seen in basic circuit theory, linear system theory, and electronics to develop a design procedure that solves a wide range of problems. The chief reason for this course is to introduce the student to a mathematically intense design methodology and to show how known concepts can be combined to develop useful design procedures.

1.1 Filter Terminology

Before proceeding further, it is helpful to review some standard filter terminology. It will help to understand specifications of filter performance requirements in forms that are historically standard. In addition, this chapter concludes with a review of filter technologies and frequency ranges over which they are usually encountered in practice.

The filter theory discussed in this course deals with linear networks. The filter is described by means of its network function, $H(s)$, or by other symbols, like $T(s)$, that relate output to input. The independent variable for the network function is the complex frequency, s, of the signal input to the network.

1.1. FILTER TERMINOLOGY

Typically, the network function is described as a ratio of polynomials as shown in (1.1).

$$H(s) = \frac{\mathbf{V}_2}{\mathbf{V}_1}(s) = \frac{b_m s^m + b_{m-1} s^{m-1} + \ldots + b_1 s + b_0}{a_n s^n + a_{n-1} s^{n-1} + \ldots + a_1 s + a_0}$$
$$= \frac{\sum_{k=0}^{m} b_k s^k}{\sum_{k=0}^{n} a_k s^k} \quad (1.1)$$

Normally, the network function is measured by applying a sinusoidal input to the network and measuring the steady-state output response that occurs in response to the input. This procedure yields a phasor representation for the input and output, respectively, at the complex frequency $s = j\omega$. Equation (1.2) gives the ratio of output to input for a sine wave measurement.

$$H(j\omega) = \frac{|\mathbf{V}_2|e^{j\theta_2}}{|\mathbf{V}_1|e^{j\theta_1}} = t(s)\rfloor_{s=j\omega} \quad (1.2)$$

Then, the magnitude and phase of the network function are as follows:

$$|H(j\omega)| = \frac{|\mathbf{V}_2|}{|\mathbf{V}_1|}$$
$$\angle H(j\omega) = (\theta_2 - \theta_1) = \theta(\omega) \quad (1.3)$$

Note that they are both a function of frequency.

A common quantity used to specify filter response is known as the attenuation function. This is a dBv ratio of the input to output voltage magnitudes, and it is positive when $\mathbf{V}_1 > \mathbf{V}_2$, which represents a loss in voltage magnitude. Equation (1.4) shows this relationship:

$$a(\omega) = 20\log_{10}|\mathbf{V}_1/\mathbf{V}_2| \equiv -20\log_{10}|H(j\omega)| \quad \text{dBv} \quad (1.4)$$

Historically, filters were passive and always displayed loss, or attenuation, so the design objectives were defined to be positive for convenience. With the advent of the active filter, it is possible for filters to supply gain, and so, both gain and attenuation are used in this text. The important thing to understand is that either leads to a description of the magnitude and phase of the network function versus the input frequency, $s = j\omega$.

So far, the discussion has been directed at a general input-output function. Note that a network, N, may be designed to include both the source and the load impedance so that practical sources and loads are also included in the discussion. Figure 1.2 illustrates how this can be done for the network shown in Figure 1.1. Note that attenuation is also a power ratio consistent with the definition of a dB (or Bell) unit of power measure. Let (1.5) represent average powers in response to sinusoidal excitation for a network, as in Figure 1.2:

$$P_{in} = \frac{|\mathbf{V}_1|^2}{2R_{in}}$$
$$P_{out} = \frac{|\mathbf{V}_2|^2}{2R_L} \quad (1.5)$$

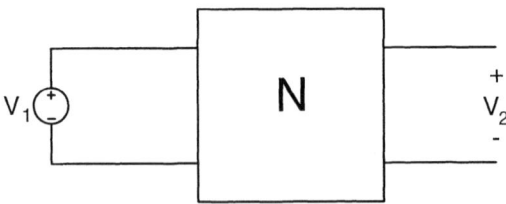

Figure 1.1. Circuit model for network function, $H(s)$.

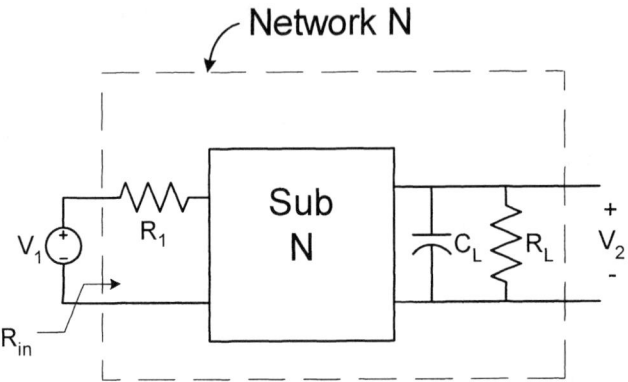

Figure 1.2. Circuit model to include source and load resistance.

1.1. FILTER TERMINOLOGY

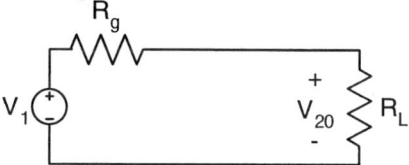

Figure 1.3. Power to load without the filter.

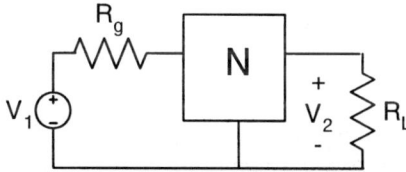

Figure 1.4. Power to load with the filter inserted.

The power ratio, in decibels, then shows that power loss is exactly equal to the attenuation function whenever R_{in} and R_L are equal. Normally, the real part of the input impedance, R_{in}, is a function of frequency. This makes the power loss function in (1.6) more complicated than the attenuation function, $a(\omega)$, by itself. Since $a(\omega)$ is sufficient to obtain the network function, $T(s)$, it is usually the preferred method for describing a network.

$$\text{Power loss} = 10\log_{10}\left(\frac{P_{in}}{P_{out}}\right) \equiv 10\log_{10}\left(\frac{|\mathbf{V}_1|^2 R_L}{R_{in}|\mathbf{V}_2|^2}\right)$$
$$\equiv a(\omega) + 10\log_{10}(R_L/R_{in}), \text{ dB} \qquad (1.6)$$

Insertion loss is a classical filter response function of frequency. It is defined as the decibel ratio of power delivered to the load with the filter removed to the power delivered to the load with the filter inserted. Figures 1.3 and 1.4 illustrate the measurements for forming the insertion loss ratio. From Figure 1.3, the power to the load without the network is as developed in (1.7):

$$P_{L0} = \frac{R_L I_L^2}{2} = \frac{R_L}{2}\left(\frac{\mathbf{V}_1}{R_L + R_g}\right)^2 = \frac{R_L \mathbf{V}_1^2}{2(R_L + R_g)^2} \qquad (1.7)$$

And the power to the load with the network connected is developed in (1.8):

$$P_L = \frac{\mathbf{V}_L^2}{2R_L} = \frac{|H(j\omega)|^2 \mathbf{V}_1^2}{2R_L} \qquad (1.8)$$

Then, the insertion loss is obtained as follows:

$$\text{Insertion loss} = 10\log_{10}\left(\frac{P_{L0}}{P_L}\right)$$

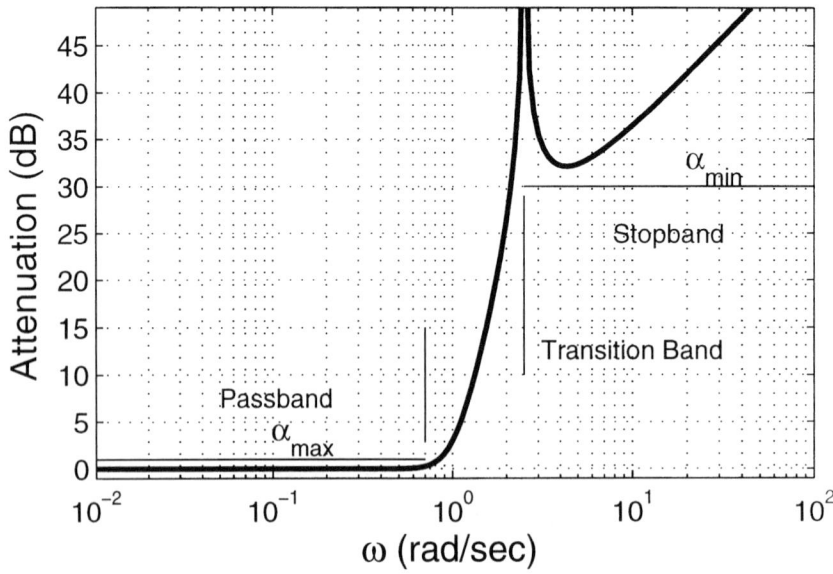

Figure 1.5. A lowpass filter response.

$$\begin{aligned}
&= 10\log_{10}\left(\frac{R_L \mathbf{V}_1^2}{2(R_L+R_g)^2} \cdot \frac{2R_L}{|H(j\omega)|^2 \mathbf{V}_1^2}\right) \\
&= 20\log_{10}\left(\frac{R_L}{R_L+R_g}\right) - 20\log_{10}|H(j\omega)| \\
&\equiv 20\log_{10}\left(\frac{R_L}{R_L+R_g}\right) + a(\omega) \quad\quad (1.9)
\end{aligned}$$

This last result shows that insertion loss differs from the power loss function by a constant (flat loss) term and that the frequency-dependent behavior comes from the attenuation function, $a(\omega)$, defined earlier.

Four basic filter types are illustrated with their attenuation functions in Figures 1.5 through 1.8. These are the lowpass (LP), highpass (HP), bandpass (BP), and bandstop (BS) filters, respectively. Each filter is characterized by a nominal function, or curve, passing through a passband region, allowing a maximum attenuation, α_{max}. These are just constants for the basic filter types shown, but they may be functions of frequency for more complex filters. The region between a passband and a stopband is referred to as the transition band. The transition band is usually nonzero in extent in order to provide physically realizable filter requirements. The variable α_{min} specifies a minimum attenuation in the stopband region.

The response boundaries given by α_{max} and α_{min} are normally worst case specifications. The response for the required filter should fit inside the tolerance bands for all operating temperatures and for the full peak-to-peak signal range required for all frequencies.

1.1. FILTER TERMINOLOGY

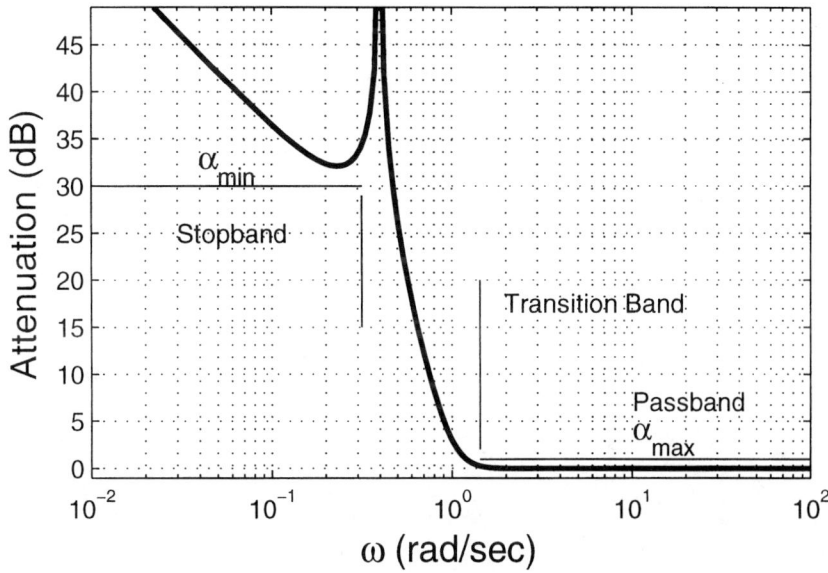

Figure 1.6. A highpass filter response.

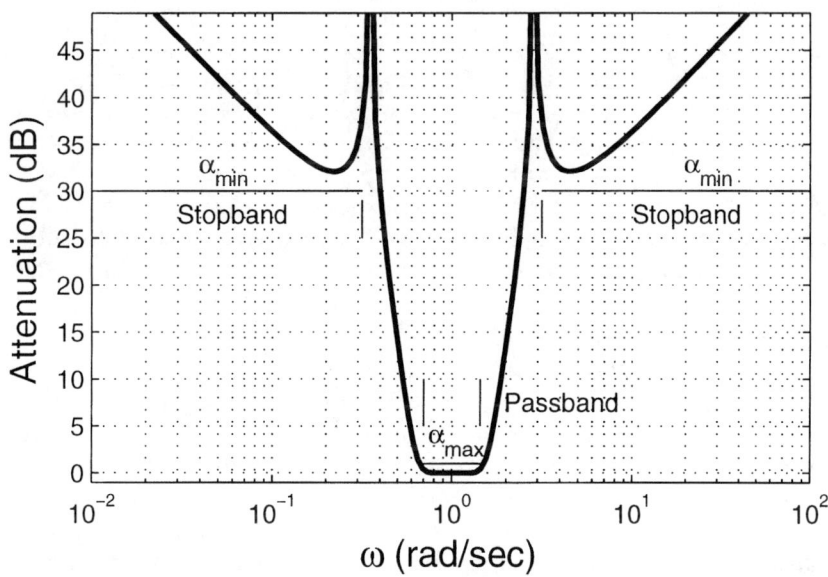

Figure 1.7. A bandpass filter response.

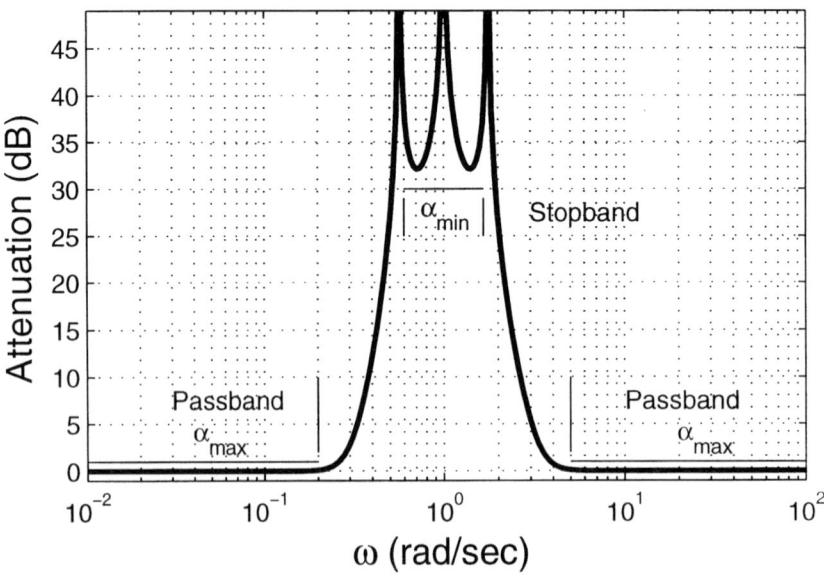

Figure 1.8. A bandstop filter response.

1.1.1 Filter Component Values

Designing a filter to meet worst-case attenuation specifications requires high-quality components. It is common to build filters with component values specified to a fraction of a percent of their nominal design values. This requirement of exactness will be a new experience for most students, but everyone is forewarned to be careful about rounding off numbers in the filter design process. If care is exercised, a good design will always check theoretical curves exactly. Vendors will sort components, for a fee, and are happy to supply any value to any tolerance whenever substantial volume is involved. This means that it is sometimes feasible to implement filters that require precision components in order to meet design specifications.

Another factor related to filter performance is the range of element values encountered in the final design for a filter realization. Whenever component values must have a large ratio of minimum to maximum, it is quite possible that the design is poorly conditioned. This is reflected in the filter having difficulty meeting specifications for all frequencies and operating temperatures. In addition, element values should not be too small or too large. Nominal values that are too small are vulnerable to parasitic effects of packaging, and values that are too large are usually also physically large or costly and difficult to obtain. For example, in wirewrap or printed circuit packages, a connecting node easily has a capacitance of 5 pF to the ground plane. This parasitic capacitance adds in parallel to any element connected from node to ground. A second example is that plain wire conductors, or printed circuit paths, exhibit on the order of 8 to 10 nH/cm, so small inductance values

1.1. FILTER TERMINOLOGY

are inserted with a series connection to other elements. Finally, it is noted that contact resistance in connectors is in the range of 0.1 to 0.3 Ω, while surface contamination provides parallel leakage paths on the order of 20 MΩ or larger. In view of the above, useful guidelines for the normal range of RCL component values are $\Omega < R < M\Omega$, $\mu H < L < H$, and $pF < C < \mu F$. These listed ranges are useful for most applications.[1]

In each case the specific amount may be interpreted as "a few" (e.g., capacitance could range between a few pF to a few μF). The practical range of element values is about 1 to 10^6 in each case.

A word on temperature stability is also appropriate. Whenever a circuit design requires that component values possess very small tolerances, behavior with respect to temperature becomes important. Most components that the student has used for previous laboratory experiments are not of sufficient quality to use for exact filter design. The bypass ceramic capacitors have both a poor tolerance and a poor temperature coefficient. So do electrolytic capacitors and carbon film resistors. Exact filters use capacitors made from polystyrene or mylar films, mica, or specially treated ceramics that yield a temperature coefficient near zero. Standard high-quality resistors are available through the use of various metal oxides. A typical filter component should aim for a temperature coefficient between +100 ppm/°C (ppm = parts per million). This means that the component would change by 1 part in 100 (1%) over a 100°C operating temperature range. It is common to have a wide operating temperature range requirement for many rather ordinary applications. Electronic equipment gets put in small boxes in out of the way places where they get very hot (underneath equipment in a factory) or very cold (in a relay station in Alaska). The operating environment for the filter is always a serious consideration for the selection of components to use in its fabrication. This course will not address the selection of components directly, but a filter topology will be developed that provides minimal sensitivity to component variation. The role of component sensitivity will be developed in detail at that time.

1.1.2 An Active Filter Definition

The next topic to consider is the definition of an active filter. At least two concepts are involved in the answer to this question. One involves the meaning of flat loss, or gain. The second involves the concept of using amplifiers and RC networks to build network functions equivalent to those for RLC circuits (i.e., to eliminate inductance from the circuit).

Whenever the incremental signal average power delivered to a load exceeds that delivered to the filter input, it is clear that some amplification has occurred. It is not possible for a purely passive network to deliver more power than it receives, as it will normally provide some loss due to nonideal or parasitic losses in the LC elements. This incidental loss is often represented by means of a flat loss, as illustrated in Figure 1.9 for a bandpass filter function. The nominal design has zero loss in the

---

[1] Practical values for resistance and capacitance on very large scale integrated (VLSI) chips are considerably different, and these values will be discussed later.

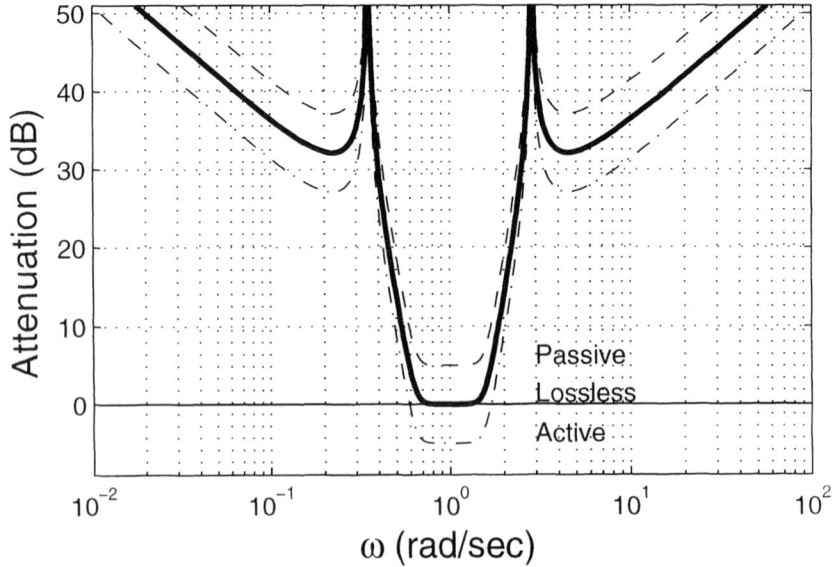

Figure 1.9. Illustrating the concept of flat loss or gain.

passband, but an actual passive realization shows that the curve is displaced by a constant value from the nominal design curve. The positive displacement on an attenuation curve represents a flat loss for the filter realization. It is called flat loss due to its being a constant across the frequency band.

If an amplifier is inserted between the passive filter and its load, then the loss may be recovered, and gain may even be possible. If the amplifier has a constant frequency response across the band of interest, then it adds a flat gain to the response as shown in Figure 1.9. The combination of passive filter plus amplifier could be called an active filter simply because all of the components are not passive. While such filters do occur frequently in practice, they will not be of much interest in this course. Basically, the amplifier in such designs is used to restore signal levels and to provide isolation. Telephone repeater-equalizer networks have incorporated such designs in the past.

Modern filters avoid the use of inductors wherever possible. This becomes feasible because RC networks combined with amplifiers and feedback are sufficient to provide complex poles and zeros to realize a network function. Current technology can readily fabricate resistance and capacitance in integrated circuit forms, as is also possible for operational amplifiers. This technology makes the integrated RC-active filter an attractive filter design solution, so long as dc operating voltages are available to bias the filter, and the filter will handle the signal ranges required. Generally, the inductor does not lend itself to integrated circuit technology. It requires highly permeable materials to concentrate magnetic flux in areas required to obtain integrated circuit inductance. So far, such materials have not been easily

1.1. FILTER TERMINOLOGY

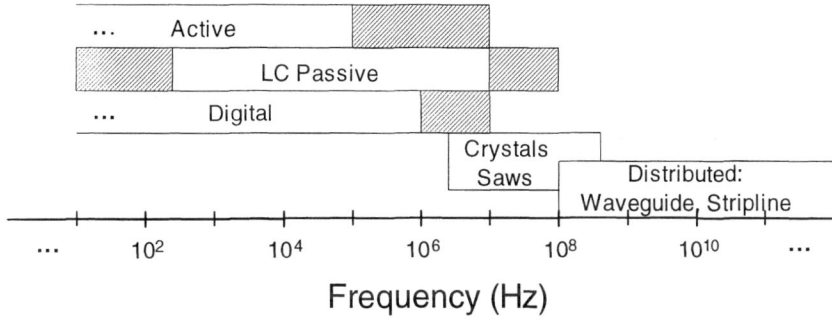

Figure 1.10. Filter component technology versus frequency.

fabricated or found to be compatible with the semiconductor materials required for transistor junctions and other devices.

In summary, this course considers an active filter to be any RC-amplifier topology that does not use inductors to obtain a filter response that has complex poles. In addition, the active filters considered here will be based on currently available operational amplifiers. The frequency range over which good filter designs may be obtained is found to be limited somewhat, as shown in Figure 1.10. There, the range of filter implementation is compared to other technologies versus frequency. The cross-hatched regions are gray areas where an implementation may be possible, depending upon the requirement or the problem. In some cases, a particular technique may not be feasible due to required power levels, size, weight, cost, and so forth. Such problem-specific conditions always dictate an approach to finding a suitable filter.

1.1.3 Conclusion

The following factors are usually considered somewhere in making the choice of component technology or other design trade-offs.

1. Frequency range of application;
2. Operating temperature range;
3. Signal levels;
4. Aging, mean time before failure (MTBF);
5. Size, weight;
6. Cost, production volume;
7. Complexity, production line manual labor, yield.

The items in the list are self-explanatory. Active filters always score well on size, weight, MTBF, and complexity. They are straightforward to tune and can provide very stable behavior when properly designed. The purpose of this course is to describe techniques for doing just that.

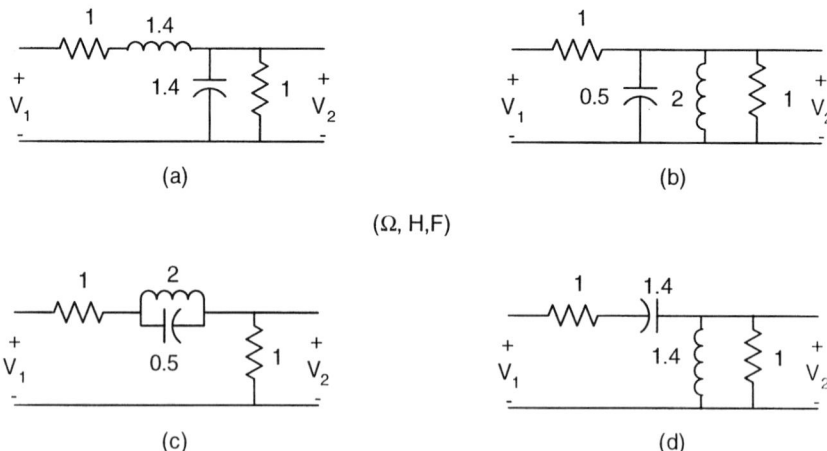

Figure 1.11. (a–d) Circuits for Problem 1.1.

This chapter has provided an introduction to some filter terminology, raised the question of component values and stability, and defined the active filter. The next chapter reviews network analysis procedures, the idea of a prototype design, network scaling, and other issues, before we launch into the active filter design problem.

1.2 Problems

1.1 Find the network function, $\mathbf{V}_2/\mathbf{V}_1$, and plot the attenuation versus frequency, ω, for each circuit (a,b,c,d) shown in Figure 1.11. Identify the amount of flat loss and determine the type (LP, BP, and so forth) for each filter. Specify the passband and stopband frequencies in each case. The different bands may be determined as follows: Compared to the flat loss, consider any frequency to be passed if the relative attenuation is within 3 dB. Also, any frequency may be considered not to be passed if the relative attenuation is 12 dB or greater.

1.2 Two 10-kΩ resistors are connected in parallel on a printed circuit board. A circuit model that includes parasitic elements is shown in Figure 1.12 for this connection. Determine the input impedance function, and plot its magnitude versus frequency, ω. Over what frequency range does the connection provide a reasonably good approximation to the desired 5 kΩ of the connection?

1.3 In the circuit in Figure 1.13, the capacitor has a temperature coefficient (TC) of $+150$ ppm/$°$C, and the resistor has a TC of $+350$ ppm/$°$C. The values given in the circuit are for a $+25°$C operating temperature. Find the change in the location of the pole of the input impedance when the operating temperature changes to $+100°$C.

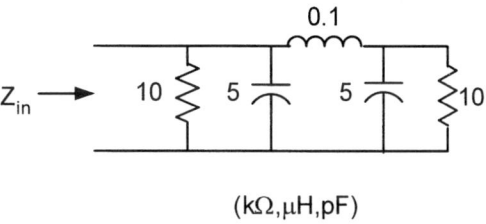

Figure 1.12. Circuit for Problem 1.2.

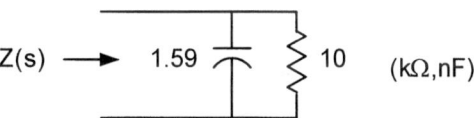

Figure 1.13. Circuit for Problem 1.3.

1.4 Find the equivalent inductance, L_{eq}, as a function of operating frequency, ω, for the equivalent reactance of an ideal inductance, L, placed in parallel with a parasitic capacitance, C_p, on a printed circuit board. Over what range of frequency, in terms of L and C_p, is $L_{eq}/L \leq 1.10$?

1.5 What is the maximum value of resistance, R, placed in parallel with a parasitic capacitance, $C_p = 1$ pF, so that the phase difference between the voltage across the combination, **V**, and the current delivered to the combination, **I**, is less than $\tan^{-1}(0.1)$ at a frequency of 10 MHz?

Chapter 2

Review of Circuit Analysis Concepts

Everyone has probably heard that design (synthesis) is the reverse of analysis, but most undergraduate courses spend more time on analysis than on design. Consequently, design is not a very familiar, or comfortable, procedure for most students. The saying that design is the reverse of analysis sounds straightforward, but there is a hidden assumption in the saying that implies that the student really understands analysis so that properties of analysis results can be used to synthesize solutions to a problem. This chapter reviews several properties of circuit analysis that are not so well known but are useful to the network design process. Concepts covered here include loop and node parameter matrices, the prototype network concept, impedance and frequency scaling, and some simple network transformation procedures.

2.1 Network Parameter Matrices

Whenever loop current variables are used to analyze a circuit, it is possible to arrange the equations into a matrix form that looks like the analysis of a series RLC circuit. This form is shown in (2.1):

$$\{s[L] + [R] + \frac{1}{s}[C^{-1}]\}[\ell] = [\mathbf{V}] \qquad (2.1)$$

where s is complex frequency, $[\ell]$ is a column vector of loop current variables, $[\mathbf{V}]$ is a column vector of loop source voltages, and $[L], [R]$, and $[C^{-1}]$ are loop parameter inductance, resistance, and reciprocal capacitance matrices, respectively.[1] The circuit shown in Figure 2.1 interprets the set of equations in series form for a one-dimensional set of variables.

Alternately, when node voltage variables are used to analyze a circuit, it is possible to arrange the analysis into a matrix form that looks like a parallel GCL

---
[1] The voltage and current variables are phasors, or complex frequency components.

16   CHAPTER 2. REVIEW OF CIRCUIT ANALYSIS CONCEPTS

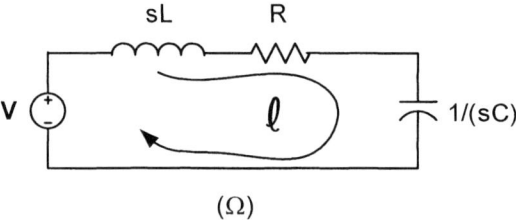

Figure 2.1. The series RLC loop-variable form.

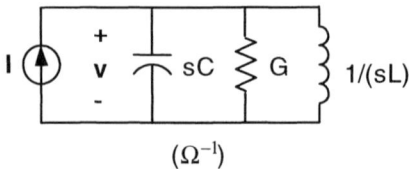

Figure 2.2. The parallel GLC node-variable form.

circuit. This form is shown in (2.2):

$$\{s[C] + [G] + \frac{1}{s}[L^{-1}]\}[\mathbf{v}] = [\mathbf{I}] \qquad (2.2)$$

where s is complex frequency, $[\mathbf{v}]$ is a column vector of node voltage variables, $[\mathbf{I}]$ is a column vector of node source currents, and $[C], [G]$, and $[L^{-1}]$ are node parameter capacitance, conductance, and reciprocal inductance matrices, respectively. The circuit shown in Figure 2.2 interprets (2.2) in a parallel form for a one-dimensional set of variables.

Example 1 - Find the loop and node parameter matrices for the circuit given in Figure 2.3. Use the results to construct circuit models from the parameter matrices.

Solution - The given circuit has three independent node voltage variables and three independent loop current variables as identified by $\mathbf{v}_{1,2,3}$ and $\ell_{1,2,3}$, respectively, on the circuit schematic.

a - Nodal Analysis

$$\text{KCL}_1 : \quad \frac{\mathbf{v}_1 - \mathbf{V}_{s1}}{10} + \frac{s}{5}\mathbf{v}_1 + \frac{\mathbf{v}_1 - \mathbf{v}_2}{20} = \mathbf{I}_{s1}$$

$$\text{KCL}_2 : \quad 24\mathbf{i}_x + \frac{\mathbf{v}_2 - \mathbf{v}_3}{2s} = 0,$$

$$\text{but} \quad \mathbf{I}_x = \frac{\mathbf{v}_1 - \mathbf{v}_2}{20}, \quad \text{so}$$

2.1. NETWORK PARAMETER MATRICES

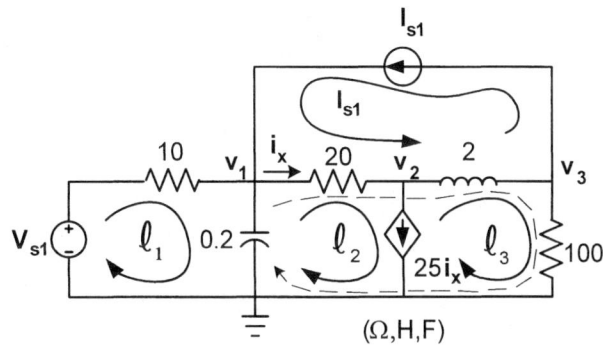

Figure 2.3. Example circuit for parameter matrix analysis.

$$\frac{24(\mathbf{v}_1 - \mathbf{v}_2)}{20} + \frac{\mathbf{v}_2 - \mathbf{v}_3}{2s} = 0$$

KCL$_3$:
$$\frac{\mathbf{v}_3}{100} + \frac{\mathbf{v}_3 - \mathbf{v}_2}{2s} = -\mathbf{I}_{s1} \quad (2.3)$$

These three equations combine to yield (2.4):

$$\{s[C] + [G] + \frac{1}{s}[L^{-1}]\} \begin{bmatrix} \mathbf{v}_1 \\ \mathbf{v}_2 \\ \mathbf{v}_3 \end{bmatrix} = \begin{bmatrix} \mathbf{I}_{s1} + 0.1\mathbf{V}_{s1} \\ 0 \\ -\mathbf{I}_{s1} \end{bmatrix}$$

where
$$[C] = \begin{bmatrix} 0.2 & 0 & 0 \\ 0 & 0 & 0 \\ 0 & 0 & 0 \end{bmatrix} \text{F}$$

$$[G] = \begin{bmatrix} 0.15 & -0.05 & 0 \\ 1.2 & -1.2 & 0 \\ 0 & 0 & 0.01 \end{bmatrix} \Omega^{-1}$$

$$[L^{-1}] = \begin{bmatrix} 0 & 0 & 0 \\ 0 & 0.5 & -0.5 \\ 0 & -0.5 & 0.5 \end{bmatrix} \text{H}^{-1} \quad (2.4)$$

Equation (2.4) can be implemented in reverse to obtain the circuit shown in Figure 2.4. When interpreting KCL$_2$, at $\mathbf{v}_2$, it should be noted that $1.2(\mathbf{v}_1 - \mathbf{v}_2) \equiv (1.25 - 0.05)\mathbf{v}_1 - (1.25 - 0.05)\mathbf{v}_2$, so that the 20-$\Omega$ resistor on Node 1 can be accounted for. This circuit model, obtained by nodal analysis on the given circuit, is different, as now all sources are current sources, and the floating current source, $\mathbf{I}_{s1}$, is changed into two sources connected to ground. The important result is that both circuits yield identical responses for the node voltages, $\mathbf{v}_{1,2,3}$.

b - Loop Analysis
KVL$_1$:

$$10\ell_1 + \frac{5}{s}(\ell_1 - \ell_2) = \mathbf{V}_{s1}$$

18   CHAPTER 2. REVIEW OF CIRCUIT ANALYSIS CONCEPTS

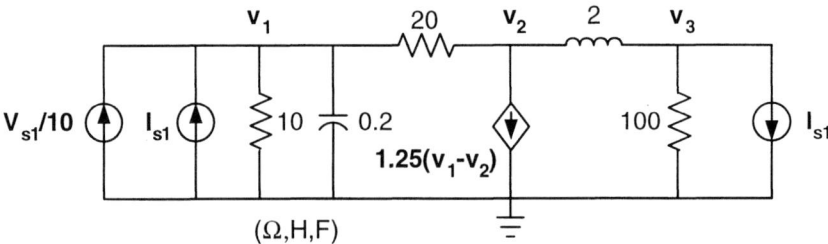

Figure 2.4. Circuit model obtained from nodal analysis.

KVL$_2$ (a super-loop):

$$\frac{5}{s}(\ell_2 - \ell_1) + 20(\ell_2 + \mathbf{I}_{s1}) + 2s(\ell_3 + \mathbf{I}_{s1}) + 100\ell_3 = 0$$

Dependent Source Constraint:

$$25i_x = 25(\ell_2 + \mathbf{I}_{s1}) = \ell_2 - \ell_3 \text{ or}$$
$$24\ell_2 + \ell_3 = -25\mathbf{I}_{s1} \quad (2.5)$$

These three equations combine to yield (2.6):

$$\{s[L] + [R] + \frac{1}{s}[C^{-1}]\} \begin{bmatrix} \ell_1 \\ \ell_2 \\ \ell_3 \end{bmatrix} = \begin{bmatrix} \mathbf{V}_{s1} \\ -(2s+20)\mathbf{I}_{s1} \\ -25\mathbf{I}_{s1} \end{bmatrix}$$

$$\text{where} \quad [L] = \begin{bmatrix} 0 & 0 & 0 \\ 0 & 0 & 2 \\ 0 & 0 & 0 \end{bmatrix} \text{ H}$$

$$[R] = \begin{bmatrix} 10 & 0 & 0 \\ 0 & 20 & 100 \\ 0 & 24 & 1 \end{bmatrix} \Omega$$

$$[C^{-1}] = \begin{bmatrix} 5 & -5 & 0 \\ -5 & 5 & 0 \\ 0 & 0 & 0 \end{bmatrix} \text{ F}^{-1} \quad (2.6)$$

Equation (2.6) can be implemented in reverse to obtain the circuit model shown in Figure 2.5. Again, this circuit model is different from the given circuit, as here the sources are all transformed to voltage sources. The circuit of Figure 2.5 provides loop current responses, $\ell_{1,2,3}$, that are the same as those for the original circuit.

This example serves to show that circuits are not unique, and many circuits exist that provide identical loop current or node voltage response. The difficult part of design is deciding which answer is best for a given problem.

2.2. NETWORK SCALE FACTORS

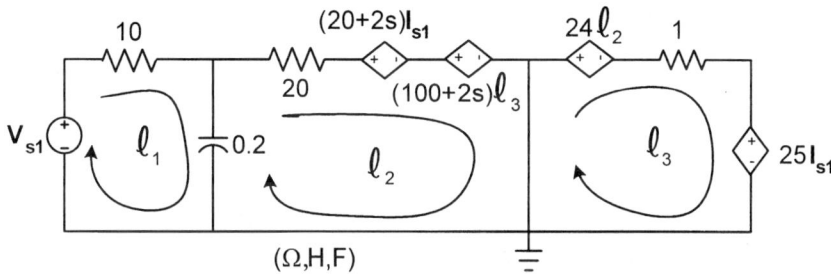

Figure 2.5. Circuit model obtained from loop analysis.

2.2 Network Scale Factors

You have probably already noticed that 1 Ω-1 rad/sec circuit values are easier to use for analysis than are values that have a wide range of decimal multipliers. Unfortunately, the 1 Ω-1 rad/sec circuits do not represent practical values for real circuit applications. It turns out, however, that there is no loss in generality in studying circuits with convenient element values. This is because we can always scale a linear circuit to accommodate practical impedance and frequency values for applications once the design or analysis has been completed.

Many books list lowpass prototype filter designs. They are listed in terms of 1 Ω-1 rad/sec element values because the designer has no way of knowing ahead of time what impedance and passband frequencies are required for specific applications. The tabulated designs are referred to as prototypes because they are models that can be scaled to accommodate specific application requirements. Let us first consider how we can change all branch impedances of a circuit by a specific factor, and then we will consider frequency-scaling operations.

The loop parameter impedance formulation provides us with the means to determine how scaling operations can be applied to any linear network. The use of loop current variables provides an analysis approach that can be used on any linear network and is thus a general analysis procedure. Consider the two circuits shown in Figure 2.6. The prototype circuit in Figure 2.6(a) is to be impedance scaled by a factor, k_R, so that each branch in the scaled circuit will have an equivalent impedance k_R times that of the prototype circuit, while the voltage sources are left unchanged. Using loop parameter matrices, we have the following results:

$$[Z_L][\mathbf{I}] = [\mathbf{V}]$$
$$\text{and} \quad [Z'_L][\mathbf{I'}] = [\mathbf{V}] \qquad (2.7)$$

$[Z_L] = j\omega[L] + [R] + [C^{-1}]/(j\omega)$ is the loop parameter matrix for the circuit. The scaling provides a relationship between Z_L and Z'_L as shown in (2.8):

$$[Z'_L] = k_R[Z_L] \equiv j\omega k_R[L] + k_R[R] + \frac{1}{j\omega}k_R[C^{-1}]$$

CHAPTER 2. REVIEW OF CIRCUIT ANALYSIS CONCEPTS

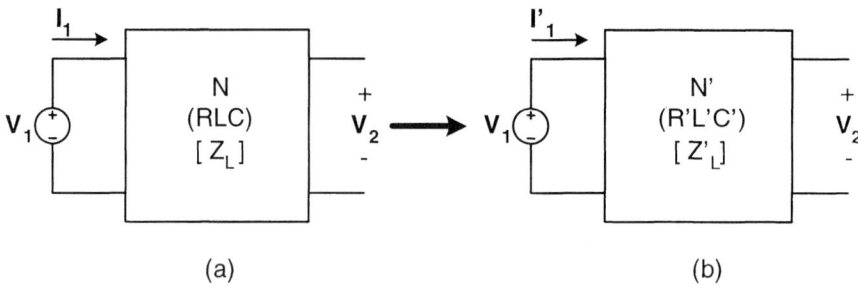

Figure 2.6. A prototype network, (a), is scaled to yield a practical impedance scaled network, (b).

$$\equiv j\omega[L'] + [R'] + \frac{1}{j\omega}[C'^{-1}] \qquad (2.8)$$

The condition required in (2.8) tells us how to change each element in the prototype to obtain the desired scaled impedance behavior. Note, the prime is used here to note the new element value, and the unprimed is the old, or original, element value.

$$\begin{aligned} \text{All} \quad L'_{jk} &= k_R L_{jk} \\ R'_{jk} &= k_R R_{jk} \\ C'_{jk} &= C_{jk}/k_R \end{aligned} \qquad (2.9)$$

Note that capacitance is inversely proportional to impedance, and so, capacitance is scaled inversely by the impedance scale factor.

So impedance scaling the RLC elements is straightforward, as shown above. We should note some details before leaving this derivation. The first conclusion is that the loop currents in a scaled network will also be scaled inversely by the impedance scale factor when the voltage excitation is held constant for both the prototype and scaled networks. This conclusion is seen by looking at the inverse of the loop impedance formulation:

$$\begin{aligned} [\mathbf{I}] &= [Z_L]^{-1}[\mathbf{V}] \\ [\mathbf{I'}] &= [Z'_L]^{-1}[\mathbf{V}] = [k_R Z_L]^{-1}[\mathbf{V}] \\ \therefore [\mathbf{I'}] &\equiv \frac{1}{k_R}[\mathbf{I}] \end{aligned} \qquad (2.10)$$

All currents in the scaled network are reduced by the impedance factor, k_R. If impedances are scaled by 1000, then all currents will be in mA. We should observe that voltage and current divider ratios are unaffected by the impedance scaling operation because the scale factor cancels for impedance or admittance ratios.

As far as dependent sources are concerned, their effect is included in the loop parameter matrices, and their scaling depends upon the units of the source function. Voltage-to-voltage and current-to-current dependent sources are unaffected by

2.2. NETWORK SCALE FACTORS

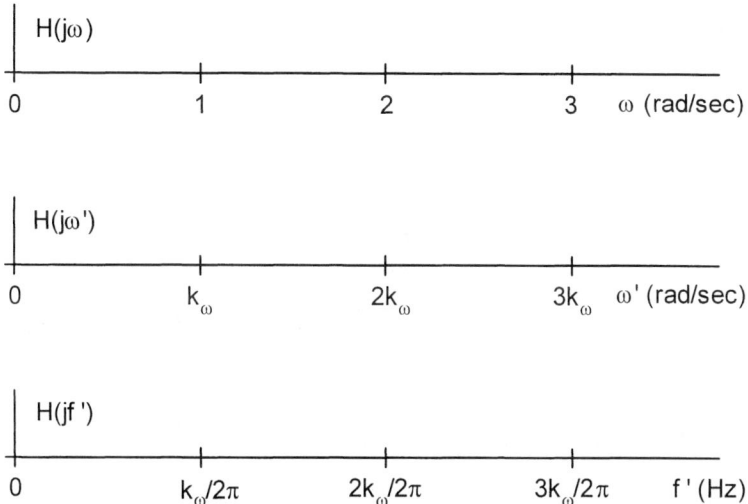

Figure 2.7. A desired mapping of prototype frequency to a practical frequency scale.

impedance scaling. A current controlled voltage source, however, has a resistance proportionality factor, and so it is scaled proportionally to the factor, k_R. A voltage controlled current source has an admittance proportionality, so it is inversely scaled by the k_R factor. You can always check the loop parameter matrices whenever in doubt about the effect of any particular dependent source.

Frequency scaling is just a mapping of the prototype ω-value to a different scale as shown in Figure 2.7. The prototype frequencies are scaled by the factor k_ω, so that some point of interest on the prototype ω-axis is scaled to occur at a practical frequency on the ω', or f', practical frequency scale. The network function value is required to remain the same function of ω; only the frequency changes scale. We can see how this works by again using the loop parameter matrix formulation for any circuit:

$$[Z_L(j\omega)][\mathbf{I}] = [\mathbf{V}] \text{ at } \omega$$
but we want
$$\omega' = k_\omega \omega$$
so
$$[Z_L(j\omega'/k_\omega)][\mathbf{I}] = [\mathbf{V}] \text{ at } \omega = \omega'/k_\omega \quad (2.11)$$

Equation (2.11) just shows that using ω in terms of ω' yields the desired mapping while maintaining the network function as an unaltered function of the original prototype frequency, ω. Substitution of $j\omega'/k_\omega$ into the loop impedance matrix yields the required frequency scaling operations as follows:

$$\begin{aligned} Z'_L(j\omega') &= Z_L(j\omega'/k_\omega) \\ &= \frac{j\omega'}{k_\omega}[L] + [R] + \frac{k_\omega}{j\omega'}[C^{-1}] \end{aligned}$$

$$= j\omega'[L'] + [R'] + \frac{1}{j\omega'}[C'^{-1}] \qquad (2.12)$$

By associating matrix elements in (2.12), we can deduce the procedure for frequency scaling the network:

$$\begin{aligned} L'_{jk} &= L_{jk}/k_\omega \\ R'_{jk} &= R_{jk} \\ C'_{jk} &= C_{jk}/k_\omega \end{aligned} \qquad (2.13)$$

Not surprisingly, only energy storage elements are affected by the frequency-scaling operation. For k_ω factors greater than one (usually true), the prototype LC values are always lowered to smaller values for real applications.

Finally, the two scalings may be combined into composite scale factors for the RLC elements just by performing the two procedures in sequence. The result is shown in (2.14):

$$\begin{aligned} k_L &= k_R/k_\omega \\ k_R &= k_R \\ k_C &= 1/(k_R k_\omega) \end{aligned} \qquad (2.14)$$

The completely scaled circuit values will then be given in (2.15):

$$\begin{aligned} L'_{jk} &= k_L L_{jk} \\ R'_{jk} &= k_R R_{jk} \\ C'_{jk} &= k_C C_{jk} \end{aligned} \qquad (2.15)$$

Example 2 - The circuit shown in Figure 2.8 is an example of a prototype filter design. The element values are at a 1 Ω level, and the frequency response is at a 1 rad/sec range. Scale the circuit of Figure 2.8 so that it will have a load resistance value of 1 kΩ and the parallel LC branch will resonate at 20 kHz.

Solution - Clearly, the impedance scale factor is $k_R = 10^3$, by inspection, since the load resistance needs to scale from 1 Ω to 1 kΩ. However, in order to determine the frequency scale factor, k_ω, we need to use the prototype frequency value at which the parallel LC branch is designed to resonate. Equation (2.16) gives this result, which is also shown with the circuit schematic:

$$\omega_0 = \frac{1}{\sqrt{LC}} = \frac{1}{\sqrt{1.68 \times 0.09524}} = 2.50 \qquad (2.16)$$

Now, k_ω is found by scaling to the desired value for ω'_0 in the scaled circuit:

$$\begin{aligned} \omega'_0 &= 2\pi 20 \times 10^3 = k_\omega \omega_0 = k_\omega 2.50, \text{ or} \\ k_\omega &= \frac{40\pi \times 10^3}{2.50} = 16\pi 10^3 \end{aligned} \qquad (2.17)$$

2.2. NETWORK SCALE FACTORS

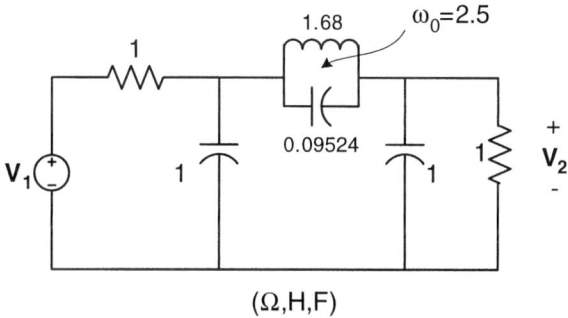

Figure 2.8. A prototype passive network.

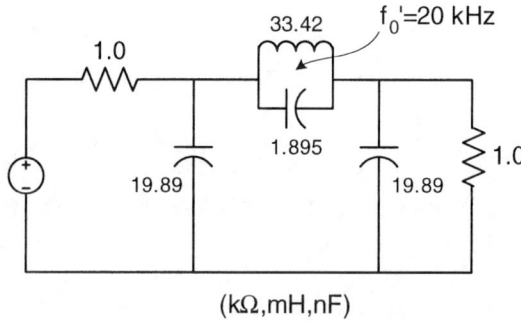

Figure 2.9. Scaled network for practical element values and a specified branch resonance.

The required network factors are then obtained by substitutions into (2.14) and (2.15):

$$
\begin{align}
k_R &= 10^3 \ \Omega/\Omega \\
k_L &= \frac{k_R}{k_\omega} = \frac{1}{16\pi} = 19.894 \text{ mH/H} \\
k_C &= \frac{1}{k_R k_\omega} = \frac{1}{16\pi 10^6} = 19.894 \text{ nF/F}
\end{align}
\tag{2.18}
$$

The resulting scaled network is shown in Figure 2.9.

This example shows one way to determine scale factors for prototype networks. With two unknowns, k_R and k_ω, two independent specifications are always required in order to find a unique scaling of element values.

Example 3 - What are k_R and k_ω for the prototype shown in Figure 2.8 so that the load capacitance will equal 100 pF and the inductor will have a value of 10 mH?

Solution - The factors are found from the scale factor relations given in (2.14) and the circuit data as follows:

$$100 \times 10^{-12} = 1 k_C = \frac{1}{k_R k_\omega}$$

$$10 \times 10^{-3} = 1.68 k_L = 1.68 \frac{k_R}{k_\omega}$$

Multiplication yields k_ω:

$$1000 \times 10^{-15} = \frac{1.68}{k_\omega^2} \text{ or } k_\omega^2 = \frac{1.68}{10^{-12}}$$

The scale factors are thus $k_\omega = 1.29615 \times 10^6$ and $k_R = 10 \times 10^{-3} k_\omega / 1.68 = 7715.17$.

This example tells us over what frequency range the circuit can be scaled in order to have a specified minimum load capacitance on the basis of a given inductor value.

2.3 Frequency Transformations of Passive Filters

Tabulated lowpass prototype designs can be transformed to alternate filter types such as highpass, bandpass, or bandstop through the use of frequency transforms. There is a wealth of tabulated filter designs already in existence and the ability to transform to different filter functions just enhances the utility of the prototype concept. In addition, we will learn later that the very stable leapfrog active filter topology is based upon passive filter topology, and so it is useful to understand how to get different filter functions from tabulated, or known, LP prototype designs.

The basic idea behind frequency transformation is that impedance is preserved between each network (original and transformed result) at different frequencies on a one-to-one basis. In the following examples we let $s = \sigma + j\omega$ be the prototype frequency and $s' = \sigma + j\omega'$ be the new, or use, frequency variable. The procedure to obtain a particular filter function with practical values is to first do the frequency transform on a prototype frequency scale (1 rad/sec), and then scale the resulting design to practical element values.

2.3.1 LP to LP

Example 2 illustrated the lowpass-to-lowpass (LP to LP) transformation procedure implicit in (2.14) and (2.15). In this section we review the procedure formally to illustrate the transformation notation and concept.

The LP to LP is characterized by the relation given in (2.19):

$$s' = k_\omega s \longleftrightarrow s = s'/k_\omega \qquad (2.19)$$

Equation (2.19) expands or contracts the scale, as shown in Figure 2.7. Invoking the preservation of branch impedances, the following results are obtained for the

2.3. FREQUENCY TRANSFORMATIONS OF PASSIVE FILTERS

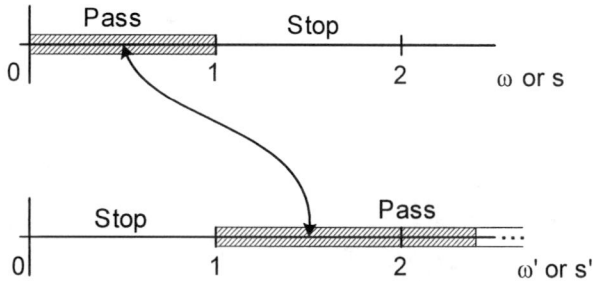

Figure 2.10. The one-to-one LP to HP transform.

transformed RLC elements:
a. Resistance is independent of frequency, so $R' = R$.
b. Inductance is inversely proportional to k_ω:

$$Ls = L's' \quad \text{so}$$
$$L' = L\frac{s}{s'} = \frac{L}{k_\omega} \quad (2.20)$$

c. Capacitance is inversely proportional to k_ω:

$$Cs = C's' \quad \text{so}$$
$$C' = C\frac{s}{s'} = \frac{C}{k_\omega} \quad (2.21)$$

The above equations are the same as those given previously in (2.13).

2.3.2 LP to HP

This formation swaps the LP passband and stopband through the use of the reciprocal frequency relationship given in (2.22):

$$s' = 1/s \longleftrightarrow s = 1/s' \quad (2.22)$$

The frequency effect of this transform is shown in Figure 2.10. The following are results of holding branch impedances invariant:

a. Resistance is independent of frequency, so $R' = R$.
b. Reciprocal inductance maps to capacitance:

$$Z(s) = Ls = Z(s') = \frac{L}{s'} = \frac{1}{C's'}$$
$$C' = \frac{1}{L} \quad (2.23)$$

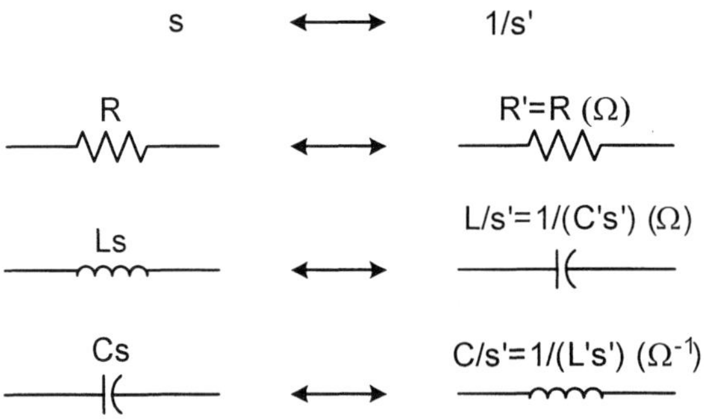

CHAPTER 2. REVIEW OF CIRCUIT ANALYSIS CONCEPTS

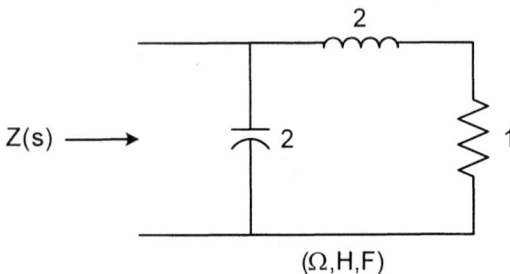

Figure 2.11. Element mappings for the LP to HP transformation.

Figure 2.12. LP circuit for Example 4.

c. Reciprocal capacitance maps to inductance:

$$Y(s) = Cs = Y(s') = \frac{C}{s'} = \frac{1}{L's'}$$

$$L' = \frac{1}{C} \quad (2.24)$$

These element transformations are illustrated in Figure 2.11.

Example 4 - Perform an LP to HP transform on the LP prototype circuit given in Figure 2.12. Calculate the input impedance for each network, and show that $Z(s')$ is equal to $Z(1/s)$.

Solution - The procedures given in (2.22) through (2.24) are applied to the circuit of Figure 2.12, yielding the result shown in Figure 2.13. The input impedances for each network are obtained by repeated series-parallel calculations (repeated fractions) as follows:

2.3. FREQUENCY TRANSFORMATIONS OF PASSIVE FILTERS

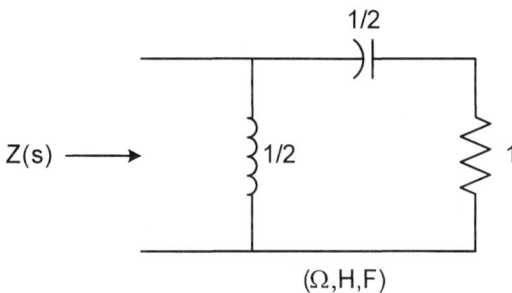

Figure 2.13. HP circuit obtained from the LP circuit of Figure 2.12.

From Figure 2.12:
$$Z(s) = \frac{1}{2s + \frac{1}{2s+1}} \quad (2.25)$$

From Figure 2.13:
$$Z(s') = \frac{1}{\frac{2}{s'} + \frac{1}{2/s'+1}} \quad (2.26)$$

Clearly, $Z(s) = Z(s')$.

Note in this transformation that a pole or zero at $s = \alpha$ goes to a pole or zero at $s' = 1/\alpha$. The same can be said for critical frequencies at infinity or zero.

Example 5 - Apply the LP to HP transform to the following network function, $H(s)$, and compare the critical frequencies for each network function:

$$H(s) = \frac{2s+1}{4s^2 + 2s + 1} = H_{LP} \quad (2.27)$$

Solution - Substitute $s = 1/s'$ into (2.27) to get the transformed network function of (2.28):

$$H(s') = \frac{2/s' + 1}{4/s'^2 + 2/s' + 1} = \frac{s'(s'+2)}{4 + 2s' + s'^2} = H_{HP} \quad (2.28)$$

The critical frequencies are as follows:

For H_{LP}: zeros, $-0.5, \infty$; poles, $0.5\angle \pm 120°$.

For H_{HP}: zeros, $-2, 0$; poles, $2.0\angle \mp 120°$.

Clearly, the critical frequencies are reciprocal to each other for the two networks.

2.3.3 LP to BP

This transform maps a single frequency in the LP frequency domain to two frequency points in the BP domain. It is a one-to-two valued transform. Although this transform is mathematically complex, it is nevertheless a widely used transform because of its inherent practicality.

The transform combines the LP to LP and the LP to HP in one operation, while including a factor to control the resulting bandwidth of the transform. First,

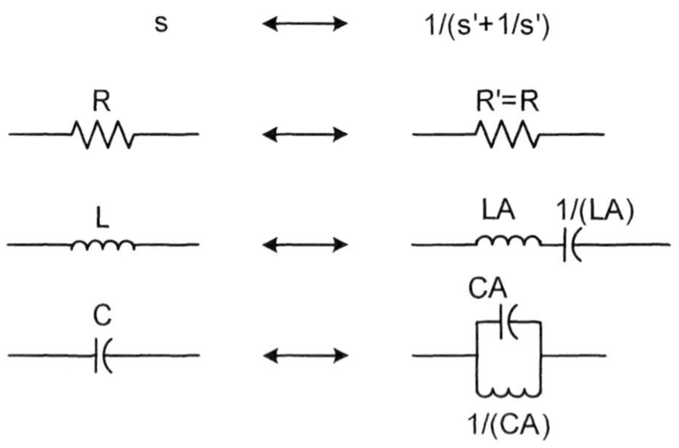

CHAPTER 2. REVIEW OF CIRCUIT ANALYSIS CONCEPTS

Figure 2.14. Element mappings for the LP to BP transformation.

we consider s in terms of s' (a quadratic relation) to gain some understanding of how the transform works; then we solve for s' as a function of s to determine the bandwidth relationships. The fundamental frequency transform is given in (2.29):

$$s = As' + A/s' \qquad (2.29)$$

The following results are obtained for RLC elements:

a. Resistance is independent of frequency, so $R' = R$.
b. Inductance maps to a series LC branch:

$$Z(s) = Ls = Z(s') = LAs' + LA/s' = L's' + \frac{1}{C's'} \quad \text{so}$$

$$L' = LA \quad \text{and} \quad C' = \frac{1}{LA} \qquad (2.30)$$

c. Capacitance maps to a parallel LC branch:

$$Y(s) = Cs = Y(s') = CAs' + CA/s' = C's' + \frac{1}{L's'} \quad \text{so}$$

$$C' = CA \quad \text{and} \quad L' = \frac{1}{CA} \qquad (2.31)$$

The element transformation results are shown in Figure 2.14. Note that each LP energy storage element transforms into two energy storage elements that resonate at $\omega_0 = 1$ rad/sec.

Example 6 - Perform the BP transformation, $s = 10(s' + 1/s')$, on the LP circuit given in Figure 2.15; find $\mathbf{V}_2/\mathbf{V}_1$ for each network (the given LP and the derived BP); and show that the networks satisfy the required frequency transform. In addition, determine the filter Q^2 for the bandpass circuit response.

---

[2]Recall that filter Q is the ratio of center frequency to 3-dB bandwidth.

2.3. FREQUENCY TRANSFORMATIONS OF PASSIVE FILTERS

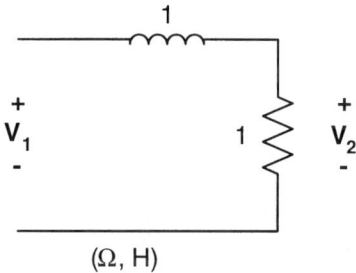

Figure 2.15. LP circuit for Example 6.

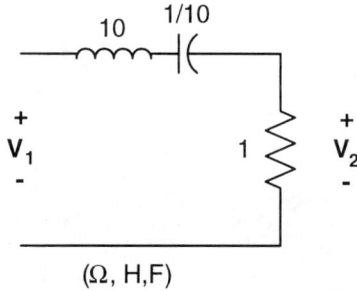

Figure 2.16. BP circuit derived from the prototype of Figure 2.15.

Solution - Using the voltage division concept, the two network functions are obtained as follows:

From Figure 2.15:

$$H(s) = \frac{V_2}{V_1} = \frac{1}{s+1} = H_{LP} \tag{2.32}$$

From Figure 2.16:

$$H(s') = \frac{1}{10s' + 10/s' + 1} = H_{BP} \equiv H_{LP}(10s' + 10/s') \tag{2.33}$$

$$= \frac{s'/10}{s'^2 + s'/10 + 1} = \frac{Ks'}{s'^2 + 2\alpha s' + \omega_0^2} \tag{2.34}$$

The simple LP function for the circuit in Figure 2.15 is transformed into the well-known BP quadratic form, (2.34), for the circuit in Figure 2.16. In standard form, $\omega_0^2 = 1$ is the resonant frequency (peak) of the steady-state sine wave response, while $2\alpha = 1/10$ is the -3-dB bandwidth of the response. The filter Q is equivalent to the complex pole Q, $\omega_0/2\alpha$, for this simple quadratic network function, and this

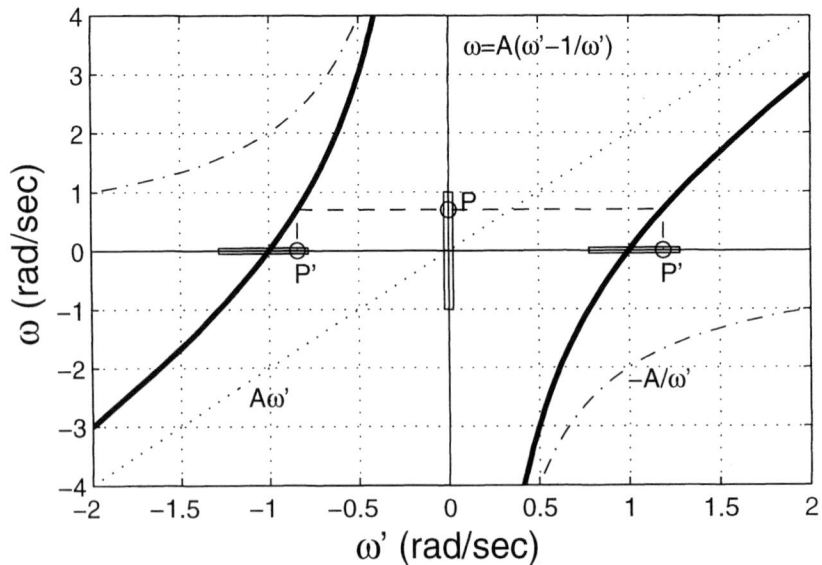

Figure 2.17. The LP to BP frequency transformation template for $s' = j\omega'$.

factor is found to equal 10. Note that this Q factor is also equal to the bandpass transformation factor, A.

We now look at the LP to BP transformation in more detail in order to derive bandwidth relationships between the LP prototype and the derived BP prototype.

Let $s' = j\omega'$. Then substitution into (2.29) yields the corresponding relation for $s = j\omega$ for steady-state sinewave excitation:

$$\omega = A(\omega' - 1/\omega') \tag{2.35}$$

A sketch of (2.35) is shown in Figure 2.17, using ω' as the independent variable (abscissa). The block of ω-values represents passband frequencies for the LP circuit, and these are mapped to corresponding points on the ω'-axis. Observe that each ω-value maps to two ω'-values for this transform as noted earlier.

The inverse of (2.29) provides the desired relationship to show how two s' points are related to a single LP complex frequency, s.

$$\begin{aligned} \text{With } s &= A(s' + 1/s') \quad \text{or equivalently} \\ 0 &= s'^2 - \left(\frac{s}{A}\right)s' + 1 \quad \text{we obtain} \\ s' &= \frac{s}{2A} \pm \sqrt{\frac{s^2}{4A^2} - 1} \end{aligned} \tag{2.36}$$

Equation (2.36) shows that for each value of s, there are two values of s' due to the quadratic relationship between s and s'. Now, let $s = j\omega$, and consider the LP

2.3. FREQUENCY TRANSFORMATIONS OF PASSIVE FILTERS

sinusoidal steady-state results. Substitution yields (2.37):

$$\text{For } s = j\omega$$
$$s' = j\omega' = \frac{j\omega}{2A} \pm j\sqrt{\frac{\omega^2}{4A^2} + 1} \text{ or}$$
$$\omega' = \frac{\omega}{2A} \pm \sqrt{\frac{\omega^2}{4A^2} + 1} \tag{2.37}$$

The pass bandwidth for the BP filter is given by $BW = \omega'_2 - \omega'_1$, where ω'_2 is the $+\omega'$-value for $\omega = 1$, and ω'_1 is the $+\omega'$-value for $\omega = -1$. The following is obtained:

$$\text{For } \omega = +1, \quad \omega'_2 = \frac{1}{2A} + \sqrt{\frac{1}{4A^2} + 1} \text{ and}$$
$$\text{for } \omega = -1, \quad \omega'_1 = \frac{-1}{2A} + \sqrt{\frac{1}{4A^2} + 1} \tag{2.38}$$

From (2.38), the following identities are obtained:

$$BW = \omega'_2 - \omega'_1 = 1/A \text{ and}$$
$$\omega_0^2 = \omega'_2 \omega'_1 = 1 \tag{2.39}$$

Thus, the BP bandwidth is the reciprocal of the constant, A, in the frequency transformation, and the passband edge frequencies are reciprocal to each other. This assumes that the LP bandwidth is unity, as is customary.

2.3.4 LP to BP Network Functions

The LP to BP transformation can be applied directly to an LP network function to obtain the corresponding BP network function for any filter that has a lowpass equivalent, or source network. Specific results are as follows.

Let $s = A(s' + 1/s')$ in the following LP network function to obtain a transformed network function:

$$H_{LP}(s) = \frac{b_m \prod_{k=1}^{m}(s - z_k)}{a_n \prod_{k=1}^{n}(s - p_k)} = \frac{b(s)}{a(s)} \tag{2.40}$$

The z_k and p_k are the LP zeros and poles, respectively. Now, substitute the LP to BP frequency transform, (2.29), to get the corresponding BP network function:

$$H(s') = \frac{b_m \prod_{k=1}^{m}(As' - z_k + A/s')}{a_n \prod_{k=1}^{n}(As' - p_k + A/s')}$$
$$= \frac{b_m}{a_n}\left(\frac{A}{s'}\right)^{m-n} \frac{\prod_{k=1}^{m}(s'^2 - z_k s'/A + 1)}{\prod_{k=1}^{n}(s'^2 - p_k s'/A + 1)} \tag{2.41}$$

This result shows that each pole or zero of the LP function gets mapped to a pair of complex poles or zeros. The resulting critical frequencies are reciprocal in the s'-plane as shown in Figure 2.18 for an example LP pole-zero diagram. Note that

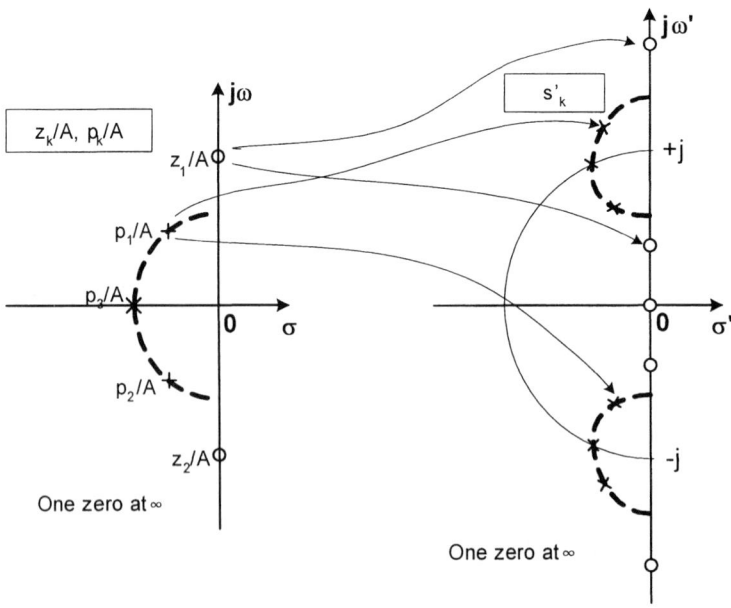

Figure 2.18. Example LP pole-zero BP transformation geometry where the LP function is scaled by the transform factor, A.

the zero at infinity in the LP function maps to a zero at infinity and a zero at the origin in the BP result. This is in accordance with the geometrical mean result for the mapping of any LP point to two BP points. As follows from (2.36), we get (2.42) for any s-value:

$$s'_2 = \frac{s}{2A} + \sqrt{\frac{s^2}{4A^2} - 1} \quad \text{and}$$

$$s'_1 = \frac{s}{2A} - \sqrt{\frac{s^2}{4A^2} - 1} \quad \text{so that}$$

$$s'_2 s'_1 \equiv 1 \tag{2.42}$$

Equation (2.42) says that the mapped BP frequencies are reciprocal to each other.

Example 7 - The network function for the LP prototype of Figure 2.8 is as follows:

$$H(s) = \frac{V_2}{V_1} = \frac{0.08(s^2 + 6.25)}{(s+1)(s^2 + 0.84s + 1)} = H_{LP}(s) \tag{2.43}$$

Transform this function to a BP function to obtain a filter Q of 5. Determine the poles and zeros for the resulting BP function, and plot the attenuation function, $\alpha(\omega)$, versus ω' over the range $10^{0.3} \leq \omega' \leq 10^{0.3}$. Check the pass bandwidth for the BP response. The 3-dB bandwidth for the LP prototype is 1 rad/sec.

2.3. FREQUENCY TRANSFORMATIONS OF PASSIVE FILTERS

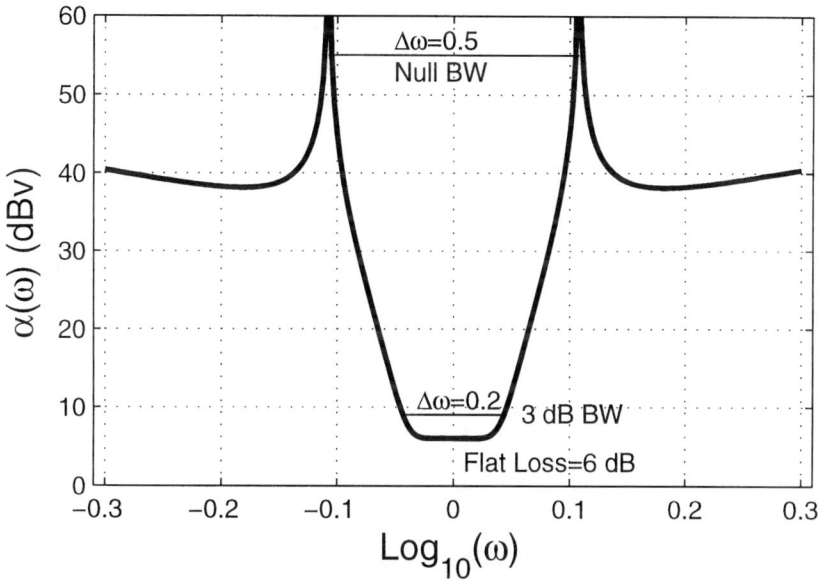

Figure 2.19. Frequency response for the BP function of (2.44).

Solution - The filter Q equals A, so we require $s = 5(s' + 1/s')$ to be substituted into $H(s)$ to obtain $H(s')$:

$$H(s') = \frac{0.08[25(s' + 1/s')^2 + 6.25]}{[5(s' + 1/s' + 1)][25(s' + 1/s')^2 + 4.20(s' + 1/s') + 1]} \times \frac{s'^3}{s'^3}$$

$$= \frac{0.08s'[25(s'^2 + 1)^2 + 6.25s'^2]}{[5(s'^2 + 1) + s'][25(s'^2 + 1)^2 + 4.20(s'^2 + 1)s' + s'^2]}$$

$$= \frac{0.08s'(25s'^4 + 56.25s'^2 + 25)}{(5s'^2 + s' + 5)(25s'^4 + 4.20s'^3 + 51.0s'^2 + 4.20s' + 25)} \qquad (2.44)$$

$$H_{BP}(s) = \frac{2.0s(s^4 + 2.25s^2 + 1)}{125(s^6 + 0.3680s^5 + 3.0736s^4 + 0.7440s^3 + 3.0736s^2 + 0.3680s + 1)}$$

The poles are $p_{1,2} = -0.0458 \pm j1.094$; $p_{3,4} = -0.1000 \pm j0.9950$; and $p_{5,6} = -0.0382 \pm j0.9125$. The zeros are $z_{1,2} = \pm j1.2808$; $z_{3,4} = \pm j0.7808$; and $z_5 = 0.0$.

The attenuation response for (2.44) is shown in Figure 2.19, plotted against $\log_{10}(\omega)$ to show the geometrical symmetry of the response. The null-bandwidth-to-3-dB-bandwidth ratio is 2.5:1, which is the same as for the LP prototype function. The 6 dB flat loss at the BP center frequency is due to the equal terminations, which provide a transmission factor of 0.5 at the dc frequency of the LP prototype.

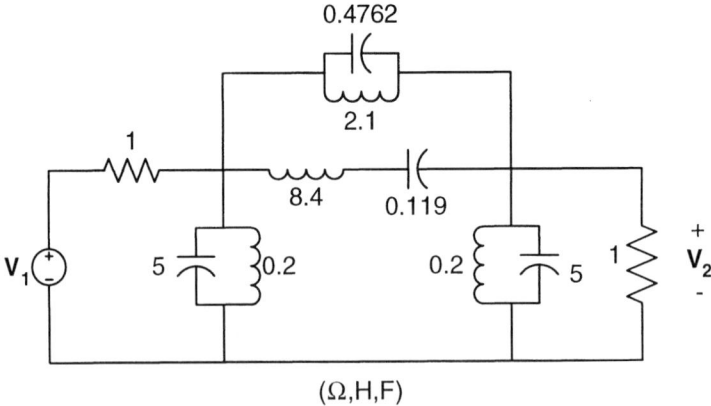

Figure 2.20. BP circuit obtained from the LP of Figure 2.8 for a filter Q of 5.

2.3.5 BP Filter Element Values

When we apply the LP to BP transformation, $s = 5(s' + 1/s')$, to the prototype circuit (see Figure 2.8), we get the circuit shown in Figure 2.20. We see that the inductance ratio for this BP circuit, $8.4/0.2 = 42.0$, is the same as that for the capacitance ratio, $5/0.1190 = 42.0$. These ratios are proportional to the transformation factor squared, A^2. Thus, as the filter Q is increased, the element ratios required to realize the BP circuit go up by Q^2. This points out a serious difficulty in using the LP to BP transform to obtain BP filters.

The problem of the element value ratio depending upon the transformation factor, A^2, is greatly reduced through a procedure that uses an equivalent circuit for the bandpassed parallel LC branch. Consider the result for a bandpassed parallel LC branch, as shown in Figure 2.21. By using element values for the BP result, shown in Figure 2.21(b), the following element ratio is obtained:

$$\frac{L_{max}}{L_{min}} = \frac{AL}{1/AC} = A^2 LC = \frac{A^2}{\omega_0^2} \equiv \frac{C_{max}}{C_{min}} \qquad (2.45)$$

The impedance for the bandpassed branch is obtained by applying the frequency transform to the LP impedance function. Let z_{LP} be the impedance function for the parallel LC branch shown in Figure 2.21(a).

$$z_{LP} = \frac{s}{C(s^2 + \omega_0^2)} \qquad (2.46)$$

By inspection, z_{LP} is changed to z_{BP} by letting $s = A(s' + 1/s')$, and then we can write the impedance, z_{BP}, for the bandpassed branch shown in Figure 2.21(b):

$$z_{BP} = \frac{A(s' + 1/s')}{C[A^2(s' + 1/s')^2 + \omega_0^2]} \times \frac{s'^2}{s'^2} = z_{LP}(A(s' + 1/s'))$$

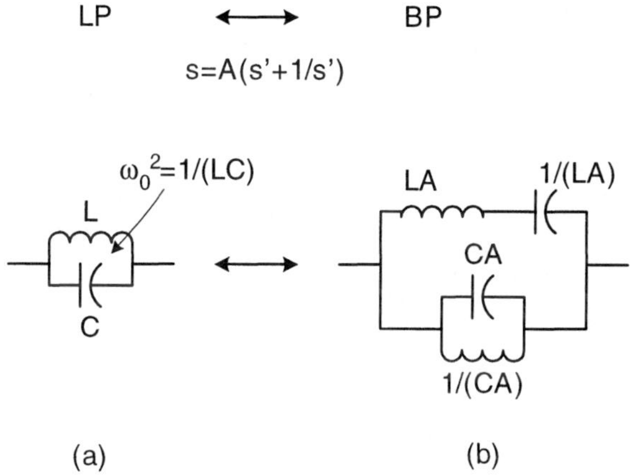

2.3. FREQUENCY TRANSFORMATIONS OF PASSIVE FILTERS

LP ⟷ BP

$$s = A(s' + 1/s')$$

(a) prototype with $\omega_0^2 = 1/(LC)$, L and C in parallel

(b) bandpass equivalent with LA and $1/(LA)$ in series with parallel combination of CA and $1/(CA)$

Figure 2.21. Result of bandpassing a parallel LC branch: (a) prototype; and (b) bandpass equivalent.

$$= \frac{A(s'^2+1)s'}{C[A^2(s'^2+1)^2 + \omega_0^2 s'^2]}, \text{ and}$$

$$= \frac{1}{AC} \frac{s'(s'^2+1)}{(s'^4 + (2+\omega_0^2/A^2)s'^2 + 1)} \quad (2.47)$$

Equation (2.47) shows that z_{BP} is zero at $s = j1$, the BP prototype center frequency, but it does not show the poles where the BP branch exhibits parallel resonance. An equivalent circuit for that of Figure 2.21(b) can be obtained by means of a partial fraction expansion on the poles of z_{BP}. Equation (2.48) illustrates this form of two parallel LC branches in series to obtain z_{BP}:

$$z_{BP} = \frac{1}{AC} \frac{s'(s'^2+1)}{(s'^2+\omega_1^2)(s'^2+\omega_2^2)} \equiv \frac{s'}{C_1(s'^2+\omega_1^2)} + \frac{s'}{C_2(s'^2+\omega_2^2)} \quad (2.48)$$

$$\text{where } \omega_1^2 = 1 + \frac{\omega_0^2}{2A^2} - \frac{\omega_0}{A}\sqrt{1 + \frac{\omega_0^2}{4A^2}} \text{ and}$$

$$\omega_2^2 = 1 + \frac{\omega_0^2}{2A^2} + \frac{\omega_0}{A}\sqrt{1 + \frac{\omega_0^2}{4A^2}} \quad (2.49)$$

Note that $\omega_2 > \omega_1$, and $\omega_1^2 \omega_2^2 = 1$. The roots given in (2.49) are obtained by factoring the denominator of z_{BP}. By using partial fraction expansion procedures, C_1 and C_2 are found as follows:

$$\frac{1}{C_1} = \frac{1}{AC}\left[\frac{s'^2+1}{s'^2+\omega_2^2}\right]_{s'^2=-\omega_1^2} = \frac{1}{AC}\left(\frac{1-\omega_1^2}{\omega_2^2-\omega_1^2}\right) \text{ and}$$

$$\frac{1}{C_2} = \frac{1}{AC}\left[\frac{s'^2+1}{s'^2+\omega_1^2}\right]_{s'^2=-\omega_2^2} = \frac{1}{AC}\left(\frac{1-\omega_2^2}{\omega_1^2-\omega_2^2}\right) \quad (2.50)$$

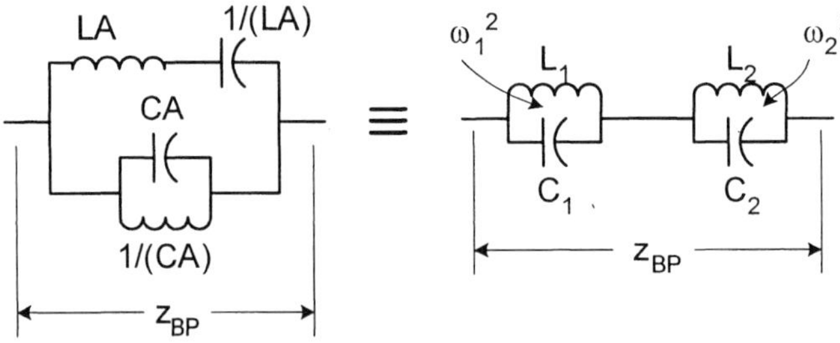

CHAPTER 2. REVIEW OF CIRCUIT ANALYSIS CONCEPTS

[Circuit diagram: Left side shows a parallel LC branch with LA, 1/(LA), CA, 1/(CA) between terminals labeled Z_{BP}. Right side shows equivalent series branches with L_1, C_1 (with ω_1^2) and L_2, C_2 (with ω_2^2) between terminals labeled Z_{BP}.]

$L_{max}/L_{min} = A^2 LC = C_{max}/C_{min}$ and $L_1/L_2 = \omega_2^2$

Where $\omega_1^2 = 1 + \omega_0^2/(2A^2) - \omega_0[1+\omega_0^2/(4A^2)]^{0.5}/A$

$\omega_2^2 = 1 + \omega_0^2/(2A^2) + \omega_0[1+\omega_0^2/(4A^2)]^{0.5}/A > \omega_1^2$

$\omega_2^2 \omega_1^2 = 1$

$C_1 = AC(\omega_2^2 - \omega_1^2)/(1-\omega_1^2)$

$C_2 = AC(\omega_2^2 - \omega_1^2)/(\omega_2^2 - 1)$

$L_1 = 1/(C_1 \omega_1^2) = 1/(AC(1+\omega_1^2))$

$L_2 = 1/(C_2 \omega_2^2) = 1/(AC(1+\omega_2^2)) < L_1$

$L_1/L_2 = (\omega_2^2 - 1)/(1-\omega_1^2) = C_1/C_2$

Figure 2.22. Design parameter summary for an equivalent form of the bandpassed parallel LC branch.

Interpreting (2.48) as $s \to 0$ yields the inductances L_1 and L_2 for the equivalent series branches:

$$L_1 = \frac{1}{C_1 \omega_1^2} \quad \text{and} \quad L_2 = \frac{1}{C_2 \omega_2^2} \tag{2.51}$$

Figure 2.22 summarizes the equivalence for the transformed LP parallel LC branch of Figure 2.21(b). This equivalent circuit provides the extremely interesting result that the element ratio is independent of A^2! This is found by using $\omega_1^2 \omega_2^2 = 1$ and evaluating the ratio for L_1/L_2:

$$\begin{aligned}
\frac{L_1}{L_2} &= \frac{C_2 \omega_2^2}{C_1 \omega_1^2} = AC \left(\frac{\omega_2^2 - \omega_1^2}{\omega_2^2 - 1} \right) \frac{1}{AC} \left(\frac{1 - \omega_1^2}{\omega_2^2 - \omega_1^2} \right) \frac{\omega_2^2}{\omega_1^2} \quad \text{and} \\
&= \frac{(1-\omega_1^2)\omega_2^2}{(\omega_2^2 \omega_1^2 - \omega_1^2)} \equiv \omega_2^2
\end{aligned} \tag{2.52}$$

2.3. FREQUENCY TRANSFORMATIONS OF PASSIVE FILTERS

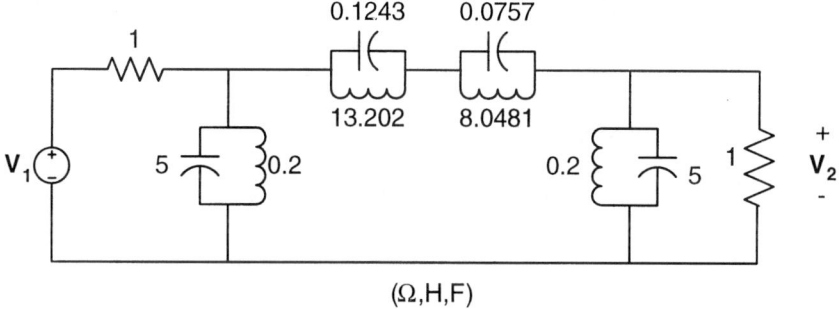

(Ω,H,F)

Figure 2.23. BP circuit obtained from Figure 2.8 for a filter Q of 5 using the equivalency shown in Figure 2.22.

Substitution also confirms that $C_1/C_2 = L_1/L_2$. So, application of this equivalency removes the A^2 component ratio dependence for this transformed branch.

Example 8 - Determine the equivalent circuit transform for the circuit in Figure 2.20, and determine element value ratios for the completed filter design.

Solution - The circuit in Figure 2.20 originated from an LP prototype for which $\omega_0^2 = 6.25$ for the parallel LC branches, $C = 1$, and the transform factor $A = 5$. Substitution then yields the following values for the equivalent transformed branch:

$$\omega_1^2 = 1 + \frac{6.25}{2(25)} - \frac{2.5}{5}\sqrt{1 + \frac{6.25}{4(25)}} = 0.6096$$

$$\omega_2^2 = 1 + \frac{6.25}{2(25)} + \frac{2.5}{5}\sqrt{1 + \frac{6.25}{4(25)}} = 1.6404$$

$$C_1 = 5(1)(\omega_2^2 - \omega_1^2)/(1 - \omega_1^2) = 13.2019$$
$$C_2 = 5(1)(\omega_2^2 - \omega_1^2)/(\omega_2^2 - 1) = 8.0481$$
$$L_1 = 1/(C_1\omega_1^2) = 0.1243$$
$$L_2 = 1/(C_2\omega_2^2) = 0.0757$$

The complete equivalent BP prototype is given in Figure 2.23. The element ratios for this circuit are $L_{max}/L_{min} = 0.2/0.0757 = 2.6404$, and $C_{max}/C_{min} = 13.2019/5 = 2.6404$, respectively. These factors are appreciably less than the 42:1 ratio found for the straight BP circuit shown in Figure 2.20.

This last is a remarkable result and serves to illustrate an important aspect of circuit design, namely, that it is usually possible to obtain equivalent circuits, and they exhibit different properties that are useful to the designer. The circuit of Figure 2.23 is superior to that of Figure 2.20 (even though they give equivalent response in theory) because it is easier to implement and to tune.

The branch equivalence developed here is well known to filter designers and used extensively, and it is applicable to prototype circuits whose network functions have finite j-axis zeros. This feature allows the resulting network design to include capacitance in parallel with each inductor appearing in the BP design. This is very practical as it allows the filter designer to account for inductor parasitic winding capacitance in parallel with each inductance. The consequence is that actual circuits exhibit theoretical response over a wider frequency range.

2.3.6 LP to BS

This transform, like the BP transform, is also a one-to-two transform. The results are analogous to the BP transform, and so, we will not spend much time here on this procedure. The scaling and equivalent impedance ideas are similar to those presented for the BP procedure, and so, it is not necessary to repeat similar examples for this case.

The basic LP to BS frequency transform is as follows. Note that the process is equivalent first to making an LP to HP prototype and then applying the BP transform to the HP circuit:

$$s = \frac{1}{A(s' + 1/s')} = \frac{s'}{A(s'^2 + 1)} \qquad (2.53)$$

The following results are obtained for RLC elements.

a. Resistance is independent of frequency, so $R' = R$.
b. Inductance maps to a parallel LC branch:

$$\begin{aligned} Z(s) &= Ls = Z(s') = \frac{Ls'}{A(s'^2 + 1)} \\ &= \frac{1}{(C's' + 1/L's')} \quad \text{so} \\ C' &= \frac{A}{L} \quad \text{and} \quad L' = \frac{L}{A} \end{aligned} \qquad (2.54)$$

c. Capacitance maps to a series LC branch:

$$\begin{aligned} Y(s) &= Cs = Y(s') = \frac{Cs'}{A(s'^2 + 1)} \\ &= \frac{1}{(L's' + 1/C's')} \quad \text{so} \\ L' &= \frac{A}{C} \quad \text{and} \quad C' = \frac{C}{A} \end{aligned} \qquad (2.55)$$

The topology of the BS result appears to switch the L and C results for the BP transform, and the transform factor A is reciprocal to the BP result. Here, A is a factor related to the stopband center frequency to the stop bandwidth, so normally this is a number greater than 1, as was the case for the BP transform. The fact

2.3. FREQUENCY TRANSFORMATIONS OF PASSIVE FILTERS

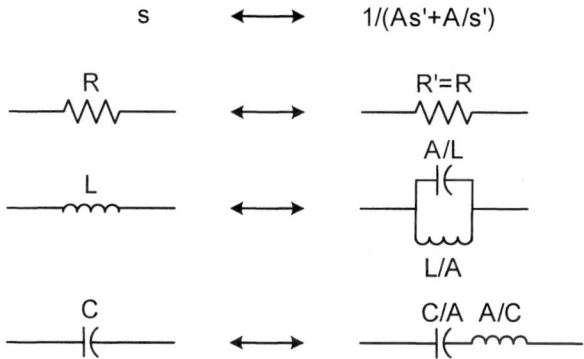

Figure 2.24. Element mappings for the LP to BS transformation.

that an inductor goes to a parallel LC branch is reasonable when it is recalled that an HP transform changes L to C and then a BP transform changes C to a parallel LC branch.

The element transformation results are shown in Figure 2.24 for the LP to BS transform. The corresponding LP to BS frequency template is developed by the following discussion and illustrated in Figure 2.25.

For $s' = jw'$, the BS frequency transform, (2.53), yields the following j-axis frequency relationship:

$$\omega = \frac{\omega'}{A(1 - \omega'^2)} \qquad (2.56)$$

Equation (2.56) is plotted in Figure 2.25 and shows the transformation of the LP pass- and stopbands. The stopband width, $\omega'_2 - \omega'_1$, can be obtained by solving (2.56) for $\omega = \pm 1$, as follows:

$$0 = \omega'^2 + \frac{\omega'}{A\omega} - 1, \quad \text{which yields}$$

$$\omega' = \frac{-1}{2A\omega} \pm \sqrt{\frac{1}{4A^2\omega^2} + 1} \qquad (2.57)$$

Then, for $\omega = +1$

$$\omega'_1 = \frac{-1}{2A} + \sqrt{\frac{1}{4A^2} + 1} \qquad (2.58)$$

and for $\omega = -1$

$$\omega'_2 = \frac{+1}{2A} + \sqrt{\frac{1}{4A^2} + 1} \qquad (2.59)$$

Thus, the stopband bandwidth, $\omega'_2 - \omega'_1$, is just $1/A$, the reciprocal transformation factor:

$$\omega'_2 - \omega'_1 = \frac{1}{A} \quad \text{and}$$
$$\omega'_2 \omega'_1 = 1 = \omega_0^2 \qquad (2.60)$$

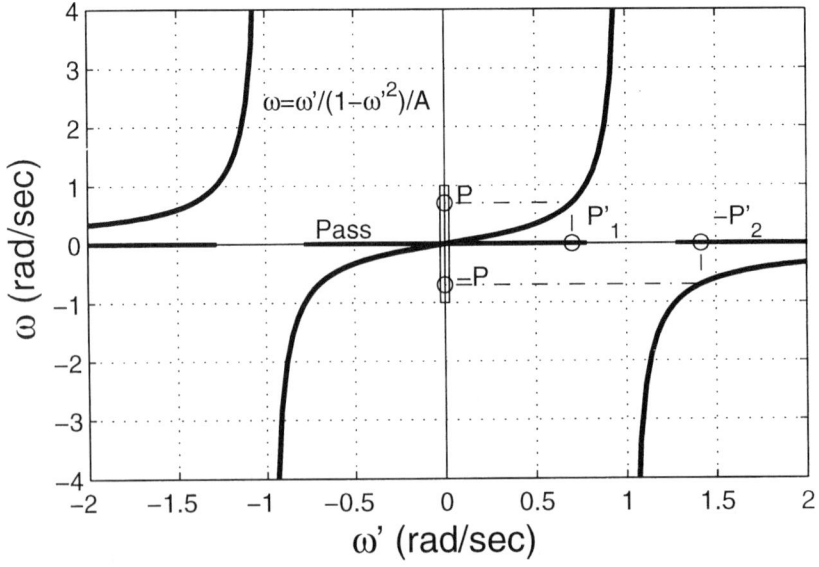

Figure 2.25. The LP to BS frequency transformation template for $s' = j\omega'$.

Equation (2.60) shows that the geometric mean of the bandedge frequencies is just the branch resonant frequency for each LC pair. A similar result is obtained for the transformed bandwidth for $\omega = \pm P$ (any point on the ω-axis) in the LP filter to the bandstop response. We get the following from (2.57) for $\omega = \pm P$:

$$\begin{aligned} P_2' &= \frac{+1}{2AP} + \sqrt{1 + \frac{1}{4A^2P^2}} \quad \text{and} \\ P_1' &= \frac{-1}{2AP} + \sqrt{1 + \frac{1}{4A^2P^2}} \quad \text{so} \\ P_2' - P_1' &= \frac{1}{AP} \quad \text{and} \\ P_2' P_1' &= \omega_0^2 = 1 \end{aligned} \qquad (2.61)$$

The result in (2.61) facilitates choosing a suitable prototype to meet required specifications.

2.4 Impedance Transformations of Passive Filters

In some cases, it is desirable to use filters whose network functions put all the zeros at infinity. These are referred to as all-pole filters. Such a filter is shown in Figure 2.26 for a third-order Butterworth function. The circuit requires a pure inductance, which is difficult to realize in practice. While these filters have implementation

2.4. IMPEDANCE TRANSFORMATIONS OF PASSIVE FILTERS

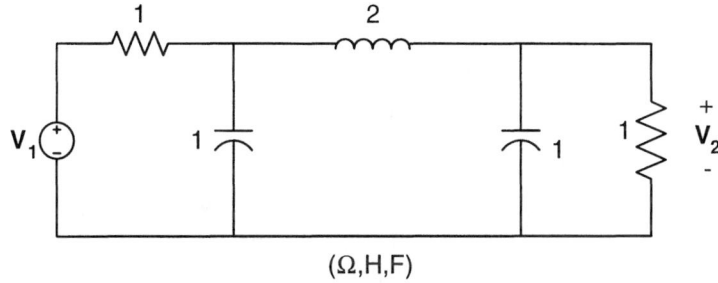

Figure 2.26. A third-order Butterworth LP prototype filter.

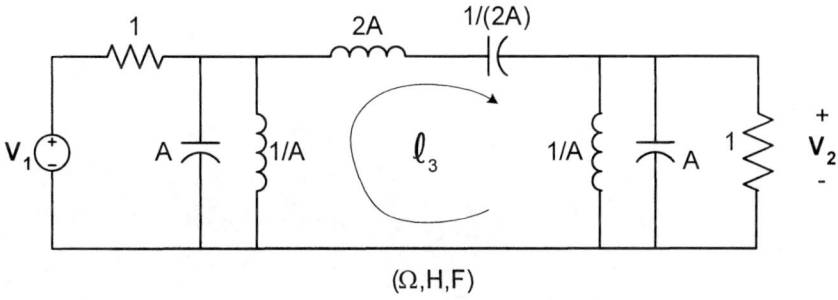

Figure 2.27. BP obtained from the LP prototype of Figure 2.26.

problems, they are used in many applications, and especially for leapfrog active implementations, they provide useful low-sensitivity realizations for the LP filter. However, their BP realizations have a serious element ratio problem, which is not easy to ignore. The next section illustrates the idea of network internal scaling techniques, which are sometimes used to alleviate element value ratios for such circuits.

2.4.1 The Internal Loop Scaling Procedure

Figure 2.27 illustrates a BP realization for the LP prototype given in Figure 2.26 and subjected to the BP frequency transform, $s = A(s' + 1/s')$. The element value ratio for this circuit is $2A^2$ for both L and C, and this can be hard to implement for moderate to high Q requirements. Unlike the last example, there are no straightforward equivalencies to help reduce the required ratio of element values. It helps here to use a concept known as internal loop scaling. This concept originates through a consideration of an ideal transformer impedance scaling property.

A series of steps indicated in Figure 2.28 provides a change in load impedance for a given two-port network. This change does not affect the input impedance at

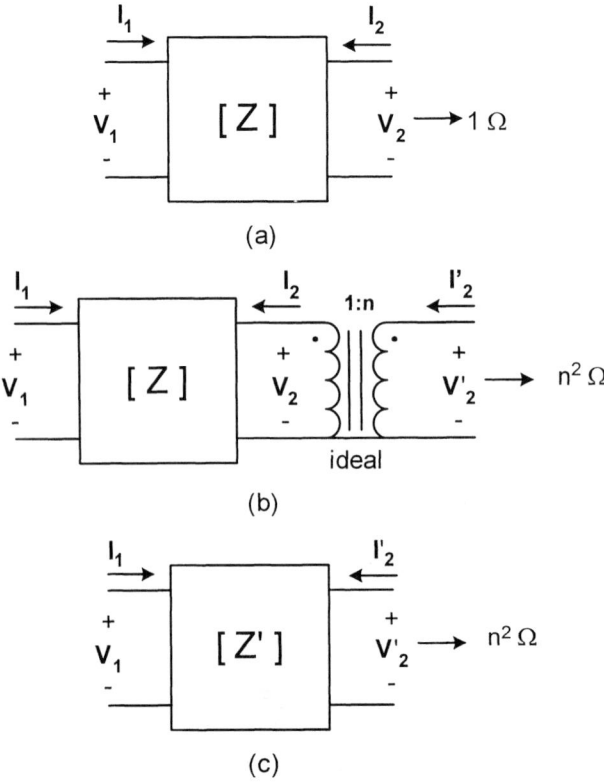

Figure 2.28. An ideal transformer is used to scale the load impedance level of a two-port network without changing the input impedance and power transfer: (a) given two-port circuit model, (b) with ideal transformer added to Port-2, and (c) transformed two-port circuit model.

Port-1, but it does require the impedance following Port-2 to go from a 1 Ω level to an n^2 Ω level. Note that ideal transformers do not absorb any power, so the same power appears at the transformed load for the circuit in Figure 2.28(c) as it does for the circuit given in Figure 2.28(a). Evaluating the effect of this change on the z-parameters shows the transformer effect. The transformer provides the following constraints on $\mathbf{V}_2$ and $\mathbf{I}_2$:

$$\begin{aligned} \mathbf{V}'_2 &= n\mathbf{V}_2 \text{ and} \\ \mathbf{I}_2 &= n\mathbf{I}'_2 \end{aligned} \quad (2.62)$$

The z-parameters relate $\mathbf{I}_1, \mathbf{I}_2$ to $\mathbf{V}_1, \mathbf{V}_2$:

$$\begin{aligned} \mathbf{V}_1 &= z_{11}\mathbf{I}_1 + z_{12}\mathbf{I}_2 = z_{11}\mathbf{I}_1 + z_{12}n\mathbf{I}'_2 \text{ and} \\ \mathbf{V}'_2 &= n\mathbf{V}_2 = nz_{21}\mathbf{I}_1 + n^2 z_{22}\mathbf{I}'_2 \end{aligned} \quad (2.63)$$

2.4. IMPEDANCE TRANSFORMATIONS OF PASSIVE FILTERS

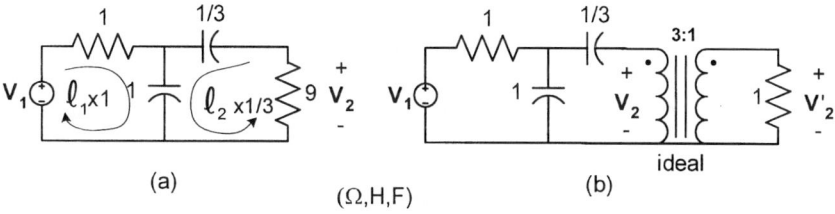

(a)   (Ω,H,F)   (b)

Figure 2.29. Using an ideal transformer to lower load resistance: (a) a given circuit, and (b) circuit (a) with a transformed load.

$[Z']$ is then given in terms of $[Z]$, as in (2.64):

$$[Z'] = \begin{bmatrix} z_{11} & nz_{12} \\ nz_{21} & n^2 z_{22} \end{bmatrix} \equiv \begin{bmatrix} 1 & 0 \\ 0 & n \end{bmatrix} \begin{bmatrix} z_{11} & z_{12} \\ z_{21} & z_{22} \end{bmatrix} \begin{bmatrix} 1 & 0 \\ 0 & n \end{bmatrix}$$

$$= \begin{bmatrix} 1 & 0 \\ 0 & n \end{bmatrix} [Z] \begin{bmatrix} 1 & 0 \\ 0 & n \end{bmatrix} \qquad (2.64)$$

The ideal transformer, placed in the output loop, has multiplied the second row and column of the y-parameter matrix. This result is a circuit example of the use, or meaning, in application of elementary transformation matrices on a set of linear equations.

Before applying loop scaling to the all-pole BP filter element ratio problem, it is helpful to see that loop scaling does in fact work equivalently to using an ideal transformer in a circuit.

Consider the circuits shown in Figure 2.29. Using ladder analysis on the circuit of Figure 2.29(a), it is easy to find the following network function for the circuit:

$$\frac{\mathbf{V}_2}{\mathbf{V}_1}(s) = H(s) = \frac{9}{9s + 13 + 3/s} \qquad (2.65)$$

The transformer, in Figure 2.29(b), converts the 1-Ω resistor to 9 Ω so that $\mathbf{V}_2/\mathbf{V}_1$ is the same for circuit (b) as it is for circuit (a). However, the transformer sets the load voltage, $\mathbf{V}'_2$, to one-third of $\mathbf{V}_2$, and so the 1 Ω load voltage is 1/3 the voltage of circuit (a). Note that the power delivered to the load resistor is the same for both circuits; in circuit (a), $P_L = \mathbf{V}_2^2/9 = H^2(s)V_1^2(s)/9$, and in circuit (b), $P_L = \mathbf{V}'^2_2/1 = H^2(s)V_1^2(s)/3^2$. So, both circuits deliver the same power to the different load resistors.

Now an equivalent circuit for Figure 2.29(b) can be derived by applying loop scaling, (2.64), to the circuit in Figure 2.29(a) as follows. By inspection, the loop parameter matrices, for the circuit in Figure 2.29(a), are as follows:

$$[Z] = [R] + \frac{1}{s}[C^{-1}] = \begin{bmatrix} 1 & 0 \\ 0 & 9 \end{bmatrix} + \frac{1}{s}\begin{bmatrix} 1 & 1 \\ 1 & 4 \end{bmatrix}$$

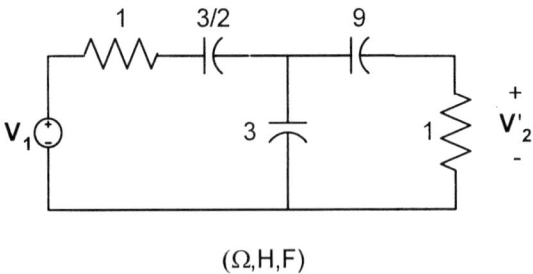

(Ω,H,F)

Figure 2.30. Circuit derived from the circuit of Figure 2.29(a) using loop scaling.

Now, apply loop scaling, with $n = 1/3$, to $[Z]$ to get $[Z']$ for an equivalent loop scaled network, where Loop 2 is scaled by $n = 1/3$:

$$
\begin{aligned}
[Z'] &= \begin{bmatrix} 1 & 0 \\ 0 & 1/3 \end{bmatrix} [Z] \begin{bmatrix} 1 & 0 \\ 0 & 1/3 \end{bmatrix} \\
&= \begin{bmatrix} 1 & 0 \\ 0 & 1/3 \end{bmatrix} \left\{ \begin{bmatrix} 1 & 0 \\ 0 & 3 \end{bmatrix} + \frac{1}{s} \begin{bmatrix} 1 & 1/3 \\ 1 & 4/3 \end{bmatrix} \right\} \\
&= \begin{bmatrix} 1 & 0 \\ 0 & 1 \end{bmatrix} + \frac{1}{s} \begin{bmatrix} 1 & 1/3 \\ 1/3 & 4/9 \end{bmatrix}
\end{aligned}
\quad (2.66)
$$

Interpretation of (2.66) for two loops yields the circuit in Figure 2.30. Simple ladder analysis of Figure 2.30 yields the following result for its network function:

$$\frac{\mathbf{V'}_2}{\mathbf{V}_1} = \frac{3}{9s + 13 + 3/s} \equiv \frac{1}{3} H(s) \quad (2.67)$$

Equation (2.67) shows that the circuit in Figure 2.30 is equivalent to the circuit in Figure 2.29(b) with the result that loop scaling is equivalent to inserting an ideal tranformer into a circuit. This is an amazing result, namely, that ideal transformer action can be obtained by loop scaling part of a circuit! The result is that a half-tee network gets changed to a full tee or the half-tee swaps sides topologically. Now let's apply loop scaling to the all-pole BP filter element value problem.

Example 9 - Use internal loop scaling on ℓ_3 for the circuit in Figure 2.27, and find an equivalent circuit for the resulting parameter matrices. Choose a value for the loop scale factor, n, so that all inductors are equal in the transformed circuit.

Solution - The given circuit, Figure 2.27, has five loops (note alternating directions) and yields the following parameter matrices:

$$[Z] = s[L] + [R] + \frac{1}{s}[C^{-1}]$$

2.4. IMPEDANCE TRANSFORMATIONS OF PASSIVE FILTERS

$$[R] = \begin{bmatrix} 1 & 0 & 0 & 0 & 0 \\ 0 & 0 & 0 & 0 & 0 \\ 0 & 0 & 0 & 0 & 0 \\ 0 & 0 & 0 & 0 & 0 \\ 0 & 0 & 0 & 0 & 1 \end{bmatrix}$$

$$[L] = \begin{bmatrix} 1/A & 1/A & 0 & 0 & 0 \\ 1/A & 1/A & 0 & 0 & 0 \\ 0 & 0 & 2A^2 & 0 & 0 \\ 0 & 0 & 0 & 1/A & 1/A \\ 0 & 0 & 0 & 1/A & 1/A \end{bmatrix}$$

$$[C^{-1}] = \begin{bmatrix} 0 & 0 & 0 & 0 & 0 \\ 0 & 1/A & 1/A & 0 & 0 \\ 0 & 1/A & 2A+2/A & 1/A & 0 \\ 0 & 0 & 1/A & 1/A & 0 \\ 0 & 0 & 0 & 0 & 0 \end{bmatrix} \qquad (2.68)$$

Then
$$[Z'] = [n][Z][n] = s[L'] + [R'] + \frac{1}{s}[C'^{-1}]$$

where
$$[n] = \begin{bmatrix} 1 & 0 & 0 & 0 & 0 \\ 0 & 1 & 0 & 0 & 0 \\ 0 & 0 & n & 0 & 0 \\ 0 & 0 & 0 & 1 & 0 \\ 0 & 0 & 0 & 0 & 1 \end{bmatrix} \qquad (2.69)$$

The internal scaling is just a single scaling on Loop 3, so the result multiplies Row 3 and Column 3 (of the parameter matrices) by the factor, n, to obtain the transformed loop parameter matrices:

$$[R'] = [n][R][n] = \begin{bmatrix} 1 & 0 & 0 & 0 & 0 \\ 0 & 0 & 0 & 0 & 0 \\ 0 & 0 & 0 & 0 & 0 \\ 0 & 0 & 0 & 0 & 0 \\ 0 & 0 & 0 & 0 & 1 \end{bmatrix}$$

$$[L'] = [n][L][n] = \begin{bmatrix} 1/A & 1/A & 0 & 0 & 0 \\ 1/A & 1/A & 0 & 0 & 0 \\ 0 & 0 & 2A^2n^2 & 0 & 0 \\ 0 & 0 & 0 & 1/A & 1/A \\ 0 & 0 & 0 & 1/A & 1/A \end{bmatrix}$$

$$[C'^{-1}] = [n][C^{-1}][n] = \begin{bmatrix} 0 & 0 & 0 & 0 & 0 \\ 0 & 1/A & n/A & 0 & 0 \\ 0 & n/A & n^2(2A+2/A) & n/A & 0 \\ 0 & 0 & n/A & 1/A & 0 \\ 0 & 0 & 0 & 0 & 0 \end{bmatrix} \qquad (2.70)$$

The transformation does not affect the load resistors while providing an arbitrary factor for scaling the inductor in Loop 3. The reciprocal capacitance (susceptance) matrix provides a constraint on the choice of n, assuming it is desirable to realize $[C'^{-1}]$ with all susceptance elements greater than or equal to zero. From Loop 2, $n < 1$ is required, and then the total susceptance around Loop 3 has to be greater than the sum of susceptance shared with other loops (ℓ_2 and ℓ_4 in this case), as given in (2.71):

$$\frac{2n}{A} \le n^2(2A + \frac{2}{A})$$

Thus, for both conditions,

$$1 > n \ge \frac{1/A}{(A + 1/A)} = \frac{1}{A^2 + 1} \tag{2.71}$$

Equation (2.72) is required in order to equalize all inductor values in $[L']$:

$$2n^2 = 1/A \quad \text{or}$$
$$n = \frac{1}{\sqrt{2A}} = \frac{0.7071}{A} \tag{2.72}$$

It is always possible to select this desired value of n since $1/\sqrt{2A}$ is always greater than $1/(A^2 + 1)$ for all real $A > 0$. Normally, A is real and greater than 1 for this procedure.

$[R']$ is unaffected by the choice of n, and for $n = 1/\sqrt{2A}$, $[L']$ is apparent by inspection. $[C'^{-1}]$ is as follows:

$$[C'^{-1}] = \begin{bmatrix} 0 & 0 & 0 & 0 & 0 \\ 0 & 1/A & 0.7071/A & 0 & 0 \\ 0 & 0.7071/A & (1/A + 1/A^3) & 0.7071/A & 0 \\ 0 & 0 & 0.7071/A & 1/A & 0 \\ 0 & 0 & 0 & 0 & 0 \end{bmatrix} \tag{2.73}$$

A circuit realization for the transformed parameter matrices is shown in Figure 2.31. The network function between $\mathbf{V}_2$ and $\mathbf{V}_1$ is identical for both circuits given in Figures 2.27 and 2.31. A big difference is that the inductor element value ratio is 1 for the circuit in Figure 2.31, whereas it is $2A^2$ for the prototype given in Figure 2.27. The capacitance ratio is not improved as much, but it has the following value depending upon the BP factor, A.

For $A > \sqrt{2}$, it is possible to show that the uncoupled capacitor in Loop 3 is larger than the uncoupled capacitor in Loop 2 or 4; and so, the following ratio is obtained:

$$\frac{C_{max}}{C_{min}} = \frac{\sqrt{2}A^2}{\sqrt{2}A^2/(\sqrt{2}A - 1)} = \sqrt{2}A - 1 \tag{2.74}$$

A different ratio is obtained for $A < \sqrt{2}$, but this is not an interesting result. The capacitance ratio in (2.74) is proportional to A rather than A^2, as it was for the original prototype. This result is straightforward to achieve for capacitors, and so,

2.4. IMPEDANCE TRANSFORMATIONS OF PASSIVE FILTERS 47

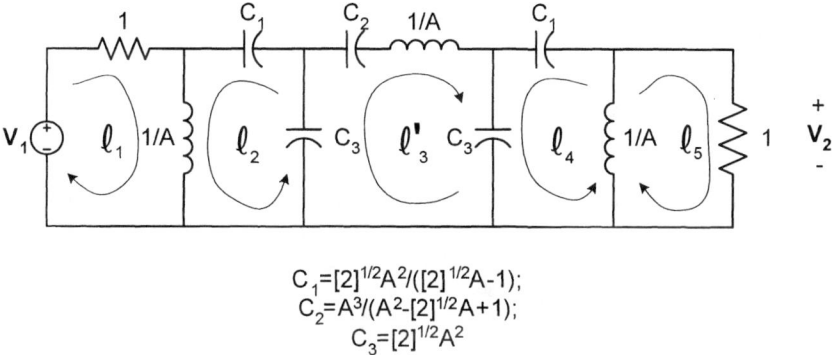

$C_1 = [2]^{1/2} A^2 / ([2]^{1/2} A - 1);$
$C_2 = A^3 / (A^2 - [2]^{1/2} A + 1);$
$C_3 = [2]^{1/2} A^2$

Figure 2.31. Loop-scaled equivalent for the circuit of Figure 2.27.

the result shown in Figure 2.31 is very practical and easier to implement than the BP prototype given in Figure 2.27. It is interesting that the capacitors act to lower the Loop 3 impedance level but preserve the desired impedance levels on both the source and load sides of the network.

Before concluding this chapter on circuit analysis concepts, it is useful to develop another equivalent circuit concept for the application of a special class of LP prototype circuits.

2.4.2 The Double Terminated Network and Its Dual

This discussion shows that any linear prototype network terminated at both source and load with 1-Ω resistors yields the same network function for its dual network as for itself. Consider such a network and its dual as shown in Figure 2.32.

The given network is described with its z-parameters, and so the network function between $\mathbf{V}_L$ and $\mathbf{V}_g$ is as follows for the circuit in Figure 2.32(a):

At Port-1:

$$\mathbf{V}_g = 1 \cdot \mathbf{I}_1 + z_{11}\mathbf{I}_1 + z_{12}\mathbf{I}_2 \quad \text{gives}$$
$$\mathbf{I}_1 = \frac{\mathbf{V}_g - z_{12}\mathbf{I}_2}{1 + z_{11}} \tag{2.75}$$

At Port-2:

$$\mathbf{V}_2 = z_{21}\mathbf{I}_1 + z_{22}\mathbf{I}_2 \quad \text{and} \quad \mathbf{I}_2 \equiv -1 \cdot \mathbf{V}_2$$
$$= \frac{z_{21}\mathbf{V}_g}{1 + z_{11}} + \left[z_{22} - \frac{z_{12}z_{21}}{1 + z_{11}}\right] \cdot (-\mathbf{V}_2) \quad \text{yields}$$
$$\frac{\mathbf{V}_2}{\mathbf{V}_g} = H(s) = \frac{z_{21}}{(1 + z_{11})(1 + z_{22}) - z_{12}z_{21}} \tag{2.76}$$

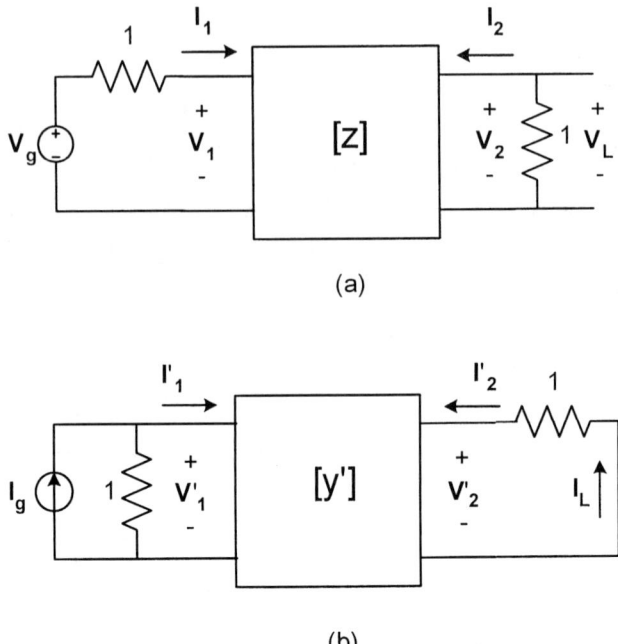

Figure 2.32. (a) A given double terminated prototype, and (b) its dual network.

2.5. SUMMARY

For the dual network in Figure 2.32(b), we have the following numerical equalities through the property of duality:

$$\mathbf{I}_g \doteq \mathbf{V}_g$$
$$\mathbf{I}_L \doteq \mathbf{V}_L$$
$$\mathbf{V'}_1 \doteq \mathbf{I}_1$$
$$\mathbf{V'}_2 \doteq \mathbf{I}_2$$
$$[y'] = \begin{bmatrix} y'_{11} & y'_{12} \\ y'_{21} & y'_{22} \end{bmatrix} \doteq \begin{bmatrix} z_{11} & -z_{12} \\ -z_{21} & z_{22} \end{bmatrix}$$

Analysis of the dual circuit then yields the following network function between $\mathbf{I}_L$ and $\mathbf{I}_g$:

At Port-1:

$$\mathbf{I}_g = 1 \cdot \mathbf{V'}_1 + y'_{11}\mathbf{V'}_1 + y'_{12}\mathbf{V'}_2 \quad \text{gives}$$
$$\mathbf{V'}_1 = \frac{\mathbf{I}_g - y'_{12}\mathbf{V'}_2}{1 + y'_{11}} \quad (2.77)$$

At Port-2:

$$\mathbf{I'}_2 = y'_{21}\mathbf{V'}_1 + y'_{22}\mathbf{V'}_2 \quad \text{and} \quad \mathbf{V'}_2 \equiv -1 \cdot \mathbf{I'}_2$$
$$= \frac{y'_{21}\mathbf{I}_g}{1 + y'_{11}} + \left[y'_{22} - \frac{y'_{12}y'_{21}}{1 + y'_{11}}\right] \cdot (-\mathbf{I'}_2) \quad \text{yields}$$
$$\frac{\mathbf{I'}_2}{\mathbf{I}_g} = \frac{y'_{21}}{(1 + y'_{11})(1 + y'_{22}) - y'_{12}y'_{21}} \doteq -\frac{\mathbf{V'}_2}{\mathbf{V}_g} \quad (2.78)$$

Clearly, the dual network gives the identical network magnitude function as the original network.

Quite often, in tabulated passive filter designs, the dual network is listed along with its original network. This gives the designer topological choices in the prototype to use when it is going to be transformed to another filter type, such as BP or BS. For example, Figure 2.33 shows the dual of the LP circuit given in Figure 2.8 and used throughout this chapter. The dual filter has used Thévenin's transformation on the parallel current source and conductance to obtain the series voltage source and resistor shown. The network function, $\mathbf{V}_L/\mathbf{V}_g$, is the same for both networks.

2.5 Summary

In this chapter we have reviewed basic circuit analysis and properties that lead to equivalent circuits. In particular, the parameter matrix forms were discussed, and the prototype filter concept was introduced, along with both impedance and frequency-scaling procedures. Frequency transformations of the prototype network were developed, and this led to considerations of element value ratio problems for BP (and BS) filters. Additional circuit equivalencies were discussed to alleviate the

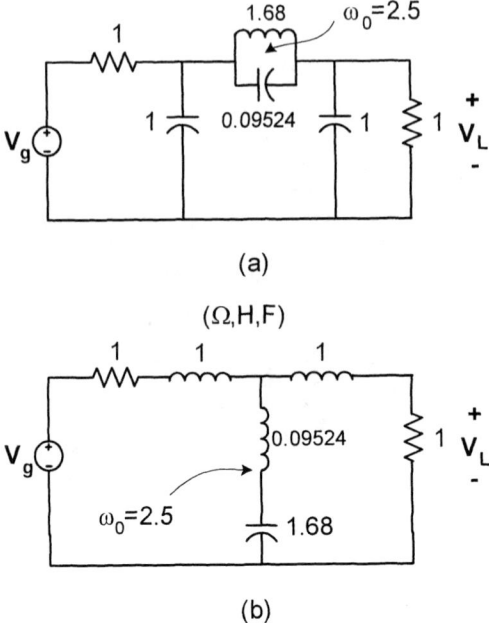

Figure 2.33. (a) An LP prototype, and (b) its dual network.

2.6. PROBLEMS

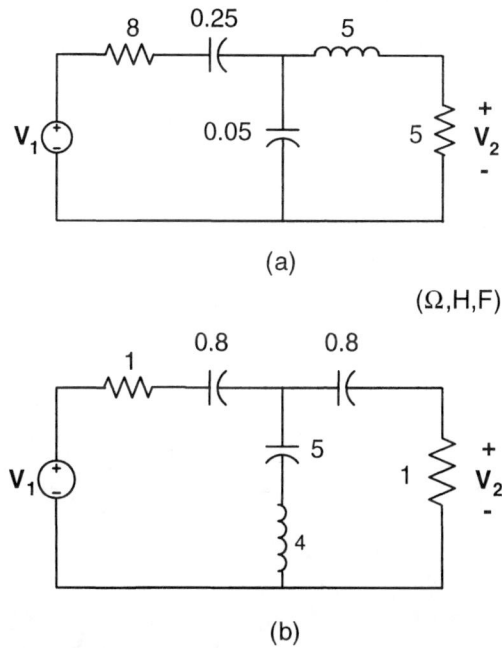

Figure 2.34. Circuits (a) and (b) for Problems 2.1 through 2.3.

element ratio problem. These ideas involved the use of equivalent impedance forms in the case of the BP transform of a parallel LC and the concept of internal loop scaling for the BP form of all-pole filters. Finally, it was shown that the doubly terminated (1 Ω) linear network provides the same network function as its dual network, thus yielding two topologies as candidates for circuit transformations and applications.

2.6 Problems

2.1 Scale each of the circuits, (a) and (b), given in Figure 2.34 so that each load resistance will equal 1 kΩ, and $\omega = 1$ rad/sec will go to 4 MHz. Obtain plots of the network function, $|V_2/V_1|_{dB}$, for each of the scaled networks via simulation, and determine the type of response (i.e., LP, BP). Specify the flat loss and the passband limits for each filter. Use +3-dB relative attenuation to estimate the passband limits.

2.2 To observe the effect of element tolerances, enter 5% tolerance for the energy storage elements in the circuits, (a) and (b), in Figure 2.34, and use the repeated runs (or cases) option to obtain several plots for random element values over the specified tolerance spread in a single frequency response graph.

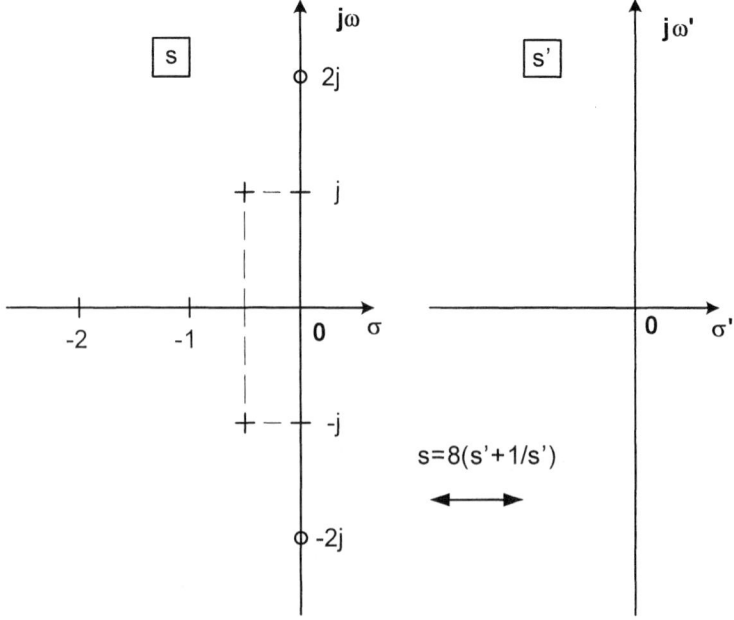

Figure 2.35. Pole-zero diagrams for Problem 2.4.

Discuss the results for each filter, and give conclusions about the effect of element value sensitivity and the network frequency response.

2.3 To observe the effect of distributed capacitance in a circuit layout, add 5 pF from each node to ground in the scaled circuits, (a) and (b), in Figure 2.34, and repeat the calculations and plots of Problem 2.1 for the parasitically disturbed circuits. In order to best display any errors, it is recommended to simulate both circuits at the same time and plot the difference between the responses for the two circuits in decibels versus frequency. Specify any changes in flat loss and passband limits for each filter.

2.4 A lowpass prototype network has the pole-zero pattern shown in Figure 2.35 for the following network function. Sketch a pole-zero pattern for the bandpass prototype of this function using a filter Q of 8. *Note:* This is a concept problem; it is not necessary to transform the function. Just indicate how poles and zeros are affected by the transform, and show expected differences (bandwidth) between the bandpass pole and zero locations:

$$H(s) = \frac{(s^2 + 4)}{3.2(s^2 + s + 1.25)}$$

2.6. PROBLEMS

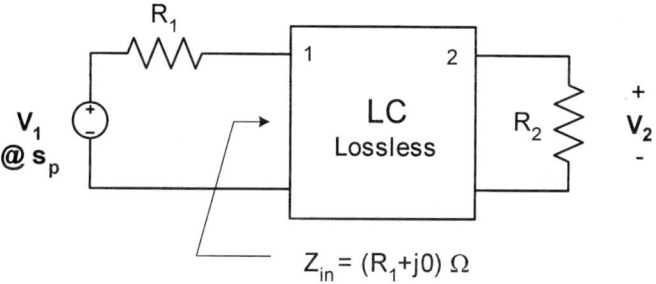

Figure 2.36. Circuit for Problem 2.6.

2.5 A lowpass prototype filter has the following network function:

$$H(s) = \frac{1}{5.06s^5 + 6.26s^4 + 10.2s^3 + 7.15s^2 + 4.1s + 1}$$

(a) Find the BP transform factor, A, required to transform the given LP filter function so as to provide a bandpass filter having a 2-kHz passband at a geometric mean frequency of 12 kHz. Specify the resulting passband frequency limits, f_1 and f_2.

(b) What is the resulting bandwidth that provides a minimum attenuation of 60 dB relative to the center frequency of the filter? Specify the frequencies, f_3 and f_4, that define this 60-dB band for the resulting filter function.

(c) Determine the network function for the prototype bandpass network function with $\omega_0 = 1$ and as a ratio of polynomials in s'. Check your answer using computer methods.

2.6 A lossless LC network is operated between unequal resistances as shown in Figure 2.36. At certain frequencies, s_p, in the filter passband, the input impedance at Port-1 is equal to $R_1 + j0$ Ω, when the network is terminated with a resistance of R_2 Ω. What is the magnitude of the transmission function squared, $|T|^2 = |\mathbf{V}_2/\mathbf{V}_g|^2$, at these frequencies?

2.7 Show how the LP prototype pass- ($\omega < 1$) and stop- ($\omega > 1$) bands are mapped for the transformation, $s = Bs'/(s'^2+1)$. Explain how the B parameter affects the transformation, and show what happens to the LP elements when they are subjected to this transform. Also, illustrate the mapping of ω versus ω', and determine the stopband bandwidth in the ω' domain.

2.8 A passive prototype filter has the loss function shown in Figure 2.37. A bandpass filter is required so that the center frequency will be at 455 kHz, and the two nulls, due to the prototype null at $\omega = 2$, should be separated by 45.5 kHz. What bandpass transformation factor, A, and frequency scale

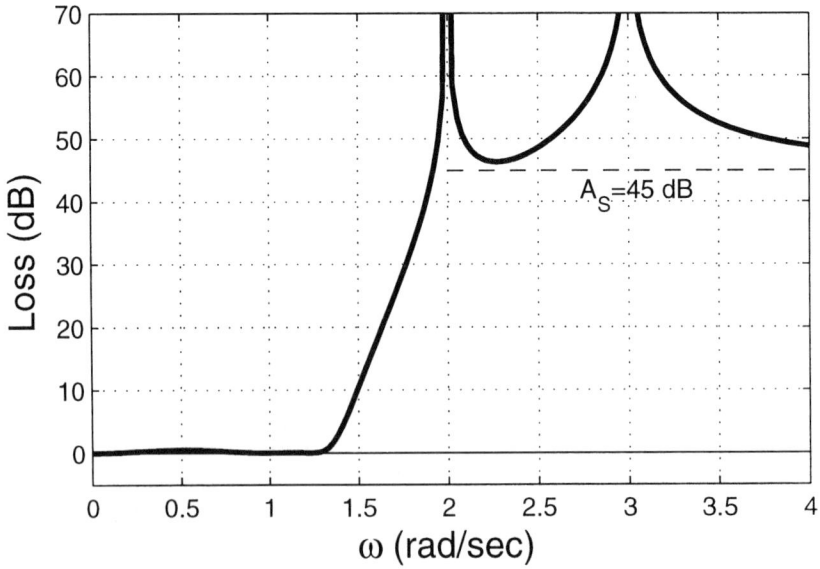

Figure 2.37. LP prototype response for Problem 2.8.

factor, k_ω, are required to obtain the desired response from the prototype design? Explain your answer, and show a sketch of the expected loss function for the resulting bandpass design.

2.9 The LP prototype shown in Figure 2.26 has a 3-dB bandwidth of 1 rad/sec.

(a) Apply an LP to BP transform to the circuit so as to obtain a BP filter with a filter Q of 5. Check the result by circuit simulation.

(b) Apply an LP to BS transform using the transform factor, $A = 0.5$. What is the resulting stopband width for the BS circuit? Check the result using circuit simulation.

2.10 The following lowpass prototype network function, $T(s)$, has a 3-dB bandwidth of 1 and a transmission zero at $\omega = 2.5$. Use the bandstop transformation, $s = 0.1s'/(s'^2 + 1)$, to convert $T(s)$ to a bandstop function. Determine the poles and zeros of the result, and plot the resulting loss function in decibels versus ω using a computer. What is the 3-dB bandwidth of the resulting stopband, and what is the bandwidth separating the nulls?

$$T(s) = \frac{0.16(s^2 + 6.25)}{s^3 + 1.84s^2 + 1.84s + 1}$$

2.11 A lowpass filter has its passband cutoff at 1 rad/sec and transmission zeros at $\omega = 2$ and $\omega = 4$ rad/sec. The filter is transformed to a bandpass filter and then scaled to have its center frequency at 10 kHz. The transformation factor is chosen so that the bandwidth between the widest pair of nulls will also be 10 kHz.

(a) What transformation factor is required?

(b) At what positive frequencies will the scaled bandpass filter have zero transmission?

2.12 Transform the lowpass network function of Problem 2.10 to obtain a highpass prototype using the LP to HP frequency transform. Determine the resulting $T(s')$ function, and sketch its pole-zero pattern in the s'-plane. Also sketch and dimension the resulting loss function versus ω'.

2.13 Confirm, using computer simulation, that the bandpass equivalent circuit, as shown in Figure 2.23, has the same response (see Figure 2.19) as the original circuit, Figure 2.20.

Chapter 3

Frequency Effects in Feedback Circuits

The purpose of this chapter is to consider the consequences involved when an operational amplifier[1] has a finite, frequency-dependent gain function. Introductory circuit courses do not discuss this effect of the opamp model. As long as the designer is not too concerned with frequency response behavior, a simple model is adequate to accomplish many useful instrumentation (low-frequency or dc) applications. However, when the goal is to design frequency-dependent network functions for filters, the opamp nonideal frequency response becomes an important issue to be aware of and to understand.

Many parameters are actually used to describe and specify the performance of a practical opamp (e.g., common mode rejection, input offset currents). The discussion presented here will not go into the different parameters, even though they are all important considerations for the proper utilization of such devices. The student is referred to other books (or courses) for information on the parameters not treated here. References [1–8] are listed at the end of the chapter. Reference [6] defines parameters used in the industry to specify opamp performance, and test circuits are provided in an appendix. James K. Roberge discusses the meaning and describes measurement techniques for important parameters in Chapter 11 of [4]. Review of these references will provide perspective on how practical opamps are modeled by designers and manufacturers.

This chapter looks at the effects of finite frequency-dependent gain for an otherwise ideal circuit model. Some examples are analyzed to show why the high-gain wide-bandwidth criteria evolved. These examples show how the functions may be altered (compensated for) to improve response in some cases. The chapter then concludes with a discussion of a current controlled amplifier and some properties of its closed-loop response function.

The goal of this chapter is to develop an understanding of certain behavior that is always present whenever an opamp is used in a circuit. This understanding

---

[1] Note: An operational amplifier is customarily referred to as an opamp.

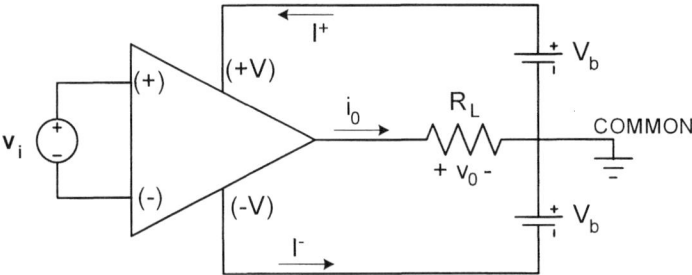

Figure 3.1. An opamp with supporting circuitry.

enables the filter designer to be aware of nonideal frequency effects that occur in a proposed design. It also provides understanding of either how to correct undesired behavior or how to change opamp design so as to obtain better response for a given application.

3.1 The Operational Amplifier

The opamp is used in many signal-processing applications for analog signals. Examples are transducers, control systems, preamplifiers, wave-shaping circuits, analog-to-digital converters, and many more. It is a versatile and common circuit element worthy of thorough study. Virtually all modern opamps are integrated circuits (ICs) packaged in a flat dual-inline-pin (DIP) ceramic or plastic package. Some sophisticated units are packaged one per package, and they have many connections brought out for adjusting or optimizing the performance and stability of the amplifier. Simpler units are packaged in dual or quadpacks and have no external connections for optimizing response. These amplifiers find application in audio and low-frequency instrumentation measurements and other devices. The simpler design is considered here as an introduction to these devices.

A simple opamp with the circuitry required for bipolar operation is shown in Figure 3.1. The dc voltage sources are required to bias the amplifier (i.e., provide power so the amplifier can do its job). The normal opamp has a balanced input and an unbalanced output, as shown. The output is delivered to a load resistor that goes to the common source connection. If multiple amplifiers share bias connections inside the IC package, then only three additional leads per opamp are required in a package of multiple opamps. The LM1458 has two opamps in an eight-pin DIP package.

When v_0 versus v_i is measured for the circuit in Figure 3.1, it looks like one of the curves shown in Figure 3.2. Note that peak output is limited near the 15-V bias voltages that are used to operate the amplifier. The biasing sets upper limits for the peak-to-peak output voltage range. In addition, the input required to drive the output from its negative to positive limits is on the order of 200 μV for a 30-V output swing. The transition represents an incremental slope (gain) of about

3.1. THE OPERATIONAL AMPLIFIER

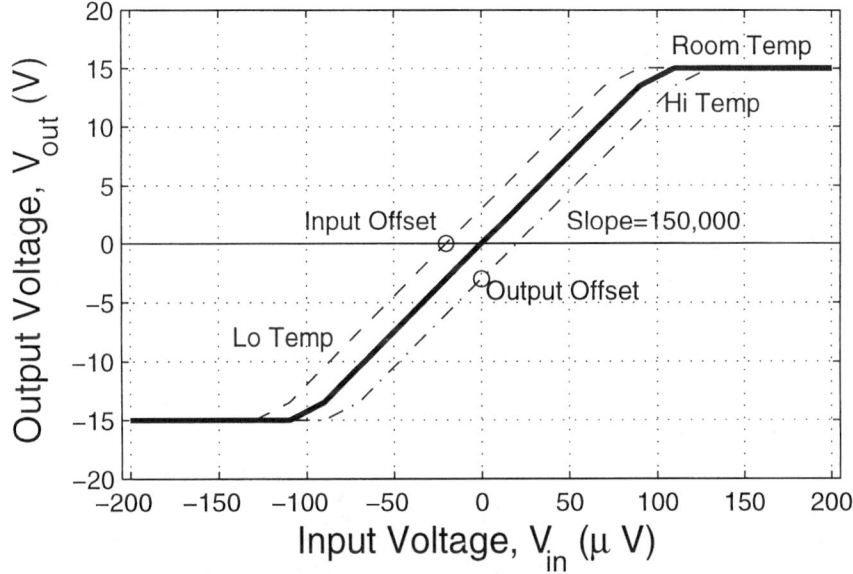

Figure 3.2. Plot of open-loop gain for a typical low-frequency opamp.

150,000:1. This is a really large number. Next, we observe that the curve does not always go through the origin as the zero crossing (input or output offset) is temperature dependent. The offsets are caused by temperature-dependent internal element mismatches. Finally, we might observe that when R_L is varied over a wide range (open circuit to very small values), the gain curve is unaffected. Adjusting R_L amounts to changing the output current, i_0, and so, the opamp output appears to act as a dynamic voltage source.

The curve of Figure 3.2 for an opamp is very difficult to measure due to the very small input signals that are required. Most multimeters do not go below a few millivolts for a full-scale reading, and so, to measure a few microvolts becomes a challenge. Here is another way to think of the problem. Suppose we have 100 μV on a 1-kΩ resistor. The power absorbed by the resistor is given in (3.1):

$$P_R = \frac{\mathbf{V}^2}{R} = \frac{10^{-8}}{10^3} = 10 \text{ pW} \tag{3.1}$$

Now, suppose a radio station 10 miles distant is radiating 50 W uniformly over a hemisphere above the Earth. The power density at your circuit (in free space) is given in (3.2):

$$S = \frac{50}{2\pi(16090)^2} \frac{\text{W}}{\text{m}^2} = 30.7 \text{ nW/m}^2 \tag{3.2}$$

If the circuit is on a 10 × 10 cm^2 board, then the board is exposed to an incident radiated power of 30.7×10^{-11} W = 307 pW. This power level is about 30 times

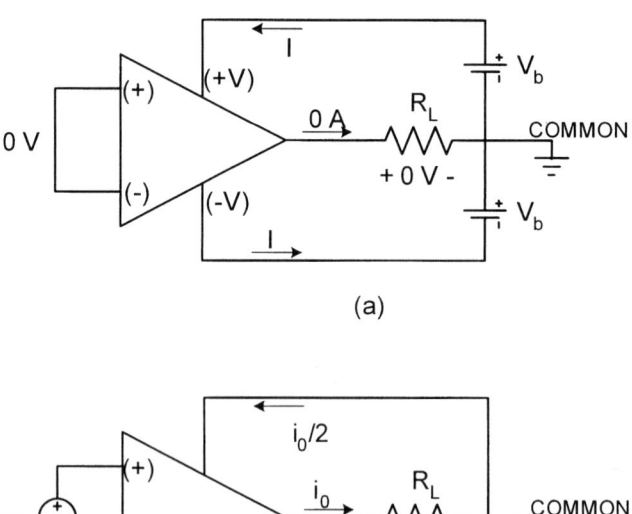

Figure 3.3. (a) Static and (b) small signal models for the opamp.

the power given in (3.1). A lot of factors are ignored in this simple estimate, such as losses of radiated power through the building, nonuniformity (directivity) of radiation, and weather losses. However, the important conclusion is that incident radiated power from TV, radio, CBs, and the like combines to yield unwanted signals that are on the order of a test signal required to measure the opamp gain. The signal that makes the opamp work is really small! That is an important conclusion. So far, the opamp is shown to have a nonlinear characteristic. The bias supplies provide dc plus incremental current. When the opamp is operated at input near zero, it is in the transition region between its upper and lower output limits. The incremental signals have a constant gain or relationship to each other. Static and small-signal circuit models are illustrated in Figure 3.3 for an opamp with zero offset, or in an ideal balanced state. The important observation here is that the small signal output, v_0 or i_0, comes indirectly from the bias supplies operating through the amplifier. The amplifier works like a valve. Steady current flows through it without going to the load when there is no call for output current or voltage. When there is a need for output, then a portion of the bias current is diverted to the load as required by external circuitry.

3.1. THE OPERATIONAL AMPLIFIER

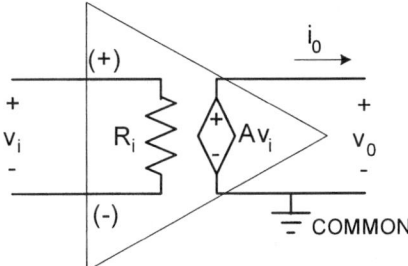

Figure 3.4. A first-order signal model for the opamp.

Applications considered in this course are interested in the amplifier's dynamic response.[2] A useful first-order model is shown in Figure 3.4. This model assumes that the input has some resistance and the input voltage controls a dependent voltage source at the output. The output voltage source appears between the amplifier output and the common connection of the bias supplies. Now, this model fits what we have described about the opamp so far, but to save time and to simplify tedious examples, the model is carried one step further.

It has been observed that for a finite output voltage, the input is quite small, on the order of a few microvolts. So for practical purposes (of engineering), it is possible to consider the input so small as to be negligible! Think of it as zero, or tending toward zero. Now consider what is happening at the output. It is obtained by multiplying v_i (a small number that tends toward zero) by a very large factor, A. Suppose we think of this factor as tending toward infinity. Then v_0 is given by (3.3), and we see that the answer might be indeterminant:

$$v_0 = A v_i \Rightarrow \infty \cdot 0 = \; ? \tag{3.3}$$

However, it is possible that a limit might exist if each tendency is controlled in a correlated fashion. Here is how it works. Let the opamp input voltage (therefore, also current) tend in a limit toward zero in such a way that as the gain tends in a limit toward infinity, there is a finite voltage developed on the opamp output. The model shown in Figure 3.5 illustrates this set of assumptions. Figure 3.5(a) shows the output voltage source with its return connection to the common point, and Figure 3.5(b) is an abbreviated symbol that is used to convey the same understanding as the circuit in Figure 3.5(a).

The validity of the model in Figure 3.5 can be compared to a result obtained using the more detailed model in Figure 3.4 via the following experiment. Consider the application of the opamp in a feedback circuit, as shown in Figure 3.6, find the voltage transfer between v_0 and v_S, and compare the results for high input-resistance (R_i) and open-loop gain (A). The constraints for the second model (zero

---
[2]Signals are considered to be dynamic whenever their relationship to each other is governed by means of a differential equation. For linear, time-invariant circuits, the equation coefficients are constants, and the equation independent variable is time.

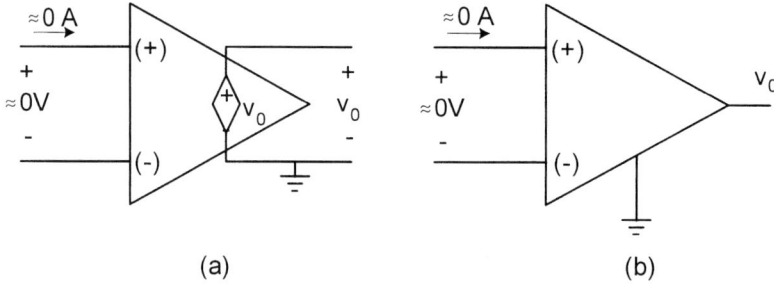

Figure 3.5. A second dynamic signal model for the opamp. (a) Model showing output voltage source and zero input, and (b) schematic model representing the circuit in (a).

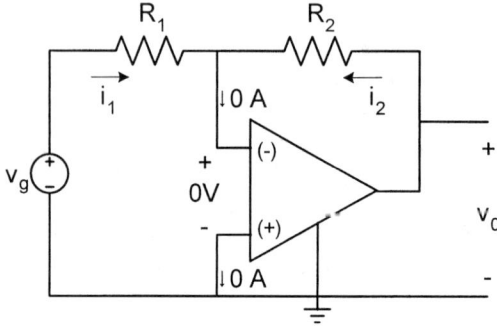

Figure 3.6. An inverting amplifier configuration.

input for current and voltage and v_0 at the output) are shown on the circuit in Figure 3.6. Analysis proceeds as follows.

Applying KVL to the rise v_0 through R_2 through the opamp input yields (3.4):

$$v_0 = R_2 i_2 + 0 \tag{3.4}$$

KVL around the source loop yields (3.5):

$$v_g = R_1 i_1 + 0 \tag{3.5}$$

KCL at the opamp input yields (3.6):

$$i_1 + i_2 = 0 \tag{3.6}$$

Substitution yields (3.7):

$$\frac{v_0}{v_g} = -\frac{R_2}{R_1} \tag{3.7}$$

3.1. THE OPERATIONAL AMPLIFIER

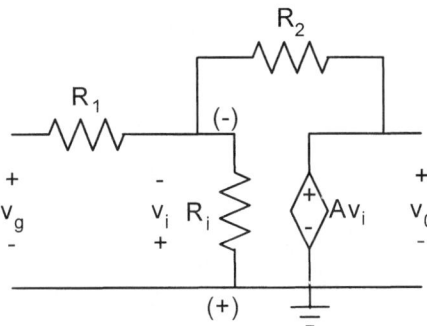

Figure 3.7. Feedback example using a finite-gain input-impedance opamp model.

This result contains no amplifier parameters in its formulation. The answer is very close to what would be measured for low-frequency, small-amplitude input signals. Measurements would confirm the validity of the simple model for the opamp.

The circuit shown in Figure 3.7 is used to analyze the application of Figure 3.6 using the more complex opamp model shown in Figure 3.4. Both R_i and A have large values. Analysis proceeds as follows:

$$v_0 = Av_i, \quad \text{by KVL, and}$$
$$0 = G_i(-v_i) + G_2(-v_i - Av_i) + G_1(-v_i - v_g)$$

by KCL at the opamp input node $(-v_i)$. Collection of terms and substitution yields (3.8):

$$v_0 = \frac{-AG_1 v_g}{[AG_2 + (G_1 + G_2 + G_i)]} \quad \text{or}$$
$$\frac{v_0}{v_g} = \frac{-G_1 R_2}{1 + [1 + R_2(G_1 + G_i)]/A} \tag{3.8}$$

Inspection of (3.8) shows that when $G_i << G_1$ and $A >> 1 + G_1 R_2$, then it goes in the limit to the result given by (3.7) for the simpler opamp model. Clearly, the simple opamp model does not predict any opamp effects on the transfer function. It is necessary to use the more complex model when it is desired to understand opamp effects.

The development of active filter design theory conventionally ignores opamp effects and uses the simpler model for filter design purposes. This course will ultimately do the same thing, but first we will look at some typical opamp effects in feedback circuits, explore reasons behind what happens, and indicate what can be done to modify or correct undesired response. Consequently, it is necessary to use the model in Figure 3.4 for these discussions.

64   CHAPTER 3. FREQUENCY EFFECTS IN FEEDBACK CIRCUITS

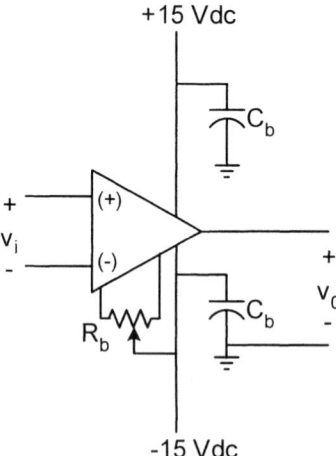

Figure 3.8. The opamp and support components for a typical circuit application.

3.1.1 Some Application Procedures

Before taking up the subject of the frequency effects of the opamp in a feedback circuit, this section provides a few words about typical procedures used in opamp applications. As shown in Figure 3.8 there are often three external components associated with each opamp installation in a circuit, namely, the two bypass capacitors, C_b, and a balance potentiometer, R_b.

The capacitors are very important as they are used to prevent oscillation of the high-gain amplifier through long round-trip paths to the dc power supplies. The capacitors should have excellent high-frequency response (electrolytics do not), they should be mounted adjacent to the opamp, and they should have terminal leads as short as possible. Ceramic capacitors are usually adequate for this job. They are typically small-value (0.01 to 0.1 μF), high-Q capacitors that do not require much space on the board. The bypass capacitors constrain incremental signal current paths to lie within a localized region of the opamp, thus avoiding, or greatly reducing, coupling with other opamp currents on the same circuit board. These capacitors are not intended to help regulate bias voltages. If that is required, then it is customary to place larger-value tantulum capacitors at terminals where power enters the board.

The balance resistor, R_b, provides for an incremental adjustment to zero the output offset voltage (see Figure 3.2). The result is a linear characteristic that is zero for zero input as long as the input-output voltages are within the linear range of the amplifier characteristic.

When the input voltage is outside the linear region, the amplifier is saturated at either its ±dc voltage limit. This saturation occurs quite often in poorly designed circuits. As the amplifier goes from the linear to a saturated region, the incremental

3.1. THE OPERATIONAL AMPLIFIER

signal gain goes from very large to zero. This significantly lower gain provides quite a different response than intended by the designer and results in holding the amplifier in the saturated state. This is referred to as latch-up and can be a difficult problem to deal with when a complex multiopamp circuit is involved. When opamps interact with each other due either to poor layout or not providing bypass capacitors for each opamp, then latch-up may occur when the circuit is powered up. Latch-up may also occur in a circuit where opamps are cascaded and an intermediate stage saturates before the last stage reaches saturation. This provides strange results that are often very subtle (e.g., wider bandwidth than expected, or not as much passband gain). Most opamp applications assume that the opamp is operating in its linear region in order to achieve the expected response. Whenever the amplifier is outside the range of the model, nonlinear analysis methods are required in order to predict the resulting response.

The design methods presented in this text are restricted to linear circuit models. Other techniques are outside the scope of this course. There is a circuit design technique that helps to keep the opamp in its linear operating region. The technique is referred to as negative feedback, and it is provided by circuit connections external to the amplifier.

A qualitative understanding of the circuit is obtained by inspection of Figure 3.7. Thus, with v_g set to zero, when v_0 is a large positive value, the control voltage, v_i, takes on a negative value (by voltage division), which in turn makes v_0 negative, or smaller. A stable value is reached when v_0 goes to zero. Thus, the circuit forces the output to be zero for zero input. Note that the negative feedback is accomplished by reversing the input to the opamp so that there is a negative gain between the output and its control input, v_i.

The result found in (3.8), referred to as the closed-loop gain, tells a lot about how the opamp can be used. First of all, the conductances G_1 and G_2 are chosen to be much greater than the amplifier input conductance, G_i. Alternately, opamp design seeks to make G_i approach zero. A typical value would be 0.5 μSiemens or μMhos.[3] Then, as the gain, A, becomes much greater than the value of $1 + G_1 R_2$, the closed-loop gain approaches the product, $-G_1 R_2$, in the limit as $|A| \to \infty$. Obtaining a gain proportional to a resistor ratio is very appealing because it is very linear, accurate, and stable. Note that if either R_1 or R_2 is a complex impedance, then the closed-loop gain becomes a function of frequency determined by the product, $-Y_1 Z_2$. Historically, the circuit in Figure 3.7 has been used to build integrators ($R_1 = $ constant, $Y_2 = Cs$), differentiators ($Y_1 = Cs$, $R_2 = $ constant), or real multipliers (both $R_1, R_2 = $ constants) for analog simulation of differential equations.

A consequence of the above concept is that the negative feedback configuration leads to the design requirement for high opamp input resistance, R_i, and a very large open-loop gain, A. As shown in Figure 3.2, a typical gain in the linear region is on the order of 150,000. One might then ask how such a large gain is obtained in practice. A typical bipolar junction transistor yields an incremental current gain on the order of 200 or so for a single device. It would require the cascade of two

---

[3] The Siemens has replaced the Mho as the conductance unit in the international system, but this text uses the historical Mho unit.

or more stages to get a transfer gain on the order of 10^5. It is also known that an individual transistor yields a frequency response that cuts off for higher frequencies. So, intuitively, we know that cascading transistor gains will cause the amplifier to provide an open-loop small-signal gain with frequency dependence somewhat like that shown in (3.9):

$$A \to A(s) = \frac{A_0}{(1+\frac{s}{p_1})(1+\frac{s}{p_2})(1+\frac{s}{p_3})\cdots} \qquad (3.9)$$

In (3.9), the form assumes that the frequency dependence remains a simple real pole for each stage of gain, but this is neither obvious nor easy to accomplish in practice. Now we want to look at what happens to the closed-loop gain when the open-loop gain, A, is assumed to be frequency dependent, as in (3.9), and the input conductance, G_i, is allowed to approach zero. The latter assumption is not necessary, as G_i is usually small compared to $(G_1 + G_2)$, but it does help to simplify the discussion that follows.

3.1.2 Single-Pole Open-Loop Gain

Consider first the case where the opamp open-loop gain is approximated using a single-pole response function as given by (3.10):

$$A \to A(s) = \frac{A_0}{1+\frac{s}{p_1}} \qquad (3.10)$$

Here, A_0 is the static slope of the opamp transfer function, and p_1 is the pole. When $A(s)$ is substituted for A in (3.8), along with $G_i = 0$, (3.11) is obtained:

$$\frac{v_0}{v_g} = \frac{-G_1 R_2}{1 + \frac{(1+G_1 R_2)}{A_0} + \frac{(1+G_1 R_2)s}{A_0 p_1}} \approx \frac{-G_1 R_2}{1 + \frac{(1+G_1 R_2)s}{A_0 p_1}} \qquad (3.11)$$

In (3.11), the approximation results from the assumption that A_0 is much greater than $(1 + G_1 R_2)$. This is normally a reasonable approximation, but note that the closed-loop gain, $G_1 R_2$, should not approach the open-loop gain, A_0. Equations (3.10) and (3.11) are compared in Figure 3.9 by means of Bode plots for $s = j\omega$. The curve with high gain is the open-loop function, while the lower-gain curve is the closed-loop response.

The curves plotted in Figure 3.9 provide a basis for observing some general behavior of the open- and closed-loop functions for an opamp that is characterized by a single-pole response. Both curves are lowpass functions with the closed-loop curve inside and asymptotic to the open-loop gain for any value of closed-loop gain. In addition, both curves have virtually the same unit-gain bandwidth. This unit-gain bandwidth is the product of the open-loop static gain, A_0, and the open-loop half-power bandwidth, p_1 (the open-loop pole). The gain bandwidth product is often used to compare different devices, but there is an implicit understanding that the devices possess single-pole frequency functions. Note that the closed-loop gain

3.1. THE OPERATIONAL AMPLIFIER

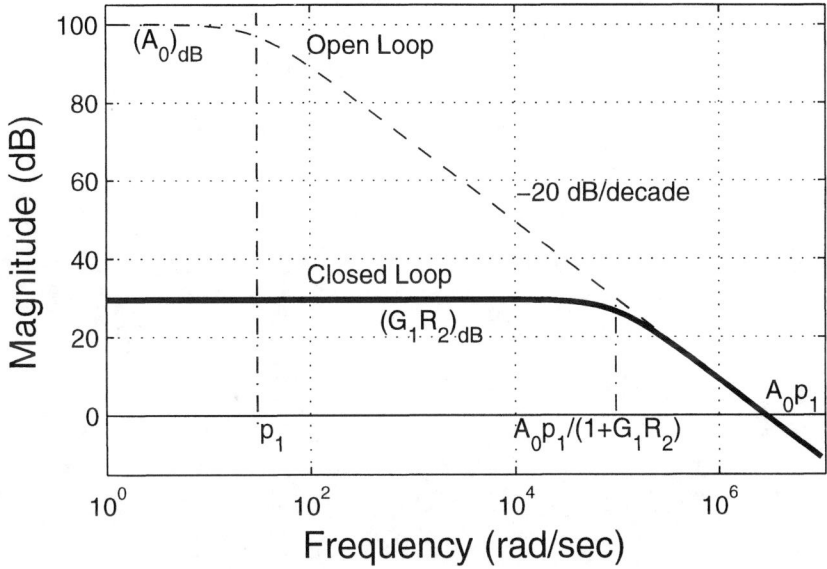

Figure 3.9. Single-pole open- and closed-loop gain comparison.

bandwidth product is slightly less than the open-loop product, $A_0 p_1$, by the factor, $G_1 R_2/(1+G_1 R_2)$. The difference is negligible for $G_1 R_2 > 1$.

This relationship between open- and closed-loop gain is very well behaved, and so, for those applications where bandwidth is not too important, opamp manufacturers strive to produce the single-pole open-loop gain curve as faithfully as they can. With increased static gain, A_0, the pole, p_1, decreases since the amplifiers are limited by the unit-gain bandwidth of their internal amplifying devices. This deduction is not obvious but can be understood through examination of a typical opamp design.

3.1.3 Double-Pole Open-Loop Gain

Now consider the case where the open-loop gain is approximated by a two-pole response function as given by (3.12):

$$A \rightarrow A(s) = \frac{A_0}{(1+\frac{s}{p_1})(1+\frac{s}{p_2})} \qquad (3.12)$$

Again, A_0 is the static gain of the opamp transfer function, and p_1 and p_2 are its poles. Substituting $A(s)$ for A in (3.8), along with $G_i = 0$, yields (3.13):

$$\frac{v_0}{v_g} = \frac{-G_1 R_2}{1 + \frac{(1+G_1 R_2)}{A_0}(1+\frac{s}{p_1})(1+\frac{s}{p_2})}$$

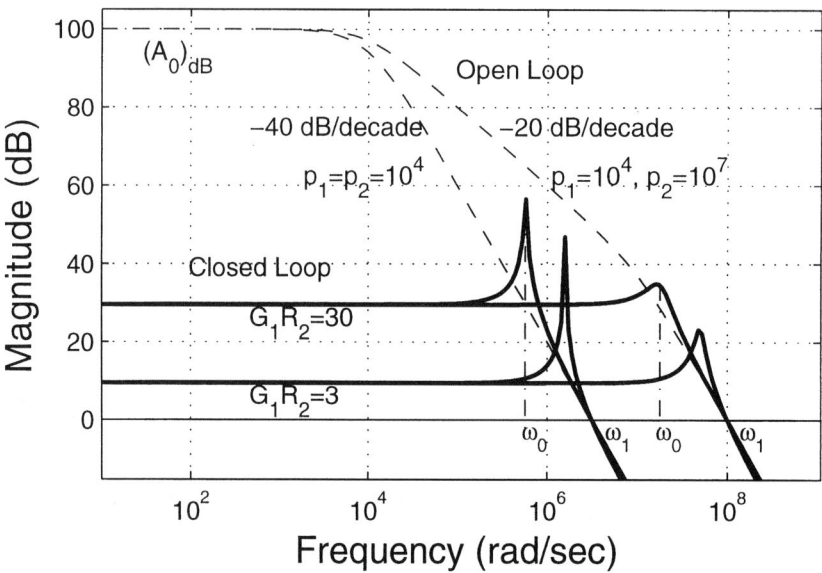

Figure 3.10. Example responses for two-pole open-loop gain.

$$= \frac{-G_1R_2}{1 + \frac{(1+G_1R_2)}{A_0} + \frac{2\alpha s}{\omega_0^2} + \frac{s^2}{\omega_0^2}} \approx \frac{-G_1R_2}{1 + \frac{2\alpha s}{\omega_0^2} + \frac{s^2}{\omega_0^2}} \quad (3.13)$$

where $2\alpha = p_1 + p_2$ and $\omega_0^2 = A_0 p_1 p_2/(1 + G_1 R_2)$.

The approximation is due to the assumption that the term $(1 + G_1R_2)/A_0$ is small compared to 1. This means that the closed-loop gain factor, G_1R_2, should not approach the static gain, A_0, in order for the assumption to remain valid. Note also that at $s = j\omega_0$, the term $(1 + G_1R_2)/A_0$ cannot be dropped from the denominator of (3.13). Equations (3.12) and (3.13) are compared in Figure 3.10 by means of Bode plots for $s = j\omega$. Parameters are assumed for two different types of open-loop functions for this comparison. The closed-loop response is plotted for two values of G_1R_2 for each open-loop function.

The results shown for this comparison are conceptually similar to the one-pole results, but there are some significant differences as well. Again, the curves display lowpass character, and the closed-loop curve is inside and asymptotic to the open-loop curve. The unit-gain bandwidth is also virtually the same for each open-loop and companion closed-loop response function. In this case, the open-loop unit-gain bandwidth $\omega_1 = (A_0 P_1 P_2)^{1/2}$, so a gain-bandwidth factor is not a useful concept for the two-pole amplifier frequency response. (*Note*: In the examples, $\omega_1 = 3.16 \times 10^6$ for the opamp with two identical poles, and $\omega_1 = 10^8$ for the other case of separated poles.)

Compared to the single-pole response, the two-pole result yields appreciable variation in its closed-loop response as it approaches, or transitions to, its high-frequency

3.1. THE OPERATIONAL AMPLIFIER

asymptote. These curves are characterized by peaked response, or resonances, due to the fact that complex poles are generated by the feedback circuit and the open-loop function acting together. The peaks are usually undesirable and are sometimes referred to as parasitic responses. A smooth, well-behaved transition is usually desired from the closed-loop lower-frequency gain to its asymptotic high-frequency response.

When the two poles are equal, the open-loop response drops off at -40 dB/dec, and high-Q complex poles are obtained in the resulting closed-loop response (pole $Q = \omega_0/2\alpha$). Separating the open-loop poles by a large amount, as in the second example, causes the response first to drop off at -20 dB/dec and then to go to -40 dB/dec when the second breakpoint frequency is reached. In this case the resulting complex poles have much less Q, and the closed-loop response shows less peaking in the transition region as the asymptote is approached. As p_2 is increased from 10^7 to higher values, the peaked response vanishes. This condition occurs for values of p_2 near the open-loop unit-gain bandwidth ($p_2 = A_0 p_1$). When p_2 is this large, it is interesting to note that the open-loop gain looks like a single-pole response for frequencies all the way to its unit-gain bandwidth. The associated closed-loop response is then well behaved for a very wide range of closed-loop gain requirements.

Generally, opamp designers try to achieve open-loop gain responses that follow single-pole behavior (-20 dB/dec) over as wide a band as possible. When the high static-gain is achieved by means of a cascade of amplifying stages, normally one stage sets the lowest pole, and then all other stages have bandwidth as wide as can readily be achieved. This procedure provides a desirable response for a majority of feedback applications.

If an amplifier design provides undesirable pole combinations, it is sometimes possible to alter the response by compensation. This procedure consists of adding a capacitor to lower a pole to a suitable value, or by cascading simple RC networks to cancel and shift an undesired pole. For example, the two-pole open-loop gain with $p_1 = p_2$ may be altered by cascading it with the RC function, $(s + p_2)/(s + p_2')$, so the pole at p_2 is moved to $p_2' > p_2$. Manufacturers usually provide application notes with suggested procedures for units that provide these compensation methods.

Before we conclude this discussion with a three-pole example, it is instructive to consider the closed-loop pole locations implied for the quadratic function of (3.13). Solving for the zero-frequencies of the denominator of (3.13) yields the closed-loop poles, $s_{1,2}$, given in (3.14):

$$\begin{aligned} s_{1,2} &= -\alpha \pm \sqrt{\omega_0^2 - \alpha^2} \approx -\alpha \pm j\omega_0 \quad \text{for } \omega_0 \gg \alpha \\ \therefore s_{1,2} &\approx \frac{-(p_1 + p_2)}{2} \pm j\sqrt{\frac{A_0 p_1 p_2}{1 + G_1 R_2}} \end{aligned} \quad (3.14)$$

This result shows that even though the closed-loop response is peaked, it is stable since the poles, $s_{1,2}$, are restricted to the left-half of the complex-frequency plane for any positive real values of A_0, p_1, and p_2. This conclusion is not valid for the three-pole example considered in the following discussion.

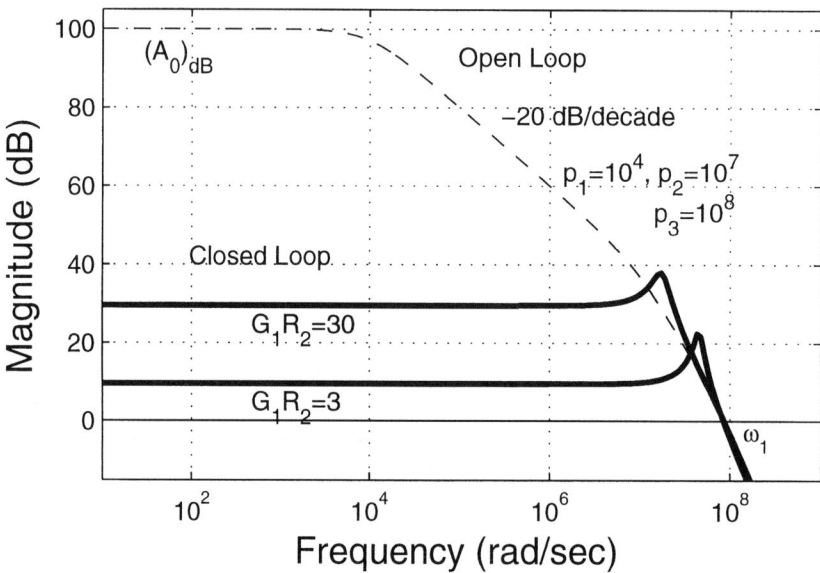

Figure 3.11. Example responses for three-pole open-loop gain.

3.1.4 Triple-Pole Open-Loop Gain

An example is shown in Figure 3.11 to illustrate closed-loop results for a three-pole opamp. The procedure is the same as was used for the previous examples. Again, it is seen that the closed-loop response has peaks (complex poles) as it approaches its high-frequency asymptote. The unit-gain bandwidth is virtually the same for both the open-loop and its companion closed-loop responses. In this case, the open-loop unit-gain bandwidth is $\omega_1 = (A_0 p_1 p_2 p_3)^{1/3}$. (Is there a trend here?) Note, with $p_3 = 10^8$, the third pole is at the unit-gain bandwidth, and so, the response looks like the two-pole example given in Figure 3.10. The closed-loop response for this case is nearly identical to the corresponding two-pole example, except that the poles for the closed-loop response are in the right half-plane for any $G_1 R_2 < 8.08$ (18.15 dB). That means the response is unstable for all closed-loop gain requirements less than 8.08 (for the given parameters). Note that the Bode plot does not show whether a response function is stable or not. It is necessary to determine the location of all poles to determine stability. This condition is found as follows. Let $K = G_1 R_2$; then, substitution into the closed-loop gain function, (3.8), yields (3.15) for $G_i \ll G_1$:

$$\begin{aligned}
\frac{v_0}{v_g} &= \frac{-K}{1 + (1+K)(1+s/p_1)(1+s/p_2)(1+s/p_3)/A_0} \\
&= \frac{-K A_0 p_1 p_2 p_3}{(1+K) Q(s)}
\end{aligned}$$

$$\text{where} \quad Q(s) = (s+p_1)(s+p_2)(s+p_3) + \frac{A_0 p_1 p_2 p_3}{1+K} \qquad (3.15)$$

3.1. THE OPERATIONAL AMPLIFIER

Solution of (3.15) for $Q(s) = 0$ yields the poles, $s_k = s_{1,2,3}$, for the closed-loop gain function. Through consideration of the polynomial, it becomes apparent that right-half-plane poles are possible as the constant term, $A_0 p_1 p_2 p_3/(1+K)$, becomes arbitrarily large, or K becomes small. This is possible through a procedure known as the root locus technique or simply by letting Matlab factor (3.15) for specified parameter values. However, at some critical value, the poles just cross the $j\omega$-axis and will have zero real parts. Let $s = j\omega_c$ for this case. Then, $Q(j\omega_c)$ reduces to the following requirement:

$$0 = (j\omega_c)^3 + (p_1 + p_2 + p_3)(j\omega_c)^2 + (p_1 p_2 + p_2 p_3 + p_3 p_1)(j\omega_c) \\ + p_1 p_2 p_3 [1 + A_0/(1+K)]_{K=K_{crit}} \quad (3.16)$$

Equation (3.16), a complex number relationship, has two parts that must be true (i.e., its real and imaginary parts are separately equal to zero). Thus, we get the conditions given in (3.17) for obtaining j-axis poles for the closed-loop gain function of (3.15):

$$\omega_c^2 = p_1 p_2 + p_2 p_3 + p_3 p_1$$
$$\frac{A_0}{1+K} = \frac{(p_1 + p_2 + p_3)\omega_c^2}{p_1 p_2 p_3} - 1 \quad (3.17)$$

ω_c^2 is from the imaginary part, and the critical maximum gain is from the real part of (3.16). For the parameters given in Figure 3.11—$A_0 = 10^5, p_1 = 10^4, p_2 = 10^7, p_3 = 10^8, K = 30$ dB—factoring the resulting $Q(s)$, (3.15), yields the poles $s_{1,2,3} = (-103.2, -34.1 \pm j16.9)10^6$ sec^{-1}. These poles are all in the left half-plane, and so the closed-loop gain function is stable. The corresponding critical feedback gain is $K_c = 18.15$ dB for the given A_0 and p_k values. When $K = 3$ dB, (3.15) yields the poles $s_{1,2,3} = (-127.6, +8.8 \pm j56.3)10^6$ sec^{-1}, and two poles are in the right half-plane. This closed-loop gain value yields an unstable network function.

The graph in Figure 3.12 shows the locus of poles [roots of $Q(s)$] for a range of closed-loop gain values, $K = G_1 \cdot R_2$. It confirms the above discussion for the stable range of closed-loop gain values. The plot is easy to obtain from Matlab as the following code indicates. The code assumes that the parameters, A_0, $p_{1,2,3}$, have been previously defined.

```
q1=p1+p2+p3;
q2=p1*p2+p2*p3+p3*p1;
q3=p1*p2*p3;
Kf=logspace(0,2,25);
for k=1:25
  q=[1 q1 q2 q3*(1+A0/(1+Kf(k)))];
  p(:,k)=roots(q)/1e6;
end
plot(real(p),imag(p),'xk'),grid
... add labels, data, etc
```

Plotting poles versus design parameters is a helpful procedure to determine network function performance. Matlab helps to make the task very simple.

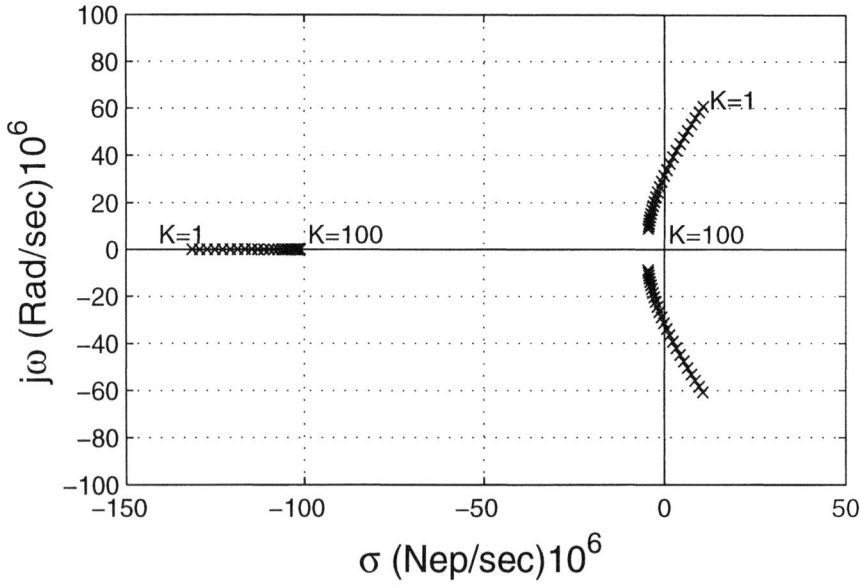

Figure 3.12. Graph of poles as a function of closed-loop gain, K.

Generally, opamp design attempts to place all additional poles (above two) at the unit-gain bandwidth or higher, if at all possible. This pole placement provides an opamp that can be used for a wider range of applications without having to compensate it for each application. Thus, the opamp user is cautioned to be aware of possible problems whenever an opamp is used in complex feedback applications. As will be seen, most active filter configurations offer the possibility for unstable operation due to feedback gain and frequency requirements.

3.2 An Operational Amplifier Dual

The previous discussion dwelled upon response obtainable from a voltage controlled voltage source (VCVS) with zero input conductance. For such a circuit, shown in Figure 3.13(a), the output voltage is controlled by the input voltage, v_x, but the signal power input to the opamp, $v_x i$, is zero. This is a remarkable condition [i.e., to be able to control the signal on a load (drawing power) with virtually no input signal power]. In addition, since the output, v_0, is a voltage source, it is possible to drive an arbitrary load without affecting the transfer function, v_0/v_g. The circuit is said to have isolation because the transfer is independent of the load.

Through consideration of duality, it is possible that a current controlled current source (CCCS) could yield a voltage transfer function while requiring zero input power if the control current is developed through a short circuit. Such a circuit is shown in Figure 3.13(b). In this circuit, the output is controlled with the input

3.2. AN OPERATIONAL AMPLIFIER DUAL

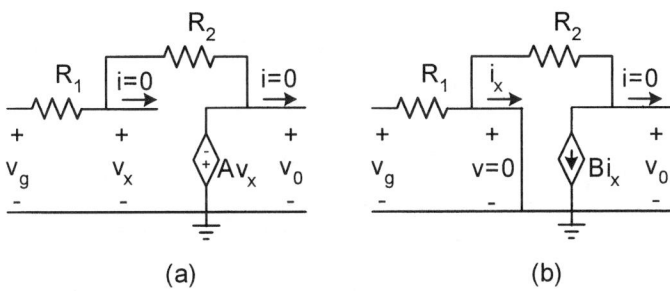

Figure 3.13. Feedback and a dual amplifier model: (a) VCVS, and (b) CCCS.

current, i_x, and the amplifier input power, vi_x, is zero. So, this circuit also yields an output voltage controlled with zero input power to the amplifier. However, a comparison of the voltage transfer function for each network yields a striking difference between the two circuits. Straightforward circuit analysis yields the transfer functions given by (3.18) and (3.19) for the circuits of Figure 3.13(a, b), respectively. For Figure 3.13(a):

$$\frac{v_0}{v_g} = \frac{-G_1 R_2}{1 + (1 + G_1 R_2)/A} \qquad (3.18)$$

For Figure 3.13(b):

$$\frac{v_0}{v_g} = \frac{-G_1 R_2}{1 + 1/B} \qquad (3.19)$$

As derived earlier, (3.18) is the same as (3.2) when $G_i = 0$, and (3.19) is the voltage transfer for Figure 3.13(b). There is a startling difference here! Note that for large A and B, both circuits yield the same feedback gain, $-G_1 R_2$, as required by the feedback ratio. But, observe that if A and B are both described by single-pole response functions, then the resulting poles in (3.18) depend upon $G_1 R_2$, but in (3.19), they do not. In other words, changing the feedback ratio in the second circuit does not alter the shape of the frequency response, whereas it does in the first circuit. What is the hitch here? Why was the first circuit analyzed so much if it is inferior to the second? We are correct to be suspicious as it is usually true that nature has fixed things so you don't get something for nothing. The first circuit was analyzed so much because the majority of opamps (in America) emulate that circuit. This is what you find in most applications. The hitch is that the second circuit does not have isolation. Connecting a load will change the transfer, v_0/v_g. So, it is not a very useful circuit in the form given in Figure 3.13(b). The next question is, can the current controlled circuit be modified to provide isolation and still preserve its feedback gain-independent frequency response?

The circuit in Figure 3.14 is one possibility for a current controlled opamp with voltage source output. A unit-gain voltage follower is used to convert from a current source output to a voltage source output. Again, a simple circuit analysis yields (3.20) for the transfer function of the circuit.

74   CHAPTER 3. FREQUENCY EFFECTS IN FEEDBACK CIRCUITS

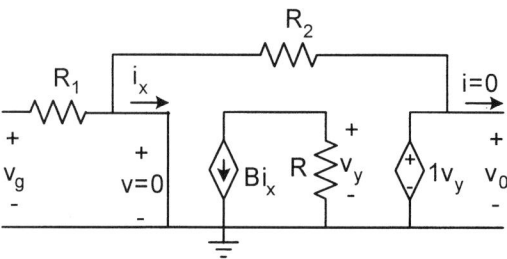

Figure 3.14. Current controlled feedback with isolation.

$$\frac{v_0}{v_g} = \frac{-G_1 R_2}{1 + R_2 G/B} \qquad (3.20)$$

In the last result, it is seen that changing R_2 has an effect on the frequency response for the closed-loop transfer function. However, changes in R_1 have no effect on the shape of the frequency response. In addition, attaching a load to the output will not affect v_0/v_g since the output is a voltage source. This result is almost as good as the result implied by (3.19), where either R_1 or R_2 could be varied without affecting the frequency response. So, we paid a little bit for the isolation. Here, for the circuit in Figure 3.14, only R_1 can be varied, while the frequency response shape is maintained.

3.2.1 Invariant Frequency Response Example

The closed-loop response of (3.20) is plotted in Figure 3.15 for different R_1, using the following parameters: $R = 1$ MΩ, $R_2 = 100$ kΩ, and $B = 10^3/[(1 + s/10^4)(1 + s/10^8)]$. Substitution into (3.20) yields (3.21):

$$\begin{aligned}
\frac{v_0}{v_g} &= \frac{-G_1 10^5}{1 + \frac{10^5}{10^3 10^6}(1 + s/10^4)(1 + s/10^8)} \\
&\approx \frac{-G_1 10^5}{1 + s/10^8 + s^2/10^{16}}
\end{aligned} \qquad (3.21)$$

The approximation ignores one part compared to 10^4. Magnitude plots for v_0/v_g are shown in Figure 3.15 versus ω for $s = j\omega$ and for different values of closed-loop gain, $G_1 R_2$. These curves are quite different from the responses shown earlier in Figures 3.10 and 3.11 for the voltage controlled closed-loop responses. In those examples the half-power bandwidth changed as $G_1 R_2$ changed, while the unit-gain bandwidth remained virtually constant. The opposite is apparently true for the isolated current controlled feedback circuit, as the half-power bandwidth is constant, and the unit-gain bandwidth changes with the feedback ratio (closed-loop gain).

Each type of feedback circuit has its own advocate and applications. The VCVS is used primarily for low-frequency audio and instrumentation applications. They

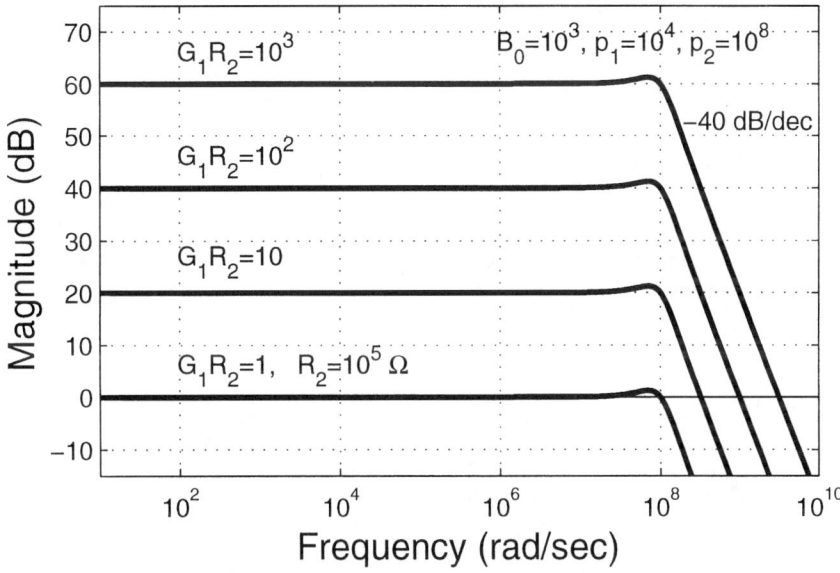

Figure 3.15. Current controlled closed-loop response example.

are widely available and commercially prolific. The CCCS is used for UHF components and other wideband applications such as radar pulse shaping circuitry and linear buffer amplifiers for high-speed analog-to-digital converters. Relatively, they are not widely available and are restricted to a few highly specialized product lines.

3.3 Summary

The previous discussions have shown that the use of an opamp in a feedback circuit provides frequency-dependent effects upon a desired transfer function. The introductory opamp circuit model, encountered in sophomore circuit courses, does not yield any prediction of frequency-dependent behavior. The examples presented here have shown that closed-loop response provides a lowpass gain equal to the desired closed-loop gain but that high-frequency behavior depends a great deal upon both the opamp response and the feedback configuration. Based on examples considered so far, it appears that VCVS feedback yields constant unit-gain bandwidth, whereas CCCS feedback yields constant half-power bandwidth.

The concepts illustrated have been presented to develop students' intuition about how an opamp interacts with a circuit to provide desirable or undesirable frequency response behavior. The problem is treated more rigorously by methods referred to as root locus techniques. These methods are concerned with stability and how the natural frequencies of a network function depend upon any given parameter of interest (e.g., static gain, feedback ratio, open-loop poles). While the root locus

3.4 Problems

3.1 Sketch (or plot), in the s-plane, the locations of the poles for the closed-loop gain, (3.8), as a function of the parameter, $K = G_1 R_2$. Assume $G_i = 0$, use a two-pole open-loop function with $A_0 = 10^5$, $p_1 = p_2 = 10^4$, and vary K from 1 to 100.

3.2 Repeat Problem 3.1 using a three-pole open-loop function, with $A_0 = 10^5$, $p_1 = 10^4$, $p_2 = 10^6$, and $p_3 = 10^7$.

3.3 Repeat Problem 3.1 using a three-pole open-loop function, with $A_0 = 10^5$, $p_1 = 10^4$, $p_2 = 10^7$, and $p_3 = 10^8$.

3.4 Determine an approximate relation for the location and value of the peaks of the closed-loop responses shown in Figure 3.10, and show why both sets differ by about 11 dB between $G_1 R_2 = 3$ and 30.

3.5 Obtain Bode magnitude plots for the closed-loop gain, (3.8), for a two-pole open-loop function. Use the parameters $A_0 = 2 \times 10^5$, $p_1 = 100$, $p_2 = 10^6$, and $G_i = 0$. Let $G_1 R_2 = 9$ and 99. Compare the results to the open-loop response as was done in the text. Are these responses better or worse than the ones obtained with $p_1 = p_2 = 10^4$?

3.6 Find an expression for the closed-loop gain of the circuit in Figure 3.7 when R_2 is changed to a capacitor, C_2. Let the desired gain be $-1/(R_1 C_2 s)$, and plot the closed-loop gain compared to the desired gain (in dB) versus ω for $G_i = 0$, $R_1 = 1$ kΩ, and $C_2 = 100$ μF. Determine the frequency band over which the circuit provides the desired response to within 1 dB. How well does the circuit perform as an ideal integrator?

3.7 This problem is focused on finding out how the integrator performance is degraded if the opamp has a nonzero output resistance, as shown in Figure 3.16. Let $A = 2 \times 10^5 / (1 + s/100)$ and the desired response be $D = -1/(RCs)$. Find the residual function, $R(s) = (\mathbf{V}_o/\mathbf{V}_g)/D(s)$, and plot $|R(j\omega)|_{\text{dB}}$ versus ω. Determine the 3 dB bandwidth for this residual response. Note that $|R| = 1$ for ideal behavior. What effect does the output resistance have on the bandwidth of R?

3.8 Repeat Problem 3.7 for $C_2 = 0.01$ μF.

3.9 Compare the input impedances for the two circuits in Figure 3.13.

3.10 Repeat the example plotted in Figure 3.15 by adjusting R_2 and holding $R_1 = 100$ Ω. Compare the results to those in Figure 3.15.

3.4. PROBLEMS

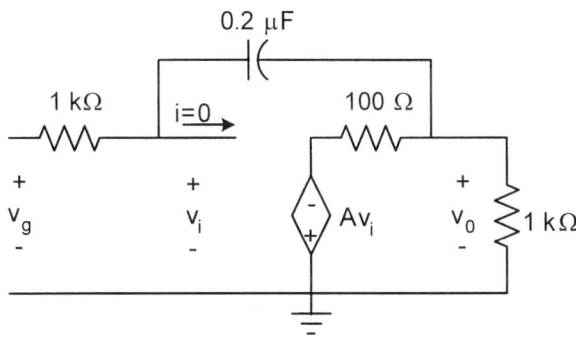

Figure 3.16. Circuit for Problem 3.7.

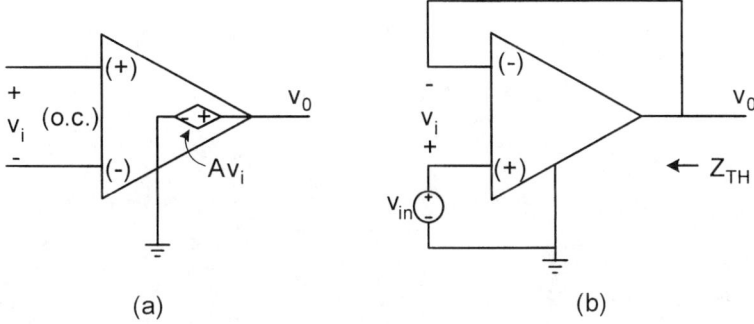

Figure 3.17. Circuits for Problem 3.11: (a) opamp model, and (b) buffer amplifier.

3.11 Use the opamp model shown in Figure 3.17(a) and find the Thévenin (output) impedance for the buffer amplifier shown in Figure 3.17(b). Use a one-pole model for the opamp open-loop gain [e.g., $A = A_0/(1 + s/\omega_0)$]. Draw an equivalent circuit model for the output impedance function that you found. What kind of circuit is the impedance? Is the output impedance RC, RL, or RLC?

3.12 Find the effective closed-loop input inpedance, Z_{in}, in Figure 3.18 for the case $G_i < G_F$. How does the open-loop gain affect the input impedance?

3.13 Find the Thévenin equivalent circuit model for the opamp circuit shown in Figure 3.19. Let $A = 10^5/(1 + s/250)$. Specify all element values for an equivalent circuit model of the output impedance. What is the effect of the output resistance, and what is the 3 dB bandwidth and low-frequency gain for the Thévenin voltage source?

3.14 In Figure 3.20, a two-port network is used to provide a negative feedback

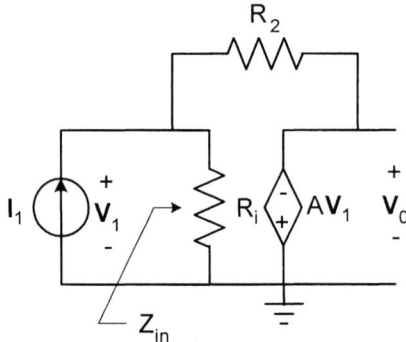

Figure 3.18. Circuit for Problem 3.12.

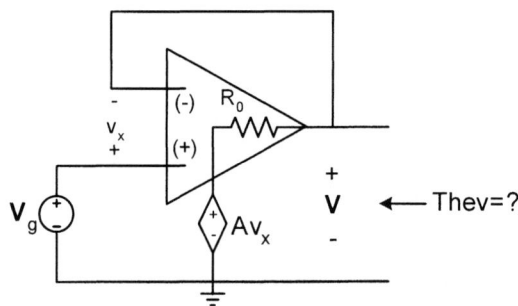

Figure 3.19. Circuit for Problem 3.13.

3.4. PROBLEMS

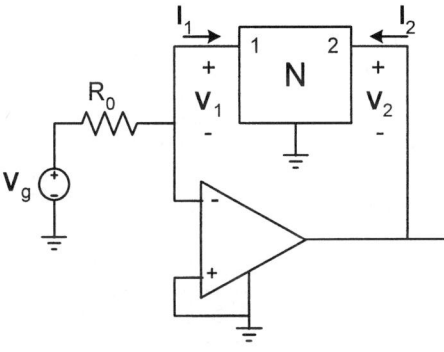

Figure 3.20. Circuit for Problem 3.14.

path for the opamp. Use y-parameters to represent the two-port, and find the network function, $\mathbf{V}_2/\mathbf{V}_g$, in terms of R_0 and the y-parameters on N; that is,

$$\begin{bmatrix} \mathbf{I}_1 \\ \mathbf{I}_2 \end{bmatrix} = \begin{bmatrix} y_{11} & y_{12} \\ y_{21} & y_{22} \end{bmatrix} \begin{bmatrix} \mathbf{V}_1 \\ \mathbf{V}_2 \end{bmatrix}$$

Why might someone try to use a circuit with j-axis transmission zeros for the network N?

3.15 The circuit in Figure 3.21 is used to obtain a negative input resistance. Use the opamp model, shown in Figure 3.17(a), with $A = A_0/(s+\alpha)$, and find the output impedance, $Z(s) = \mathbf{V}/\mathbf{I}$, in terms of parameters $R_{1,2,3,4}, A_0, \alpha$, and s.

(a) What are the poles and zeros of $Z(s)$?

(b) Determine the limit for $Z(s)$ as $s \to 0$ and A_0 becomes much larger than α. What condition must be satisfied by resistor values, $R_{1,2,3,4}$, in order to obtain a negative resistance for $Z(0)$?

(c) What type of source, voltage or current, is required for stable operation when the circuit is tuned to yield a negative input resistance, $R_{in} = \mathcal{R}e[Z(j\omega)] < 0$?

80   CHAPTER 3. FREQUENCY EFFECTS IN FEEDBACK CIRCUITS

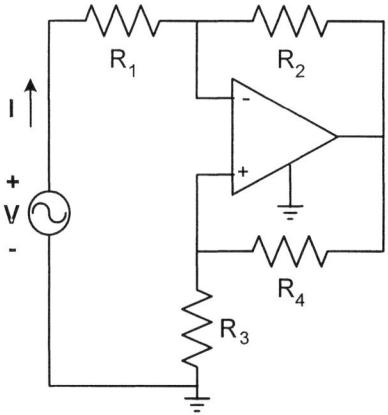

Figure 3.21. Circuit for Problem 3.15.

References

[1] Kuo, B. C., *Automatic Control Systems*, 6th ed., Upper Saddle River, NJ: Prentice Hall, 1991.

[2] Mathworks, Inc., *The Student Version of Matlab*, Upper Saddle River, NJ: Prentice Hall, 1991.

[3] Morris, J. F., *Introduction to PSPICE with Student Exercise Disk*, Boston, MA: Houghton Mifflin, 1991.

[4] Roberge, J. K., *Operational Amplifiers—Theory and Practice*, New York: Wiley, 1975.

[5] Roden, M. S., *Micro-Cap II, Electronic Circuit Analysis Program, Student Version*, Reading, MA: Addison-Wesley, 1988.

[6] Tobey, G. E., Graeme, J. G., and Huelsman, L. P., *Operational Amplifiers—Design and Applications*, New York: McGraw-Hill, 1971.

[7] Toumazou, C., Lidgey, F. J., and Haigh, D. G., *Analogue IC Design: The Current-Mode Approach*, IEEE Series, London, England: Peter Peregrinus Ltd., 1990.

[8] Truxal, J. G., *Control System Synthesis*, New York: McGraw-Hill, 1955.

Chapter 4

Some Opamp Design Considerations

The previous chapter emphasized how the opamp frequency response interacts with a simple feedback circuit to provide various effects. This chapter continues our effort to develop better understanding of opamp behavior. The emphasis is on circuit design and what is involved to actually achieve 100 dB of open-loop gain. In addition, some issues involved with controlling the bandwidth and cut off rates for the gain response are also considered. Both the voltage controlled voltage source (VCVS) and current controlled current source (CCCS) structures are studied, and it is interesting to see that there is not much difference between the two structures.

The goal of the chapter is to study a typical, or generic, opamp structure to enable the student to develop a feel for opamp design issues and see problems related to operating point stability and frequency response for high-gain circuits. As will be seen, such circuits are not trivial, and the performance limitations they offer are easier to understand when some design structures are known.

The material in this chapter is not critical to the development of concepts in this text related to active filter synthesis procedures. Consequently, it is possible to skip this entire chapter. It has been included for the curious student who wishes to know more about opamp design, terminology, and performance ideas. Whether or not this material were included in any course would depend upon available time and course objectives.

4.1 The Current Mirror

The opamp design story begins with a consideration of current mirrors. Fixed current sources are at the heart of all integrated circuit operational amplifier designs. Despite the name, the concept has nothing to do with mirrors and magic! Current sources are required to bias transistor circuits and provide stable operating conditions. The mirror is used to obtain a solid-state current source by reflecting (or mirroring) a reference current, as shown in Figure 4.1. It is usually required that

82   CHAPTER 4. SOME OPAMP DESIGN CONSIDERATIONS

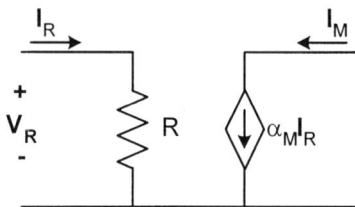

Figure 4.1. A current mirror is a current controlled current source.

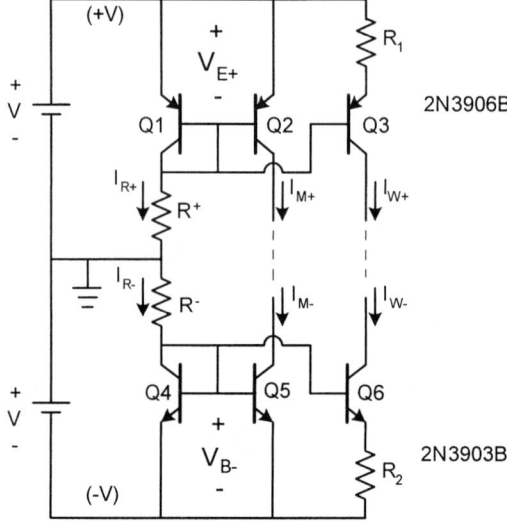

Figure 4.2. Bias current distribution for an integrated circuit.

the mirror gain factor, α_M, be as close to unity as possible. Note that the mirror looks like a current controlled current source.

A typical implementation for current sources and sinks is illustrated in Figure 4.2.[1] The mirror currents are independent of the load to which they are connected as long as the load is not an open circuit. An open circuit causes the mirror transistor to saturate, at which point it leaves the normal active region and ceases to function as a current source. The performance of individual mirrors is considered in the following discussion.

Figure 4.3(a) illustrates a basic current mirror that sinks current to the negative source in the bipolar transistor circuit. An analysis of the behavior of the circuit is obtained by assuming that both transistors are in the active region and that the base-emitter voltage, V_B, controls the currents, I_E. (See Appendix 4A for a

---

[1] See [1] for a more complete discussion of mirror theory and applications than is presented here.

4.1. THE CURRENT MIRROR

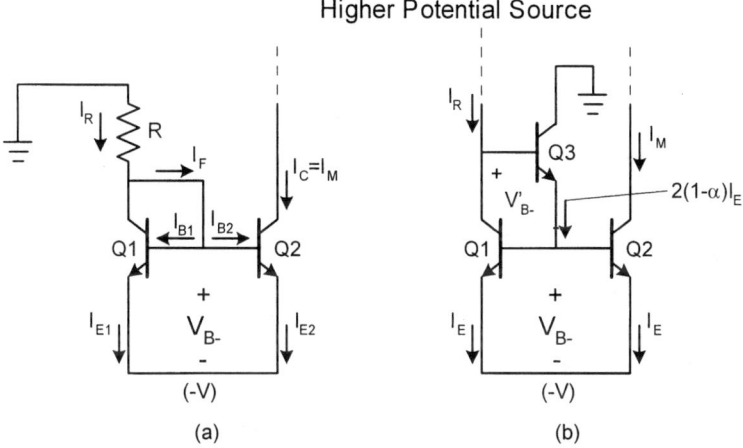

Figure 4.3. Basic unit-gain current mirrors: (a) basic mirror, and (b) EFA mirror.

discussion of polynomial modeling of the junction transistor in its active operating region.)

With both transistors in Figure 4.3(a) controlled by the same base-emitter voltage, V_B, and both transistors fabricated on the same substrate with nominally identical parameters, the following analysis yields an approximation for the mirror gain, α_M:

$$\begin{aligned} I_{B2} &= I_{B1} e^{(b_1 V_B + b_0)} = I_B \\ \therefore I_{C2} &= I_{C1} = \beta_{dc} I_B = e^{(c_2 V_B^2 + c_1 V_B + c_0)} \\ \text{and} \quad I_{E2} &= I_{E1} = (\beta_{dc} + 1) I_B \end{aligned} \tag{4.1}$$

Applying KCL to the circuit yields α_M:

$$\begin{aligned} I_M &= I_{C2} = \beta_{dc} I_B \\ I_R &= \beta_{dc} I_B + 2 I_B = (\beta_{dc} + 2) I_B \\ \alpha_M &= \frac{I_M}{I_R} = \frac{\beta_{dc}}{\beta_{dc} + 2} \approx 1 - \frac{2}{\beta_{dc}} \end{aligned} \tag{4.2}$$

For a typical β_{dc} of 100, it is seen that the mirror yields an error of about -2% in the reflection of I_R.

As will be seen shortly, it is often desirable to have the mirror gain closer to 1 than the simple circuit of Figure 4.3(a) can manage. There are alternate circuits that have been proposed to accomplish this goal, and the circuit shown in Figure 4.3(b) is one that is frequently encountered. It is referred to as an emitter follower augmented (EFA) mirror. The analysis of its behavior proceeds as follows. All transistors are assumed to have identical parameters and are fabricated on the same substrate:

$$I_M = I_{C2} = \beta_{dc} I_B$$

$$I'_B = \frac{2I_B}{\beta_{dc}+1}$$

$$I_R = I_{C1} + I'_B = \beta_{dc}I_B + \frac{2I_B}{\beta_{dc}+1} = \frac{\beta_{dc}(\beta_{dc}+1)+2}{\beta_{dc}+1}I_B$$

$$\therefore \alpha_M = \frac{I_M}{I_R} = \frac{\beta_{dc}(\beta_{dc}+1)}{\beta_{dc}(\beta_{dc}+1)+2} \approx 1 - \frac{2}{\beta_{dc}^2} \qquad (4.3)$$

Now, for a typical β_{dc} of 100, the EFA mirror yields a unit gain to within about -0.02%. This is a quite accurate reproduction of the reference current, I_R, and justifies the use of the additional transistor.

For example, let $I_M = 1$ mA for the EFA in Figure 4.3(b). Then, we get the following results using Matlab and the modeling developed in Appendix 4A:

$$\text{With } I_C = I_M, \text{ the inverse model yields}$$
$$V_B = \text{inv3903B}(10^{-3}, 3) = 0.6303 \text{ V}$$

The base and collector currents are then obtained from the base-emitter voltage forward relationship:

$$[I_B \ I_C] = \text{npn3903B}(0.6303) = [12.976 \ \mu\text{A} \ 1 \text{ mA}]$$

Application of KCL fixes I_{E3}, which in turn controls the remaining parameters for $Q3$:

$$I_{E3} = 2I_B = 25.951 \ \mu\text{A}$$
$$\text{so} \quad V'_B = \text{inv3903B}(I_{E3}, 2) = 0.49296 \text{ V which yields}$$
$$[I_{B3} \ I_{C3}] = \text{npn3903B}(V'_B) = [0.83727 \ 25.119] \ \mu\text{A}$$

Application of KCL yields I_R:

$$I_R = I_{B3} + I_C = 1.0008 \text{ mA}$$

So, finally, $\alpha_M = I_M/I_R = 0.99916$ and $V_{CE1} = V'_B + V_B = 1.1232$ V. This solution shows that the EFA provides an excellent gain factor; however, the transistors operate with quite different dc betas due to the very small current in $Q3$ as follows.

$$\beta_{dc_1} = \beta_{dc_2} = \frac{I_{C1}}{I_{B1}} = 77.1 \quad \text{while}$$
$$\beta_{dc_3} = \frac{I_{C3}}{I_{B3}} = 30.0$$

4.1.1 The Widlar Mirror

Current sources are shown in Figure 4.2 with resistance in the emitter lead of the mirror transistor. This circuit allows for obtaining mirror factors of arbitrary value less than one. The base-emitter drop of the reference transistor is used to obtain

4.1. THE CURRENT MIRROR

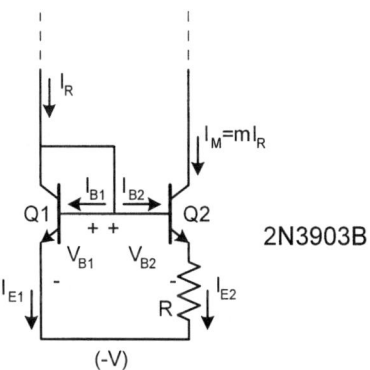

Figure 4.4. The Widlar current mirror.

a current source that is some fraction, k_R, of the reference current.[2] This circuit is referred to as a Widlar current source. Its advantage lies in the fact that quite small bias currents, on the order of a few microamps, can be obtained with small resistance values. This is an important feature for integrated circuit fabrication.

On the basis of the variables defined in Figure 4.4, the following relationships yield the resistor for a required (specified) mirror current, I_M:

$$\begin{aligned}
I_{C2} &= I_M \quad \text{gives} \\
V_{B2} &= \text{inv3903B}(I_{C2}, 3); \quad \text{then} \\
[I_{B2}\ I_{C2}] &= \text{npn3903B}(V_{B2}) \quad \text{gives} \\
I_{E2} &= I_{B2} + I_{C2}, \quad \text{and KVL requires} \\
V_{B1} &= V_{B2} + RI_{E2} \quad \text{so} \\
R &= (V_{B1} - V_{B2})/I_{E2}
\end{aligned} \qquad (4.4)$$

Everything except V_{B1} is known in order to determine R.

If the reference current, I_R, is specified to be any value greater than I_M, then a global KCL yields (4.5):

$$I_{E1} = I_R + I_M - I_{E2} \qquad (4.5)$$

The required value of emitter current, I_{E1}, then yields a required base-emitter voltage, V_{B1}:

$$V_{B1} = \text{inv3903B}(I_{E1}, 2) \qquad (4.6)$$

With all required variables thus determined, it is possible to determine a design value for R. This procedure assumes identical transistors on the same substrate.

A numerical example helps to show the value of this mirror. For a desired current, $I_M = 10\ \mu\text{A}$, and a reference current of $I_R = 1\ \text{mA}$, (4.4) yields $R = 15.965\ \text{k}\Omega$. Note that it would require 1.5 MΩ in series with 15 V to get the same 10 μA. A

---

[2]Note in IC design this may also be accomplished by reducing or increasing the emitter areas of different transistors.

CHAPTER 4. SOME OPAMP DESIGN CONSIDERATIONS

few kilohms requires much less area to fabricate using integrated circuit techniques than values on the order of a few megohms do.

Example 1 - For the circuit in Figure 4.2, find resistor values so that the following currents will be obtained for $V = 15$ V: $I_{M+} = I_{M-} = 500$ μA; $I_{W+} = 20$ μA; and $I_{W-} = 40$ μA.

Solution - The current sources are independent of the current sinks, so the design can be accomplished in two separate halves. First, complete the upper half, or source design. We start by noting that $Q1$ and $Q2$ have the same operating point, and then KCL yields I_{R+};

$$I_{R+} = I_{C1} + I_{B1} + I_{B2} + I_{B3}$$
$$\text{so that} \quad R+ = (15 - V_{E1})/I_{R+}$$

We need to find V_{E1} and I_{R+} to determine $R+$. The following results are obtained:

$$V_{E2} = \text{inv3906B}(I_{M+}, 3) = 0.6347 \text{ V}$$
$$[I_{B2} \; I_{C2}] = \text{pnp3906B}(V_{E2}) = [5.289 \; 500] \; \mu\text{A}$$
$$V_{E1} = V_{E2}$$
$$I_{B1} = I_{B2}$$
$$V_{E3} = \text{inv3906B}(I_{W+}, 3) = 0.5447 \text{ V}$$
$$[I_{B3} \; I_{C3}] = \text{pnp3906B}(V_{E3}) = [0.2972 \; 20.0] \; \mu\text{A}$$
$$I_{E3} = I_{B3} + I_{C3} = 20.2972 \; \mu\text{A}$$
$$R_1 = (V_{E2} - V_{E3})/I_{E3} = 4.4326 \text{ k}\Omega \Leftarrow$$
$$I_{R+} = I_{C1} + I_{B1} + I_{B2} + I_{B3} = 510.88 \; \mu\text{A}$$
$$R+ = (15 - V_{E1})/I_{R+} = 28.119 \text{ k}\Omega \Leftarrow$$

The design for the current sinks follows a procedure similar to that for the current sources. $Q4$ and $Q5$ have identical operating points, and we have to find the $Q6$ operating point before we can find I_{R-} and $R-$:

$$I_{C5} = I_{M-} = 500 \; \mu\text{A}$$
$$V_{B5} = \text{inv3903B}(I_{C5}, 3) = 0.6019 \text{ V}$$
$$[I_{B5} \; I_{C5}] = \text{npn3903B}(V_{B5}) = [7.372 \; 500.0] \; \mu\text{A}$$
$$I_{C4} = I_{C5}$$
$$I_{B4} = I_{B5}$$
$$V_{B4} = V_{B5}$$
$$I_{C6} = I_{W-} = 40 \; \mu\text{A}$$
$$V_{B6} = \text{inv3903B}(I_{C6}, 3) = 0.5087 \text{ V}$$
$$[I_{B6} \; I_{C6}] = \text{npn3903B}(V_{B6}) = [1.147 \; 40.00] \; \mu\text{A}$$
$$I_{E6} = I_{B6} + I_{C6} = 41.147 \; \mu\text{A}$$

$$R_2 = (V_{B5} - V_{B6})/I_{E6} = 2.2645 \text{ k}\Omega \Leftarrow$$
$$I_{R-} = I_{C4} + I_{B4} + I_{B5} + I_{B6} = 515.89 \text{ }\mu A$$
$$R- = (15 - V_{B4})/I_{R-} = 27.909 \text{ k}\Omega \Leftarrow$$

The example shows that small constant current values may be generated with relatively small resistors, provided that a reference voltage is made available. The calculations account for the difference between collector and emitter currents. These differences affect the reference current values required in order to obtain the desired mirror currents.

4.2 The Differential Amplifier Input Stage

With the current mirror concepts established, it is now possible to turn our attention to the opamp design problem. It has already been argued, in connection with (3.9), that the opamp will, of necessity, consist of cascaded stages in order to provide a large open-loop gain (e.g., 10^5). Assuming that two or more stages are employed to achieve high gain, then a little thought tells us that the first stage will be very important to the overall behavior of the result. Whatever undesired behavior is developed in the first stage is subsequently amplified by the following stages. Examples of undesired behavior include dc offset, temperature dependence, and excessive nonlinearity. The differential amplifier is a circuit that overcomes many of the undesired properties inherent to semiconductor circuits. Primarily, this circuit uses symmetry to obtain a cancellation in temperature dependence that would be present in single transistor devices. The objective of this section is to study the differential amplifier and learn about some of its useful features. Traditionally, this amplifier is used in the first stage of opamp design.

The circuit in Figure 4.5 shows that the output current, i_0, is proportional to the difference of the collector currents of transistors $Q1$ and $Q2$. These currents in turn are controlled by the difference voltage, $v_{g1} - v_{g2}$, as will be shown. In effect, this circuit provides a VCCS. Note that the circuit consists of only transistors and a bias current source, I. We are interested in how the circuit processes the desired signal voltages, v_{g1} and v_{g2}, and what undesired output is present in the current, i_0. The analysis of the circuit proceeds as follows.

All transistors are assumed to be fabricated on the same substrate with the NPN transistors having nominally identical parameters and the PNP transistors having parameters similar but not identical to the NPNs. Assume that all transistors are operating in their active region. KCL at the current source constrains the emitter currents:

$$I = i_{e1} + i_{e2} \approx i_{c1} + i_{c2} \tag{4.7}$$

Because this is the first stage of a high-gain amplifier, the differential input, $v_{g1} - v_{g2}$, is very nearly zero to within a few microvolts. Therefore, for identical transistors, the collector currents are virtually equal. From our polynomial modeling we can determine a dependence between the collector currents as follows:

$$i_{c1} = e^{c_2 v_1^2 + c_1 v_1} e^{c_0}$$

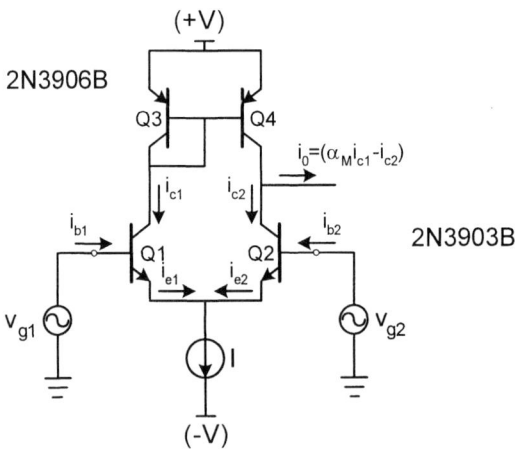

Figure 4.5. The differential amplifier (VCCS).

$$\begin{aligned}
i_{c2} &= e^{c_2 v_2^2 + c_1 v_2} e^{c_0}, \text{ so} \\
\frac{i_{c1}}{i_{c2}} &= e^{c_2(v_1^2 - v_2^2) + c_1(v_1 - v_2)} \\
&\equiv e^{c_2(v_1 + v_2)(v_1 - v_2) + c_1(v_1 - v_2)}
\end{aligned} \quad (4.8)$$

Since the variables include both static and dynamic quantities, let $i_{c1} = I/2 + i_c$ for $v_1 = V_b + v_b$ and $i_{c2} = I/2 - i_c$ for $v_2 = V_b - v_b$, since $I \approx i_{c1} + i_{c2}$. Thus, we get the following relationship for the dynamic component of the collector current:

$$\begin{aligned}
\frac{I}{2} &= e^{(c_2 V_b^2 + c_1 V_b + c_0)} \text{ for the static current} \\
v_1 + v_2 &= 2V_b + 0 \\
v_1 - v_2 &= 2v_b = (v_{g1} - v_{g2}) \equiv v_d \\
\frac{i_{c1}}{i_{c2}} &= e^{2c_2 V_b (2v_b) + c_1 (2v_b)} = K_{12}^{2v_b} \\
\therefore K_{12} &= e^{2c_2 V_b + c_1}
\end{aligned} \quad (4.9)$$

K_{12} is an operating point constant for the differential pair. It is a constant for a given static value of V_b. Substitution of i_{c1} from (4.9) into (4.7) for I then yields a dynamic relation for i_{c2}:

$$i_{c2} \approx \frac{I}{(K_{12}^{2v_b} + 1)} \quad (4.10)$$

Finally, through the active load, the output current, i_0, is obtained as in (4.11):

$$i_0 = \alpha_M i_{c1} - i_{c2} = \frac{(\alpha_M K_{12}^{2v_b} - 1)}{(K_{12}^{2v_b} + 1)} I = i_0(v_b) \quad (4.11)$$

4.2. THE DIFFERENTIAL AMPLIFIER INPUT STAGE

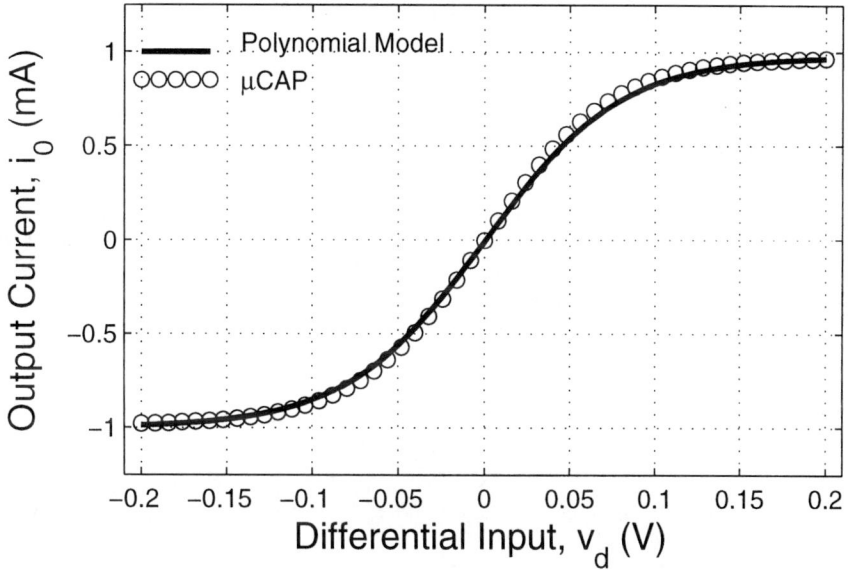

Figure 4.6. The differential amplifier gain function: polynomial model results compared to μCAP simulation results.

The derivative for i_0 with respect to the dynamic voltage, v_b, can be obtained from (4.11) (after some work) with the following result:

$$\frac{di_0}{dv_b} = 2IK_{12}^{2v_b}\log_e(K_{12})\frac{(\alpha_M + 1)}{(K_{12}^{2v_b} + 1)^2} \qquad (4.12)$$

The behavior of the output current, i_0, for small input, v_b, is then obtained from the first two terms of a Taylor expansion of (4.11) around $v_b = 0$:

$$i_0 = \frac{(\alpha_M - 1)}{2}I + 2I\log_e(K_{12})\frac{(\alpha_M + 1)}{4}v_b + \ldots \qquad (4.13)$$

where $\log_e(K_{12}) = (2c_2 V_b + c_1)$ and $\alpha_M = \beta'/(\beta' + 2)$.

For the 2N3903B, with $i_c = I/2 = 500$ μA, $V_b = \text{inv3903B}(i_c, 3) = 0.6019$ V, $\log_e(K_{12}) = 25.0857$, and $di_0/dv_b(0) = 24.8$ mA/V. This value of slope checks the μCAP simulation for the differential amplifier output current as shown in Figure 4.6.[3] Several properties of the amplifier are apparent from this result. First of all, the output is proportional to the bias current, I. This in turn is derived from a mirror of a reference current, so it is a well-known and stable value. Second, the

---

[3]The response looks much like a $\tanh(x)$ function, but the usual result, which gives the $\tanh(v_d/V_T)$, does not actually fit the slope obtained for this differential amplifier realized with the example transistors. As Figure 4.6 illustrates, the polynomial models yield a good fit to the simulated response.

gain relation between i_0 and the differential input, v_d, looks like a well-behaved function, but it is not a linear function. The transfer function for i_0 saturates at $+\alpha_M I$ for large positive differential input and at $-I$ for large negative differential input values.

Equation (4.13) is a useful form for the output function when the differential input, v_d, is very small. Here it is seen that the output term consists of a constant, referred to as offset, and a term proportional to the differential input, $v_d = (v_{g1} - v_{g2})$. The question, how small is small enough? is appropriate. The differential input is the same as v_i in the discussion relating to the opamp introduction in Chapter 3. Figure 3.2 shows that $|v_i| > 150~\mu V$ saturates the amplifier.

Now, in order to answer whether this is small enough, it is helpful to find a power series expansion for (4.11) and compare terms. Equation (4.14) gives such an expansion about $v_d = 2v_b = 0$:

$$\begin{aligned}
\frac{i_0}{I} &= \frac{\alpha_M K_{12}^{v_d} - 1}{K_{12}^{v_d} + 1} \equiv \alpha_M - \frac{(\alpha_M + 1)}{(K_{12}^{v_d} + 1)} \\
&= \frac{(\alpha_M - 1)}{2} + \frac{(\alpha_M + 1)}{4}\log_e(K_{12})v_d + (0)\frac{v_d^2}{2!} \\
&\quad - \frac{(\alpha_M + 1)}{2^3}\log_e^3(K_{12})\frac{v_d^3}{3!} + \cdots
\end{aligned} \qquad (4.14)$$

Requiring the first-order term to be much greater than the third-order term yields $v_d^2 \ll 12/\log_e^2(K_{12})$, or $|v_d| \ll 3.46/25.0857 = 0.138$ V. Clearly, the above value of v_i for a typical opamp application satisfies the test for higher-order terms to be considered negligible, and so, (4.13) is commonly used to express the small-signal behavior for this circuit.

Our objective here is to see why the differential amplifier is considered good and is used for the first stage in the opamp circuit. To achieve this understanding, it is helpful to derive similar results for a single transistor amplifier.

4.2.1 A Single Transistor Amplifier

Consider the simple circuit shown in Figure 4.7. It is a VCCS derived from a single voltage source as compared to the dual source for the differential amplifier. The constant current source is obtained from a mirror, and the dc voltage source is used to adjust the output offset current to zero. The output current is derived in (4.15):

$$\begin{aligned}
v_1 &= v_g + V_0 \\
i_c &= e^{(c_2 v_1^2 + c_1 v_1 + c_0)} \\
&\equiv e^{(c_2 V_0^2 + c_1 V_0 + c_0)} \cdot e^{(c_2(2V_0 v_g + v_g^2) + c_1 v_g)} \\
&\approx I_0 K_c^{v_g} \quad \text{for } v_g \ll V_0 \text{ and so} \\
i_0 &= I - i_c \approx I - I_0 K_c^{v_g} \qquad (4.15)
\end{aligned}$$

where $K_c = e^{(c_2 2V_0 + c_1)}$ and I_0 is the collector operating current for $v_{be} = V_0$. Note that K_c is similar to the factor, K_{12}, defined for the differential amplifier analysis.

4.2. THE DIFFERENTIAL AMPLIFIER INPUT STAGE

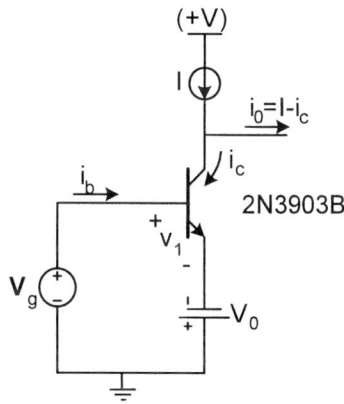

Figure 4.7. A single transistor amplifier.

The output current is then expanded into a power series dependence upon v_g for small input values:

$$i_0 = (I - I_0) - I_0 \log_e(K_c) K_c^{v_g} \left[v_g + \log_e(K_c) \frac{v_g^2}{2!} + \log_e^2(K_c) \frac{v_g^3}{3!} + \ldots \right] \quad (4.16)$$

An example transfer function, for (4.16), is shown in Figure 4.8 with $I = 100$ μA. At first glance, one might consider that this is a better amplifier since the offset current is zero, and the small-signal gain (coefficient of the linear term) is bigger by a factor of two when (4.16) is compared to (4.13).

However, this amplifier is inferior for at least two reasons. First, the dc offset has a bad temperature coefficient, and second, the linear signal range is smaller. We will now consider each of these factors.

The signal range limit for linear response is seen by inspection of (4.16). In order for linear response to dominate all higher-order terms, it is seen that $|v_g| \ll 2/\log_e(K_c)$ is required. The corresponding limit for the differential amplifier was found to be $|v_d| \ll 3.46/\log_e(K_{12})$. Therefore, the linear dynamic range of the differential amplifier exceeds that of the single transistor amplifier by $20\log_{10}(3.46/2) = 4.77$ dB, even though it appears to have less gain.

The temperature dependence for dc offset can be compared in terms of the mirror bias current temperature dependence for each amplifier. Let I_Δ be the offset current for the differential amplifier.

From (4.13):

$$I_\Delta = \frac{(\alpha_M - 1)}{2} I \text{ then}$$

$$\frac{\partial I_\Delta}{\partial T} = \frac{(\alpha_M - 1)}{2} \frac{\partial I}{\partial T} = \frac{-1}{(\beta' + 2)} \frac{\partial I}{\partial T} \quad (4.17)$$

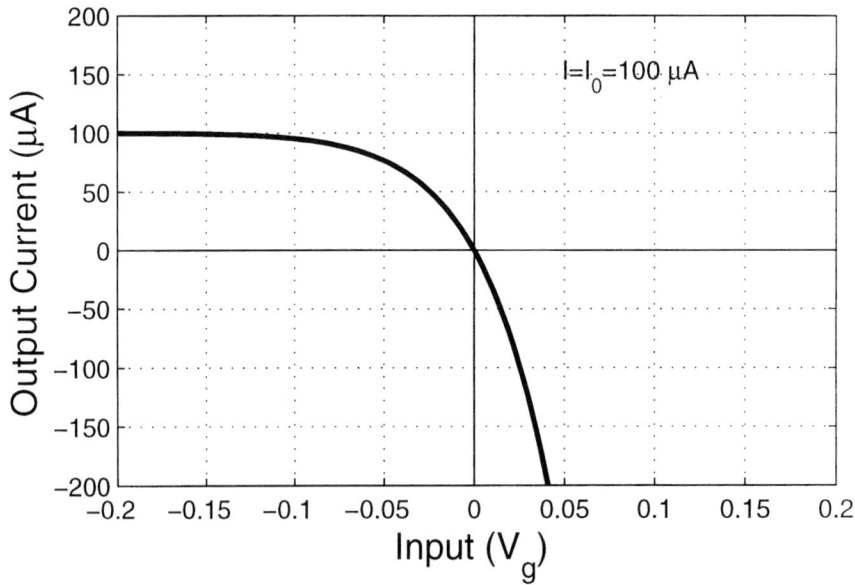

Figure 4.8. Static transfer function for the single transistor amplifier.

The temperature dependence of the mirror gain, α_M, is ignored in (4.17). $\partial I/\partial T$ is the thermal rate of change of the bias current delivered by a mirror circuit. Now let I_δ be the offset current for the single transistor amplifier with the operating point, (V_0, I_0), treated as a constant. The variation of I_0 with temperature is worse than the mirror current temperature dependence, but the result shown in (4.18) is a best-case scenario:

$$\begin{aligned} I_\delta &= I - I_0 \quad \text{then} \\ \frac{\partial I_\delta}{\partial T} &= \frac{\partial I}{\partial T} \end{aligned} \quad (4.18)$$

This last result shows that the single transistor dc offset current is directly proportional to the bias current source thermal dependence. By way of comparison, the differential amplifier dc offset current is found to be much smaller. It reduces the bias current thermal rate of change by the dc current gain, β'.

Thus, in this best case, the offset current thermal drift for the differential amplifier is on the order of two magnitudes smaller than it is for the single transistor amplifier.

Clearly, the differential amplifier is far superior to the single transistor amplifier. If the current bias source has thermal drift, $\partial I/\partial T$, the differential amplifier attenuates the effect inversely proportional to the transistor current gain, β'. In addition, the differential amplifier has a wider distortion-free input signal range. This is basically why the differential amplifier is the amplifier of choice.

4.2. THE DIFFERENTIAL AMPLIFIER INPUT STAGE

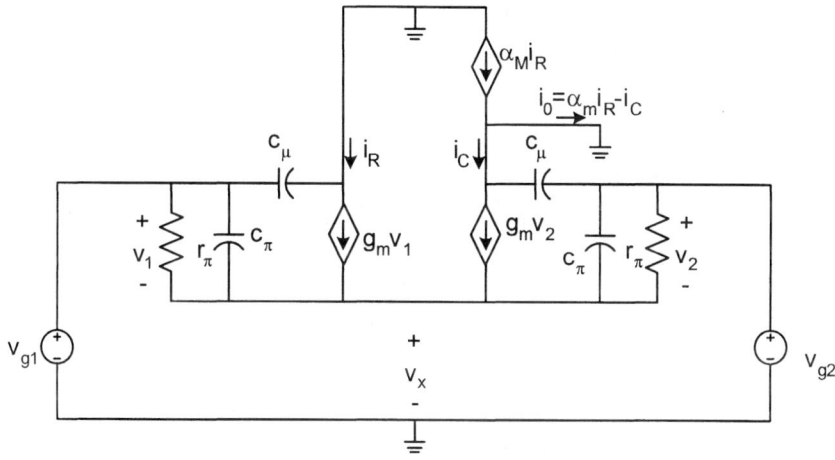

Figure 4.9. Small-signal circuit model for the differential amplifier.

4.2.2 Small-Signal Frequency Response

Before moving on to consider the second and third stages of an opamp, we need to look at the small-signal frequency response for the first stage. Comparison of the gain for the differential amplifier with that for a single transistor amplifier shows that the differential amplifier gain matches that of the single transistor for unbalanced inputs. Therefore, frequency behavior is not sacrificed by using the differential amplifier for its temperature stability and wider linear dynamic range.

A small-signal circuit model for the differential amplifier of Figure 4.5 is given in Figure 4.9. Simplified hybrid-π models are used for transistors $Q1$ and $Q2$, and a CCCS is used for the mirror pair, $Q3$ and $Q4$. See [2–5] or the μCAP manual for more details on hybrid-π models. The bias sources (constants) are replaced with short and open circuits where required. Recall that only small amplitude signals are used for the input to the first stage, and so, both $Q1$ and $Q2$ have collector currents equal to about one-half of the bias current, I. Thus, g_m, r_π, and C_μ are nominally the same for both $Q1$ and $Q2$. Since the operating points for $Q1$ and $Q2$ are not the same (i.e., the V_{CB} are not the same), then the C_μ are slightly different for the two transistors. For convenience, we assume that the C_μ are equal, but it shoud be kept in mind that they are slightly different. To make numerical comparisons, we will use the following values for the transistor small-signal parameters (see Appendix 4A).

Using a value of 1 mA for the bias current, I, the following dynamic values are obtained for the transistor hybrid models:

For the NPN transistors:

$$i_e = 500 \ \mu\text{A yields } v_{be} = \text{inv3903B}(i_e, 2) = 0.6014 \text{ V}$$

$$[i_b \ i_c] = \text{npn3903B}(v_{be}) = [7.286 \ 492.77] \ \mu\text{A}$$
$$r_\pi = 1/(b_1 i_b) = 6.875 \ \text{k}\Omega$$
$$g_m = (2c_2 v_{be} + c_1)i_c = 12.37 \ \text{mMho}$$
$$C_\pi = 17.2 \ \text{pF}$$
$$C_\mu = 1.95 \ \text{pF}$$

For the PNP transistors:
$$v_{be} = \text{inv3906B}(i_e, 2) = 0.6344 \ \text{V}$$
$$[i_b \ i_c] = \text{pnp3906B}(v_{be}) = [5.237 \ 494.80] \ \mu\text{A}$$
$$r_\pi = 1/(b_1 i_b) = 5.968 \ \text{k}\Omega$$
$$g_m = (2c_2 v_{be} + c_1)i_c = 16.61 \ \text{mMho}$$
$$C_\pi = 36.741 \ \text{pF}$$
$$C_\mu = 3.06 \ \text{pF}$$

The circuit analysis is straightforward as the node voltage, v_x, may be chosen to be the only independent variable in the circuit. Let $y_\pi = g_\pi + C_\pi s$; then, applying KCL at node v_x yields (4.19):

$$0 = y_\pi(v_x - v_{g1}) - g_m(v_{g1} - v_x) - g_m(v_{g2} - v_x) + y_\pi(v_x - v_{g2})$$
$$\text{or } v_x = (v_{g1} + v_{g2})/2 \tag{4.19}$$

The collector currents for i_R and i_C are given by (4.20) through the use of KCL at each collector node:

$$i_R = g_m(v_{g1} - v_x) - C_\mu s v_{g1}$$
$$= (g_m/2 - C_\mu s)v_{g1} - g_m v_{g2}/2$$
$$i_C = g_m(v_{g2} - v_x) - C_\mu s v_{g2}$$
$$= (g_m/2 - C_\mu s)v_{g2} - g_m v_{g1}/2 \tag{4.20}$$

The output current, i_0, is then obtained by applying KCL at the mirror output node. The result is shown in (4.21), where the mirror gain is approximated by the factor, $(1 - \epsilon_M)$:

$$i_0 = \alpha_M i_4.i_C \approx (i_4.i_C) - \epsilon_M i_R$$
$$= (g_m - C_\mu s)(v_{g1} - v_{g2}) - \epsilon_M[(g_m/2 - C_\mu s)v_{g1} - g_m v_{g2}/2] \tag{4.21}$$

The effect of the nonideal mirror gain is clearly shown in this last result. As $\epsilon_M \to 0$, the small-signal output goes to a simple result that depends only on the difference of the two input voltage sources. This would represent a response to a purely balanced signal. The result obtained in (4.21) is now compared for three possible inputs to the differential amplifier.

a. Common Mode Input: For this mode, $v_{g2} = +v_{g1}$, and we get (4.22):

$$i_0 = +\epsilon_M C_\mu s v_{g1} \tag{4.22}$$

4.2. THE DIFFERENTIAL AMPLIFIER INPUT STAGE

Here the output current is directly proportional to the mirror-gain error, ϵ_M, and it also goes to zero as $|s| \to 0$.

b. *Differential Mode (Balanced) Input*: For this mode, $v_{g2} = -v_{g1}$, and we get (4.23):

$$\begin{aligned} i_0 &= (2 - \epsilon_M)(g_m - C_\mu s)v_{g1} \\ &\approx 2(g_m - C_\mu s)v_{g1} \end{aligned} \quad (4.23)$$

Here, the output current has a dc value, as $|s| \to 0$. This last result is due to the simplicity of the model. Source resistance and base spreading resistance, r_{bb}, limits the actual high-frequency current response for this circuit.

c. *Unbalanced Mode Input*: This input is common for many applications. For this mode, $v_{g2} = 0$, and we get (4.24):

$$\begin{aligned} i_0 &= [(1 - \epsilon_M/2)(g_m - (1 - \epsilon_M)C_\mu s)]v_{g1} \\ &\approx 1(g_m - C_\mu s)v_{g1} \end{aligned} \quad (4.24)$$

In this result, the output current is approximately one-half that obtained for the differential mode input. However, frequency-dependent mirror gain error can affect the actual difference in the shape of the two frequency responses that are obtained.

Equations (4.22) through (4.24) are plotted and compared in normalized form in Figure 4.10. The curve plots the magnitude in a dB ratio for the transfer function $|\mathbf{I}_0/\mathbf{V}_{g1}|$ compared to the transconductance, g_m. The complex frequency, $s = j\omega$, is normalized by the transmission-zero bandwidth, g_m/C_μ. The curves show typical behavior associated with each input mode that may be used to drive a differential amplifier. For the parameters of the amplifier biased at 500 μA, the transmission-zero bandwidth, g_m/C_μ, is found to be a very large value:

$$BW = g_m/C_\mu = 12.37 \times 10^{-3}/1.95 \times 10^{-12}/(2\pi) = 1.01 \text{ GHz}$$

Thus, it is seen that the differential amplifier provides a wideband response for the output current transfer function.

It is useful to compare the differential amplifier gain to that for the single transistor amplifier. The small-signal circuit model is shown in Figure 4.11 for the amplifier of Figure 4.7. Note that the bias current should equal $I/2 = 50$ μA in order to get the same parameters for g_m, r_π, and so forth, as was used for the differential amplifier example. All voltages are known in the circuit of Figure 4.11, and so KCL yields the output current, i_0, as in (4.25):

$$i_0 = -(g_m - C_\mu s)v_{g1} \quad (4.25)$$

It is seen that this current transfer function equals the result for the *unbalanced* mode input to the differential amplifier when the collector bias currents are equal for both circuits. The single transistor circuit can only be operated with unbalanced input, and so it is seen that the differential amplifier does not lose any frequency response performance when compared to the single transistor amplifier.

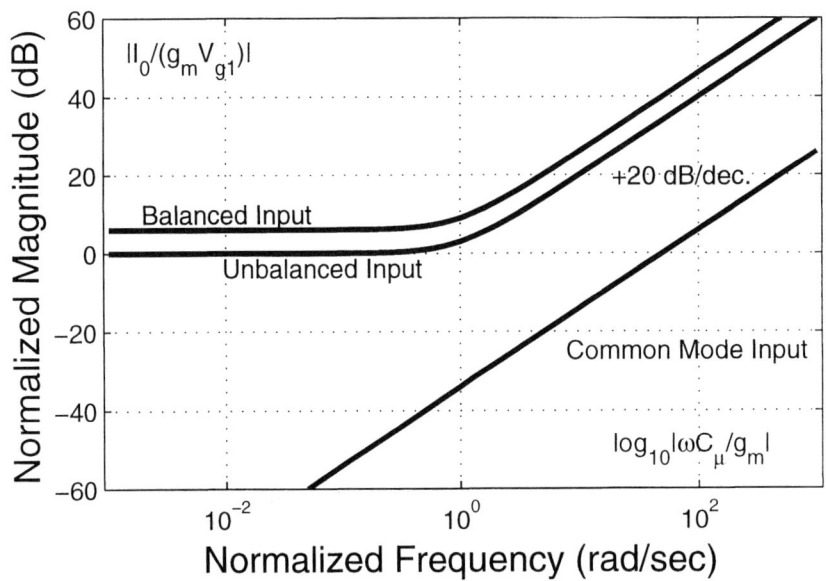

Figure 4.10. Differential amplifier gain comparisons.

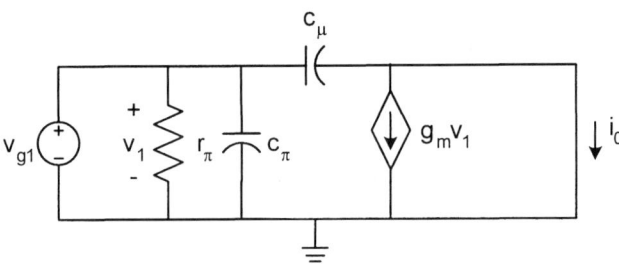

Figure 4.11. Small-signal circuit model for the single transistor amplifier.

4.2. THE DIFFERENTIAL AMPLIFIER INPUT STAGE

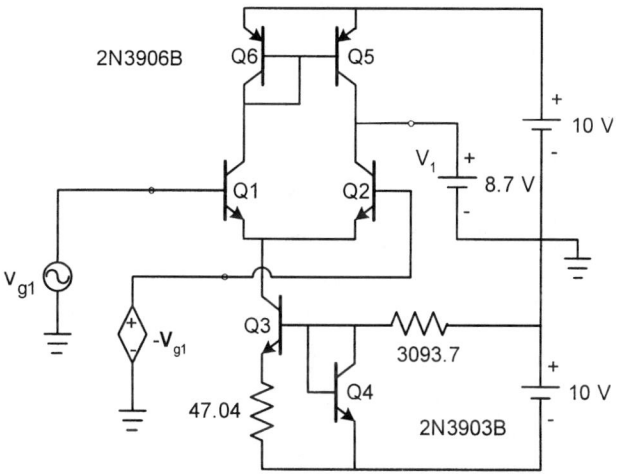

Figure 4.12. μCAP circuit for the differential amplifier test.

A final point to this discussion is illustrated by the μCAP simulation circuit shown in Figure 4.12. A differential amplifier uses discrete transistors, is driven with a balanced differential signal, and is used to simulate the output current transfer function. The 8.7-V battery is used to model bias imposed by a second stage to be discussed later. The incremental signal frequency response is plotted in Figure 4.13 where it is compared to the result found in (4.23). Note that $2g_m 10^3 = 24.74 \rightarrow 27.87$ dB for the low-frequency asymptote of the curve and that the low frequency responses compare well. The difference between the two curves is due to the fact that the collector of $Q1$ does not remain at ac signal ground as assumed in the above analysis. This effect is discussed later in more detail. However, it is still possible to conclude that the output current has an inherently wide bandwidth that increases at $+20$ dB/dec above some frequency that is on the order of g_m/C_μ rad/sec. It should be clear that the overall bandwidth of practical opamps does not have to depend upon the bandwidth of the first stage.

4.2.3 Signal Representation

It is instructive to note at this point that arbitrary input signals to a differential amplifier are often represented, or referred to, in terms of the *differential mode* voltage, v_d, and the *common mode* voltage, v_c. Let v_{g1} and v_{g2} be two amplifier inputs (see Figure 4.5). Then (4.26) is given by definition:

$$\begin{aligned} v_d &\doteq v_{g1} - v_{g2} \\ v_c &\doteq \frac{v_{g1} + v_{g2}}{2} \end{aligned} \qquad (4.26)$$

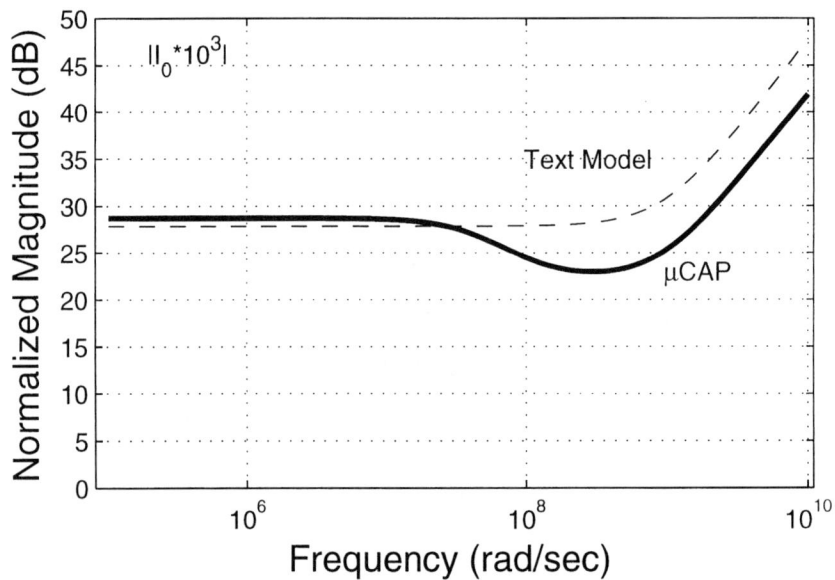

Figure 4.13. µCAP frequency response comparison.

The inverse of (4.26) is given by (4.27):

$$v_{g1} = \frac{v_d}{2} + v_c$$
$$v_{g2} = -\frac{v_d}{2} + v_c \qquad (4.27)$$

Note that the node voltage, v_x, in (4.19) found in the analysis of the circuit of Figure 4.9 is clearly equal to the common mode voltage of the input to the differential amplifier. Thus, inputs with a common mode component will stimulate common mode response.

Now, consider the representations for the three input combinations considered for the gain plots of Figure 4.10.

a. $v_{g2} = +v_{g1}$: Substitution into (4.26) yields the following:

$$v_d = 0$$
$$v_c = v_{g1}$$

Only a common mode voltage is input to the circuit for this test.

b. $v_{g2} = -v_{g1}$: Substitution into (4.26) yields the following:

$$v_d = 2v_{g1}$$
$$v_c = 0$$

4.2. THE DIFFERENTIAL AMPLIFIER INPUT STAGE

Only a differential mode voltage is input to the circuit for this test.

c. $v_{g2} = 0$: Substitution into (4.26) yields the following:

$$v_d = v_{g1}$$
$$v_c = 0.5 v_{g1}$$

Both mode voltages are input to the circuit for this test. Since this unbalanced signal is a very commonly desired input to the opamp, good opamp design requires that the common mode gain be zero or negligible, when compared to the differential mode gain. Looking at Figure 4.10, we see that the common mode gain, for this simple amplifier, is 66 dB below the differential mode gain for all frequencies less than one-tenth of the transmission-zero bandwidth of the output current transfer function. Typical common mode gain is often specified at greater than 80 dB below the differential mode gain, but this is for static gain behavior or for frequencies very close to dc. Clearly, this specification depends upon the current mirror gain error and to some extent upon the bias current stability with respect to v_x.

4.2.4 Second-Order Effect on Amplifier Model

This section is for the student who likes to get it right. It can be skipped without losing the continuity of the material being developed. In the small-signal frequency response just completed, the μCAP response of Figure 4.13 was different from the response developed for the circuit of Figure 4.9. The reason is that the circuit model used for the current mirror is a little too simple. Sometimes in circuit analysis, it is useful to oversimplify at first just to get a feel for what the circuit does. Then, if substantial differences are found between predicted and measured responses, it may be appropriate to examine and correct the models used to make the prediction. A feature of linear circuit design is that models can almost always be developed to predict or explain observed response properly to within a decibel or so of measured results. μCAP or SPICE simulations thus help to save a lot of time and money by accurately predicting a proposed circuit response before it is built. This section briefly shows how the previous analysis may be altered to better explain the frequency response for the first-stage differential amplifier.

The circuit of Figure 4.9 is modified in Figure 4.14 by adding input impedance, Z_M, for the mirror used on the collectors of the difference amplifier. When this impedance is added, the collector voltage of $Q1$, denoted as v_R, moves off of ground potential and is no longer 0 V. The circuit now has two independent node voltage variables, which are selected to be v_x and v_R. Analysis proceeds as follows.

KCL at v_x is unchanged and yields the same result as in (4.19). v_x is unaltered. Then, KCL at v_R yields (4.28):

$$i_R = -Y_M v_R$$
$$0 = Y_M v_R + g_m v_1 + C_\mu s(v_4.v_{g1}) \quad \text{reduces to}$$
$$-(Y_M + C_\mu s)v_R = i_4.C_\mu s v_R$$
$$= (g_m/2 - C_\mu s)v_{g1} - g_m v_{g2}/2 \qquad (4.28)$$

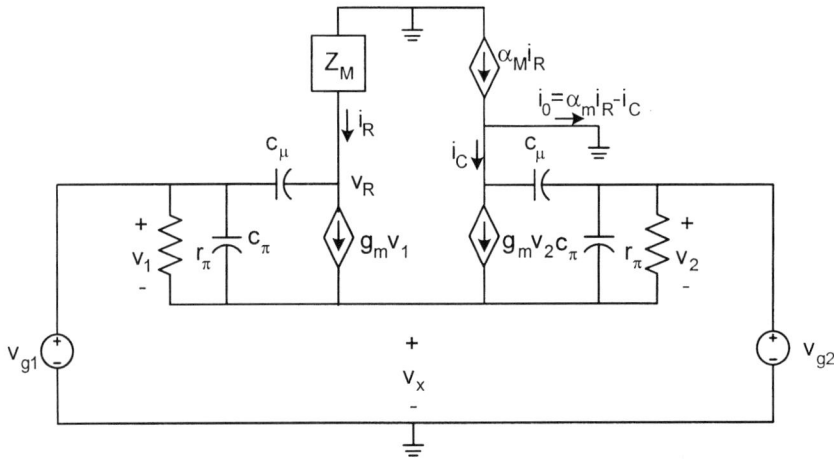

Figure 4.14. Small-signal circuit with improved mirror model.

Note that the right-hand side of (4.28) is the same as the first equation for i_R in (4.20). This would be the situation for the limiting case where $v_R \to 0$ and $Y_M \to \infty$ for the present circuit model. The KCL for the current, i_c, is unchanged and is given by the second half of (4.20). Then, when i_0 is formed by KCL at the output node, (4.29) is obtained:

$$\begin{aligned} i_0 &= \alpha_M i_4 . i_c \\ &= \alpha_M Y_M \frac{(g_m/2 - C_\mu s)v_{g1} - g_m v_{g2}/2}{Y_M + C_\mu s} \\ &\quad - [(g_m/2 - C_\mu s)v_{g2} - g_m v_{g1}/2] \end{aligned} \quad (4.29)$$

This last result may be used to obtain the response function, $i_0(s)$, for any of the input conditions described earlier. Note that the mirror model is clearly evident in this last result.

To compare results with the μCAP response of Figure 4.13, we need to consider the differential input mode where $v_{g2} = -v_{g1}$. Substituting this constraint into (4.29) yields (4.30):

$$\begin{aligned} i_0]_{DM} &= i_0]_{v_{g2}=-v_{g1}} \\ &= \left[1 + \frac{\alpha_M Y_M}{Y_M + C_\mu s}\right](g_m - C_\mu s)v_{g1} \end{aligned} \quad (4.30)$$

Equation (4.30) yields the answer obtained earlier if we let $Y_M \to \infty$ (a short circuit) and $\alpha_M \to 1$. This is as far as we can go without some knowledge of the frequency-dependent mirror parameters. Fortunately, this has been left as a homework exercise (see Problem 4.3), where both Y_M and α_M have to be found. It should be noted that the primed parameters are for PNP transistors:

$$Y_M = 2g'_\pi + g'_m + (2C'_\pi + C'_\mu)s$$

4.3. THE SECOND STAGE

$$\alpha_M = (g'_m - C'_\mu s)/Y_M \tag{4.31}$$

Inspection of these relationships shows that the mirror input admittance is a minimum for low frequency, and it gets larger, in proportion to s, as the frequency increases. So, the mirror input admittance does not seem to be a serious problem. The mirror gain, α_M, however, is severely affected with increasing frequency. It starts out at $\beta'/(\beta' + 2)$, as found earlier for the static case, and goes to $-C'_\mu/(2C'_\pi + C'_\mu)$ for large frequencies. This latter result is a small negative number. So, not only is α_M not close to 1; it becomes a negative number for high frequencies. It is this transition that causes the discrepancy between the results shown in Figure 4.13.

The analysis will now be completed with the evaluation of (4.30) using the improved mirror parameter results. Substitution of (4.31) into (4.30) yields (4.32):

$$\begin{aligned} i_0 \rfloor_{DM} &= \left[1 + \frac{g'_m - C'_\mu s}{2g'_\pi + g'_m + (2C'_\pi + C'_\mu)s + C_\mu s} \right] (g_m - C_\mu s) v_{g1} \\ &= \left[\frac{g'_\pi + g'_m + (C'_\mu + 0.5C_\mu)s}{2g'_\pi + g'_m + (2C'_\pi + C'_\mu + C_\mu)s} \right] 2(g_m - C_\mu s) v_{g1} \end{aligned} \tag{4.32}$$

The bracketed term in (4.32) contains the frequency-dependent effect of the mirror, and the balance of the equation is the same as the simpler model result [see (4.23)] plotted in Figure 4.13. Substituting values for all of the dynamic parameters, as determined earlier for this amplifier example, yields the response shown in Figure 4.15 for the evaluation and comparison of (4.32).

The analysis of the revised model clearly agrees with the μCAP simulation. We may conclude that this modified analysis adequately represents the first-stage differential amplifier frequency-dependent behavior. More complex models are always necessary in order to accurately predict high-frequency response.

4.3 The Second Stage

So far, the discussion has been directed at understanding the behavior of the first stage in the opamp architecture. We found that the differential amplifier had better temperature stability and dynamic range than the single transistor can produce. The bandwidth is also fairly wide, especially for most audio and instrumentation applications. The analysis also showed that the differential amplifier yields a current source response, i_0, to the difference input, $v_d = v_{g1} - v_{g2}$ [see (4.14)]. The dynamic transconductance between i_0 and v_d is given approximately by (4.33) as $\alpha_M \to 1$:

$$\frac{i_0}{v_d} \to \frac{I}{2}(2c_2 V_b + c_1) \tag{4.33}$$

I is the bias current for the differential amplifier, and V_b is the static base-emitter voltage for the differential transistors. A voltage gain is usually required, and so, if the current, i_0, is delivered to a load resistor, R_L, then (4.34) shows the load resistance required to obtain an incremental voltage gain of 10^5 (100 dB):

$$R_L = \frac{10^5 v_d}{i_0} = \frac{2 \times 10^5}{I(2c_2 V_b + c_1)} \approx 8 \text{ M}\Omega \tag{4.34}$$

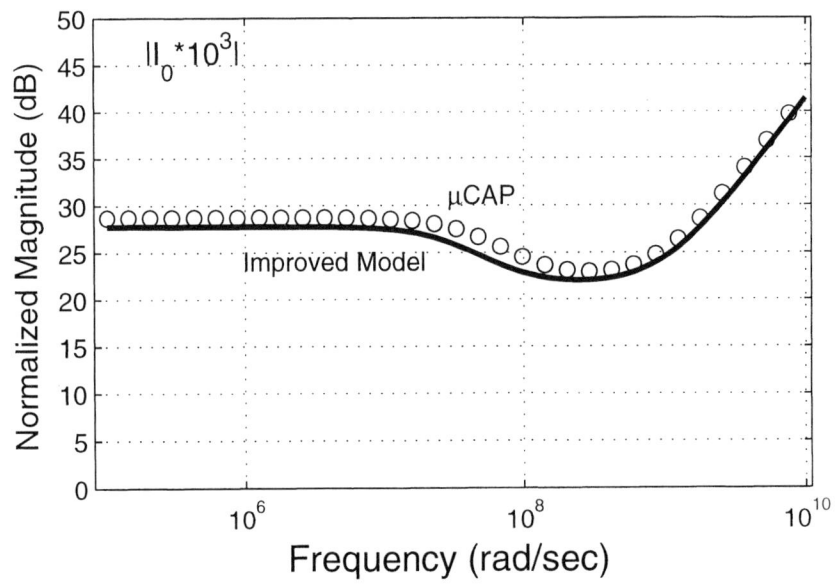

Figure 4.15. Revised analysis frequency response comparison.

This result tells us that R_L is about 8 MΩ for a 1 mA bias current. Such a value is rather large to be very practical, and so, it is conventional to provide a second stage of current gain in order to require smaller values of load resistance.

Conventional opamp design normally uses three stages. The second stage provides additional gain, and the third stage converts the current source output to a voltage source output. The voltage source output helps to provide a gain function that is not dependent on load variations.

A simple circuit is shown in Figure 4.16 for a second-stage amplifier. The circuit consists of two PNP transistors and a current sink, I_2. R_L is the load resistance, and R_S is a resistor used to provide a stable operating point for $Q6$. It is usually large but not as large as the transistor collector-emitter resistance, r_0, attributed to the Early voltage. Typically, R_S would be less than 2 MΩ. $Q5$ is used to provide an adequate collector-to-emitter voltage for the mirror output transistor of the source, i_0. This guaranteed voltage range helps to preserve the high-output impedance for the current source to the second stage. The bias, I_2, and load, R_L, are used to set the gain and the peak-to-peak signal range for the opamp. These various details will now be obtained through circuit analysis for the given amplifier.

The analysis proceeds in two steps. First, nominal values are determined using static considerations, and then, a small-signal circuit model is analyzed to determine bandwidth relationships. The incremental signal gain is determined first.

When the output voltage is at its positive limit, $V_L = V+ = +V_0$, which yields an upper limit for the load current, $i_{L1} = V_0 G_L$. The collector current for $Q6$ then

4.3. THE SECOND STAGE

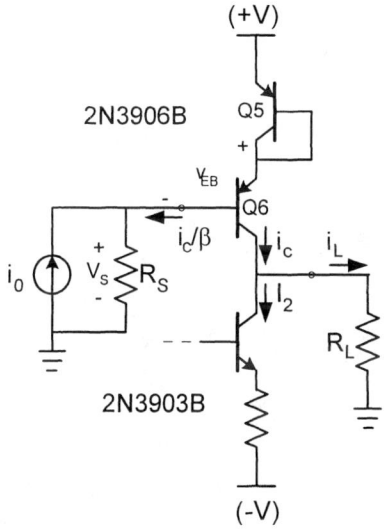

Figure 4.16. A second-stage amplifier circuit.

has the value, $i_{c1} = I_2 + i_{L1}$, by KCL. The current required from the differential amplifier is then $i_{o1} = V_S G_S - i_{c1}/\beta'_{dc}$. When the output voltage is at its negative limit, $V_L = V- = -V_0$, and the lower limit for the load current, $i_{L2} = -V_0 G_L$. Q6 collector current is then $i_{c2} = I_2 + i_{L2}$. The load current required from the differential amplifier is then $i_{o2} = V_S G_S - i_{c2}/\beta'_{dc}$. Incrementally, for Q6,

$$\Delta v_L = 2V_0$$
$$\Delta i_o = i_{o1} - i_{o2} = -\frac{2V_0 G_L}{\beta'_{dc}} \approx \frac{I}{2}(2c_2 V_b + c_1)\Delta v_d \qquad (4.35)$$

where c_1 and c_2 are NPN parameters, and β'_{dc} is a PNP parameter. The static voltage gain is then given by (4.36):

$$\frac{\Delta v_L}{\Delta v_d} = -\frac{R_L \beta'_{dc} I}{2}(2c_2 V_b + c_1) = -A_0 \qquad (4.36)$$

This last result is equivalent to $\Delta v_L = -\beta'_{dc} R_L \Delta i_o$, which might be seen by inspection. It is also the nominal unbalanced input gain. Note that in (4.36) the bias current, I_2, does not affect the gain. It helps to set the operating point for Q6 as shown in (4.35). With $\beta'_{dc} = 100$, $I = 1$ mA, and a desired gain, $A_0 = 10^5$, (4.36) shows that $R_L \approx 80$ kΩ would be required. Essentially, the second stage reduces the required load resistance by the current gain factor, β'_{dc}. Sometimes Darlington pairs are used to obtain current gain proportional to β'^2, thereby further reducing required load resistance. There is a corresponding sacrifice in bandwidth that goes with this choice.

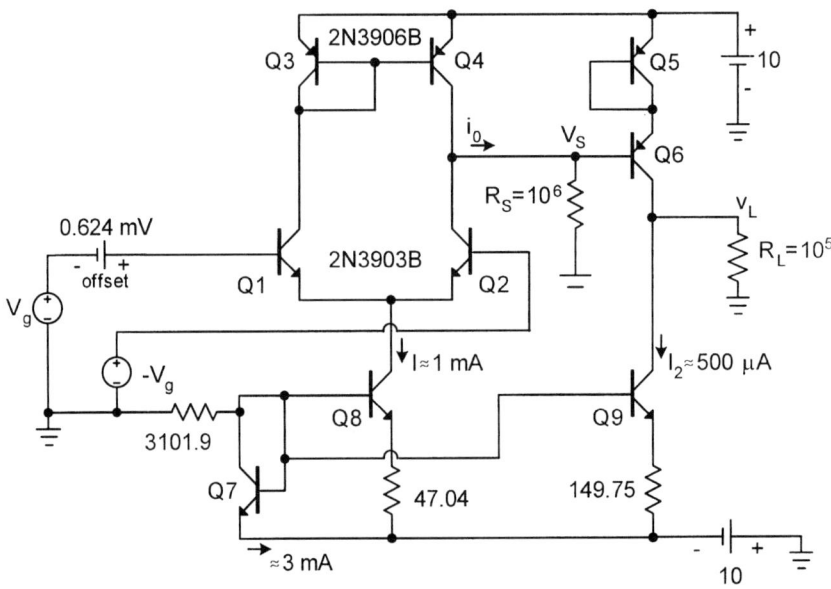

Figure 4.17. μCAP circuit for the two-stage amplifier simulation.

The second stage introduces an input dc offset requirement due to its biasing arrangement. The input offset is found as follows. When the output voltage is zero, the load current, i_L, is also zero, and so, the collector current is equal to the bias current, I_2. Then, applying KCL on the input to $Q6$ provides (4.37):

$$i_0 \rfloor_{v_L=0} \doteq i_{offset} = V_S G_S - I_2/\beta'_{dc}$$
$$\text{where } V_S = V_0 - 2V_{eb} \qquad (4.37)$$

A differential input offset voltage is required to obtain an output offset current from the first stage. Note that the offset, $(\alpha_M - 1)I/2$, due to the differential amplifier is not negligible compared to the current in (4.37). The required input offset voltage is then given by (4.38):

$$i_{offset} \approx I\frac{(\alpha_M - 1)}{2} + I\frac{(\alpha_M + 1)}{4}\log_e(K_{12})(v_d)_{offset}$$
$$\therefore (v_d)_{offset} = \frac{4}{\log_e(K_{12})}\left[\frac{V_S G_S - I_2/\beta'_{dc}}{I(\alpha_M + 1)} - \frac{(\alpha_M - 1)}{2(\alpha_M + 1)}\right] \qquad (4.38)$$

The offset voltage is important to consider when doing μCAP simulations for an amplifier in open-loop calculations. An example will help to clarify what is meant. The following parameters are used to determine gain and offset for the two-stage amplifier shown in Figure 4.17. Refer to Figure 4.17 for notation and to derive the following parameters: $R_L = 10^5$ Ω, $R_S = 1$ MΩ, $I = 1$ mA, and $I_2 = 0.5$ mA. The reference source is adjusted from its value of 3093.7 Ω in Figure 4.12 to a new

4.3. THE SECOND STAGE

value of 3101.9 Ω in order to supply both mirror currents, I and I_2. When the load voltage, v_L, is zero, then,

$$
\begin{align*}
i_{c6} &= I_2 = 500 \; \mu\text{A} \\
v_{eb6} &= \text{inv3906B}(i_{c6}, 3) = 0.6347 \text{ V} \\
[i_{b6} \; i_{c6}] &= \text{pnp3906B}(v_{eb6}) = [5.289 \; 500.0] \; \mu\text{A} \\
\beta'_{Q6} &= i_{c6}/i_{b6} = 94.50
\end{align*}
$$

Both transistors, $Q5$ and $Q6$, have identical emitter currents, so their emitter-base voltages are also identical. The above data then determine the offset current required to establish zero output, $v_L = 0$:

$$
\begin{align*}
V_S &= 10 - 2v_{eb6} = 8.7306 \text{ V, and then} \\
(i_0)_{offset} &= V_S G_S - i_{b6} = 3.4418 \; \mu\text{A}
\end{align*}
$$

Substitution of these parameters into (4.38) yields an estimate for the input differential offset voltage as follows:

$$(v_d)_{offset} = 1.0665 \text{ mV}$$

A μCAP simulation circuit, with differential mode input, is shown in Figure 4.17 for the two-stage amplifier with the above parameters. Note, in the differential mode, with $v_{g2} = -v_{g1}$,

$$(v_{g1})_{offset} = 0.5(v_d)_{offset} = 0.533 \text{ mV}$$

The dc offset voltage in the circuit is determined by trial and error, starting from the predicted 0.533 mV, so that the three static gain curves shown in Figure 4.18 will all cross the axis for zero input. Note that if the offset had not been included, the amplifier would saturate at the upper output level for the full input range shown. Consequently, the incremental signal gain as estimated by μCAP would be virtually zero. In terms of differential input, the response in Figure 4.18 shows that the offset temperature drift coefficient is approximately 0.9 μV/K.

The next question asks about the open-loop bandwidth of the resulting two-stage amplifier. The μCAP solution at 27°C is plotted in Figure 4.19 for the circuit of Figure 4.17, which includes the dc offset source. If the dc source were to be omitted, the gain would be quite small (try it and see what you get). Here we see that the low-frequency differential-input gain of $\approx$104.1 dB is about 2.7 dB different from the estimate obtained from (4.36) for the circuit with the above parameters. This is a fair prediction, and the result is sensitive to the offset value used at the input. The half-power bandwidth appears to be about 10 kHz for the first pole, and a second pole appears around 2 MHz. Recall that the region between 10 and 100 MHz is where the current mirror begins to affect the gain of the current source, i_0, that drives the second stage. We will now briefly analyze a small-signal circuit model for this second stage to see how the two-stage amplifier bandwidth can be predicted.

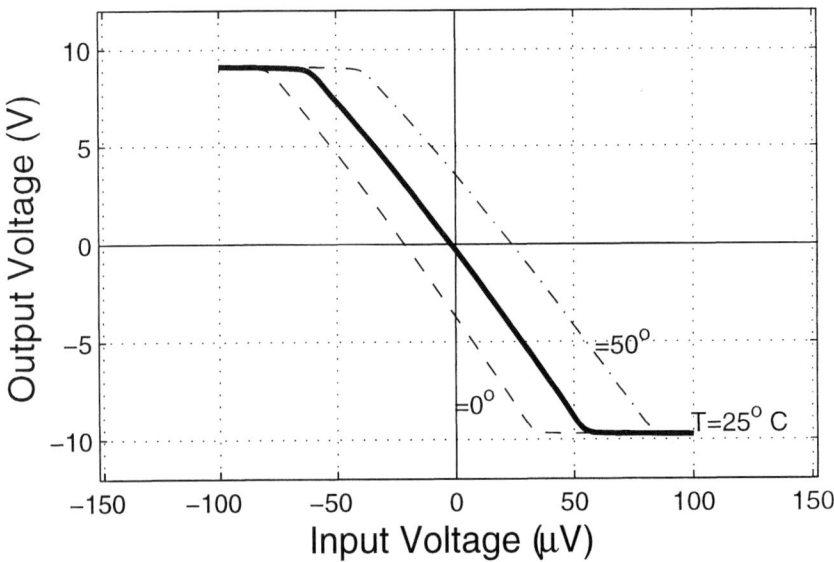

Figure 4.18. Two-stage amplifier static transfer function as a function of temperature.

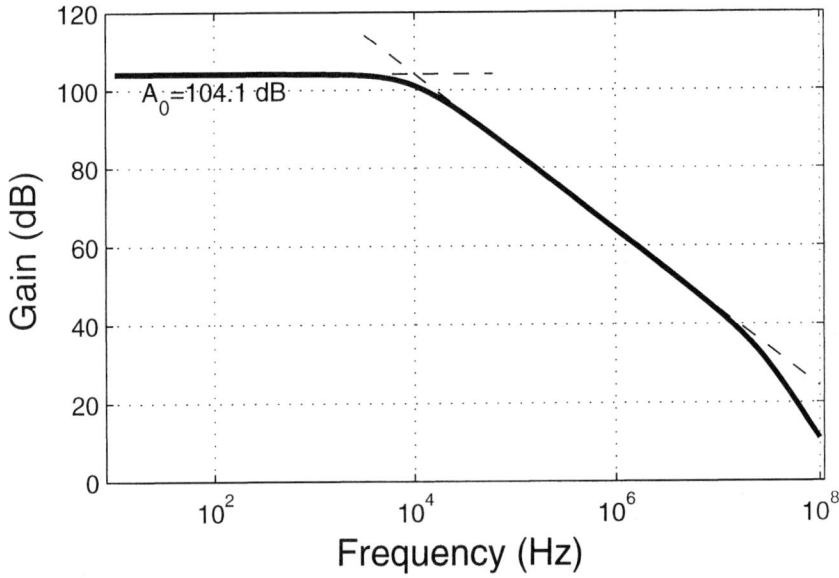

Figure 4.19. Differential-gain frequency response for the two-stage amplifier.

4.3. THE SECOND STAGE

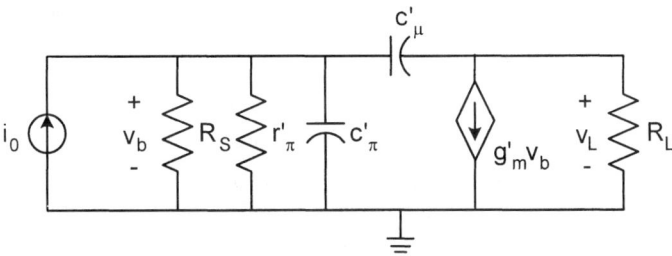

Figure 4.20. Small-signal circuit model for the second stage.

A small-signal circuit is shown in Figure 4.20 for the second-stage amplifier. It assumes that the input is a current source; that the diode connected transistor, $Q5$, behaves as a short circuit; and that the bias source, I_2, is an open circuit. The result is the hybrid-π model for a grounded emitter PNP transistor. The circuit has two independent voltage variables, so the analysis proceeds from two KCL equations as follows. Applying KCL at the output yields (4.39):

$$0 = G_L v_L + g'_m v_b + C'_\mu s(v_L - v_b) \quad \text{yields}$$
$$v_b = -\frac{(G_L + C'_\mu s)}{(g'_m - C'_\mu s)} v_L \tag{4.39}$$

Application of KCL at the input yields (4.40):

$$y'_\pi = g'_\pi + C'_\pi s$$
$$i_0 = (G_S + y'_\pi) v_b + C'_\mu s(v_b - v_L) \quad \text{yields}$$
$$\frac{v_L}{i_0} = \frac{(C'_\mu s - g'_m)}{C'_\mu C'_\pi s^2 + (g'_m C'_\mu + (C'_\pi + C'_\mu) G_L) s + G_L (g'_\pi + G_S)} \tag{4.40}$$

When typical parameter values are substituted into (4.40), it is found that the poles are both real and widely separated due to the relations shown in (4.41):

$$\frac{g'_m}{C'_\pi} > \frac{(C'_\pi + C'_\mu) G_L}{C'_\pi C'_\mu} \quad \text{and}$$
$$\left(\frac{g'_m}{C'_\pi}\right)^2 > \frac{4 G_L (g'_\pi + G_S)}{C'_\pi C'_\mu} \tag{4.41}$$

The poles are then given by the following approximate functions:

$$\sigma_2 \approx \frac{g'_m}{C'_\pi} > \sigma_1$$
$$\sigma_1 \approx \frac{G_L (g'_\pi + G_S)}{C'_\mu g'_m} = \frac{1 + r'_\pi G_S}{R_L C'_\mu \beta'} \tag{4.42}$$

The above second-stage amplifier design yields the following parameters and values for the poles:

$$G_S = 1\ \mu\text{Mho} \text{ and } G_L = 10\ \mu\text{Mho}$$
$$r'_\pi = 5.968\ \text{k}\Omega,\quad g'_m = 16.61\ \text{mMho},\quad \text{and } \beta' = 99.1$$
$$C'_\pi = 36.74\ \text{pF} \text{ and } C'_\mu = 3.06\ \text{pF}$$
$$\therefore \sigma_1 \approx 33.17\ \text{kNepers/sec} \to 5.28\ \text{kHz}$$
$$\sigma_2 \approx 45.21\ \text{GNepers/sec} \to 71.95\ \text{MHz}$$

This low pole estimation compares well with the μCAP simulation shown in Figure 4.19.

The result obtained in (4.42) is often used to set the first pole (bandwidth) of the opamp. Note that the current gain, β', multiplies the base-collector capacitance, C'_μ, so that a much larger effective capacitance interacts with R_L to set this pole value. This result is sometimes referred to as the Miller effect, which is really just a manifestation of feedback. The advantage of the β-multiplication is that small capacitance values are adequate to set a desired pole value. For example, the LF351 has its first pole at about 40 Hz. Then, with $R_L = 100$ kΩ and $\beta' = 100$, the required capacitance would be as follows:

$$C = (1 + r'_\pi G_S) G_L / \beta' / 80\pi = 400\ \text{pF}$$

This value of capacitance is achievable in IC fabrication, or it may be added to the chip externally, but since it is much larger than the C'_μ provided by the junction, the added capacitance (collector-to-base on $Q6$) provides a good unit-to-unit production repeatability for the first pole of the opamp. If the pole is not set with capacitance added at this point, then much larger values may be required at other points in the circuit.

In summary, we have seen how a single transistor can be used to provide additional gain to bring the required load resistance to practical values in order to achieve an open-loop gain on the order of 100 dB. The second stage provides a controllable relationship for setting the bandwidth of the opamp with relatively small capacitance values added to the base-collector of the amplifying transistor. A dc offset is introduced during this stage, and its effect may be cancelled at the opamp input. This offset is developed entirely at the second stage, and consequently, its stability is only affected by the gain contributed by additional gain stages rather than the full open-loop gain. This stage has a current source output, or high impedance, and so, we now need to look at the third stage, which interfaces this gain with practical load applications by converting it to a voltage source output.

4.4 The Third Stage

We are getting close to the end of our journey to understand opamp performance and design issues. So far, we have developed the required open-loop gain and bandwidth control. The raw gain is developed by means of a voltage controlled

4.4. THE THIRD STAGE

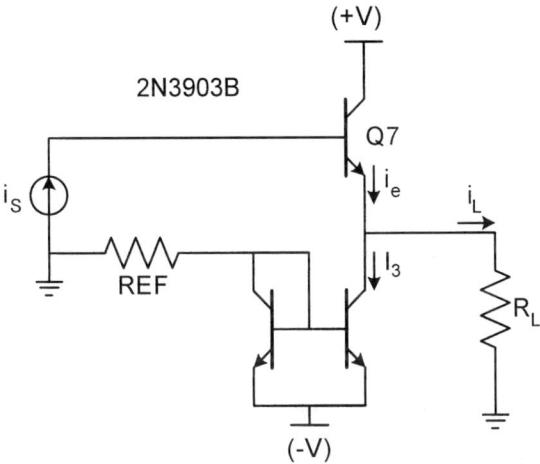

Figure 4.21. A single transistor emitter follower amplifier.

current source. We now seek to convert the output to a voltage source so that the gain may be isolated from the load. When we do this, the major consideration will be that we not undo what the first stages have achieved. So, we will want to check the dc offset performance, the dynamic range of the overall transfer function, and the small-signal frequency response. Because of these requirements, the third stage is almost always some variation of a voltage follower (common collector) circuit. Voltage followers are characterized by having high input, low output impedances, and unit voltage gain. They also have wide bandwidth, and so, they should have very little effect on the open-loop network function of the opamp.

The emitter follower circuit shown in Figure 4.21 illustrates some of these ideas for the output stage. The bias, I_3, is used to improve the efficiency of the stage and, with the minimum value of R_L, is used to set the peak-to-peak voltage output of the amplifier. The efficiency of the output stage is an issue because this stage may dissipate a lot of power in order to handle large voltage swings and load current values. These ideas will be considered first. For minimum output voltage, $i_L = -I_3$ ($Q7$ is cut off, so $i_e = 0$) and $v_L = -R_L I_3 = V^-$ when the bias current source saturates. At this negative output level, neither the bias source nor $Q7$ is dissipating any average power while the load is receiving maximum power. When the output voltage is zero, then $i_e = I_3$, and $Q7$ dissipates $V^+ I_3$ W while the bias source dissipates $|V^-|I_3$ W; so, for this load condition, the amplifier is dissipating a total of $(V^+ + |V^-|)I_3$ W, while the load receives zero power, or is idling. To obtain a positive output voltage equal in magnitude to the negative output (equal voltage swings) requires $i_L = I_3$, $i_e = 2I_3$, and $V_L = R_L I_3 = V^+$ (if $Q7$ is saturated). Again, $Q7$ dissipates negligible power when it is saturated, and the bias source dissipates $(V^+ + |V^-|)I_3$ W. Normally, $|V^-| = V^+ = V_0$, and so, we get an output

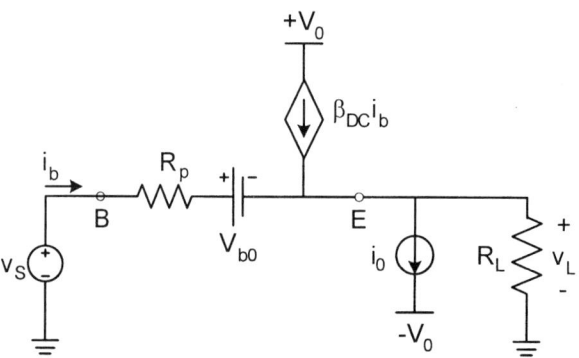

Figure 4.22. A static circuit model for the single transistor output stage.

amplifier idle-power to maximum load-power ratio of 2:1.

$$\frac{2V_0 I_3}{V_0^2/R_L} = \frac{2V_0 I_3}{V_0 I_3} = 2 \qquad (4.43)$$

If a resistor is used in place of the bias source, I_3, then the idle-power ratio can easily approach 10:1 or higher. Normally, it is desirable to choose I_3 slightly larger than the required maximum load current so that $Q7$ does not have to cut off in order to obtain the minimum output voltage. This arrangement keeps the transistor operating in the active region continuously in what is sometimes referred to as Class A operation. The advantage is that the stage will have more bandwidth (though not obviously) and also provide a better transfer function between input and output. Some of these relationships will now be found from an analysis of the static circuit model shown in Figure 4.22.

The forward base-emitter $i - v$ characteristic is crudely modeled as a voltage source, V_{b0}, in series with a resistance, R_p. Use of KCL yields the conditions given in (4.44):

$$\begin{aligned} (\beta_{dc} + 1)i_b &= I_3 + G_L v_L \quad \text{and} \\ i_b &= G_p(v_S - V_{b0} - v_L) \end{aligned} \qquad (4.44)$$

These are combined to obtain relationships for v_L and i_1 as given in (4.45). Note that $G_m \approx G_p(\beta_{dc} + 1)$ has been used to obtain the results for the static circuit.

$$\begin{aligned} v_L &= \frac{G_m(v_S - V_{b0}) - I_3}{G_m + G_L} \\ i_b &= G_p \frac{G_L(v_S - V_{b0}) + I_3}{G_m + G_L} \end{aligned} \qquad (4.45)$$

Equation (4.45) tells us that when $v_L = 0$, the input voltage is $V_{b0} + I_3/G_m$ (positive), and so, this stage introduces appreciable offset. However, when this

4.4. THE THIRD STAGE

offset is referred to the input, it is reduced by the open-loop gain factor.[4] The slope of the output versus input voltage is seen to be $G_m/(G_m + G_L)$, and since G_m is normally $>> G_L$, for this stage, this is a gain very nearly equal to 1. Note from the relation for i_b that the transistor cuts off ($i_b = 0$) when $(v_S - V_{b0}) = -R_L I_3$, so the cutoff provides a lower input voltage limit for the amplifier. Finally, we note from (4.45) that the slope between the input voltage and current is the low-frequency input resistance for the amplifier. Equation (4.46) gives this result:

$$R_{in} = \beta_{dc} R_L + R_p \tag{4.46}$$

This input impedance is what generates the input voltage to the last stage in response to the output current of the second stage. As R_L becomes large, so does R_{in}, and therefore, the opamp's open-loop gain also increases. Equation (4.46) then shows that there has to be a minimum R_L to limit the range of R_{in} to achieve a required open-loop gain. The message here is that the load resistance has an effect on the open-loop gain obtainable from the opamp.

There are different ways to reduce the effects just discussed, but they are difficult to eliminate. Some opamp designs put modest power output stages on the IC and leave it to the user to boost the load drive capability by adding external circuitry. This arrangement gets the high-power consumption off chip and eases the layout constraints for the IC design. The user has to be aware of the drive capability for any opamp selected for a job. Output load constraints can place serious limitations on the utility of any opamp, so the warning is to pay attention to the load limit performance specifications.

Before considering a different output circuit, let us look at the small-signal performance for the amplifier in Figure 4.21. A small-signal circuit model is shown in Figure 4.23. In order to interpret the results, it is useful to estimate parameter values for this stage. The operating point is taken at zero output voltage, and so, the emitter current equals the bias current, I_3.

Let $R_{Lmin} = 1250\ \Omega$, or $G_L = 0.8$ mMho. With $I_3 R_{Lmin} > 10$ V, $I_3 = 10$ mA is satisfactory. Then, for the NPN 2N3903B with $i_e = 10$ mA,

$$\begin{aligned}
v_b &= \text{inv3903B}(i_e, 2) = 0.7366 \text{ V} \\
[i_b\ i_c] &= [0.1085\ 9.893] \text{ mA} \\
\beta_{dc} &= i_c/i_b = 91.22 \\
r_\pi &= 1/(b_1 i_b) = 461.9\ \Omega \\
g_m &= \beta_{dc}/r_\pi = 197.48 \text{ mMho (This is huge!)} \\
C_\pi &\approx 222.09 \text{ pF (by interpolation for } v_b \text{ value)} \\
C_\mu &\approx 1.97 \text{ pF}
\end{aligned}$$

For analysis, let v_S and v_L be the independent voltage variables for the circuit.

---

[4]It is a usual procedure to make the opamp specifications appear in their best light (e.g., 10 μV input offset sounds a lot better than 1 V output offset). Beware of product sales literature: it may be correct or just incomplete, but some companies do engage in "specmanship" to sell products.

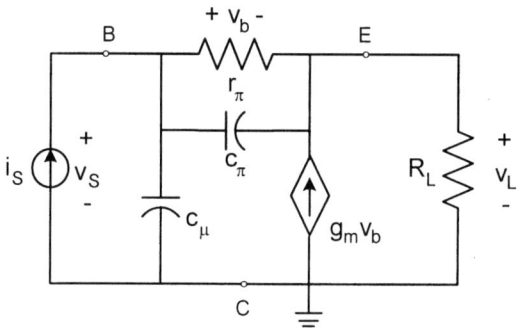

Figure 4.23. Small-signal circuit model for the single transistor output stage.

Then, applying KCL at the base and emitter, plus some algebra, yields (4.47):

$$\begin{aligned}
y_\pi &= g_\pi + C_\pi s \\
i_S &= C_\mu s v_S + y_\pi(v_S - v_L) \quad \text{and} \\
0 &= G_L v_L - (y_\pi + g_m)(v_S - v_L) \quad \text{yield} \\
\frac{v_L}{v_S} &= \frac{y_\pi + g_m}{y_\pi + g_m + G_L} \\
\frac{v_S}{i_S} &= \frac{y_\pi + g_m + G_L}{C_\mu s(y_\pi + g_m + G_L) + y_\pi G_L}
\end{aligned} \quad (4.47)$$

With $G_L \ll g_m$, as is usual for this stage, the last result shows that $v_L \approx v_S$ for all $s = j\omega$. Thus, the output voltage follows the input voltage extremely well over a very wide frequency band. The input voltage is developed by the input impedance in response to the current source. A rearranged form is shown in (4.48):

$$\begin{aligned}
\frac{v_S}{i_S} &= \frac{s + (g_\pi + g_m + G_L)/C_\pi}{C_\mu [s^2 + (g_\pi + g_m + G_L)s/C_\pi + G_L s/C_\mu + g_\pi G_L/(C_\pi C_\mu)]} \\
&\approx \frac{(s + 9.0255 \times 10^8)}{C_\mu(s + 13.056 \times 10^8)(s + 3.0321 \times 10^6)} \\
&= \frac{1}{C_\mu \left[s + \frac{G_L(s + g_\pi/C_\pi)}{C_\mu(s + (g_\pi + g_m + G_L)/C_\pi)} \right]} \quad \text{by division} \\
&\approx \frac{1}{C_\mu \left[s + \frac{1}{C_\mu(R_L(\beta_{dc} + 1) + r_\pi)} \right]}
\end{aligned} \quad (4.48)$$

Equation (4.48) is reached for the case where $|s| < (g_\pi + g_m + G_L)/C_\pi = 9.0255 \times 10^8$. The pole in (4.48), $1/C_\mu/(R_L(\beta_{dc} + 1) + r_\pi) = 4.386 \times 10^6$, is only crudely close to the actual pole at 3.032×10^6 Nep/sec.

In this last result, we see again the β-multiplication of C_μ to set the bandwidth of a simple amplifier. With the chosen gain parameters, the bandwidth is seen to

4.4. THE THIRD STAGE

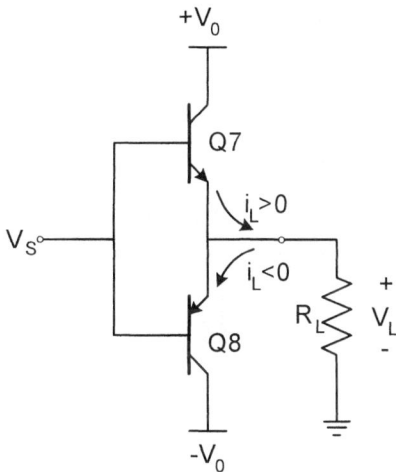

Figure 4.24. A complementary voltage follower output circuit.

be $3.032 \times 10^6/2\pi = 482.6$ kHz, which is two orders of magnitude greater than the second-stage bandwidth of about 5.3 kHz. The bandwidth of this output stage should not seriously affect the bandwidth for the composite opamp. Note that the bandwidth may be increased by using a smaller R_L or β-factor. However, the input impedance would also be reduced, and so, a gain-bandwidth trade-off is involved.

We have taken the time to study this output circuit in some detail to get a feel for some of the desired and actual properties of the output stage for an opamp. We will now take a look at a commonly used output stage that combines two emitter followers in complementary fashion in order to achieve even better efficiency and peak load capability. Some of the properties just developed will extend directly to our understanding of this next circuit.

4.4.1 Complementary Emitter Follower Output Stage

The circuit shown in Figure 4.24 is known as a complementary (NPN-PNP) emitter follower amplifier. $Q8$ has replaced the bias current, I_3, of Figure 4.21, and the idea is that $Q7$ will supply positive current to the load and that $Q8$ will supply negative current. The arrangement provides an efficient output circuit since it does not use any power for zero-load current (i.e., both transistors cut off for zero output). Thus, the idle-to-maximum load-power ratio is zero, making this a highly attractive circuit. However, as is usually the case, we don't get something for nothing, and so, we find that this circuit has some problems with its transfer function. How one goes about fixing up the problem determines how well the circuit behaves in a practical situation. We now look at the static transfer function for the circuit and some ways to compensate for its undesired behavior.

The static transfer function is easily estimated from the results of the previous

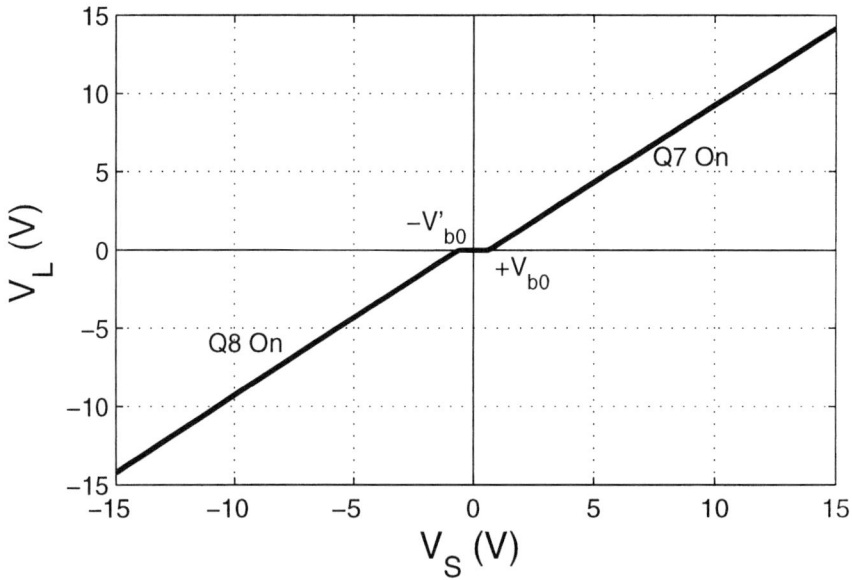

Figure 4.25. Static transfer function for the complementary voltage follower.

discussion. When the load voltage is positive, the circuit model of Figure 4.22 is used to estimate the static response (set $I_3 = 0$). Then, by applying KCL at v_L we get (4.49):

$$v_L = \frac{G_m}{(G_m + G_L)}(v_S - V_{b0}) \quad \text{for } v_L > 0 \quad (4.49)$$

Now, when the load voltage is negative, a circuit similar to Figure 4.22 applies, but V'_{b0} is negative with respect to the input. Equation (4.50) is then obtained for negative load currents:

$$v_L = \frac{G'_m}{(G'_m + G_L)}(v_S + V'_{b0}) \quad \text{for } v_L < 0 \quad (4.50)$$

The resulting transfer function is plotted in Figure 4.25, where we can see that the response is not linear. In fact, there is virtually no output over the interval between $-V'_{b0}$ and $+V_{b0}$ where the input has to overcome the diode drop of the base-emitter junction for each transistor. So, even though the circuit is highly efficient and has no voltage offset, its transfer function is not linear for midrange values of the output signal. When this "dead zone" is referred to the opamp input, it is only a few microvolts wide, and so, some designs do nothing to correct this problem other than to rely on negative feedback to cover it up. Such opamps must be used carefully in real applications because the nonlinearity presents dynamic response problems that are difficult to model in closed-loop applications.

4.4. THE THIRD STAGE

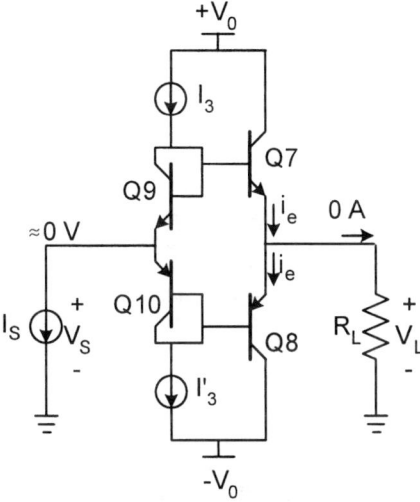

Figure 4.26. Diode offset bias for the complementary follower circuit.

A common approach to fixing the problem is to use a second set of diode junctions to create a difference voltage much like what is done on the input of a differential amplifier. One such circuit is shown in Figure 4.26. The voltages are shown for v_L and v_S grounded (set to 0). In the circuit, it is important that $I_3 = I_3'$ and $\beta = \beta'$ as closely as possible. Circuit analysis then yields (4.51):

$$\begin{aligned} I_3 &= i_b + i_e = i_b + (\beta+1)i_b = (\beta+2)i_b \\ i_e &= (\beta+1)i_b = \frac{\beta+1}{\beta+2}I_3 \doteq \frac{\beta'+1}{\beta'+2}I_3' \end{aligned} \quad (4.51)$$

This correction works nicely when the complementary transistor parameters can be closely matched, which is actually hard to achieve. The bias current, I_3, is small, on the order of tens of μA, and so a Widlar mirror is usually used rather than resistors of a few megohms to build I_3 and I_3'. This also yields a higher input impedance for the circuit. The size of I_3 affects the idle power since each transistor emitter current equals I_3 when the load voltage is zero. The idle power will thus be measured in units of μW, which is still much more efficient than the single-sided circuit considered earlier.

A stronger consideration for I_3 is the required maximum load current. When v_S becomes sufficiently positive, $Q9$ cuts off, and I_3 is the base current to $Q7$. At this point, the $Q7$ emitter current is $(\beta+1)$ times the base current, or $(\beta+1)I_3$. Thus, I_3 sets the maximum positive and negative currents that can be delivered to the load, and so, it indirectly determines, or limits, the peak-to-peak output voltage that can be achieved by the follower circuit for a given load condition. A μCAP simulation is shown in Figure 4.27 for the static transfer function, V_L versus V_S, for the circuit in Figure 4.26. Parameters were set at: $V_0 = 15$ V, $R_L = 1$ kΩ, $I_3 = I_3' = 200$ μA.

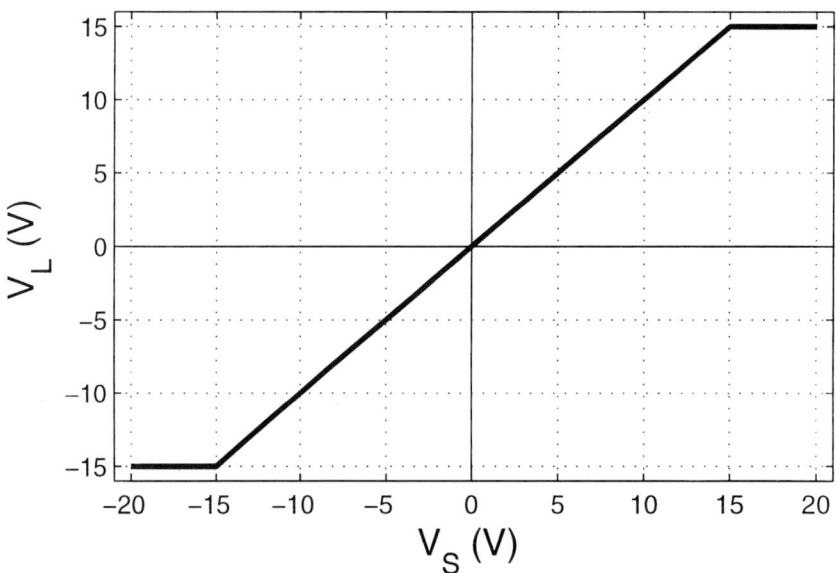

Figure 4.27. Static transfer for the diode-biased complementary follower.

The dc offset is negligible for this circuit, and the dead zone is not discernible on this graph.

The small-signal behavior of the complementary follower is not easy to predict. Normally, only one transistor is really active at one instant in time. The other is either cut off or in its idle state and, so, appears as two reverse-biased diodes. The bias diodes, $Q9$ and $Q10$, are always on for linear operation, and their small-signal impedance is small enough to be considered negligible when compared to the impedance of $Q7$ or $Q8$. Figure 4.28 illustrates a small-signal circuit model for the complementary follower in accordance with this discussion. When $\beta' = \beta$, and $g'_m = g_m$, the controlled source appears to flip back and forth between $Q7$ and $Q8$, whichever is active. But note that as far as v_L is concerned, the source does not appear to change since the two transistor circuits are virtually in parallel by considering the bias transistors to be small-signal short circuits.

Further study of Figure 4.28 reveals it to be the same as Figure 4.23 with C_μ replaced with $(C_\mu + C'_\mu)$ and C_π replaced with $(C_\pi + C'_\pi)$. Appropriate modification of (4.47) then yields (4.52) for the input impedance function:

$$\frac{v_S}{i_S} = \frac{1}{(C_\mu + C'_\mu)} \frac{s + (g_\pi + g_m + G_L)/(C_\pi + C'_\pi)}{s^2 + \left(\frac{g_\pi + g_m + G_L}{C_\pi + C'_\pi} + \frac{G_L}{C_\mu + C'_\mu}\right)s + \frac{g_\pi G_L}{(C_\pi + C'_\pi)(C_\mu + C'_\mu)}} \quad (4.52)$$

The following parameters are obtained for the 2N3903B and 2N3906B transistors when a 200 μA quiescent emitter current is assumed for both the transistors and I_3.

4.4. THE THIRD STAGE

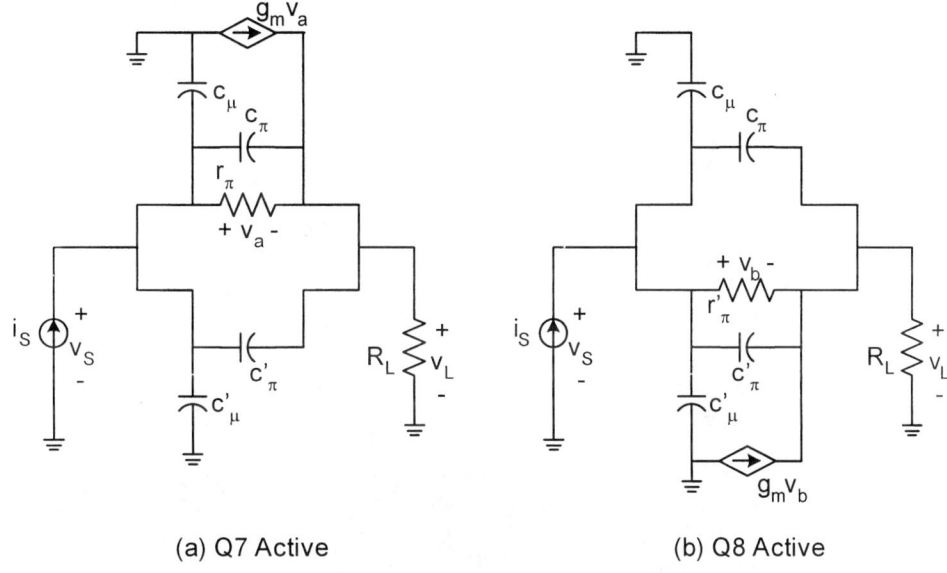

(a) Q7 Active (b) Q8 Active

Figure 4.28. Small-signal circuit models for the circuit of Figure 4.27 with $\beta' = \beta$: (a) $Q7$ active, and (b) $Q8$ active.

NPN 2N3903B:

$$\begin{aligned}
i_e &= 200 \ \mu\text{A} \quad \text{yields} \quad v_b = 0.5658 \text{ V} \\
\beta_{dc} &= 54.8 \\
g_\pi &= 71.557 \ \mu\text{Mhos} \\
g_m &= 3.922 \text{ mMhos} \\
C_\pi &= 11.43 \text{ pF} \\
C_\mu &= 1.95 \text{ pF}
\end{aligned}$$

PNP 2N3906B:

$$\begin{aligned}
i_e &= 200 \ \mu\text{A} \quad \text{yields} \quad v_e = 0.6076 \text{ V} \\
\beta'_{dc} &= 89.0 \\
g'_\pi &= 71.108 \ \mu\text{Mhos} \\
g'_m &= 6.329 \text{ mMhos} \\
C'_\pi &= 19.80 \text{ pF} \\
C'_\mu &= 3.05 \text{ pF}
\end{aligned}$$

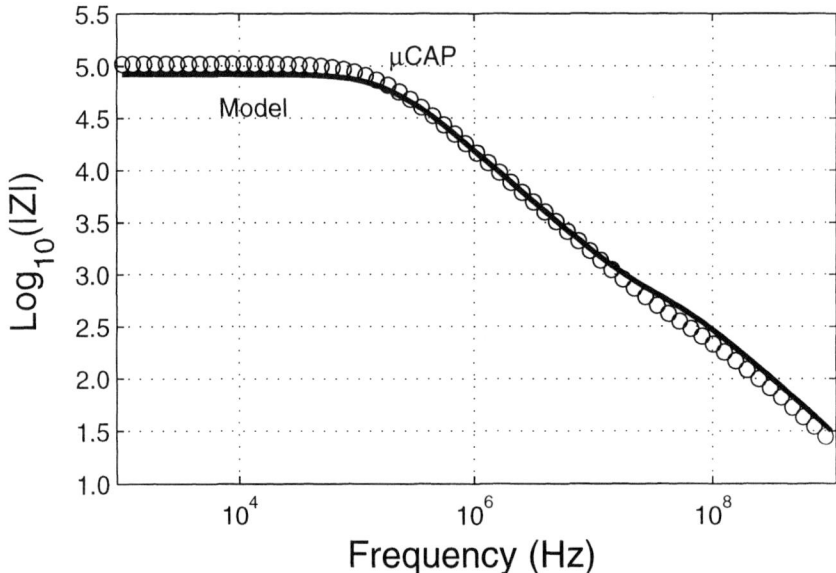

Figure 4.29. Third-stage input impedance magnitude: simulation versus theory.

Substitution of these into (4.52) yields (4.53):

$$\frac{v_S}{i_S} = \frac{1}{(C_\mu + C'_\mu)} \frac{(s + 1.535 \times 10^8)}{(s^2 + 3.135 \times 10^8 s + 3.666 \times 10^{14})} \quad (4.53)$$

This function has its poles at 186.8 kHz and 49.7 MHz with its zero at 24.4 MHz. With this circuit, the g_m/C_π term does not dominate, and it is not straightforward to estimate the low pole value, as was done for (4.48). Nevertheless, this simplified circuit model does predict the frequency response fairly well. This is shown in Figure 4.29, where the above model results are compared to the μCAP simulation for the circuit of Figure 4.26 and the input impedance of Figure 4.28. The 3 dB bandwidths are nearly identical, but there is a difference in the static gain for the two results. The difference depends strongly upon the quiescent, or effective, bias current used to estimate the hybrid-π parameters for the small-signal model. The key idea shown here is that only one of the complementary transistors is active at any instant. That is what is proven by the comparison of these simulated and modeled circuit results.

It should be noted that these results have assumed ideal current sources for the bias currents, I_3, I'_3, and the source, i_S. When nonideal sources are used, the interactions are complex and difficult to predict, but results obtained in the next section show that the bandwidth of the last stage has little or no observable effect upon the overall opamp response. Its effect appears to be flat over the whole band. The limiting bandwidth effect is basically contributed by the second stage. We will

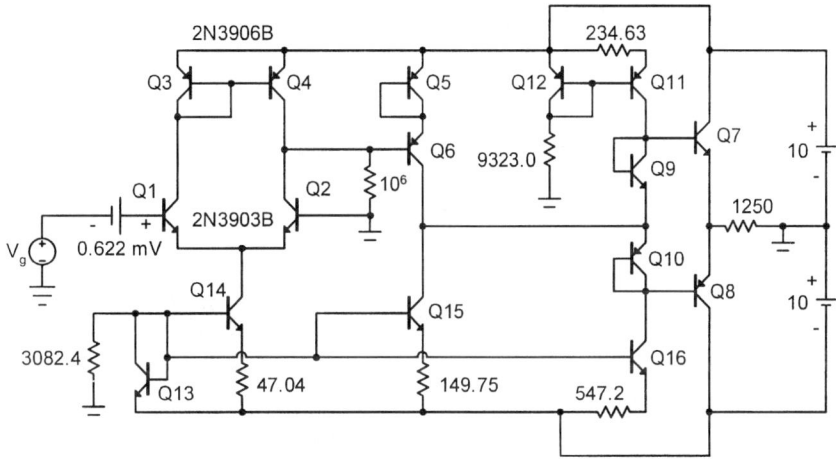

Figure 4.30. μCAP simulation circuit for the three-stage opamp.

now finally take a look at the performance of the full opamp circuit that is a result of all the previous discussion.

4.5 All Together Now—A Three-Stage Opamp

In this section, we briefly look at the performance obtained when we put the three stages together to obtain a candidate opamp circuit. The μCAP simulation circuit shown in Figure 4.30 is configured for the unbalanced mode input with $V_{g2} = 0$. Note that the output offset voltage is corrected at the input with the 0.622 mV dc source. The value for this source has been determined from the two-stage results discussed earlier.

The static transfer function is shown in Figure 4.31 for a 50°C operating temperature range. Several features are evident from this response. First, for temperatures above 25°C, the output range is limited to 17.4 V_{pp} for the 1.25 kΩ load. This limit is caused by the saturation of the output transistors, $Q7$ and $Q8$, as they are driven by the mirror bias sources at these limit values. Second, the output stage has no apparent dead zone as viewed on this microvolt scale, and so, we may conclude that the diode biasing of $Q9$ and $Q10$ with I_3 and I'_3 has solved that problem. Third, the output stage has not apparently contributed additional offset voltage, and it also has not contributed any observable temperature drift to the output offset. The drift shown in Figure 4.31 is approximately +1.8 μV/K for the unbalanced mode. This is about double the result obtained earlier for the two-stage amplifier in the differential mode. The addition of the third stage has successfully converted the near 100 kΩ load requirement of the second stage to a 1250-Ω load at the output without adding any detrimental behavior to the static transfer function.

The small-signal gain function is shown in Figure 4.32 for both open- and closed-

120   CHAPTER 4. SOME OPAMP DESIGN CONSIDERATIONS

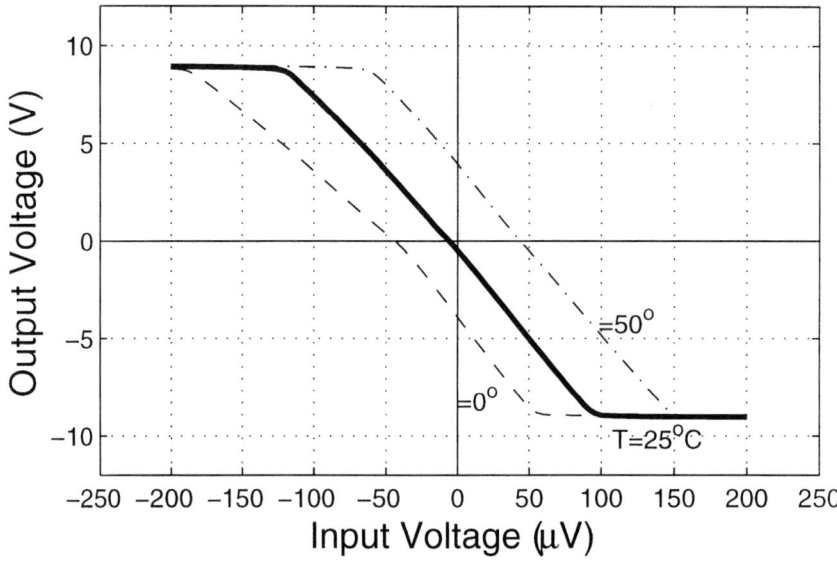

Figure 4.31. Three-stage static transfer function with unbalanced input mode.

loop gains with the unbalanced input condition. The first pole is at approximately 7 kHz ($\approx$10 kHz was determined earlier for the two-stage amplifier), and the second pole is at about 20 MHz, as seen earlier. The open-loop gain is close to 100 dB ($+10^5$) as required by the design objective. The closed-loop response is obtained by replacing the offset correction source with 125 Ω and connecting the 1250 Ω load to the base of $Q1$ rather than ground. We see that the amplifier behaves like the examples considered in Chapter 3, as the closed-loop gain, $R_L/R_{in} = 10$, is held over a wide bandwidth until it becomes asymptotic to the open-loop gain. The peak may be lowered by reducing the open-loop bandwidth (add capacitance to the collector-base of $Q6$). This effect is considered in Problem 4.14(c).

So, the results show that we have accomplished a basic design for a VCVS opamp. There are some features to improve if we want to have a better opamp. One feature commonly encountered is current limiting circuitry on the output so that short circuit loads cannot damage the output stage. This is done in several devious ways, and we will not go into that problem here. The student may refer to power supply design and regulation for further details on this subject, or just look at published circuits for general purpose opamps, such as the LM741. A second feature is the open-loop dependence upon R_L. This effect may be reduced by splitting R_L and then requiring the external load to be greater than some minimum value. For example, put $R = 2500$ Ω from output to ground internally; then $R_{Lmin} = 2500$ Ω to get the responses shown in Figures 4.31 and 4.32. Subsequently, as R_L is varied from 2500 Ω to an open circuit, the gain only varies by 6 dB, while the bandwidth only reduces by one-half. Recall that the second-stage low-frequency pole depends

4.5. ALL TOGETHER NOW—A THREE-STAGE OPAMP

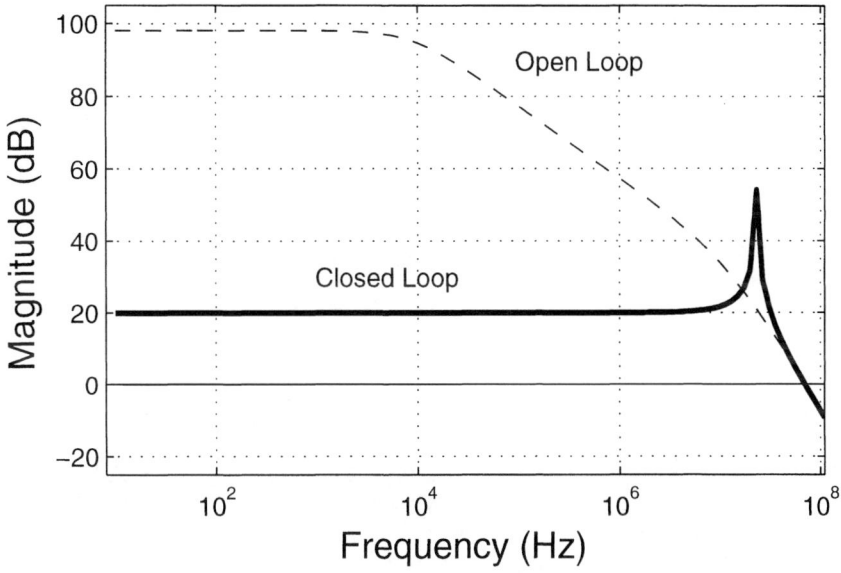

Figure 4.32. Open- and closed-loop gain response for the three-stage opamp circuit.

directly upon the load resistance to that stage. Also note that incorporating a resistor ($2 \times R_{Lmin}$) in parallel with the output does not affect the idle-power ratio for the output stage since that resistor draws zero idle power. Thus, a parallel load resistance, internal to the opamp, helps to stabilize the opamp open-loop gain and bandwidth with respect to external load resistance variations. Third, some applications require very high input impedance, and so, voltage followers may be added to the input to increase the input impedance. Using smaller bias current on the first stage yields higher input impedance, but it also gives up open-loop gain, which must be recovered in the second stage. Fourth, circuitry is sometimes added to provide means for zeroing the output offset voltage for the composite amplifier.

A final feature that it is useful to correct is the temperature drift of the output offset voltage. It is possible to correct the drift (not the offset) for this circuit design, and so, we will now investigate this problem.

4.5.1 Temperature Compensation of Output Offset

The static response in Figure 4.31 shows that the open-loop, input, dc-offset voltage drifts at about 90 μV/50°C, or 1.8 μV/K. The curves show that with the fixed 0.622 mV offset correction at the input, the incremental gain (slope of static curve at zero output voltage) will display a large variation versus temperature. It is helpful to correct for this drift, whenever possible, so that the incremental gain will not have a strong temperature dependence at the output. Since the voltage offset

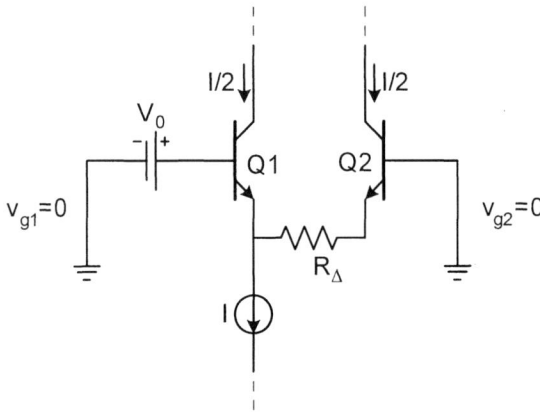

Figure 4.33. Adding a small resistance to compensate for output offset voltage drift.

has been found to depend inversely upon the first-stage bias current, I [see (4.38)], it is possible to use this same current to compensate for the drift. A scheme is shown in Figure 4.33.

Because the first-stage amplifier has very small incremental signal variation due to the large open-loop gain, the emitter currents of $Q1$ and $Q2$ are both virtually equal to one-half the bias current, I. The addition of a small resistance, R_Δ, as shown in Figure 4.33, reduces the offset voltage value required for V_0 to the value given in (4.54); v_{offset} is the same as developed in (4.38) for the two-stage amplifier:

$$V_0 = v_{offset} - R_\Delta \frac{I}{2} \qquad (4.54)$$

Then, since we have determined a value for the temperature drift of v_{offset}, it is possible to determine R_Δ if we know the temperature dependence of the bias current, I. This is shown in (4.55):

$$\frac{\Delta V_0}{\Delta T} = 0 = \frac{\Delta v_{offset}}{\Delta T} - \frac{R_\Delta}{2}\frac{\Delta I}{\Delta T} \quad \text{or}$$
$$R_\Delta = 2\frac{\Delta v_{offset}/\Delta T}{\Delta I/\Delta T} \qquad (4.55)$$

$\Delta I/\Delta T$ can be determined in several ways; the simplest is to run a μCAP simulation versus temperature to obtain the estimate, $\Delta I/\Delta T \approx 2.2$ μA/K (at T=25°C), for the current mirror used to develop the current, I. Substitution into (4.55) then yields a value for R_Δ:

$$R_\Delta \approx 2\frac{1.98 \ \mu V/K}{2.2 \ \mu A/K} = 1.64 \ \Omega \qquad (4.56)$$

4.5. ALL TOGETHER NOW—A THREE-STAGE OPAMP

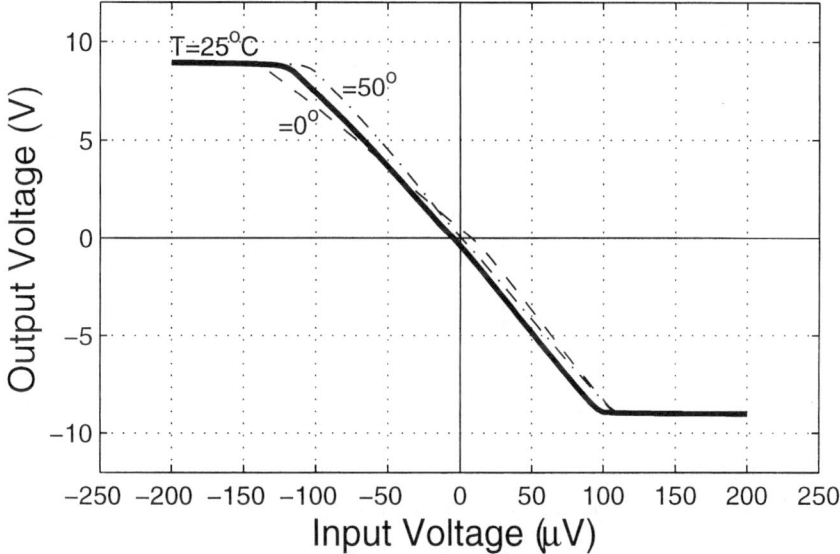

Figure 4.34. The three-stage static transfer functions with temperature compensation for the output offset drift.

The fixed offset voltage is then changed as follows:

$$V_0' = v_{offset} - R_\Delta \frac{I}{2} = 0.622 - 1.64(0.5) = -0.1989 \text{ mV} \qquad (4.57)$$

When R_Δ is inserted in series with the emitter of $Q2$, in the circuit of Figure 4.30, and the fixed input bias voltage, V_0, is set equal to V_0', static transfer responses are obtained as shown in Figure 4.34. The result shows that the temperature offset drift has been reduced from $\approx 90/50$ μV/K to less than $13/50$ μV/K. This is a reduction of about seven times in the temperature dependent effect of the offset voltage. Note that if the temperature drift had been negative, a resistor in series with the emitter of $Q1$ could have been employed to compensate for the drift.

This compensation method has yielded an extremely practical result for the opamp circuit. The curves in Figure 4.34 show that the open-loop gain has been stabilized considerably over both the 50°C temperature range and most of the linear output range of the opamp. Such performance is important enough to warrant inclusion of the compensation resistor in the final design for the circuit. Some designs place a small resistor in both emitters, $Q1$ and $Q2$, and adjust all the offset with an appropriate difference between the two resistors. This can be a dangerous thing to do since the wrong R_Δ can just as easily overcompensate for the temperature drift of the opamp output. The particular opamp designed here will always have a positive input offset due to the second-stage biasing. Therefore, the inclusion of the R_Δ, as developed, is always on the correct side to reduce input

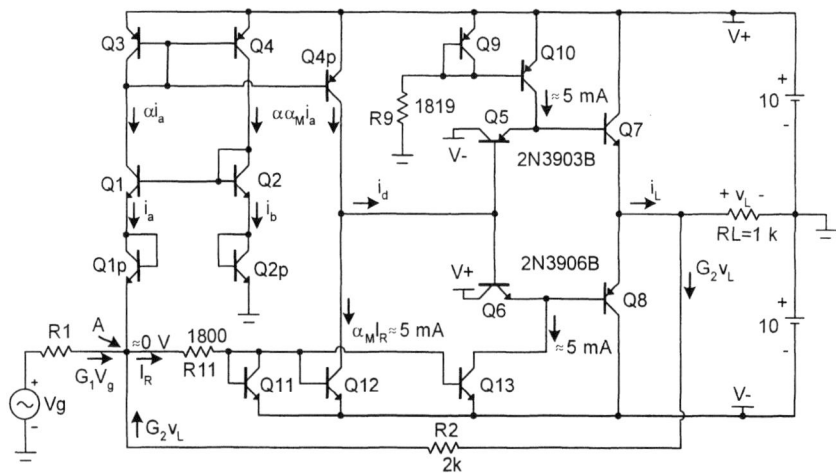

Figure 4.35. A current controlled opamp example circuit.

offset while compensating for the temperature drift whenever the correct R_Δ value is used.

These results bring us to a satisfactory conclusion for an introduction to conventional opamp design and performance issues. The material presented in this chapter should enable the student to look at any data sheet schematic for a candidate opamp and deduce how the amplifier is designed and how well it may perform for a proposed application. We will test this ability by taking a look at a current controlled opamp design to see how it works and to compare its performance with that of a voltage controlled opamp. The concept and motivation for such an amplifier were developed in Chapter 3, and considering the difference between both types of opamp structures is an objective of this chapter.

4.5.2 Case Study—A Current Controlled Opamp

In this section, we analyze a current controlled opamp feedback circuit. This will complete our study of opamp design and performance considerations.

A schematic is shown in Figure 4.35, along with some circuit variable definitions, to facilitate our analysis of the amplifier performance. This design is typical of circuits that are found in [1]. It is interesting to note that there are only two resistors internal to the amplifier design. R_1, R_2, and R_L are external to the amplifier although it is feasible to incorporate a maximum value of R inside the amplifier package since it is not normally varied to change the closed-loop gain.[5]

The next apparent feature of the amplifier is that it has a lot of current mirrors. Transistors $Q9$ through $Q13$ form three mirrors that each develop a nominal 5 mA bias current. $Q10$ and $Q13$ deliver bias current to a complementary output follower

---
[5] Recall the discussion in Chapter 3 for this type of amplifier.

4.5. ALL TOGETHER NOW—A THREE-STAGE OPAMP

stage as discussed in the previous design. The transistors $Q5$ and $Q6$ are connected to bias out the dead zone diode drops of the output transistors, but in this circuit, they also amplify the incoming current, i_d, as will now be shown.

Assume that $Q7$ delivers the positive load current, i_L, and that $Q8$ delivers the negative current. This agrees with our understanding of the complementary output circuit. Now, consider the case for $i_L > 0$. For convenience, we assume that all transistors have the same β-factor. The emitter currents to $Q5$ and $Q6$ are then given by (4.58):

$$\begin{aligned} i_{eQ5} &\approx 5 \text{ mA} - \frac{i_L}{\beta+1} \\ i_{eQ6} &\approx 5 \text{ mA} - 0 \end{aligned} \quad (4.58)$$

These currents are then referred to the base of $Q5$ and $Q6$ to form the current input, i_d, to this output stage. This result is given in (4.59):

$$\begin{aligned} i_d + \frac{5 \text{ mA} - i_L/(\beta+1)}{\beta'+1} &= i_{bQ6} \approx \frac{5 \text{ mA}}{\beta+1} \text{ or} \\ i_d &\approx \frac{i_L}{(\beta+1)^2} = \frac{(G_L + G_2)v_L}{(\beta+1)^2} \end{aligned} \quad (4.59)$$

The approximation is made on the assumption that $\beta' \approx \beta$. For $\beta \approx 100$, this result represents a gain of 80 dB between the load and input currents for this stage. So, we see substantial gain developed in the output stage rather than the input stage. Finally, the same gain relation is found by assuming a negative load current and working through $Q8$.

Now consider the input stage. The emitters of $Q1$ and $Q2$ represent differential (current) input ports to this amplifier. The circuit of Figure 4.35 thus shows an unbalanced input to the opamp since $Q2$ is grounded. Transistors $Q1$ through $Q4$ form a set of current mirrors that are connected in a cross-coupled mode. The result is that the current, i_b, in $Q2$ is a mirror of the current, i_a, in $Q1$. This will be shown first. With i_a equal to the emitter current of $Q1p$, the reference current input to mirror $Q3-Q4$ is approximately αi_a. That current is then mirrored to the collector of $Q4$ (and $Q4p$) by the mirror factor, α_M, derived earlier. Applying a global KCL to $Q1$ and $Q2$ yields i_b:

$$\begin{aligned} i_b &= (\alpha \alpha_M + \alpha - 1)i_a \text{ with} \\ \alpha &= \frac{\beta}{\beta+1} \text{ and } \alpha_M = \frac{\beta}{\beta+2} \\ i_b &= \frac{\beta-2}{\beta+2}i_a \end{aligned} \quad (4.60)$$

This last result shows that i_a and i_b are nearly equal. Therefore, the base-emitter drops on $Q1$ and $Q2$ are virtually identical, as are the drops on $Q1$ and $Q2p$. This also makes the input sum-point, A, virtually at ground potential, or 0 V. Input point A looks like a short circuit and is the current controlling sum-point used in the model of Chapter 3.

It should be noted that (4.60) was derived using a convenient assumption that the NPN and PNP transistor gains are the same. In practice, this is difficult to achieve, but the result obtained yields a gain not much different from that obtained by assuming different gains. For example, (4.60) becomes $i_b = i_a(\beta_N \beta_P - \beta_P - 2)/[(\beta_N + 1)(\beta_P + 2)])$ when a harder analysis is used to get the result. Then, with $\beta_N = 90$ and $\beta_P = 150$, (4.60) yields $i_b = 0.957 i_a$ for $\beta = 90$, and the result using different gains yields $i_b = 0.965 i_a$. The difference between the two results only amounts to 8 parts in 960. It is often practical to use simplifying assumptions for design purposes.

Now, close the loop with KCL at the sum-point A:

$$I_R \approx 5 \text{ mA with point } A = 0 \text{ V}$$
$$i_a = I_4 . G_1 v_g - G_2 v_L \tag{4.61}$$

With i_a known, we can now determine the input current, i_d, to the output stage. The transistor $Q4p$ delivers the same collector current as $Q4$, and so, i_d is found by applying KCL at the collectors of $Q4p$ and $Q12$.

$$\begin{aligned} i_d &= \alpha_M(\alpha i_a - I_R) \\ &= \alpha_M[(\alpha - 1)I_4.\alpha(G_1 v_g - G_2 v_L)] \end{aligned} \tag{4.62}$$

Now combine (4.59) and (4.62) to get the transfer function for the amplifier:

$$\left[\alpha G_2 + \frac{G_L + G_2}{\alpha_M(\beta+1)^2}\right] v_L = -\alpha G_1 v_g - (1-\alpha)I_R$$

$$v_L \equiv -\frac{G_1 R_2 v_g + R_2 I_R/\beta}{1 + (1 + G_L R_2)/(\alpha_M \beta(\beta+1))} \tag{4.63}$$

This last result shows that the output voltage has a dc offset determined by the reference current and the fixed feedback resistor, R_2. The small-signal gain is in a form identical to that derived in Chapter 3. The closed-loop gain may be adjusted by changing only the input resistor, R_1, and the half-power bandwidth should not change.

The results shown in Figure 4.36 were obtained from a μCAP simulation of the circuit of Figure 4.35 with R_2 equal to a constant 2 kΩ. The half-power frequency is on the order of 4 MHz (very large) and appears constant for $R_1 > 5$ Ω. The fact that the output is also a good voltage source is illustrated by driving a 1 kΩ load for the 50 dB gain condition. There is no visible effect on the gain as the load goes from 100 kΩ to 1 kΩ. Note that if gain approximates $-G_1 R_2$ for this circuit, R_2 appears to be approximately 1700 Ω for this circuit. (See the 0-dB closed-loop response curve for $R_1 = 1700$ Ω in Figure 4.36.)

We could consider more details about this opamp, but we will leave them and move onto the next phase of this course.

4.6 Conclusion

This chapter has provided a rather lengthy introduction to the opamp structure and some of its design issues. Many topics have still not been covered or introduced.

4.7. PROBLEMS

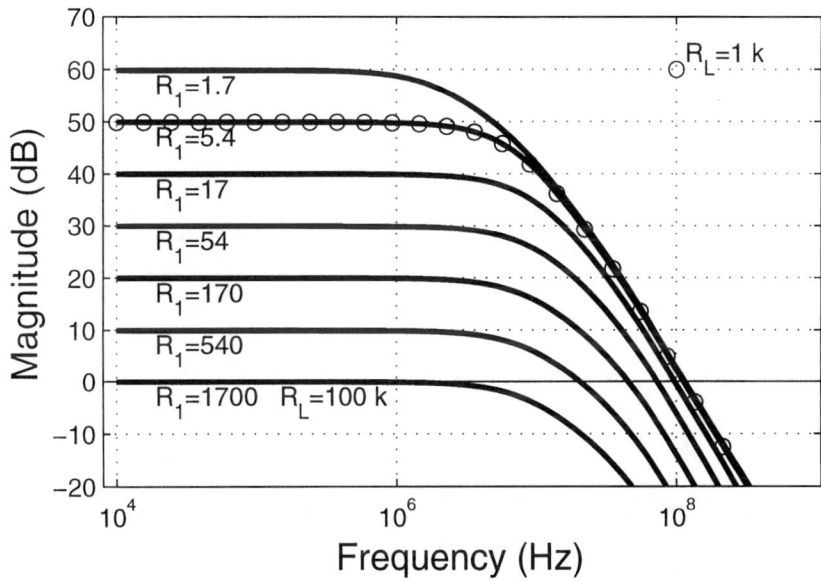

Figure 4.36. Closed-loop gain responses for the current controlled opamp example.

You might be able to make a career out of designing these circuits! There are a few companies like Analog Devices that have large groups devoted to these problems on a full-time basis. We could talk about increasing the input impedance, providing offset correction circuitry, signal limiting on the front end, overload protection on the output, single bias supply design, lead/lag frequency compensation, and so forth. The list goes on, and so shall we. This introduction has been presented so that you will have some idea of how amplifiers work, develop some respect for getting 100 dB of gain to work as a stable amplifier, and have a feel for the signal and frequency limits of such devices.

For the rest of the book, we treat the opamp as an ideal device to use in the design of frequency-dependent circuits (filters). The application requirements of operating frequency and signal level will be warning factors about whether it may be practical to design an active filter or not. The background from this chapter should help you to be aware of such limitations.

4.7 Problems

4.1 Determine the small-signal output impedance at $Q2$ in the mirror circuit of Figure 4.3(a). Plot the magnitude response versus frequency.

4.2 In the power series expansion, (4.14), for the output current, i_o, of the differential amplifier, find the maximum amplitude for v_d that will limit fundamental

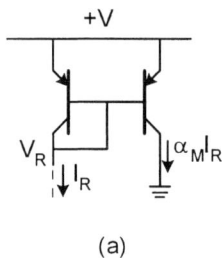

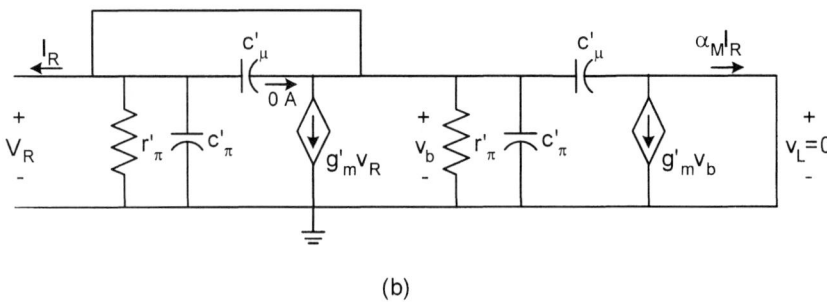

Figure 4.37. Circuit for Problem 4.3: (a) active load, and (b) hybrid-π small-signal model.

distortion to less than 1 part per 1000 from the v_d^3 term of the series when v_d is a sinusoid. Compare the result to that for a single transistor amplifier developed in (4.16).

4.3 Find Y_M and α_M for the small-signal circuit given in Figure 4.37 for a mirror used as an active load.

4.4 The purpose of this problem is to initiate a study of the first-stage differential amplifier circuit shown in Figure 4.38. The circuit is similar to the one discussed in the text, but you may try different operating current bias values and other parameter variations through the use of μCAP simulation. For example, it might be interesting to compare the response for different transistors and/or EFA on the current mirrors.

(a) Determine R for the transistors you select and then R_1 for $I_1 = 20$ μA and $I_1 = 200$ μA.

(b) Use μCAP to simulate the circuit and obtain plots for the dc transfer between v_0 and v_g for the two values of I_1.

(c) Compare the simulated results with equations developed in class. Confirm that your R and R_1 values yield the proper currents via the dc operating points determined from the μCAP simulations.

4.7. PROBLEMS

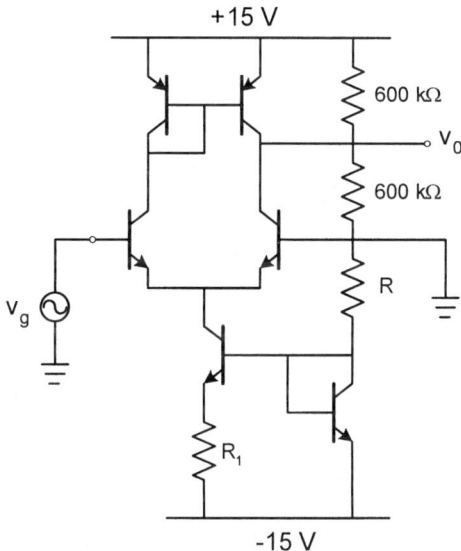

Figure 4.38. Circuit for Problem 4.4.

4.5 Replace I_3 with resistor, R_E, in Figure 4.21 and determine the idle-power ratio in terms of R_L and R_E for $|V^-| = V^+ = V_0$. How is the maximum peak-to-peak output affected by R_L and R_E?

4.6 Sometimes a V_{be} multiplier biasing circuit is used to bias complementary transistors in a follower output stage, as shown in Figure 4.39. Explain how the circuit works and find a suitable value for K.

4.7 Design your own three-stage opamp and check its performance with μCAP or SPICE. The overall open-loop gain should be at least 90 dB at low frequency for unbalanced mode input and exhibit a -20 dB/dec rolloff until unity open-loop gain is achieved. The minimum load resistance should be at least 2 kΩ.

4.8 Use your opamp (Problem 4.7) with feedback to check how it will respond to closed-loop applications. With $R_1 = 1$ kΩ, check the frequency response for different values of R_2 to see how well the amplifier responds.

4.9 Repeat Problem 4.8 by replacing R_2 with a capacitor, C_2. Determine the response for $R_1 = 1$ kΩ and $C_2 = 0.1$ μF and specify the bandwidth over which your opamp approximates ideal integration.

4.10 Obtain a schematic diagram for the LM741 and explain its design.

   (a) Choose appropriate or typical element values and simulate its response. Compare the results to the manufacturer's specifications. Are there any apparent problems with the design?

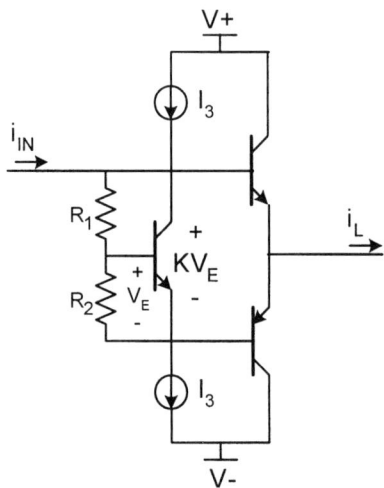

Figure 4.39. Circuit for Problem 4.6.

(b) Repeat (a) for an LH0024.

(c) Repeat (a) for an LF444.

4.11 The mirror design shown in Figure 4.40 is used to bias the differential amplifier in the LH0024 wideband opamp. Analyze the circuit to determine its mirror gain factor and explain its operation. In what way does the mirror offer better performance than the Widlar mirror of Figure 4.3(b)?

4.12 Let the transistor current gain, α, be a single-pole function of frequency, $\alpha(s) = \alpha_0/(1 + s/\omega_\alpha)$, where f_α is the alpha cut off frequency. Find $\beta(s)$ and the beta cut off frequency, f_β in terms of α_0 and β_0.

4.13 Apply the result of Problem 4.12 to the static gain, (4.63), for the current controlled opamp. How well does the use of a frequency-dependent $\beta(s)$ to predict the gain response function for $|v_L/v_g|_{s=j\omega}$ compare with the simulation results of Figure 4.36?

4.14 Answer the following questions for the circuit given in Figure 4.41. It is known that the first pole of the amplifier is at 800 Hz, and $\beta = 200$ for each transistor.

 (a) What value is required for I_1 in order to obtain an unbalanced mode gain of 100 dB for the 1-kΩ load?

 (b) Find the input dc offset voltage for this amplifier if $R_1 = 100$ μA.

 (c) Assume the first pole is due to an effective resistance, R_{L2}, loading the second stage. What is the approximate value of C_μ for $Q2$?

 (d) Assuming $C_\mu = 5$ pF, what value of capacitance added in parallel with C_μ will move the pole to 100 Hz?

4.7. PROBLEMS

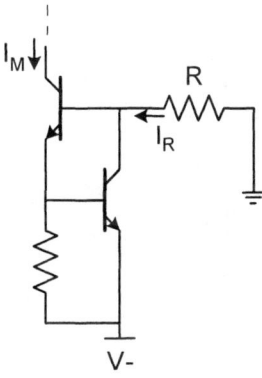

Figure 4.40. Circuit for Problem 4.11.

(d) Assuming $C_\mu = 5$ pF, what value of capacitance added in parallel with C_μ will move the pole to 100 Hz?

4.15 Set $I_1 = 100$ μA in the opamp circuit of Figure 4.41 and use μCAP to simulate and plot the open-loop ac voltage gain versus frequency from low frequency to -10 dB past the unit-gain frequency.

4.16 Confirm that the current controlled opamp of Figure 4.35 has dc offset, and propose a circuit modification to correct for the offset. Can the apparent dependence upon the closed-loop gain be eliminated?

4.17 Use μCAP to measure and plot the small-signal output admittance for the NPN mirror circuit given in Figure 4.42. Assume the transistors are type 2N3903.

(a) What does the simulation tell us about output impedance for a simple mirror circuit?

(b) Try to determine/explain why the 3-dB bandwidth is so dependent upon the reference current (through the 14.4-kΩ resistor).

4.18 In the circuit of Figure 4.43, the 1.3-V bias is large enough that both transistors are always on (i.e., $i_1 = i_o + i_L$ and $i_2 = i_o$ for $i_L > 0$, or $i_1 = i_o$ and $i_2 = i_o - i_L$ for $i_L < 0$). Use an effective β and a base current control model for each transistor to estimate the offset between i_L and i_{in} for these assumptions. *Note*: Your answer should be in terms of i_0, I_2, β_1, and β_2.

4.19 All questions pertain to the amplifier circuit given in Figure 4.44. Assume that $\beta = 200$ for each transistor and that the nominal value for R_L is 2 kΩ. The transistors are complementary pairs designed to have similar parameter values.

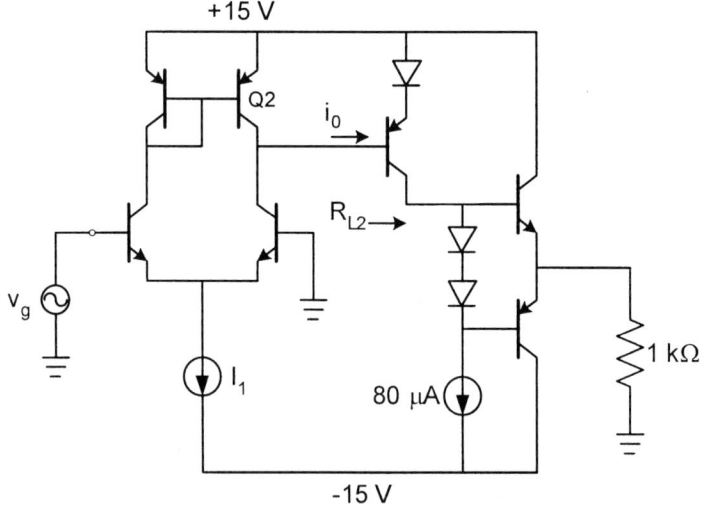

Figure 4.41. Circuit for Problems 4.14 and 4.15.

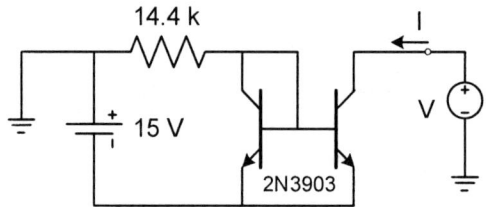

Figure 4.42. Circuit for Problem 4.17.

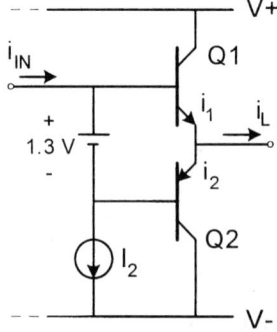

Figure 4.43. Circuit for Problem 4.18.

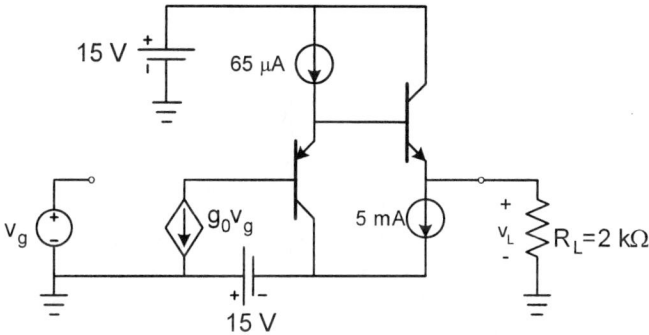

Figure 4.44. Circuit for Problem 4.19.

(a) What is the output range (in V) for the amplifier? Specify the *min* and *max* values for v_L, and determine the idle-to-maximum load-power ratio for the output stage of the amplifier (the transistor and the 5-mA source).

(b) Find an expression for the low-frequency incremental gain between the input, v_g, and the output, v_L. The answer should be in terms of parameters, β, g_0, R_L, and any other small-signal parameters that you want to use. Be sure to define any parameters that you use. Use the given nominal values, and determine a value for g_0 that will provide a gain of 100 dB for the amplifier.

(c) Find an expression, in terms of the given parameters, for the input offset voltage required for zero output, and specify the value required when $g_0 = 10^{-3}$ Mho.

(d) Design mirror circuits for each of the required bias current sources. Supply a schematic for each design and specify all element values.

4.20 The circuit given in Figure 4.45 is a candidate for the second stage of an operational amplifier. It is sometimes called a current source loaded Darlington transistor after the inventor of the configuration. Let $I_0 = 20$ μA, $R_L = 250$ kΩ, $T_1 = T_2 = $ 2N6515. Your objective is to evaluate the performance of the amplifier using methods similar to those used in the text to study a typical first-stage amplifier.

(a) Use the simple model to determine low-frequency behavior for V_L versus V_{in} and the small-signal input impedance of the amplifier for the given bias current. What effect does this circuit have on the drift and dc offset properties of the output signal?

(b) Use μCAP (or any SPICE software) to simulate incremental signal response versus frequency. The indicated transistors are optional, and you are free to choose any preferred transistors for the design. However, you should use the indicated bias current of 20 μA.

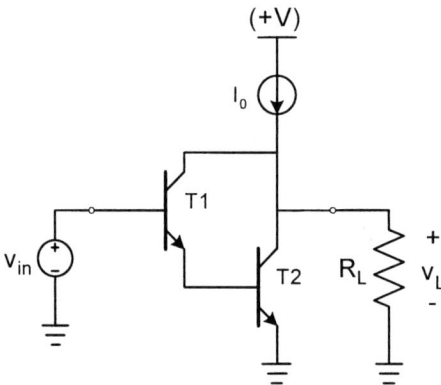

Figure 4.45. Circuit for Problem 4.20.

(c) Use the hybrid-π circuit model and determine the dominant components that control the breakpoints for the voltage transfer function obtained from the simulation.

(d) Predict/discuss/estimate the resulting behavior when this stage is cascaded with or driven by the first-stage differential amplifier circuit analyzed in the text.

References

[1] Toumazou, C., Lidgey, F. J., and Haigh, D. G., *Analogue IC Design: The Current-Mode Approach*, IEEE Series, London, England: Peter Peregrinus Ltd., 1990.

[2] Neudeck, G. W., *The Bipolar Junction Transistor*, Vol. III, Reading, MA: Addison-Wesley, 1989.

[3] Getreau, I., *Modeling the Bipolar Transistor*, Beaverton, OR: Tektronix, Inc., 1976.

[4] Gray, P. E., et al., "Physical Electronics and Circuit Models of Transistors (PCM)," *SEEC Notes*, Vol. 2, New York: Wiley, 1964.

[5] Streetman, B. G., *Solid State Electronic Devices*, Upper Saddle River, NJ: Prentice Hall, 1990.

[6] Baker, R. J., Li, H. W., and Boyce, D. E., *CMOS Circuit Design, Layout, and Simulation*, New York: IEEE Press Inc., 1998.

Appendix 4A Matlab and Transistor Modeling

It is well known that SPICE, μCAP, and other programs exist to simulate the behavior of nonlinear devices such as bipolar junction transistors (BJTs) and other semiconductor devices. The various models, in the programs, contain several tens of parameters in order to provide better simulations for real devices. This is an intimidating situation for trying to do design, and so, this appendix illustrates how a Matlab model can be created for static behavior of general purpose BJTs in their active region of operation. This discussion illustrates the development of a simple model that handles both forward and inverse calculations for the device. The procedure illustrates an approach that can be applied to any device by means of appropriate modification.

4A.1 Modeling BJT Static Response

The procedure begins by using μCAP, with a full-level model, to collect simulated calibration data for an NPN transistor, as shown in Figure 4A.1.

With V_{ce} set at a constant 2 V, data are collected for $\log_e(i_c)$ and $\log_e(i_b)$ over $0.4 < v_{be} < 0.8$ V. These are calibration data. As shown in Figure 4A.2, using circles for each data point, the natural logarithm of the base current with respect to the base-emitter voltage appears as a virtual straightline over the active region of this transistor.

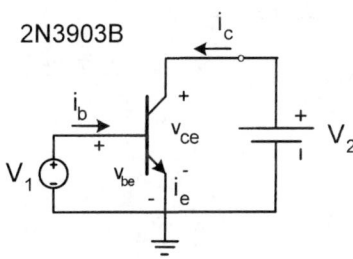

Figure 4A.1. Circuit for calibrating an NPN transistor.

Using Matlab to find a least-squares fit to the $\log_e(i_b)$ data, the following coefficients are found:[6]

$$\mathbf{b} = [b_1 \ b_0] = \text{polyfit}(v_{be}, \log_e(i_b), 1) = [19.9625 \ -23.8339] \tag{4A.1}$$

so that

$$i_b = e^{b_1 v_{be} + b_0} = e^{\mathbf{b} \cdot \mathbf{v}_b^{(1)}}$$

$$\text{where} \quad \mathbf{v}_b^{(1)} = [v_{be}^1 \ v_{be}^0] \tag{4A.2}$$

---

[6]In this case, the boldfaced lowercase letter represents a vector.

136   CHAPTER 4. SOME OPAMP DESIGN CONSIDERATIONS

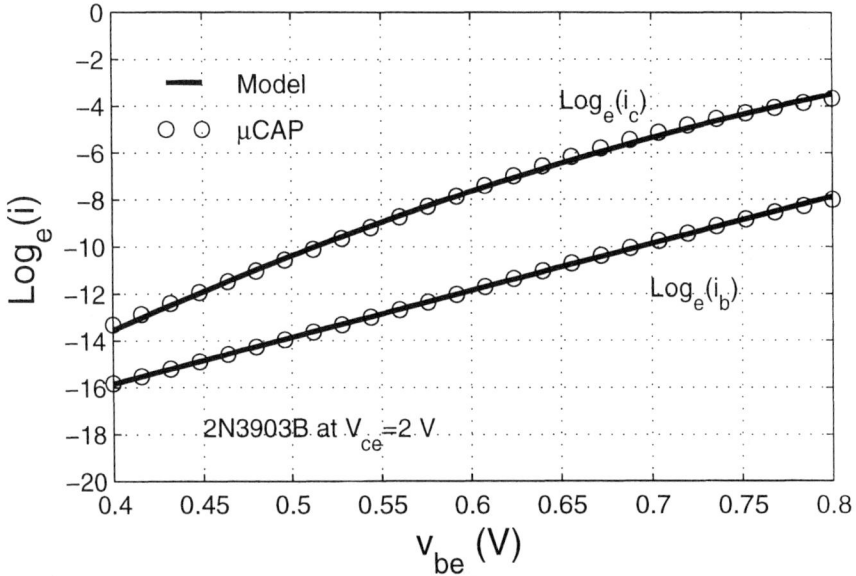

Figure 4A.2. Calibration and polynomial data for the 2N3903B.

A second-order polynomial provides a good fit to the $\log_e(i_c)$ data:

$$\mathbf{c} = [c_2\ c_1\ c_0] = \text{polyfit}(v_{be}, \log_e(i_c), 2) = [-21.7009\ 51.2093\ -30.5626] \quad (4A.3)$$

so that

$$i_c = e^{(c_2 v_{be}^2 + c_1 v_{be} + c_0)} = e^{\mathbf{c} \cdot \mathbf{v}_b^{(2)}}$$
$$\text{where}\quad \mathbf{v}_b^{(2)} = [v_{be}^2\ v_{be}^1\ v_{be}^0] \quad (4A.4)$$

The circuit of Figure 4A.3 is used to calibrate a PNP transistor using the above procedure with V_{ec} set at a constant 2 V over $0.4 < v_{eb} < 0.8$ V. The polynomials are given in (4A.5) for the 2N3906B data shown in Figure 4A.4.

$$\begin{aligned}
\mathbf{b} &= [31.9993\ -32.4605]\ \text{and} \\
\mathbf{c} &= [-24.7648\ 64.9871\ -38.8726]\ \text{so that} \\
i_b &= e^{b_1 v_{be} + b_0}\ \text{and} \\
i_c &= e^{(c_2 v_{be}^2 + c_1 v_{be} + c_0)}
\end{aligned} \quad (4A.5)$$

Matlab functions, `npn3903B.m` and `pnp3906B.m`, are written as follows to implement the forward calculation of current as a function of base-emitter voltage:

APPENDIX 4A. MATLAB AND TRANSISTOR MODELING 137

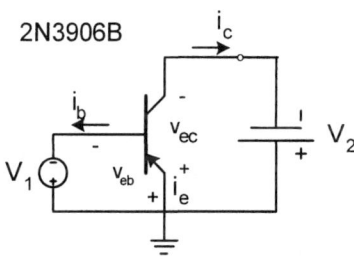

Figure 4A.3. Circuit for calibrating a PNP transistor.

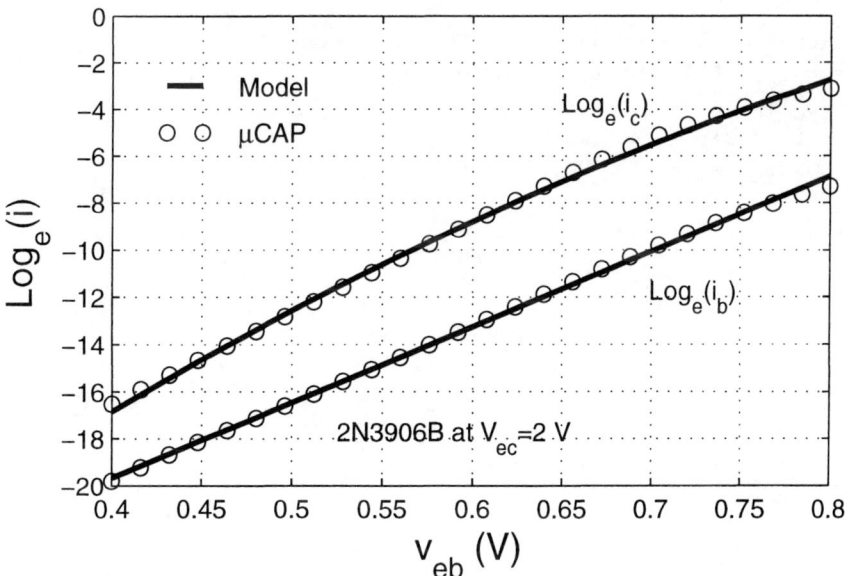

Figure 4A.4. Calibration and polynomial data for the 2N3906B.

```
% function [ib,ic]=npn3903B(vbe)
%
% calculates 2n3903B npn transistor currents
% from specified base-emitter voltage, vbe
% Model measured at Vce=2 V
% Use ic' =ic*(1+(Vce-2)/100)
% to extend model for 2<Vce<12 V
% The Early effect on ib is negligible

function [ib,ic]=npn3903B(vbe)
b1=19.9625;
b0=-23.8339;
c2=-21.7009;
c1=51.2093;
c0=-30.5626;
pb=[b1 b0];
pc=[c2 c1 c0];
ib=exp(polyval(pb,vbe));
ic=exp(polyval(pc,vbe));

% function [ib,ic]=pnp3906B(veb)
%
% calculates 2n3906B pnp transistor currents
% from specified emitter-base voltage, veb
% Model measured at Vec=2 V
% Use ic'=ic*(1+(Vec-2)/100)
% to extend model for 2<Vec<12 V

function [ib,ic]=pnp3906B(veb)
b1=31.9993;
b0=-32.4605;
c2=-24.7648;
c1=64.9871;
c0=-38.8726;
pb=[b1 b0];
pc=[c2 c1 c0];
ib=exp(polyval(pb,veb));
ic=exp(polyval(pc,veb));
```

The validity of the Matlab functions is checked by calculating and plotting the logarithm of the currents as a function of the base-emitter voltages. The result is shown in Figures 4A.2 and 4A.4, where the computed values (shown as solid lines) are compared to the μCAP calibration data (shown as circles). The result is a very good (least-square) fit to the data over the full range of v_{be} or v_{eb}. These results are for the calibrating emitter voltage $V_{ce} = V_{ec} = 2$ V.

APPENDIX 4A. MATLAB AND TRANSISTOR MODELING

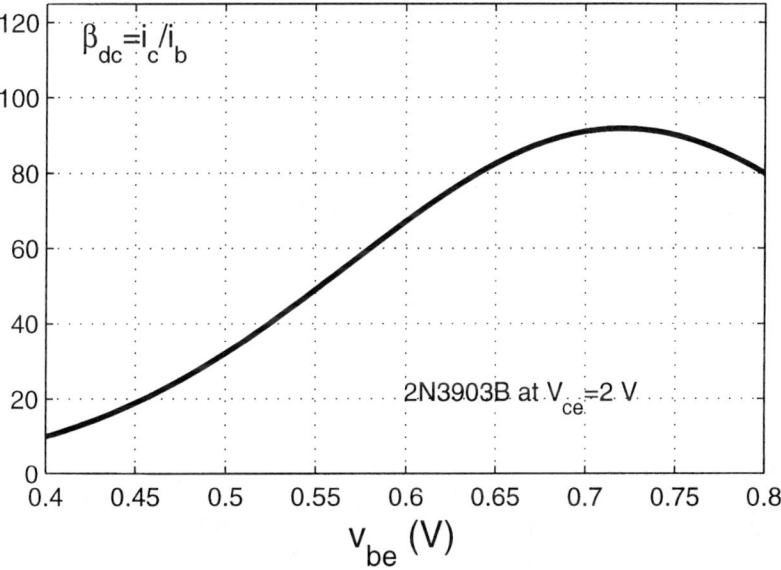

Figure 4A.5. Plot of static current gain, β, for the 2N3903B.

Note that this polynomial model allows us to model the nonlinear gain, β, between collector and base currents for the transistor. A typical curve for the static, $\beta = i_c/i_b$, is shown in Figure 4A.5. Observe that over the operating range, $0.65 < v_{be} < 0.75$, the average dc gain is approximately 90 for this transistor.

The use of models allows us to account for variation of current gain in the design of various circuits. Note that in (4A.2), for the 2N3903B NPN transistor, a conventional diode model is implied as follows for the base-current-to-base-voltage relationship:

$$i_b = e^{b_1 v_{be} + b_0} = e^{b_0} \cdot e^{b_1 v_{be}} \approx I_s e^{v_{be}/\gamma V_T}$$

where $I_s = e^{b_0} = e^{-23.8339} = 44.57$ pA and

$$\gamma = \frac{1}{b_1 V_T} = \frac{1}{(19.9625)0.026} = 1.9267 \qquad (4A.6)$$

We get the following results for the 2N3906B PNP transistor:

$$I_s = e^{b_0} = e^{-32.4605} = 7.99 \text{ fA and}$$

$$\gamma = \frac{1}{b_1 V_T} = \frac{1}{(31.9993)0.026} = 1.2019 \qquad (4A.7)$$

These are typical values for small-signal general-purpose silicon transistors.

In dealing with the design of transistor circuits and solving Kirchhoff's equations, it is usually required to find the inverse relationship for a transistor (i.e., to find the base-emitter voltage for a specified branch current). The solution for v_{be}

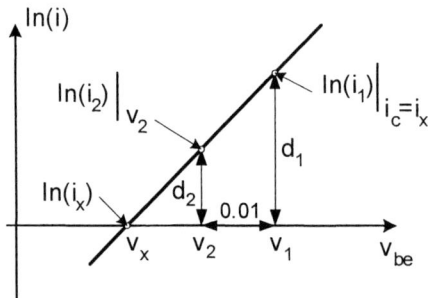

Figure 4A.6. The interpolative procedure for estimating the base voltage for a specified emitter current.

(or v_{eb}) from either a specified base or collector current is straightforward. It is not so easy if the emitter current is specified. The following illustrates a solution procedure from the models developed thus far. Let i_x be a desired current and v_x be the corresponding unknown base-emitter voltage. The model yields the following results for the range, $0.4 < v_x < 0.8$ V:

i_x *is the base current*:

$$\log_e(i_x) = b_1 v_x + b_0 \quad \text{or}$$
$$v_x = (\log_e(i_x) - b_0)/b_1 \quad (4A.8)$$

i_x *is the collector current*:

$$\log_e(i_x) = c_2 v_x^2 + c_1 v_x + c_0 \quad \text{or}$$
$$0 = c_2 v_x^2 + c_1 v_x + (c_0 - \log_e(i_x)) \quad \text{so}$$
$$v_x = \left[-c_1 \pm \sqrt{c_1^2 - 4c_2(c_0 - \log_e(i_x))}\right]/(2c_2) \quad (4A.9)$$

i_x *is the emitter current*: It requires two calculations and an interpolation to determine an estimate for this condition. First, let $i_c = i_x$ be the collector current and use (4A.9) to get v_1 for i_c. Then, $i_1 = i_e = i_c + i_b$ is greater than i_x. Next, use $v_2 = v_1 - 0.01$ to get $i_2 = i_b + i_c$ for v_2. This may be greater or less than i_x; however, i_2 is always less than i_1, as shown in Figure 4A.6. Solving the triangle relation shown in Figure 4A.6 yields (4A.10):

$$v_x = \frac{d_1 v_2 - d_2 v_1}{d_1 - d_2}$$
$$\text{where} \quad d_1 = \log_e(i_1) - \log_e(i_x)$$
$$d_2 = \log_e(i_2) - \log_e(i_x) \quad (4A.10)$$

The Matlab functions, inv3903B.m and inv3906B.m, for both transistors are as follows:

APPENDIX 4A. MATLAB AND TRANSISTOR MODELING

```
% Inverse transistor model calculator for the 2N3903B
%     vb=inv3903B(ix,fl)
%The flag fl sets the branch current, ix, from which
%it is desired to find the base-emitter voltage.
%Calculations exclude the Early effect
%for extended range of Vce>2 V cal voltage
% fl=1 for base current
% fl=2 for emitter current
% fl=3 for collector current

function vb=inv3903B(ix,fl)
b1=19.9625;
b0=-23.8339;
c2=-21.7009;
c1=51.2093;
c0=-30.5626;
pb=[b1 b0];
pc=[c2 c1 c0];

if fl<2
  vb=(log(ix)-b0)/b1;
elseif fl>2
 v=roots([c2 c1 c0 log(ix)]);
 u=(0.4<v & v<0.8);
 vb=u'*v;
else
 v=roots([c2 c1 c0-log(ix)]);
 u=(0.4<v & v<0.8);
 v1=u'*v;
 [iu iv]=npn3903B(v1);
 d1=log(iu+iv)-log(ix);
 v2=v1-0.01;
 [iu iv]=npn3903B(v2);
 d2=log(iu+iv)-log(ix);
 vb=(v2*d1-v1*d2)/(d1-d2);
end

% Inverse transistor model calculator for the 2N3906B
%     ve=inv3906B(ix,fl)
%The flag fl sets the branch current, ix, from which
%it is desired to find the base-emitter voltage.
%Calculations exclude the Early effect
%for the extended range of Vec>2 V = cal
% fl=1 for base current
% fl=2 for emitter current
% fl=3 for collector current
```

142   CHAPTER 4. SOME OPAMP DESIGN CONSIDERATIONS

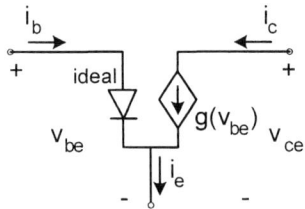

Figure 4A.7. The basic NPN model has no v_{ce} dependence for $V_{ce} \leq 2$ V.

```
function ve=inv3906B(ix,fl)
b1=31.9993;
b0=-32.4605;
c2=-24.7648;
c1=64.9871;
c0=-38.8726;
pb=[b1 b0];
pc=[c2 c1 c0];

if fl<2
  ve=(log(ix)-b0)/b1;
elseif fl>2
  v=roots([c2 c1 c0-log(ix)]);
  u=(0.4<v & v<0.8);
  ve=u'*v;
else
  v=roots([c2 c1 c0-log(ix)]);
  u=(0.4<v & v<0.8);
  v1=u'*v;
  [iu iv]=pnp3906B(v1);
  d1=log(iu+iv)-log(ix);
  v2=v1-0.01;
  [iu iv]=pnp3906B(v2);
  d2=log(iu+iv)-log(ix);
  ve=(v2*d1-d2*v1)/(d1-d2);
end
```

So far, the modeling does not include any dependence upon the collector-emitter voltage. This dependence is often ignored, and we can use a circuit model, as shown in Figure 4A.7, for the NPN transistor. The base current is modeled using an ideal diode, and the collector current is controlled by a transductance function of the diode voltage. A corresponding circuit model for the PNP transistor is shown in Figure 4A.8.

APPENDIX 4A. MATLAB AND TRANSISTOR MODELING

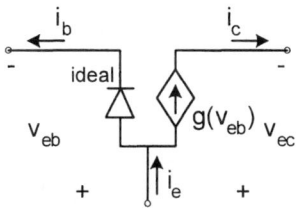

Figure 4A.8. The basic PNP model has no v_{ec} dependence for $v_{ec} \leq 2$ V.

The calibration test (simulation) can be repeated to determine the effect of collector voltage on the transistor currents. μCAP data are shown in Figure 4A.9 for the NPN transistor. The graph shows that for $2 < V_{ce} < 12$ V, there is virtually no effect upon the base current and only a very small effect upon the collector current. Traditionally, this effect upon the collector current is modeled by adding a nonlinear current source whose current is proportional to the collector junction power divided by the Early effect voltage,[7] as shown in Figure 4A.10. There is no correction for the base current. Note that the added current is for $V_{ce} > V_{cal} = 2$ V and that the polynomial models include the effect for $V_{ce} = V_{cal}$.

It is usually necessary to consider the Early effect in transistor circuits so as to find operating points where transistors are placed back-to-back on their collectors. The small additional current gives collector current dependence upon collector voltage, thus yielding limiting values for the collector voltages in such otherwise indeterminant situations. For the transistor family modeled here, the Early voltage, V_A, is on the order of 100 V.

4A.2 Modeling Dynamic Response

When we want to solve for the response of a transistor in an actual circuit application, it is customary to break the problem into two parts, namely, its static and dynamic behavior. The static part provides what is called the operating point of the transistor as discussed in the previous section. The dynamic part solves for incremental changes away from the operating, or static, solution.

The concept is illustrated in Figure 4A.11. In this figure, the total response is shown in Figure 4A.11(a), while Figure 4A.11(b, c) show the separation of the total response variables into static and dynamic parts, respectively. Both circuits [Figure 4A.11(b, c)] satisfy all of Kirchhoff's laws separately, but individual branch $i - v$ relations usually differ.

Figure 4A.12 illustrates models for the specific example of Figure 4A.11(a), which is sometimes referred to as a common emitter amplifier. The static circuit, used to find the operating point, has been discussed previously. The dynamic circuit, shown in Figure 4A.12(b), is a circuit based on physical considerations used to model small-signal (incremental) response around the operating point. It is sometimes referred to as a hybrid-π model.

---

[7]See [6], p. 118, for a $v - i$ plot of the Early effect.

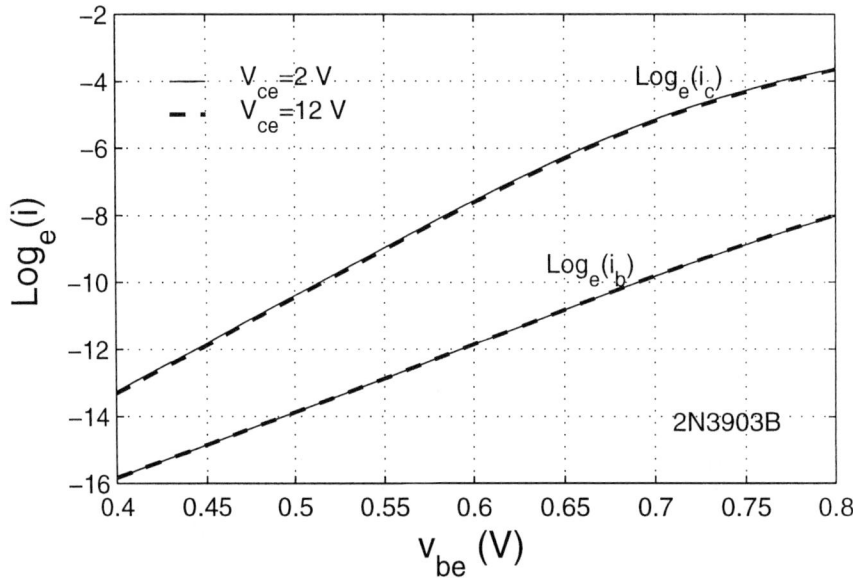

Figure 4A.9. Simulation showing the effect of V_{ce} on transistor current.

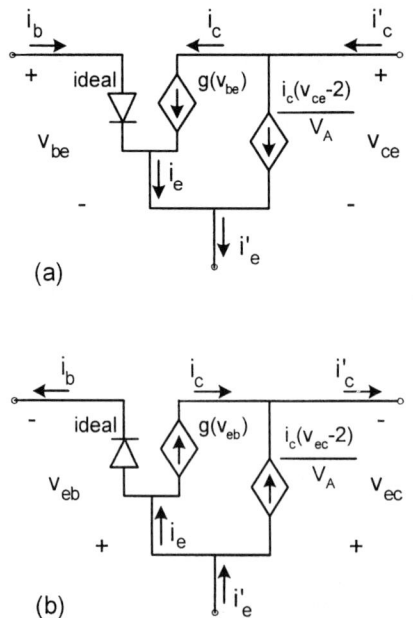

Figure 4A.10. Extended model for $V_{ce} > V_{cal} = 2$ V: (a) NPN, and (b) PNP.

APPENDIX 4A. MATLAB AND TRANSISTOR MODELING

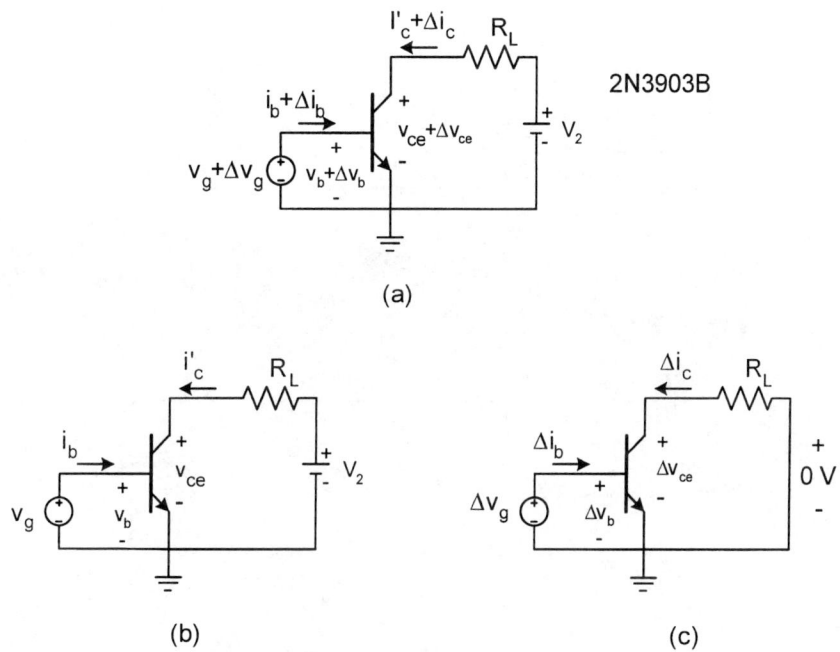

Figure 4A.11. Static and dynamic signal circuits: (a) total response circuit variables, (b) static signal circuit, and (c) dynamic signal circuit.

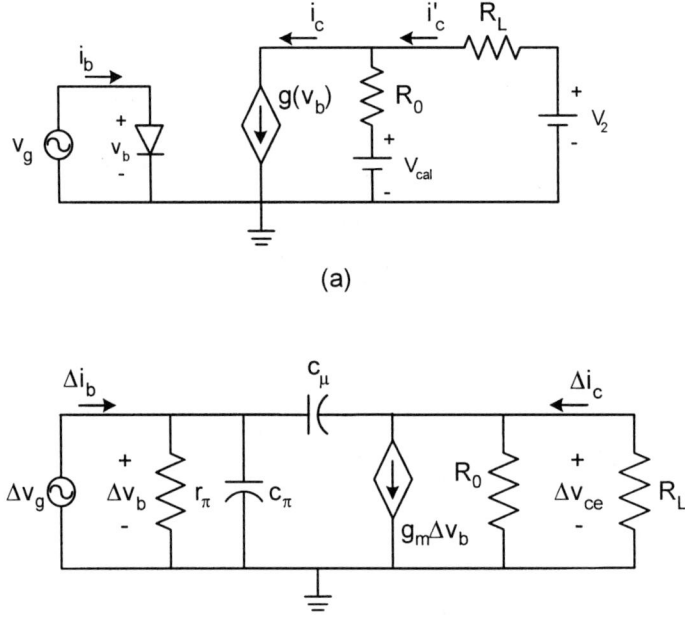

Figure 4A.12. Static and dynamic circuit models for the NPN circuit of Figure 4A.11: (a) static signal model, and (b) dynamic signal model.

APPENDIX 4A. MATLAB AND TRANSISTOR MODELING

Using values found by solving the static circuit model, dynamic model resistances are estimated as follows (at low frequency):

$$r_\pi \approx \frac{\Delta v_b}{\Delta i_b} = \frac{\Delta v_b}{b_1 e^{(b_1 v_b + b_0)} \Delta v_b} = \frac{1}{b_1 i_b} \qquad (4A.11)$$

$$g_m \approx \frac{\Delta i_c}{\Delta v_b} = \frac{(2c_2 v_b + c_1)\Delta v_b e^{(c_2 v_b^2 + c_1 v_b + c_0)}}{\Delta v_b}$$

$$= (2c_2 v_b + c_1) i_c \qquad (4A.12)$$

$$R_0 = \frac{V_A}{i_c} \qquad (4A.13)$$

where V_A is the Early voltage.

The capacitances come from transistor physical dimensions and electrical parameters of the fabrication process. If these are not known, then estimates can be made for C_π and C_μ by point matching the model to a measured response of the circuit.

To illustrate obtaining estimates for C_π and C_μ, we consider two network functions for the dynamic circuit model, Figure 4A.12(b). One is the voltage transfer, $H = \Delta v_{ce}/\Delta v_b$, and the other is the input admittance, $Y_{in} = \Delta i_b/\Delta v_b$. First, we determine the network functions and then apply them to measured (or simulated) responses in order to obtain estimates for the capacitances.

For the voltage gain function, take KCL at Δv_{ce}:

$$0 = (G_L + G_0)\Delta v_{ce} + g_m \Delta v_b + C_\mu s(\Delta v_{ce} - \Delta v_b)$$

but $\Delta v_b = \Delta v_g$ so

$$(C_\mu s + G_L + G_0)\Delta v_{ce} = (C_\mu s - g_m)\Delta v_g \text{ or}$$

$$H(s) = \frac{\Delta v_{ce}}{\Delta v_g} = \left(\frac{C_\mu s - g_m}{C_\mu s + G_L + G_0}\right) \qquad (4A.14)$$

With known conductances, C_μ can be estimated from a measure of $H(s)$ at a single frequency. For example, at $s = j\omega_1 = j2\pi f_1$, let $|H(j\omega_1)| = h_1$. Then,

$$h_1^2 = \frac{g_m^2 + C_\mu^2 \omega_1^2}{(G_L + G_0)^2 + C_\mu^2 \omega_1^2} \text{ yields}$$

$$C_\mu \omega_1 = (G_L + G_0)\left[\frac{(g_m/(G_L + G_0))^2 - h_1^2}{h_1^2 - 1}\right]^{1/2} \qquad (4A.15)$$

The behavior of $H(j\omega)$ can be deduced from (4A.15) to yield the following conditions on h_1:

$$1 < h_1 < g_m/(G_L + G_0) \text{ for } g_m > G_L + G_0$$
$$g_m/(G_L + G_0) < h_1 < 1 \text{ for } g_m < G_L + G_0 \qquad (4A.16)$$

The conditions in (4A.16) are derived from the fact that C_μ is always greater than zero and real (i.e., it is a positive real number).

148   CHAPTER 4. SOME OPAMP DESIGN CONSIDERATIONS

Table 4A.1. Table of Dynamic Parameters for the NPN 2N3903B Transistor

v_b (V)	0.50	0.55	0.60	0.65	0.70	0.75	0.80
R_π (kΩ)	53.40	19.30	6.76	2.33	0.811	0.301	0.123
G_μ (mMho)	0.83	3.44	13.60	47.20	126.0	260.0	455.0
C_π (pF)	6.43	8.77	17.20	45.60	124.0	285.0	513.0
C_μ (pF)	1.94	1.95	1.95	1.96	1.96	1.97	1.97
β_{ac}	44.3	66.2	92.1	110.0	102.0	78.2	56.0
R_0 (kΩ)	3690	885.0	218.0	58.60	19.10	8.01	4.11

Likewise, with known conductances and an estimate for C_μ, C_π can be estimated from a measure of the input admittance, $Y_{in}(s)$.

To obtain Y_{in}, evaluate KCL at Δi_b:

$$\Delta i_b = g_\pi \Delta v_g + C_\pi s \Delta v_g + C_\mu s (\Delta v_g - \Delta v_{ce})$$

Substitute (4A.14) for Δv_{ce}, and then

$$\Delta i_b = [(C_\mu + C_\pi)s + g_\pi]\Delta v_g - C_\mu s \left(\frac{C_\mu s - g_m}{C_\mu s + G_L + G_0}\right)\Delta v_g$$

$$= (C_\pi s + g_\pi)\Delta v_g + C_\mu s \left[1 - \frac{C_\mu s - g_m}{C_\mu s + G_L + G_0}\right]\Delta v_g$$

$$Y_{in} = \frac{\Delta i_b}{\Delta v_g} = C_\pi s + g_\pi + \frac{(G_L + G_0 - g_m)C_\mu s}{(C_\mu s + G_L + G_0)} \quad (4A.17)$$

Again, at $s = j\omega_1 = j2\pi f_1$, let $Y_{in}(j\omega_1) = u_1 + jv_1$. Then,

$$u_1 + jv_1 = C_\pi j\omega_1 + g_\pi + \frac{(G_L + G_0 + g_m)C_\mu j\omega_1}{(C_\mu j\omega_1 + G_L + G_0)} \quad (4A.18)$$

Equating imaginary parts in (4A.18) yields the following estimate for C_π:

$$C_\pi \omega_1 = v_1 - C_\mu \omega_1 \frac{(G_L + G_0 + g_m)(G_L + G_0)}{[(G_L + G_0)^2 + C_\mu^2 \omega_1^2]} \quad (4A.19)$$

Clearly, the hybrid-π parameters are a function of the static operating point, hence v_{be}.

Figure 4A.13 shows a plot of dynamic model parameter variables versus the static base-emitter voltage for the 2N3903B NPN transistor. The circuit of Figure 4A.11(a) was used with $R_L = 0$ and $V_2 = 6$ V. The logarithm to base 10 is plotted for each parameter. C_μ and β_{ac} show the least dependence upon v_{be}, with g_m and r_π following dependence upon i_c and i_b, respectively. Sample dynamic parameter values are listed in Table 4A.1. Tabulated values are used to interpolate values for specific values of v_{be} (e.g., in Chapter 4).

APPENDIX 4A. MATLAB AND TRANSISTOR MODELING

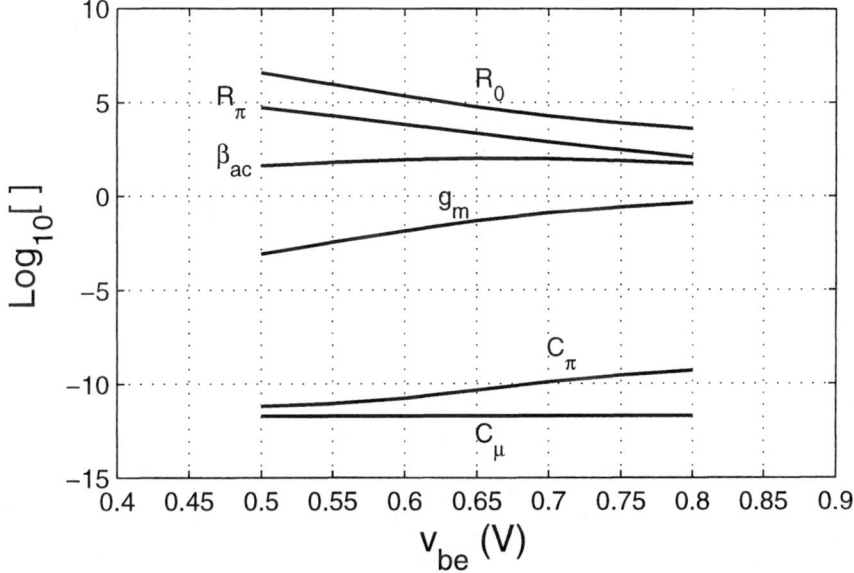

Figure 4A.13. Typical small-signal parameters for the 2N3903B.

Note that g_m and r_π are related to β_{ac} as follows:

$$\beta_{ac} = \frac{\Delta i_c}{\Delta i_b}$$
$$\text{where } \Delta i_c = g_m \Delta v_b$$
$$\Delta i_b = \Delta v_b / r_\pi$$
$$\text{and so } \beta_{ac} = g_m r_\pi \qquad (4A.20)$$

These parameters were obtained directly from μCAP simulations through measures on network functions for Y_{in} and $H(s)$.

The graph in Figure 4A.14 compares parameter values derived from the polynomial models developed here to the μCAP estimates. The comparison shows a good fit over a wide working range for the polynomial model.

Similar results, shown in Figures 4A.15, 4A.16, and Table 4A.2, are obtained for the PNP model by following the above procedures.

4A.3 Summary

This appendix on Matlab transistor modeling is included to demonstrate how the complex models used in modern simulation tools can be adapted to facilitate designing at your desk. The models are employed liberally in this chapter where a case study for designing a high-gain opamp is provided. These modeling procedures

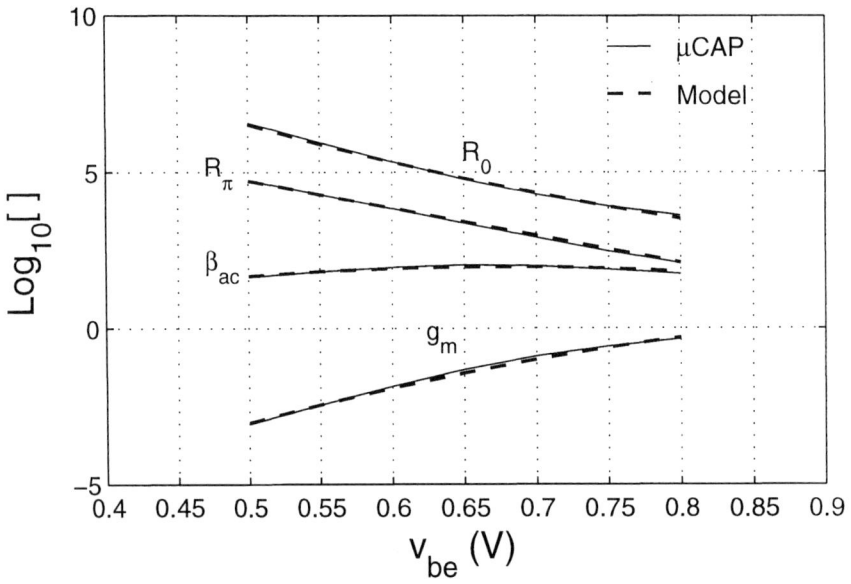

Figure 4A.14. Model parameters compared to μCAP estimates for the 2N3903B.

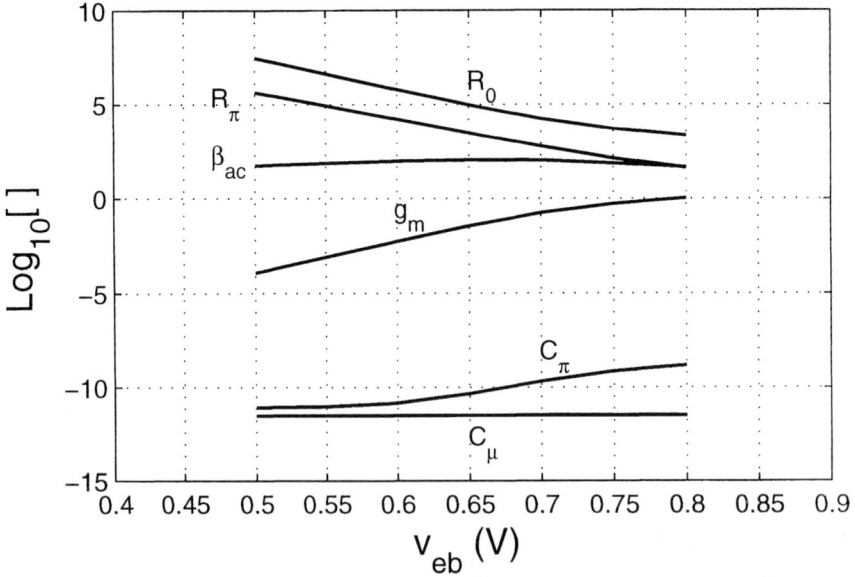

Figure 4A.15. Typical small-signal parameters for the 2N3906B.

APPENDIX 4A. MATLAB AND TRANSISTOR MODELING

Table 4A.2. Table of Dynamic Parameters for the PNP 2N3906B Transistor

v_e (V)	0.50	0.55	0.60	0.65	0.70	0.75	0.80
R'_π (kΩ)	446.0	87.40	16.50	3.03	0.578	0.134	0.041
G'_μ (mMho)	0.122	0.839	5.72	36.80	181.0	533.0	1080
C'_π (pF)	8.70	9.75	15.0	46.6	211.0	718	1500
C'_μ (pF)	3.03	3.04	3.05	3.06	3.07	3.08	3.10
β'_{ac}	54.3	73.3	94.4	111.0	104.0	71.5	44.4
R'_0 (kΩ)	29800	4320	629.0	95.50	17.50	4.94	2.11

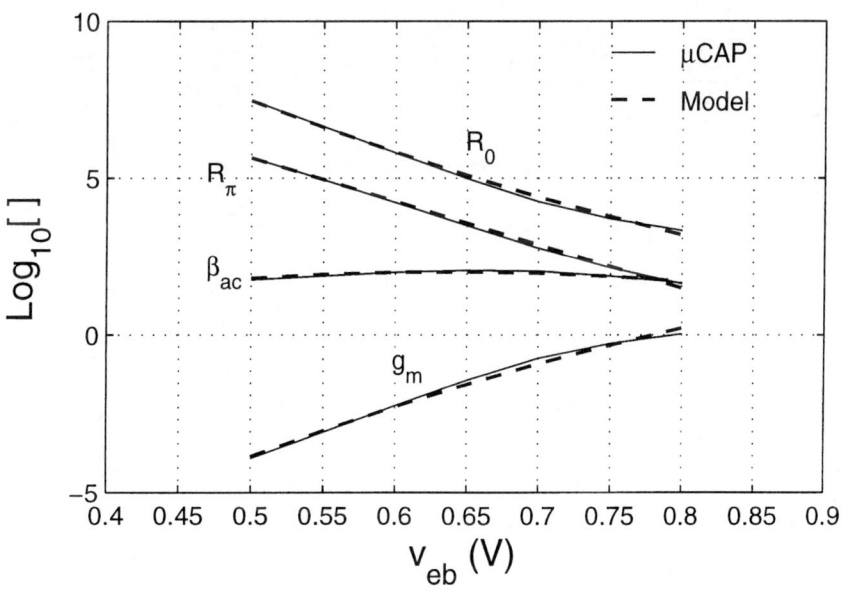

Figure 4A.16. Model parameters compared to μCAP estimates for the 2N3906B.

can be adapted to other transistors of interest just by finding suitable polynomial coefficients from simulation (or measured) data and changing the example routines appropriately to construct different models. The procedure is straightforward and allows transistor operating points to be estimated very closely prior to simulation. The procedure actually saves time over having to iterate design values and get pulled into a trial and error design procedure that is prevalent with the use of simulation tools. Models are convenient tools to use for aiding design, and the presence of a tool as powerful as Matlab cooperates to make the models easy to construct. May you enjoy many good models over your design career!

Chapter 5

Operational Design of Active Filters

The purpose of the previous chapters was to provide background on the design and use of opamps for negative feedback applications. We learned that the high-gain opamp is a complex circuit with nonideal behavior. The knowledge of opamp design and performance helps the filter designer to be aware of limitations in the use of opamps in terms of intended signal ranges, signal bandwidths, and operating temperatures. Most filters carry specifications in all of these areas, so ultimately, the filter designer must test and guarantee performance for a final design.

However, in order to simplify concepts for the initial feasibility part of a design, it is customary to use the simple opamp linear small-signal model shown in Figure 5.1(a). This model provides no information about any of the above operating parameters, but it allows simple circuit analysis based on the limit—large gain (infinity) times a vanishingly small input (zero)—that yields a finite output voltage, V_0. The result is that the opamp imposes a constraint on the connected feedback circuit so as to force 0 V across the amplifier input. At the same time the opamp provides whatever load current and voltage the feedback circuit requires in order to satisfy the zero input requirement of the opamp. This opamp model has no frequency-, thermal-, or dynamic-range-dependent information.

The model of Figure 5.1(b) is used as a second-order model to estimate opamp frequency-dependent effects for finite gain and bandwidth. Chapter 3 used this model to illustrate feedback effects. This model does not predict signal dynamic range effects or temperature behavior, but it is a practical model for studying opamp frequency-dependent effects.

When the designer has reached a proposed design and wishes to check signal range and temperature effects, then the full opamp circuit has to be used in a good simulator. This is a nontrivial step, for a modest filter can easily have a dozen or more opamps in the design. If each opamp has 20 nodes, then there are several hundred nodes involved in order to simulate a complete design. This usually means that it is necessary to use a professional (not student) software package in order to

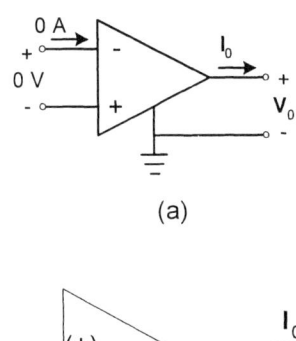

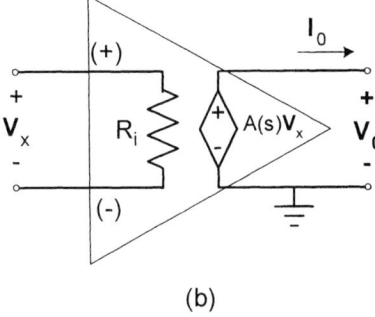

Figure 5.1. Simplified opamp models for linear dynamic signal analysis: (a) conventional opamp model, and (b) second-order dynamic model.

accomplish the simulation. If a professional package is not available, then one can always revert to assembling a prototype circuit for the design and subjecting it to the required performance testing. Testing simple designs in this way costs a lot less than buying a professional software package, but if the process has to be performed on a regular basis, there eventually comes a point when it is more economical to purchase a professional simulator package.

5.1 The Opamp as a Signal Processor

The opamp derives its name from "operational amplifier" as historically it originated as a basic analog computer component. It was a fundamental building block around which one could add resistors and capacitors to construct analog circuits to model differential equations. A few of the basic signal-processing blocks are briefly described in the following discussion.

5.1.1 The Buffer Voltage Follower

Shown in Figure 5.2, this circuit provides isolation (zero load current) so that the output voltage source, $\mathbf{V}_0 = \mathbf{V}_1$, can apply voltage to an arbitrary number of loads without loading down, or affecting, the source variable, $\mathbf{V}_1$. The operation on signal flow is illustrated by the flow graph shown to the right of the circuit.

5.1. THE OPAMP AS A SIGNAL PROCESSOR

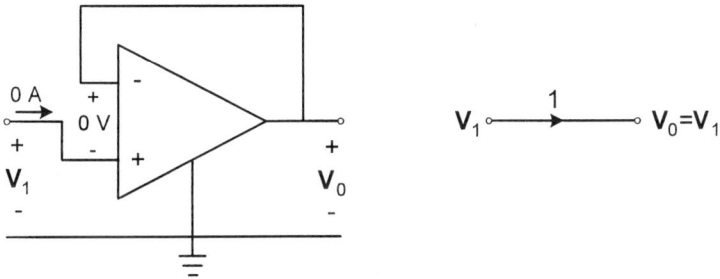

Figure 5.2. A voltage follower circuit.

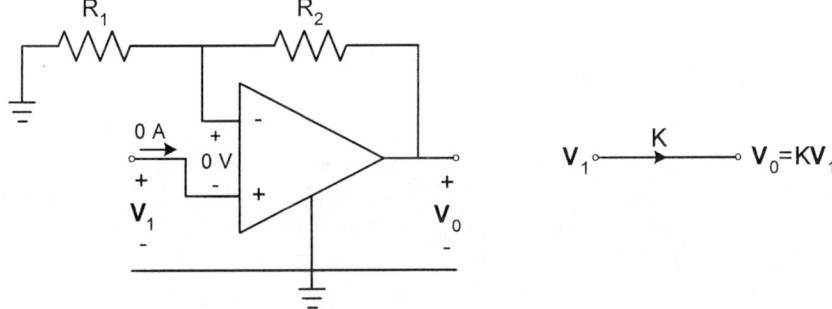

Figure 5.3. A noninverting multiplying buffer.

5.1.2 The Noninverting Multiplying Buffer

Figure 5.3 shows a noninverting multiplier circuit. Applying the opamp constraint and using KVL, we can easily show that $\mathbf{V}_0 = (1 + G_1 R_2)\mathbf{V}_1$ for this circuit. The analog multiplying factor is $K = 1 + G_1 R_2$.

5.1.3 The Noninverting Summing Multiplier

The multiple-input circuit shown in Figure 5.4 provides a buffered output signal, $\mathbf{V}_0$, that is proportional to a weighted sum of inputs, $\mathbf{V}_k$. Equation (5.1) is obtained by applying KCL at the $\mathbf{V}_\Sigma$ node.

$$\mathbf{V}_0 = K\mathbf{V}_\Sigma = K\frac{G_1\mathbf{V}_1 + G_2\mathbf{V}_2 + \ldots + G_N\mathbf{V}_N}{G_1 + G_2 + \ldots + G_N} \qquad (5.1)$$

Letting

$$\Sigma G = \sum_{k=1}^{N} G_k = G_1 + G_2 + \ldots + G_N \quad \text{and} \quad g_k = G_k/\Sigma G$$

$$\text{then} \quad \mathbf{V}_0 = K(g_1\mathbf{V}_1 + g_2\mathbf{V}_2 + \ldots + g_N\mathbf{V}_N) \qquad (5.2)$$

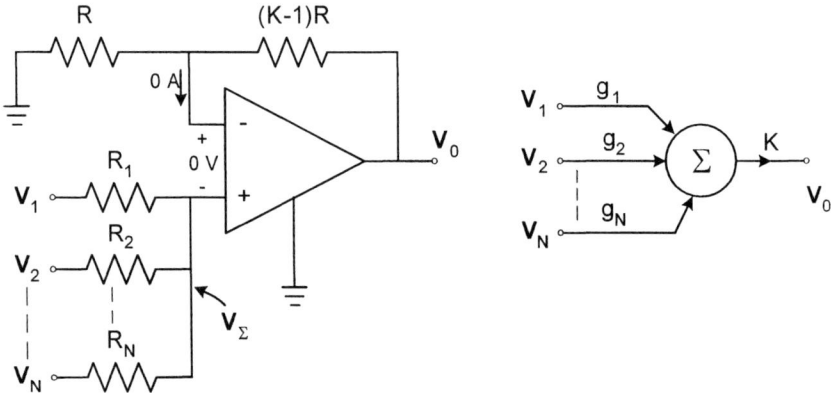

Figure 5.4. A noninverting summing multiplier.

Both K and all g_k are independent of any subsequent impedance scale factors that might be applied to the circuit since they are derived from impedance ratios. If the G_k are values for a 1 Ohm 1 rad/sec prototype circuit, then choosing $K \doteq \Sigma G$ yields the following convenient result:

$$\mathbf{V}_0 = G_1 \mathbf{V}_1 + G_2 \mathbf{V}_2 + \ldots + G_N \mathbf{V}_N = \sum G_k \mathbf{V}_k$$
$$\text{for} \quad K \doteq \Sigma G \tag{5.3}$$

The circuit isolates the resulting weighted sum from other loads, but the individual source voltages, $\mathbf{V}_k$, are clearly not isolated from each other. This coupling between the $\mathbf{V}_k$ is sometimes undesirable, and so, the inverting summing multiplier is used to reduce undesirable intermodulation effects between the $\mathbf{V}_k$ source voltages.

5.1.4 The Inverting Summing Multiplier

The circuit shown in Figure 5.5 provides a response proportional to the negative weighted sum of the $\mathbf{V}_k$ inputs. Applying KCL at the amplifier ($-$)input node yields $\mathbf{I}_\Sigma$ and the output, $\mathbf{V}_0$.

$$\mathbf{V}_0 = -R_0 \mathbf{I}_\Sigma = -R_0 (G_1 \mathbf{V}_1 + G_2 \mathbf{V}_2 + \ldots + G_N \mathbf{V}_N)$$
$$= -\sum_{k=1}^{N} R_0 G_k \mathbf{V}_k \tag{5.4}$$

Note again that the weight factors, $R_0 G_k$, are independent of any subsequent impedance scale factors applied to the circuit.

In both multiplier circuits, it is important that each $\mathbf{V}_k$ be sourced from a voltage source; otherwise, the finite loads, R_k, will load and change the $\mathbf{V}_k$ value, depending upon both its source and load resistor values. The circuit in Figure 5.5

5.1. THE OPAMP AS A SIGNAL PROCESSOR

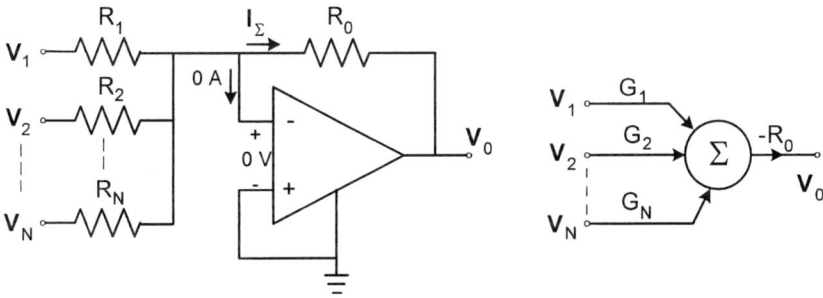

Figure 5.5. An inverting summing multiplier.

provides isolation between currents drawn from each V_k as follows. Let I_k equal the current flowing in the summing resistor, R_k. Then, for the circuit in Figure 5.5, we get the following input branch currents:

$$I_k = G_k V_k \tag{5.5}$$

Clearly, the kth branch current depends only upon the kth input voltage. This isolation is caused by the virtual ground (0 V) at the opamp summing input node. However, the input current, I_k, for the circuit in Figure 5.4 is quite different from that for the inverting circuit. Here, we get the result shown in (5.6):

$$\begin{aligned} I_k &= G_k(V_k - V_\Sigma) \text{ or} \\ R_k I_k &= V_k - V_\Sigma \\ &= V_k - \frac{G_1 V_1 + G_2 V_2 + \ldots + G_N V_N}{G_1 + G_2 + \ldots + G_N} \end{aligned} \tag{5.6}$$

where $k = 1, 2, \ldots, N$

Equation (5.6) tells us that each I_k depends upon each and every V_k value for the circuit in Figure 5.4. This result shows how intermodulation occurs between the V_k sources.

Normally, the inverting summing circuit is preferred in order to avoid intermodulation problems between sources, but there are times when the sign inversion is awkward. Consequently, the noninverting summing circuit is sometimes used, especially when each V_k is sourced from good voltage sources.

Finally, we have the last, but most important, signal-processing component: the integrator.

5.1.5 The Inverting Integrator

The RC-opamp circuit shown in Figure 5.6 provides an integral relation between V_1 and V_0 in the complex frequency domain as given in (5.7):

$$V_0 = -\frac{1}{Cs} I_C = -\frac{1}{RCs} V_1 \tag{5.7}$$

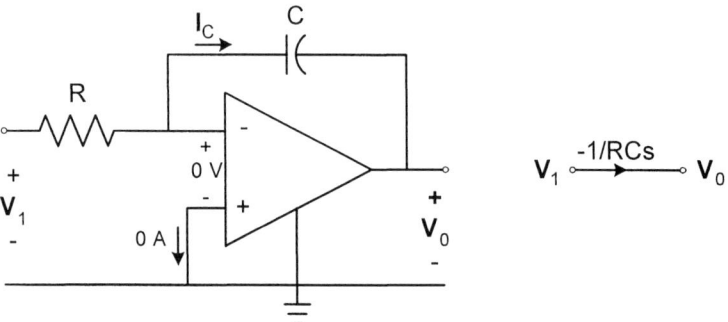

Figure 5.6. An inverting integrator.

In the time domain, the voltages are related as given by (5.8):

$$v_0(t) = v_C(0) - \frac{1}{C} \int_0^t dt' \frac{v_1(t')}{R} \qquad (5.8)$$

In old analog computers, switches and voltage sources were provided so as to control the capacitor voltage at $t = 0$.

If the opamp has either dc offset voltage at the output, or dc offset at the input, the feedback capacitor will charge until it reaches an upper or lower level of the amplifier, at which point the amplifier ceases to amplify. It sets at a rail value until the capacitor is discharged. This is a fundamental problem with the integrator as shown, and normally, it cannot be used as a stand-alone integrator. As we will now see, normally in operational circuits, the integrator is stabilized by means of negative feedback in a larger controlling circuit, as the following example illustrates.

5.2 Analog Operational Circuit Example

It is desired to obtain an operational analog circuit realization for the following prototype describing function:

$$1\frac{d^2v}{dt^2} + 4\frac{dv}{dt} + 104v = 5v_g(t) \qquad (5.9)$$

The analog circuit allows one to study the response, $v(t)$, to any forcing function, $v_g(t)$.

A solution is found by solving or rearranging (5.9) to obtain d^2v/dt^2 (the highest derivative term):

$$1\frac{d^2v}{dt^2} = 5v_g(t) - 4\frac{dv}{dt} - 104v \qquad (5.10)$$

Letting $\dot{v} = dv/dt$ and $\ddot{v} = d^2/dt^2$, an analog signal flow diagram is shown in Figure 5.7. The circuit shows that two integrators are required to obtain the second-order response.

5.2. ANALOG OPERATIONAL CIRCUIT EXAMPLE

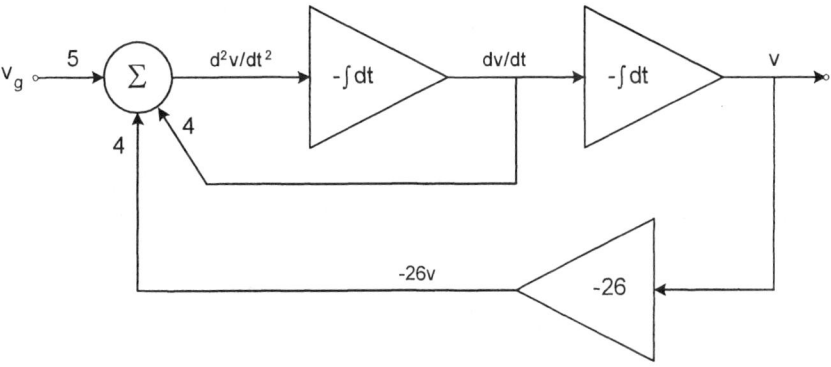

Figure 5.7. Operational circuit solution for a second-order describing function.

Now using the building blocks described above, it is straightforward to convert the conceptual signal flow diagram of Figure 5.7 to the RC-opamp circuit of Figure 5.8. Note that this circuit is designed using the s-domain equivalence for (5.10) [hence, (5.9)].

$$s^2\mathbf{V}(s) = 5\mathbf{V}_g(s) - 4s\mathbf{V}(s) - 104\mathbf{V}(s) \tag{5.11}$$

The validity of the circuit shown in Figure 5.8 can be checked by noting that the output of the summing multiplier, $s^2\mathbf{V}(s)$, is as given by (5.12) and that at the input, $\Sigma G = 4 + 4 + 5 = 13$.

$$\begin{aligned} s^2\mathbf{V} &= K\frac{5\mathbf{V}_g - 4s\mathbf{V} + 4(-26)\mathbf{V}}{\Sigma G} = 13\frac{5\mathbf{V}_g - 4s\mathbf{V} - 104\mathbf{V}}{13} \\ &= 5\mathbf{V}_g - 4s\mathbf{V} - 104\mathbf{V} \end{aligned} \tag{5.12}$$

Clearly, the circuit response, (5.12), matches the requirement given in (5.11).

The circuit shown in Figure 5.8 has been obtained in a simple, direct fashion from a specified describing function. It has the network function given by (5.13):

$$H(s) = \frac{\mathbf{V}(s)}{\mathbf{V}_g(s)} = \frac{5}{s^2 + 4s + 104} \tag{5.13}$$

$H(s)$ has poles at $s_{1,2} = -2 \pm j10$. There is an infinity of possible solutions to obtain the desired $H(s)$, so how do we determine whether the circuit in Figure 5.8 is an appropriate circuit or not? This is a design question answered by determining what required criteria are met by a candidate design.

With active circuits, there are two key criteria to meet for a good design: required element value ratios and peak gain distribution versus frequency for each amplifier output in the design.

In this example, we have four amplifiers, but only one is seen at the output of the circuit. In order to maintain linear operation, it is essential that the output (observable) amplifier have the (global) peak gain of the entire circuit. If this is not

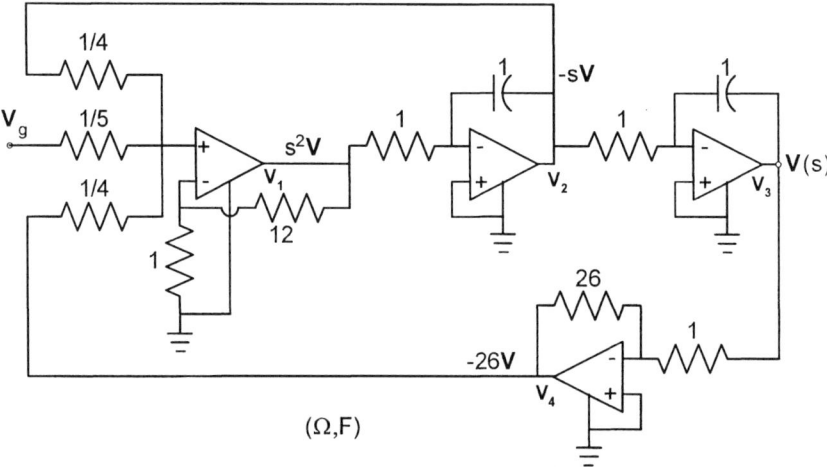

Figure 5.8. RC-opamp realization for the operational circuit in Figure 5.7.

true, then it is possible to overdrive an internal amplifier before the output amplifier saturates or cuts off. Whenever internal amplifiers saturate before the output does, it is possible to subvert the nominal response so that the circuit appears to be working, but it will exhibit an incorrect frequency response at the output. For this design, the four amplifier responses, in terms of $H(s)$, are as follows:

$$\frac{\mathbf{V}_1}{\mathbf{V}_g} = \frac{s^2 \mathbf{V}}{\mathbf{V}_g} = s^2 H(s)$$
$$\frac{\mathbf{V}_2}{\mathbf{V}_g} = \frac{-s\mathbf{V}}{\mathbf{V}_g} = -sH(s)$$
$$\frac{\mathbf{V}_3}{\mathbf{V}_g} = \frac{\mathbf{V}}{\mathbf{V}_g} = H(s)$$
$$\frac{\mathbf{V}_4}{\mathbf{V}_g} = \frac{-26\mathbf{V}}{\mathbf{V}_g} = -26H(s) \quad (5.14)$$

$\mathbf{V}_3$ provides the desired output, while $\mathbf{V}_4$ provides an internal signal at 26 times $\mathbf{V}_3$ (28 dB above the output). $|H(j\omega)|$ has its peak near its pole at $s = j10$, and the output at $\mathbf{V}_1$ at this frequency is $|s|^2 = 100$ times (40 dB) greater than the output at $\mathbf{V}_3$. Clearly, the candidate circuit does not provide its peak gain on the output amplifier.

With regard to element values, the circuit has a range of 130 to 1 on resistance values and 1 to 1 on capacitance. This is a modest range but can be considered acceptable in some applications. When the pole is scaled from $s_{1,2} = -2 + j10$ to $s'_{1,2} = 2\pi(-20+j100)$, then $k_\omega = 20\pi$. Then, for $k_R = 10^3$, we get the 1 F capacitor scaling to $1/(k_R k_\omega) = 50/\pi$ μF. The $50/\pi$ μF is rather large and expensive for a linear circuit (an electrolytic is no problem, but it is not a good choice for linear

5.2. ANALOG OPERATIONAL CIRCUIT EXAMPLE

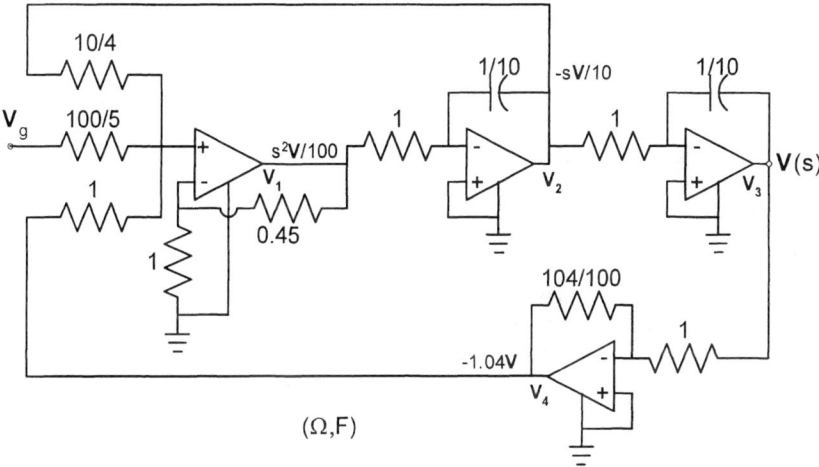

Figure 5.9. A gain-scaled version for the circuit in Figure 5.8.

behavior). The capacitor values can be reduced with a consequent increase in gain for the integrators, thus lowering the ratio between the desired voltage and its second derivative. Since $s^2\mathbf{V}$ is a factor of 100 bigger than $\mathbf{V}$ at the peak of $H(s)$, consider designing the operational circuit so as to provide $s^2\mathbf{V}/100$. This equalizes the gains between $\mathbf{V}_1$ and $\mathbf{V}_3$. Equation (5.11) thus becomes as follows:

$$\frac{s^2\mathbf{V}(s)}{100} = \frac{5}{100}\mathbf{V}_g(s) - \frac{4}{100}s\mathbf{V}(s) - \frac{104}{100}\mathbf{V}(s) \tag{5.15}$$

A candidate design to provide a circuit for (5.15) is shown in Figure 5.9. Both circuits, Figures 5.8 and 5.9, provide identical response between input and output, but the peak gain distribution and element values are remarkably different. The amplifier output functions for the circuit in Figure 5.9 are as follows:

$$\begin{aligned}
\frac{\mathbf{V}_1}{\mathbf{V}_g} &= \frac{s^2\mathbf{V}}{100\mathbf{V}_g} = \frac{s^2}{100}H(s) \\
\frac{\mathbf{V}_2}{\mathbf{V}_g} &= \frac{-s\mathbf{V}}{10\mathbf{V}_g} = -\frac{s}{10}H(s) \\
\frac{\mathbf{V}_3}{\mathbf{V}_g} &= \frac{\mathbf{V}}{\mathbf{V}_g} = H(s) \\
\frac{\mathbf{V}_4}{\mathbf{V}_g} &= \frac{-104\mathbf{V}}{100\mathbf{V}_g} = -1.04H(s)
\end{aligned} \tag{5.16}$$

We can use Matlab to check the nominal gain distribution for each amplifier response. If μCAP is used with a specific opamp, then it is recommended that the circuit be impedance and frequency scaled so as to provide satisfactory impedance values to the opamps. A Matlab analysis for the first design proceeds as follows:

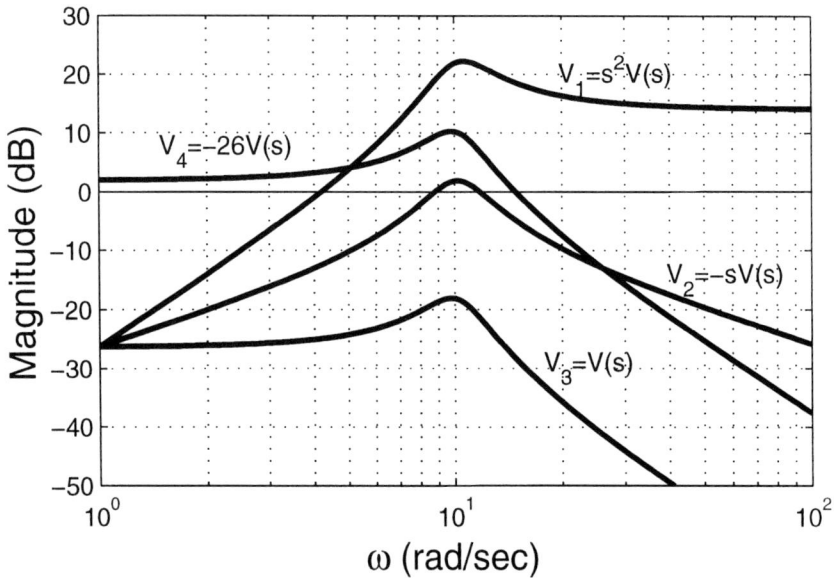

Figure 5.10. Amplifier gain-magnitude responses for the circuit in Figure 5.8.

```
w=logspace(0,2,150);
jw=j*w;
h1=[1 0 0];
h2=[0 -1 0];
h3=[0 0 1];
h4=[0 0 -26];
hn=5*[h1;h2;h3;h4];
hd=[1 4 104];
q=polyval(hd,jw);
for k=1:4
  h(k,:)=20*log10(abs(polyval(hn(k,:),jw)./q));
end
[mxgn,i]=max(h');
wmax=w(i);
```

The resulting magnitude responses (in decibels) are shown in Figure 5.10 for the first candidate design. The desired output amplifier has the lowest peak gain of all the amplifiers, which seriously limits the input range for linear response. For example, assume each amplifier has a linear output dynamic range of 20 V_{pp}. Then, with $|\mathbf{V}_3|_{max} = 10^{-18/20}|\mathbf{V}_g|_{max}$, we get $|\mathbf{V}_g|_{max} = 20 \times 10^{+18/20} = 159$ V_{pp} only if all the other amplifiers are linear for that input. Since Amplifier 1 has the global maximum gain (+22 dB), it saturates first and provides the limit on the input for linear operation. The result is that $|\mathbf{V}_g|_{max} = 20 \times 10^{-22/20} = 1.59$ V_{pp}. At

5.2. ANALOG OPERATIONAL CIRCUIT EXAMPLE

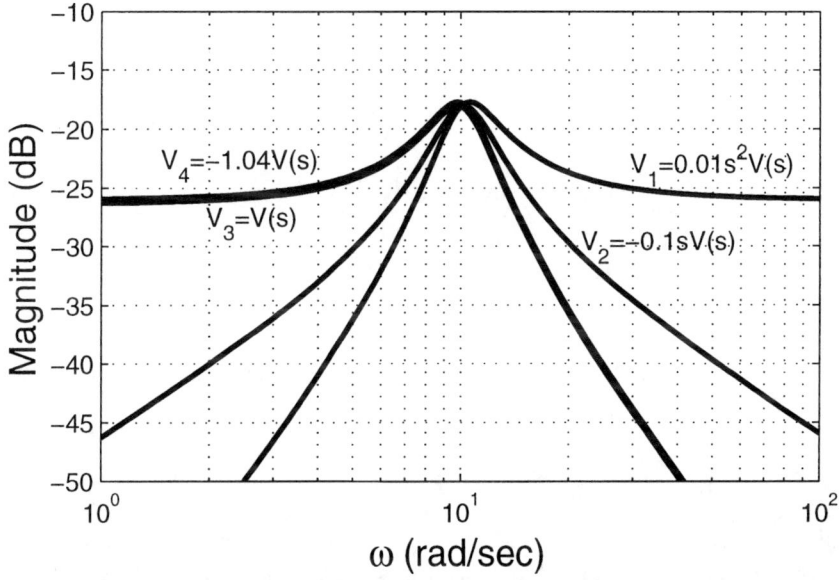

Figure 5.11. Amplifier gain-magnitude responses for the circuit in Figure 5.9.

this maximum input signal level, the maximum output on Amplifier 3 becomes $|V_3|_{max} = 10^{-18/20} \times 1.59 = 0.20$ V_{pp}, or 1/100th of the available peak output.

To evaluate the second design, the network function polynomials are changed to the following values:

```
h1=[0.01 0 0];
h2=[0 -0.1 0];
h3=[0 0 1];
h4=[0 0 -1.04];
hn=5*[h1;h2;h3;h4];
hd=[1 4 104];
```

The magnitude responses are plotted in Figure 5.11 and show the remarkable result that all amplifiers have virtually the same peak gain (−18 dB). The response of Amplifier 4 is 1.04 times (+0.34 dB) that of Amplifier 3, and this is close enough that $|V_3|$ can be used to set the dynamic range. Allowing $|V_4|_{max} = 20$ V_{pp}, we get $|V_g|_{max} = 20 \times 10^{(-18+0.34)/20} = 153$ V_{pp} so that this circuit can be specified to use the full 20 V_{pp} output range. This is a pretty good reason to spend a little design time to find a better solution. Note also that the second solution provides prototype capacitor values 10 times smaller than the first design. The *max/min* resistor ratio is 20:1, which represents a 6.5-fold reduction in the range of the resistor values. On both key issues, the second design is far superior to the first.

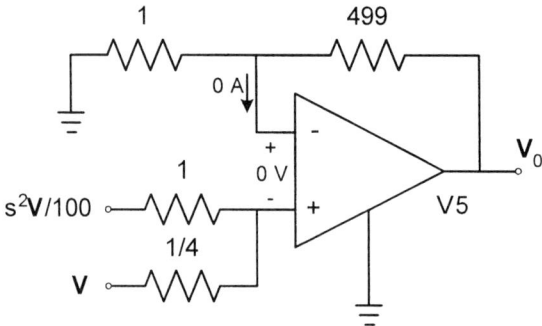

Figure 5.12. Illustrating the formation of transmission zeros.

5.2.1 Adding Transmission Zeros

An all-pole circuit, such as that in the previous example, can be used to build network functions that have finite transmission zeros. This is accomplished by summing and weighting the existing amplifier outputs to form the desired numerator polynomial. For example, suppose the function given in (5.17) is required:

$$\frac{\mathbf{V}_0}{\mathbf{V}_g} = \frac{5(s^2 + 400)}{s^2 + 4s + 104} \qquad (5.17)$$

This function has the same poles as the previous example but adds finite zeros at $z_{1,2} = \pm j20$. Equation (5.17) can be rewritten in terms of the network function, $H(s)$, of the previous example as follows:

$$\frac{\mathbf{V}_0}{\mathbf{V}_g} = (s^2 + 400)H(s) = 100\left(\frac{s^2}{100} + 4\right)\frac{\mathbf{V}}{\mathbf{V}_g} \qquad (5.18)$$

As shown in Figure 5.9, $s^2\mathbf{V}/100$ and $\mathbf{V}$ are provided at the outputs $\mathbf{V}_1$ and $\mathbf{V}_3$, respectively, so we need only to sum and weight these in order to obtain the desired $\mathbf{V}_0/\mathbf{V}_g$ network function. The result is shown in Figure 5.12. Note that the weighted summation yields $\Sigma G = 5$, so then, $K/\Sigma G = 100$ requires $K = 500$ for summing Amplifier 5. So, the extra zeros are easy to form, but the gain distribution gets messed up and needs to be repaired. A quick Matlab run tells us that the output amplifier gain exceeds all other gains by 49.5 dB (a factor of 300), and so, it will be the limiting factor on setting the dynamic range, or maximum input signal.

While technically this should be allowable, a better signal-to-noise ratio (SNR) is obtainable if all amplifiers have nearly the same peak gain so that they are all the same distance from the noise floor of the circuit.[1] Thus, in order to equalize the amplifier gains, the desired network function is rewritten as in (5.19):

$$\frac{\mathbf{V}_0}{\mathbf{V}_g} = \frac{5(s^2 + 400)}{s^2 + 4s + 104} \equiv \frac{1500(s^2/300 + 4/3)}{(s^2 + 4s + 104)} \qquad (5.19)$$

---

[1] If the peak response of one amplifier exceeds another by x dB, then its SNR is x dB higher than the amplifier having the smaller response.

5.2. ANALOG OPERATIONAL CIRCUIT EXAMPLE

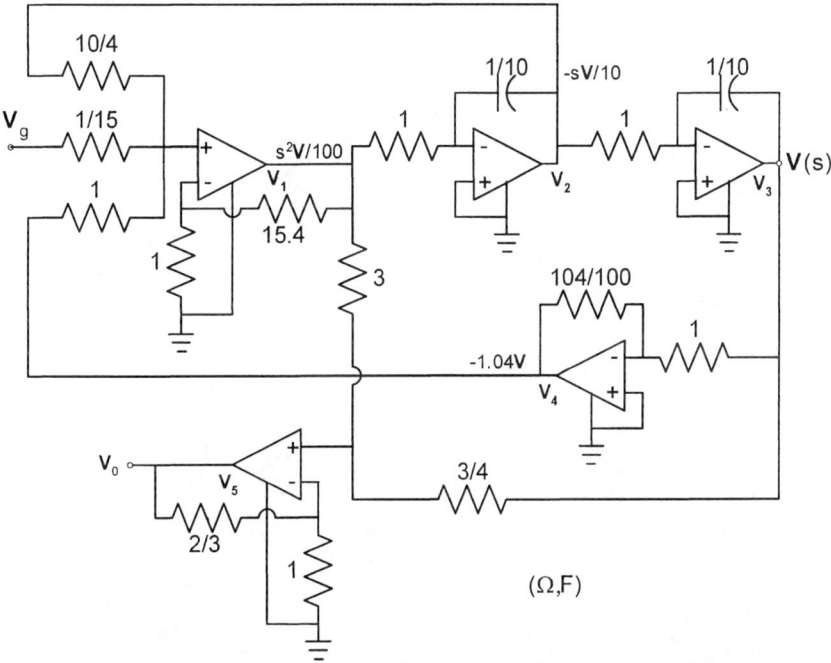

Figure 5.13. Transmission zeros obtained by summing outputs of an all-pole circuit.

Then, an all-pole function can be selected as in (5.20):

$$\frac{s^2 \mathbf{V}}{100} = \frac{1500 \mathbf{V}_g}{100} - \frac{4s}{100}\mathbf{V} - \frac{104}{100}\mathbf{V} \qquad (5.20)$$

The resulting network is given in Figure 5.13. The gain response for this circuit is checked in Matlab with the following polynomial definitions for each amplifier:

```
h1=[0.01 0 0];
h2=[0 -0.1 0];
h3=[0 0 1];
h4=[0 0 -1.04];
h5=[0.01 0 4]/3;
hn=1500*[h1;h2;h3;h4;h5];
hd=[1 4 104];
```

The resulting frequency responses are shown in Figure 5.14. All amplifiers have virtually the same peak gain ($\approx$32 dB) over the desired frequency range of the filter. It should be stressed that it is usually always worth the effort to equalize the peak gains in these types of filters. The procedure, which only requires a very small effort, maximizes dynamic range and provides improved SNR for the composite filter

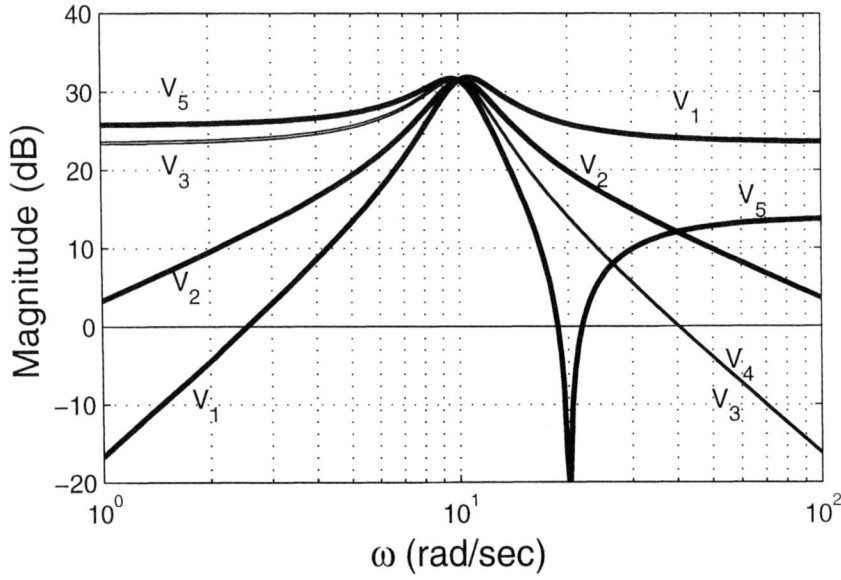

Figure 5.14. Equalized gain response for the circuit in Figure 5.13.

function. We should also note that we don't really know that a design is correct until it is built and tested, since it is easy to make an algebraic error in any analysis we do. We will use μCAP (which runs the SPICE engine) to simulate building and testing in all of our future examples.

5.3 State Variable Filters

In the previous example, the integrator outputs represented by $s^2\mathbf{V}(s)$, $s\mathbf{V}(s)$, and $\mathbf{V}(s)$ are often referred to as state variables. The implementation requires an integrator (one opamp) for each pole (in the order of the network characteristic equation), plus one summing amplifier and a sign-changing amplifier to realize an arbitrary-order all-pole function. Adding an output summing amplifier and sometimes an inverting amplifier will synthesize a network function having finite transmission zeros. The state variable filter realizes a required system function directly, but you always have to worry about gain distribution through the filter structure to optimize dynamic range for linear operation.

There is no clear-cut way to start the design so that all amplifiers have the same peak output, but one way is to start by assuming that each state variable is realized times a constant equal to its coefficient in the nth-order polynomial that specifies the required all-pole function.

5.3. STATE VARIABLE FILTERS

Example 1 - The following is an example of the design procedure applied to the example fourth-order bandpass function specified by (5.21):

$$\frac{\mathbf{V}_{BP}}{\mathbf{V}_g}(s) = \frac{6s^2}{(s^2+2s+10)(s^2+2s+17)}$$
$$= \frac{6s^2}{s^4+4s^3+31s^2+54s+170} \quad (5.21)$$

Solution - First, the all-pole function, as specified in (5.22), is selected to synthesize the state variables for the bandpass function:

$$\frac{\mathbf{V}_a}{\mathbf{V}_g} = \frac{1}{s^4+4s^3+31s^2+54s+170} \quad \text{or}$$
$$s^4 \mathbf{V}_a = \mathbf{V}_g - (4s^3+31s^2+54s+170)\mathbf{V}_a \quad \text{then}$$
$$\mathbf{V}_{BP} = 6s^2 \mathbf{V}_a(s) \quad (5.22)$$

As a first-cut design, assume that the state variables are realized with integrator gains so as to provide the polynomial coefficients of the all-pole function denominator, except for the s^2 term, which is desired to be at $6s^2\mathbf{V}_a(s)$. A Matlab run then easily provides a peak gain result for each amplifier output as follows:

```
w=logspace (-1,2,150);
jw=j*w;
h1=[1 0 0 0 0];
h2=[0-4 0 0 0];
h3=[0 0 6 0 0];
h4=[0 0 0 -54 0];
h5=[0 0 0 0 170];
h6=-[0 0 31 0 170];
hn=[h1;h2;h3;h4;h5;h6];
hd=[1 4 31 54 170];
g=polyval(hd,jw);
for k=1:6
  h(k,:)=20xlog10(abs(polyval(hn(k,:),jw)./q));
end
[mxgn,i]=max(h');
wmax=w(i);
```

There are six amplifiers in the candidate design: in the code, h1 is the set of vector coefficients that provides $s^4\mathbf{V}_a$; h2 provides $-4s^3\mathbf{V}_a$; h3 provides $6s^2\mathbf{V}_a$; h4 provides $-54s\mathbf{V}_a$; h5 provides $170\mathbf{V}_a$; and h6 provides sign inversion for the positive gains to give $-(31s^2+170)\mathbf{V}_a$ back to the summing input. The peak gains and frequencies of occurrence are listed in Table 5.1. Amplifier 3 provides the desired response at $6s^2\mathbf{V}_a(s)$, and all other outputs are on the order of 8 to 10 dB above the desired output signal. Consequently, all amplifier outputs need to be lowered with respect to Amplifier 3. Table 5.2 shows a second-cut assessment of the state variable factors so as to bring all amplifier peak gains closer to, but less than, the

Table 5.1. First-Cut Gain Distribution for the Design of a Required Filter Function

Amplifier No.	ω_{max} (rad/sec)	Peak Gain (dBv)
1	4.23	10.24
2	4.00	10.03
3	3.79	1.78
4	3.48	9.65
5	3.21	9.10
6	4.00	12.14

Table 5.2. Second-Cut Gain Distribution Obtains a Satisfactory Result for the Required Filter Design

Amplifier Number	1st-Cut Gain (dBv)	÷ Factor	2nd-Cut Gain (dBv)	Amplifier Output
1	10.24	3	0.70	$s^4 \mathbf{V}_a/3$
2	10.03	3	0.49	$-4s^3 \mathbf{V}_a/3$
3	1.78	1	1.78	$6s^2 \mathbf{V}_a$
4	9.65	3	0.11	$-54s\mathbf{V}_a/3$
5	9.10	3	-0.44	$170\mathbf{V}_a/3$
6	12.14	4	0.10	$-(31s^2 + 170)\mathbf{V}_a/4$

desired output function. Note that a factor of 3 is equivalent to +9.54 dB and of 4 is equivalent to +12.04 dB. Table 5.2 shows that all amplifier peak responses are 1 to 2 dB lower than the desired amplifier output, so the second-cut brings the peak gain distribution to desirable values.

To synthesize the circuit, we need to provide summation for $s^4 \mathbf{V}_a/3$ so that we can obtain the amplifier output factors as listed in Table 5.2. The terms needed for the summing input are given in (5.23):

$$\frac{s^4}{3}\mathbf{V}_a = \frac{1}{3}\mathbf{V}_g - \frac{1}{3}(4s^3 + 31s^2 + 54s + 170)\mathbf{V}_a \qquad (5.23)$$

The prototype design for this second cut is shown in Figure 5.15 for a μCAP simulation of the design. Note, the element values are specified times impedance scale factors, k_R and k_C. For this example, k_R was chosen so the minimum resistor load on any given opamp would be greater than 2 kΩ. Since the smallest prototype value is $1/4$, we get $k_R = 8 \times 10^3$. The frequency scale factor was arbitrarily set at $k_\omega = 2\pi 10^3$, which shifts the poles at $p_{1,2} = -1 \pm j3$ to $p'_{1,2} = (-1 \pm j3)2\pi 10^3$, and so forth. The capacitor scale factor $k_C = 1/(k_R k_\omega)$ then becomes $k_C = 125/2\pi$ nF/F. Note that a nice thing about the design is that all capacitor values are equal. This result is very hard for passive filters to meet.

5.4. CASCADE METHODS

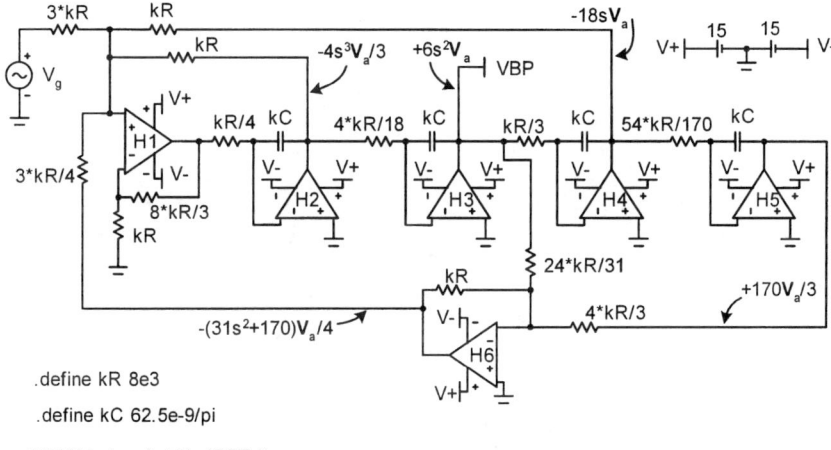

Figure 5.15. Prototype circuit in μCAP for the BP filter example.

The results of the μCAP simulation for the response of each amplifier are plotted in Figure 5.16 and compared to the nominal design computed in Matlab. The performance is excellent for these nominal design values and over this frequency range for the application. If the frequency scale factor should be increased by an order of magnitude, it would be possible to start seeing discrepancies between measured and theoretical values. For the purposes of this discussion, the results confirm the validity of the design procedure to obtain a filter directly from a specified network function.

In this relatively short section, we have introduced a very effective design procedure to obtain an arbitrary network function. If the procedure to equalize peak gain response is carefully followed, and if element values and component qualities are properly selected, the method works very well for those applications within the frequency range of the opamp. Historically, there are many other approaches to the active filter design problem, and we will review a few for their interest and use where appropriate, but this state variable approach is fundamental and always works. After the review of a few other methods, we will consider filter sensitivity issues caused by element value errors and temperature effects.

5.4 Cascade Methods

Quite often it is desired to realize a Bode characterization of a frequency response for an application such as compensation (lead-lag networks) for servo controls, speech equalizers, and instrumentation devices. A Bode characterization is based on an

170   CHAPTER 5. OPERATIONAL DESIGN OF ACTIVE FILTERS

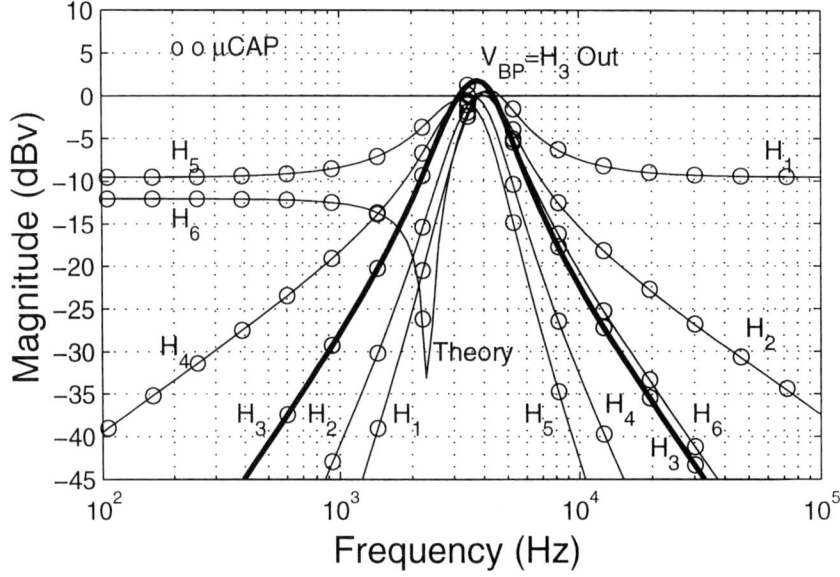

Figure 5.16. Comparison of μCAP simulation results to theory for the BP filter example.

asymptotic representation of network function. Such networks are most often realized as a cascade of poles and zeros generated by simple first-order circuits to provide the breakpoints of the Bode form. While the breakpoint poles and zeros can always be combined to form a single network function and then synthesized using the state variable form, this section illustrates the cascade approach, which offers some alternate considerations for a design approach, such as fewer amplifiers and control of or sensitivity to certain desired parameters.

5.4.1 The Cascade Concept

An overall voltage transfer is obtained from a product of transfer function stages connected in cascade. The individual transfer functions do not change as long as there is sufficient isolation between stages. Generally, this procedure allows for a single component (usually a resistor) to control the location of a pole or zero. This is in contrast to the state variable circuit, where each component affects every pole and zero of a desired network function. The isolation in a cascade realization is normally obtained from voltage sources, as shown in Figure 5.17. The transfer function per stage is $\mathbf{V}_k = H_k \mathbf{V}_{k-1}$ so that the overall transfer function works out as follows:

$$\begin{aligned} \mathbf{V}_n &= H_n \mathbf{V}_{n-1} = H_n \cdot H_{n-1} \mathbf{V}_{n-2} \\ &= H_n \ldots H_2 \cdot H_1 \mathbf{V}_0 \quad \text{or} \end{aligned}$$

5.4. CASCADE METHODS

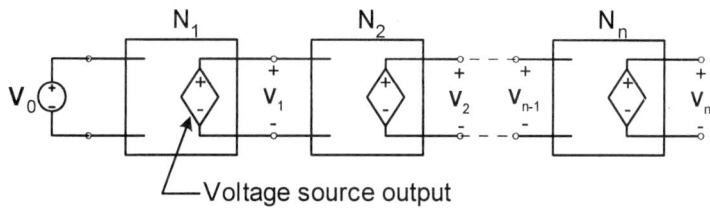

Figure 5.17. The cascade network topology.

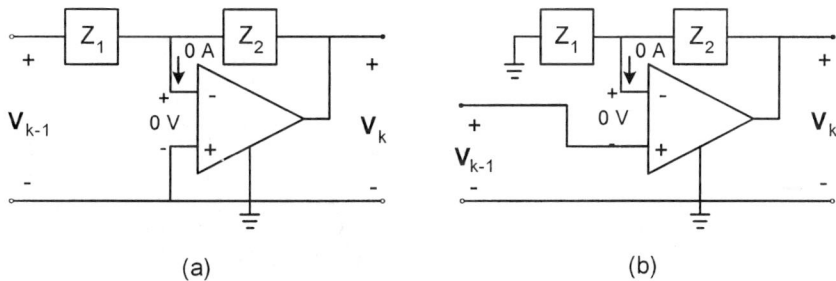

Figure 5.18. Cascade sections: (a) inverting, and (b) noninverting.

$$\frac{\mathbf{V}_n}{\mathbf{V}_0} = H = H_n \ldots H_2 \cdot H_1 \tag{5.24}$$

Typical opamp implementations are shown in Figure 5.18. For normal applications, the opamp output approximates a voltage source very well. The quality depends upon feedback, frequency, and the loading in each case. For the inverting section of Figure 5.18(a) we obtain the network function given in (5.25):

$$\frac{\mathbf{V}_k}{\mathbf{V}_{k-1}} = -\frac{Z_2}{Z_1} = -Y_1 Z_2 \tag{5.25}$$

And for the noninverting section, Figure 5.18(b), the network function is given by (5.26):

$$\frac{\mathbf{V}_k}{\mathbf{V}_{k-1}} = 1 + \frac{Z_2}{Z_1} = 1 + Y_1 Z_2 \tag{5.26}$$

Either section is usable, depending upon the application. The noninverting stage provides better isolation for its input source, but it is conceptually more awkward to use than the inverting stage. We will work examples using both types.

Before turning to some examples, it is helpful to prepare a short table of simple RC impedance functions as candidates to use for Z_1 or Z_2 in either cascade section style. A set of impedances with design parameters is shown in Figure 5.19. Impedance Type 1 is the simplest to use in the inverting stage, as it can provide uncoupled tuning for a zero and/or a pole in the resulting network function, $-Z_2/Z_1$.

No.	Circuit	Z(s)	Comments
1.	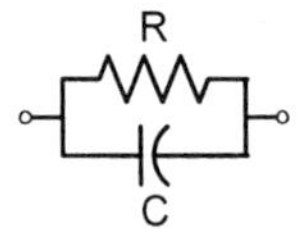 R ∥ C	$\dfrac{R}{1+s/a}$	$a = G/C$ $Z \to aR/s$ as $s \to \infty$
2.	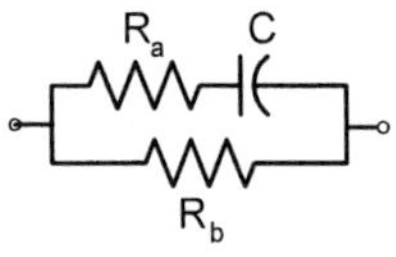 R_a in series with C, parallel with R_b	$\dfrac{R_b(1+s/a)}{1+s/b}$	$a = G_a/C$ $b = 1/(R_a+R_b)C < a$ $Z \to 1/(G_a+G_b)$ as $s \to \infty$
3.	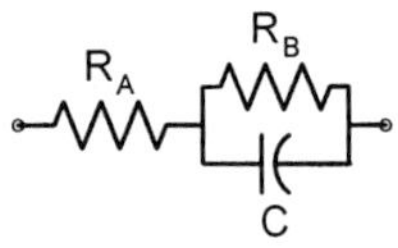 R_A in series with ($R_B \parallel C$)	$\dfrac{(R_A+R_B)(1+s/a)}{1+s/b}$	$a = (G_A+G_B)/C$ $b = G_B/C < a$ $Z \to R_A$ as $s \to \infty$
4.	 R in series with C	$\dfrac{1+s/a}{Cs}$	$a = G/C$

Figure 5.19. Simple RC impedance types for $Z_{1,2}$ in cascade sections.

5.4. CASCADE METHODS

However, when the parallel RC is used as Z_1 on an inverting stage, the resulting high-frequency short circuit presents a serious loading problem to the input source to the inverting stage. Thus, the Type 2 and 3 impedance functions are often used to ease the high-frequency loading problem of the parallel RC. The trade-off is that the resulting zeros and poles are not independent in the tuning/adjustment phase of production assembly.

Example 2 - A Compensation Network Function
A compensation network function is specified by the following function derived from its Bode asymptotic breakpoints for a decibel magnitude plot on the variable, $x = log_{10}(\omega)$:

$$\begin{aligned} h_{dB} &= 20(x - 0.602)u(x - 0.602) - 40(x - 0.903)u(x - 0.903) \\ &\quad +20(x - 1.505)u(x - 1.505) \end{aligned} \quad (5.27)$$

Find the corresponding network function required and use cascade synthesis methods to design an appropriate circuit. Discuss design considerations.

Solution - The given h_{dB} function consists of a series of ramp functions in the variable, x. Two ramps have a slope of $+20$ dB/dec (each integer range of x is a decade in ω) and one with a slope of -40 dB/dec. Recognizing that the positive sloped ramps are zero factors in $|H(j\omega)|$ and that negative slopes are pole factors, (5.27) is interpreted in reverse (using only real poles and zeros) to give the following network magnitude function. Note that $10^{0.602} = 4, 10^{0.903} = 8$, and $10^{1.505} = 32$.

$$|H(j\omega)| = 1 \cdot \frac{|1 + j\omega/4||1 + j\omega/32|}{|1 + j\omega/8|^2} \quad (5.28)$$

An appropriate network function for (5.28) is given by (5.29), where the variable, $j\omega$, is extended to the complex frequency variable, $s = j\omega$:

$$H(s) = \pm \frac{(1 + s/4)(1 + s/32)}{(1 + s/8)^2} \quad (5.29)$$

a. Inverting Two-Stage Solution
Using parallel RC impedances (Type 1 in Figure 5.19), the circuit of Figure 5.20 provides the desired $H(s)$ with a + sign as follows:

$$\begin{aligned} H_a(s) &= -Y_1 Z_2 \cdot (-)Y_3 Z_4 \\ \text{where} \quad Z_k &= R_k/(1 + s/a_k) \quad \text{and} \quad a_k = G_k/C_k \\ \text{so that} \quad H_a(s) &= +\frac{(1 + s/a_1)}{R_1} \frac{R_2}{(1 + s/a_2)} \frac{(1 + s/a_3)}{R_3} \frac{R_4}{(1 + s/a_4)} \end{aligned} \quad (5.30)$$

The problem now is to specify values for the eight RC components in the proposed circuit. There are two ways we can proceed. One is to go through a lengthy analysis for an underspecified set of constraints. The other is basically to set the values by inspection and then recalculate the answer to meet peak gain criteria,

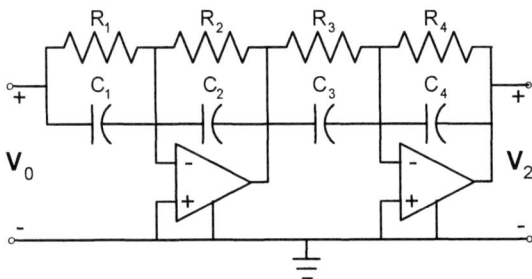

Figure 5.20. Topology for two-stage solution, a, for (5.29).

and so forth. We will elect the second method as being more direct and easier to understand, but first, note that for the eight unknowns, there are three conditions: four breakpoints are specified (a_k's); there is the network function multiplier, $R_2 R_4/(R_1 R_3) = 1$; and the peak output of Stage-1 should not exceed the peak output of Stage-2 in order to avoid internal clipping. This amounts to six constraints for the eight unknowns, thus allowing at least two arbitrary choices. The peak gain specification decides which numerator breakpoint, 4 or 32, gets assigned to Stage-1. So, a direct procedure is to set all the poles of the Z_k and then use impedance scaling to accommodate the peak gain requirement if necessary.

The constant multiplier is handled automatically in (5.30) by setting all $R_k = 1$ so that $R_2 R_4/(R_1 R_3) = 1$. Then, clearly, $a_2 = a_4 = 8 = 1/(1 \cdot C_2) = 1/(1 \cdot C_4)$.

$$C_2 = C_4 = 0.125 \text{ F} \qquad (5.31)$$

Now, we have to decide whether $a_1 = 4$ or 32 is appropriate. In (5.29), the desired network function, we have that $|H(j\omega)|$ is given by the output of Stage-2, which varies from 1 at $\omega = 0$ to 0.5 as $\omega \to \infty$. So, the gain of Stage-1 should not exceed 1. Consider the selection, $a_1 = 4$, for the Stage-1 network function given in (5.32):

$$h_1(s) = \frac{1(1 + s/4)}{(1 + s/8)} \qquad (5.32)$$

$|h_1(j\omega)|$ varies from 1 at $\omega = 0$ to 2 as $\omega \to \infty$ for $s = j\omega$. Therefore, $a_1 = 4$ is not a good choice.

Now, consider $a_1 = 32$ to get h_1, as in (5.33):

$$h_1(s) = \frac{1(1 + s/32)}{(1 + s/8)} \qquad (5.33)$$

This magnitude, for $s = j\omega$, varies from 1 at $\omega = 0$ to $8/32 = 0.25$ as $s \to \infty$. Thus, it is appropriate to choose $a_1 = 32$ for the first stage breakpoint.

$$a_1 = 1/(1 \cdot C_1) = 32 \text{ and}$$
$$a_3 = 1/(1 \cdot C_3) = 4 \text{ yield}$$

5.4. CASCADE METHODS

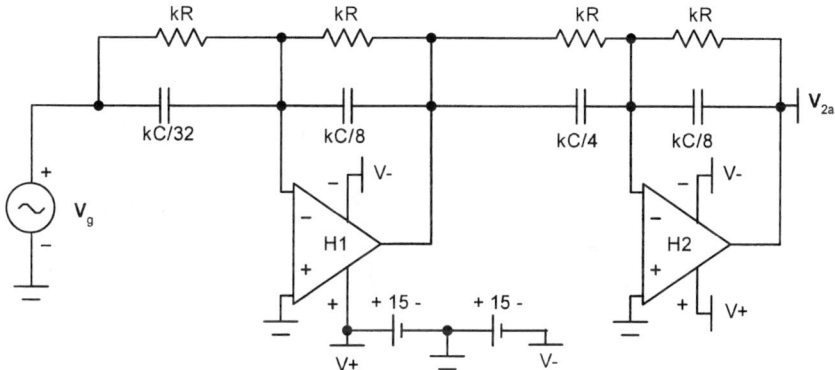

```
.define kR 2e3
.define kC 2.5e-6/pi

*** Wide bandwidth JFET i/p opamp

    .MODEL LF351 OPA (LEVEL=2 TYPE=3 C=10P A=100K ROUTAC=50 ROUTDC=75
         VOFF=5M IOFF=25P SRP=13MEG SRN=13MEG IBIAS=50P
         VEE=-15 VCC=15 VPS=14 VNS=-14 GBW=4MEG)
```

Figure 5.21. Prototype Design a for the network function in (5.29).

$$\begin{aligned} C_1 &= 1/32 \\ C_3 &= 1/4 \end{aligned} \quad (5.34)$$

The completed two-stage prototype design is shown in Figure 5.21 for this solution architecture. Note that amplifier, $H1$, has a capacitive load, which goes to a short circuit for high frequency. Such a load often causes undesired divergence from a specified network function and limits the range of useful frequency scaling (application) of the design. This will be shown later when several different design solutions are compared.

b. An Improved Two-Stage Solution

One way to eliminate the capacitive short circuit load on the opamp (in the cascade design) is to place a resistor in series with the capacitor. This leads to impedances of Type 2 or 3 as shown in Figure 5.19. In addition, the added series resistor generates a finite zero in the resulting impedance which must be accounted for in the design of the filter requirements.

The impedance is of the form $(s+a)/(s+b)$, where $b < a$ for an RC network. If the impedance is used as feedback (Z_2) around the opamp, it keeps the same form, whereas if the impedance is used on the input to the opamp (Z_1), it gets turned upside down. This means that if we use Type 2 impedances, one branch is used as feedback and another as input in order to meet the $H(s)$ of (5.29). An example architecture is shown in Figure 5.22. The circuit realizes the desired $H(s)$ with a

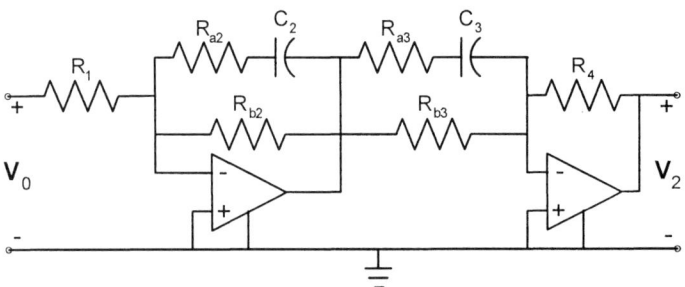

Figure 5.22. Topology for improved two-stage network.

+ sign.

$$H_b(s) = \frac{(-)1}{R_1} \cdot \frac{R_{b2}(1+s/a_2)}{(1+s/b_2)} \frac{(1+s/b_3)}{(-)R_{b3}(1+s/a_3)} R_4 \qquad (5.35)$$

where a_k and b_k are specified in Figure 5.19. Element values are found as follows. First, the $H(s)$ multiplicative constant, $R_{b2}R_4/R_1R_{b3} = 1$, is obtained by setting $R_1 = 1 = R_{b2} = R_{b3} = R_4$. Then, the previous discussion on the form for Stage-1 gain, h_1 [see (5.33)], requires $a_2 = 32$ and $b_2 = 8$. That leaves $b_3 = 4$ and $a_3 = 8$. These conditions translate to the following element values:

$$\frac{a_2}{b_2} = \frac{32}{8} = 4 = \frac{G_{a2}(R_{a2} + R_{b2})C_2}{C_2} = 1 + G_{a2}R_{b2} = 1 + G_{a2} \cdot 1$$

$$G_{a2} = 3 \to R_{a2} = 1/3$$

$$a_2 = 32 = \frac{G_{a2}}{C_2} = \frac{3}{C_2} \to C_2 = 3/32$$

$$\frac{a_3}{b_3} = \frac{8}{4} = 2 = \frac{G_{a3}(R_{a3} + R_{b3})C}{C} = 1 + G_{a3}R_{b3} = 1 + G_{a3} \cdot 1$$

$$G_{a3} = 1 \to R_{a3} = 1$$

$$a_3 = 8 = \frac{G_{a3}}{C_3} = \frac{1}{C_3} \to C_3 = 1/8$$

The completed prototype, Design b, is shown in Figure 5.23. This circuit has several advantages over the first design. First, there are no capacitive HF short circuit loads on the opamps; second, only two capacitors are used; and third, the input impedance is constant. A resistive input impedance is usually preferred as it is easier to drive and isolate from the source. A comparative graph shows that the response for this design is much better than the first design. However, the design can still be improved.

c. A Single-Amplifier, One-Stage Solution
The previous results (Figures 5.22 or 5.23) strongly suggest that two amplifiers are not required to accomplish the required design. The four breakpoints specified in the Bode form of (5.29) can actually be obtained from a single amplifier when

5.4. CASCADE METHODS

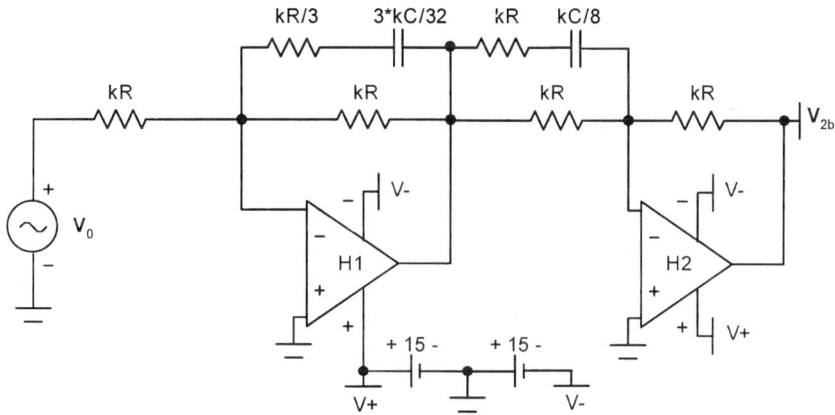

```
.define kR 2e3
.define kC 2.5e-6/pi

*** Wide bandwidth JFET i/p opamp
    .MODEL LF351 OPA (LEVEL=2 TYPE=3 C=10P A=100K ROUTAC=50 ROUTDC=75
        VOFF=5M IOFF=25P SRP=13MEG SRN=13MEG IBIAS=50P
        VEE=-15 VCC=15 VPS=14 VNS=-14 GBW=4MEG)
```

Figure 5.23. Prototype Design b for the network function in (5.29).

impedances of Type 2 or 3 are used for Z_1 and Z_2, as shown in Figure 5.24. The desired network function is provided with a $-$ sign as in (5.36):

$$H_b(s) = \frac{-Z_2}{Z_1} = \frac{-(1+s/b_1)}{R_{b1}(1+s/a_1)} \cdot \frac{R_{b2}(1+s/a_2)}{(1+s/b_2)} \quad (5.36)$$

where $b < a$ is required, $a = G_a/C$, and $b = 1/(R_a + R_b)C$ for each impedance, $Z_{1,2}$.

In order to fit (5.29), we have the constant multiplier, $R_{b2}/R_{b1} = 1$, and poles $a_1 = 8$ and $b_2 = 8$. In order for $b_1 < a_1$ to be true, we choose $b_1 = 4$, leaving $a_2 = 32$, which turns out to be greater than $b_2 = 8$ as required. For this design, we have six components with unknown values with conditions for the four critical frequencies, plus the network function multiplier, for a total of five independent constraints. Thus, one element value is arbitrary, and so, one design is obtained by choosing $R_{b2} = 1$, whence $R_{b2}/R_{b1} = 1$ yields $R_{b1} = 1$. From Figure 5.19 we form the following results:

$$\frac{a_1}{b_1} = \frac{8}{4} = 2 = \frac{G_{a1}(R_{a1}+R_{b1})C_1}{C_1} = 1 + G_{a1}R_{b1}$$
$$\therefore G_{a1} = 1 \quad \text{or} \quad R_{a1} = 1/G_{a1} = 1$$
$$\frac{a_2}{b_2} = \frac{32}{8} = 4 = \frac{G_{a2}(R_{a2}+R_{b2})C_2}{C_2} = 1 + G_{a2}R_{b2}$$
$$\therefore G_{a2} = 3 \quad \text{or} \quad R_{a2} = 1/G_{a2} = 1/3$$

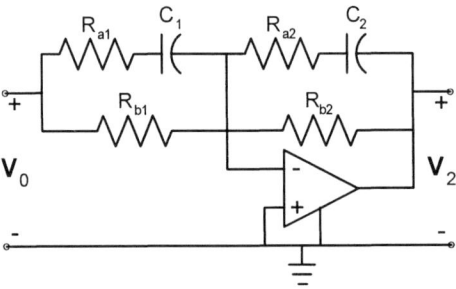

Figure 5.24. Topology for single-stage solution c in (5.29).

$$\begin{aligned}
a_1 &= 8 = \frac{1}{R_{a1}C_1} = \frac{1}{1 \cdot C_1} \to C_1 = \frac{1}{8} \\
a_2 &= 32 = \frac{1}{R_{a2}C_2} = \frac{3}{1 \cdot C_2} \to C_1 = \frac{3}{32}
\end{aligned} \quad (5.37)$$

This solution is fairly simple, and the circuit (shown in Figure 5.25) is well behaved versus frequency, since the amplifier sees basically a resistive load over the full operating range of $0 < \omega < \infty$. The response for the three designs is now compared in a single graph, as shown in Figure 5.26.

The circuits are all simulated in μCAP by using an impedance scale factor, $k_R = 2000$, and a frequency factor, $k_\omega = 2000\pi$, for each circuit. The opamp is an LF351, which has a dc gain of 100 dB, and its first pole at 40 Hz. The unit-gain bandwidth is therefore approximately $A_0 p = 4 \times 10^6 = 4$ MHz.

As mentioned above, Design a has a very poor result for frequencies above 20 kHz, as shown on the graph in Figure 5.26 marked with circles. The undesired peaks in the stopband are due to the capacitive load resonating with the opamp output impedance at frequencies near the amplifier's unit-gain bandwidth. Design b (marked with x's) fits the theoretical response (black line) to about 300 kHz before it begins to drop below the required stopband asymptote. Design c, the single-stage solution (marked with asterisks), is close to theory to about 1 MHz, where it drops off from the requirement. The difference between Designs b and c is essentially the fact that a cascade of two stages, with bandwidth limited by the opamp open-loop gain, has less bandwidth than a single stage. This is due to a squaring effect of the cascade of similar network functions. In any event, the ultimate dropping away from a theoretical stopband asymptote (for arbitrarily large frequency values in a limiting sense) is always to be expected for opamps with finite bandwidth. It is therefore important always to have specifications for any design to include frequency limits over which a desired function must be met.

As far as the poor performance of Design a, this is due to the capacitive load in combination with two equivalent inductances. One is at the output of $H1$, and the other is at the input to $H2$. The fact that these inductances are present, for frequencies in the vicinity of the opamp gain-bandwidth product, was worked in Problem 3.11. The primary lesson from this example is that capacitive loads should

5.4. CASCADE METHODS

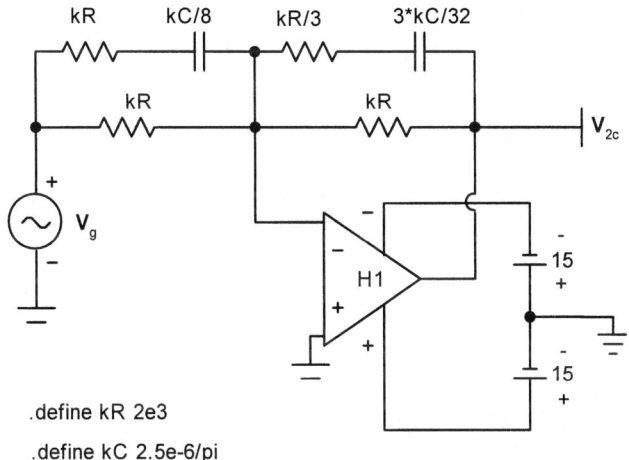

.define kR 2e3
.define kC 2.5e-6/pi

*** Wide bandwidth JFET i/p opamp
.MODEL LF351 OPA (LEVEL=2 TYPE=3 C=10P A=100K
ROUTAC=50 ROUTDC=75 VOFF=5M IOFF=25P
SRP=13MEG SRN=13MEG IBIAS=50P VEE=-15
VCC=15 VPS=14 VNS=-14 GBW=4MEG)

Figure 5.25. Prototype Design c for the network function in (5.29).

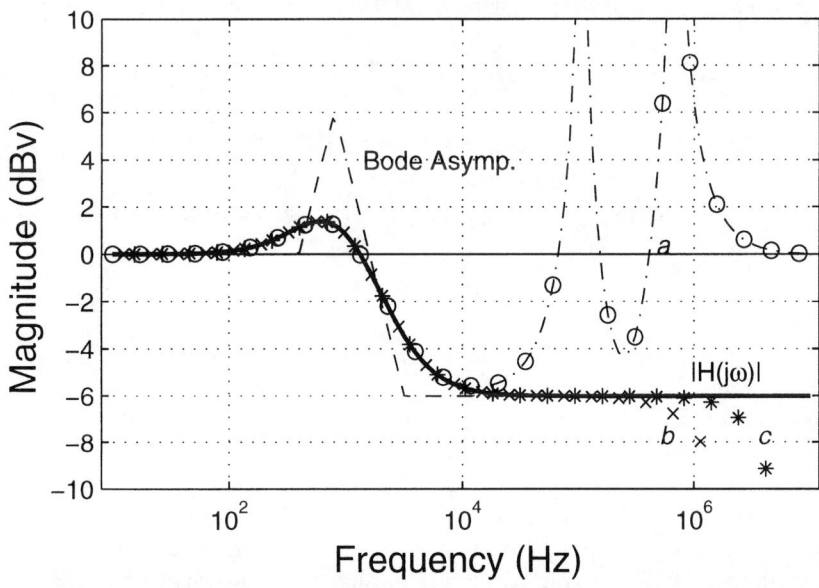

Figure 5.26. Comparison of frequency response for Designs a, b, and c.

be avoided in the cascade design approach. It is usually possible to choose design requirements so that both poles and zeros are required with the result that Type 2 or 3 loads can be used on the opamp. As the example demonstrates, being able to include series resistance on a capacitive load results in near-ideal response for the desired network functions.

Example 3 - Design a prototype network to provide the following network function and maximize K for linear operation. Determine the 3 dB bandwidth for the response.

$$H(s) = K \frac{s^2}{(s+1)^4} \tag{5.38}$$

Solution - A candidate design is shown in Figure 5.27. It uses impedances of Type 1 and 4 to realize the fourth-order pole requirement. Impedance scale factors K_1 and K_2 are used to determine the overall system multiplier. Let H_D be the network function for the circuit of Figure 5.27.

$$\begin{aligned} H_D(s) &= \frac{-Z_2^{(1)}}{Z_1^{(1)}} \cdot \frac{-Z_2^{(2)}}{Z_1^{(2)}} = + \frac{s}{(s+1)} \frac{K_1}{(s+1)} \frac{s}{(s+1)} \frac{K_2}{(s+1)} \\ &= K_1 K_2 \frac{s^2}{(s+1)^4} \end{aligned} \tag{5.39}$$

where $K_1 K_2 = K$ is required.

The network function for Stage-1 is $H_1(s)$, and it has its peak magnitude at $s = j1$.

$$H_1(s) = \frac{K_1 s}{(s+1)^2} \tag{5.40}$$

This is deduced from (5.41), which is a known functional form:

$$\begin{aligned} |H_1(j\omega)|^2 &= H_m^2 = \frac{K_1^2 \omega^2}{(\omega^2 + 1)^2} = \frac{K_1^2}{(\omega + 1/\omega)^2} \\ &= \frac{K_1^2}{2^2} \text{ at } \omega = 1 \end{aligned} \tag{5.41}$$

For $K_1 = 1$, $(H_m)_{dB} = -6$ dB, the frequency-dependent part of the function inserts 6 dB of loss below the decibel equivalent of K_1. Likewise, the second stage being identical to the first also inserts 6 dB of loss below the dB equivalent of the gain, K_2. If $K_2 = 2$ (+6 dB) is chosen for Stage-2, then both $H2$ and $H1$ will have identical peak outputs, thus leaving K_1 completely arbitrary.

The answer is $K = 2K_1$ with K_1 arbitrary. The customer has to specify a maximum input signal, and then K_1 can be chosen to give maximum output for the opamp used in the design. A specific design is checked using μCAP for $K_1 = 20$, $K_2 = 2$, $k_R = 2000$, and $k_\omega = 2000\pi$. The magnitude response is shown in Figure 5.28 in comparison to the theoretical response. The response is quite good over a 70 dB range (+20 to −50 dBv), or to about 200 kHz. The response

5.4. CASCADE METHODS

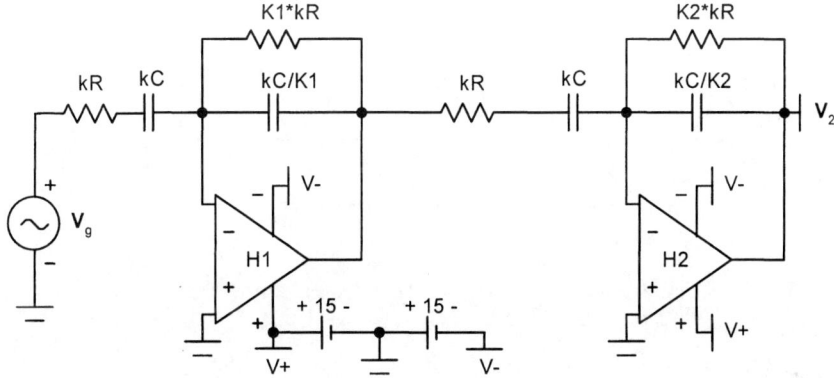

.define K1=20
.define K2=2
.define kR 2e3
.define kC 250e-9/pi
*** Wide bandwidth JFET i/p opamp
.MODEL LF351 OPA (LEVEL=2 TYPE=3 C=10P A=100K ROUTAC=50
ROUTDC=75 VOFF=5M IOFF=25P SRP=13MEG SRN=13MEG
IBIAS=50P VEE=-15 VCC=15 VPS=14 VNS=-14 GBW=4MEG)

Figure 5.27. Prototype design for the network function in (5.38).

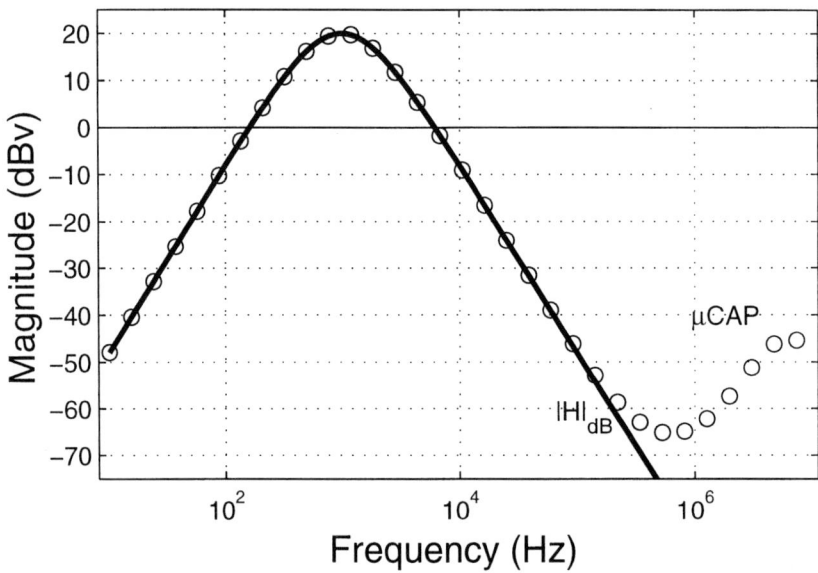

Figure 5.28. Frequency response for the prototype design in Figure 5.27, $K = 40$.

deteriorates in the region where open-loop gain decreases to unit gain. The problem statement requested the 3-dB bandwidth for the response. This bandwidth is at the points where the total response is 3 dB ($1/\sqrt{2}$) below the peak gain. The bandwidth is thus found as follows:

$$|H(j\omega)|^2 = K^2 \left|\frac{(j\omega)^2}{(1+j\omega)^4}\right|^2$$

$$H_{max}^2 = K^2 \left|\frac{j^2}{(1+j)^4}\right|^2 = \frac{K^2}{(\sqrt{2})^8} = \frac{K^2}{2^4}$$

$$|H(j\omega_{1,2})|^2 = \left(\frac{H_{max}}{\sqrt{2}}\right)^2 = \frac{H_{max}^2}{2}$$

$$= \frac{K^2 \omega_{1,2}^4}{(1+\omega_{1,2}^2)^4} = \frac{K^2}{2^5}$$

$$\therefore 2^5 \omega_{1,2}^4 = (1+\omega_{1,2}^2)^4$$

$$2^{5/4} \omega_{1,2} = (1+\omega_{1,2}^2) \text{ or}$$

$$\omega_{1,2}^2 - 2^{1.25}\omega_{1,2} + 1 = 0 \text{ whence}$$

$$\omega_{1,2} = 1.8328, 0.5456 \text{ so that}$$

$$3 - \text{dB } BW = \omega_1 - \omega_2 = 1.2872$$

5.4. CASCADE METHODS

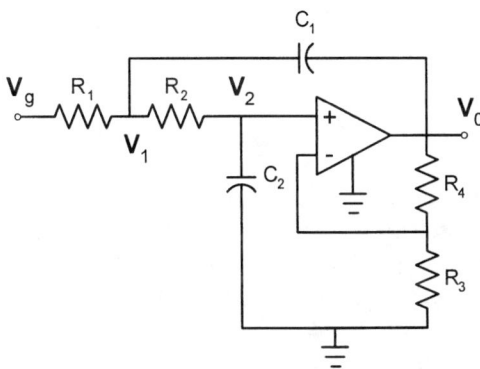

Figure 5.29. The Sallen-Key complex-pole circuit.

5.4.2 A Single-Amplifier Quadratic Factor Circuit

Quite often, in the cascade design approach, it is desirable to have a simple circuit capable of providing a complex-pole quadratic factor. This section describes the Sallen and Key (S-K), circuit which uses a single opamp and two independent capacitors to provide a desired complex pole pair [1, p. 11].

Consider the circuit given in Figure 5.29. The network function is found by taking KCL equations at $\mathbf{V}_1$ and $\mathbf{V}_2$ and noting $\mathbf{V}_0 = \mu \mathbf{V}_2$, where $\mu = 1 + R_4/R_3$.

KCL_1:
$$G_1(\mathbf{V}_1 - \mathbf{V}_g) + G_2(\mathbf{V}_1 - \mathbf{V}_2) + C_1 s(\mathbf{V}_1 - \mu \mathbf{V}_2) = 0$$

KCL_2:
$$G_2(\mathbf{V}_2 - \mathbf{V}_1) + C_2 s \mathbf{V}_2 = 0 \text{ from which}$$
$$(1 + R_2 C_2 s)\mathbf{V}_2 = \mathbf{V}_1$$

is back substituted into KCL_1 to obtain (5.42):

$$\frac{\mathbf{V}_0}{\mathbf{V}_g} = \frac{\mu \omega_0^2}{(s^2 + 2\alpha s + \omega_0^2)} \equiv \frac{\mu}{1 + \frac{2\alpha}{\omega_0^2} s + \frac{s^2}{\omega_0^2}} \tag{5.42}$$

The second form in (5.42) is the form used for Bode plots, and it provides a -40 dB/dec factor at the breakpoint, $\omega = \omega_0$. The pole parameters, 2α and ω_0^2, are defined as follows:

$$\omega_0^2 = 1/(R_1 R_2 C_1 C_2)$$
$$2\alpha = \frac{G_2(1-\mu)}{C_2} + \frac{(G_1 + G_2)}{C_1}$$
$$\text{where} \quad \mu = 1 + R_4/R_3 \tag{5.43}$$

The pole Q is $\omega_0/2\alpha$, and $Q > 1$ provides peak gain for the circuit at frequencies near $s = j\omega_0$. For example, at $s = j\omega_0$, the transfer function, (5.42), evaluates to the following value:

$$\left|\frac{\mathbf{V}_0}{\mathbf{V}_g}(j\omega_0)\right| = \frac{\mu}{2\alpha/\omega_0} = \mu Q \tag{5.44}$$

This last result points out a disadvantage of the circuit, namely, that its gain depends upon the pole-Q setting. It is therefore awkward to keep the peak gain signal on the output amplifier in a cascade circuit that uses the S-K network.

Equation (5.43) shows that 2α can be adjusted independently of ω_0 via the output amplifier gain, μ. As μ increases, 2α decreases, and it can be made to go to zero (or become negative). The value of μ that just causes $2\alpha = 0$ (infinite Q) is given by (5.45):

$$\mu_c = 1 + \frac{C_2}{C_1}(1 + R_2 G_1) \tag{5.45}$$

Example 4 - A Fourth-Order Lowpass Filter

Use a cascade of complex pole circuits to provide the following fourth-order lowpass filter requirement. Specify the constant multiplier for the function, and use μCAP to evaluate the design by scaling 1 rad/sec to 1 kHz and 1 Ω to 2 kΩ.

$$H(s) = \frac{K}{(s^2 + 0.765s + 1)(s^2 + 1.848s + 1)} \tag{5.46}$$

Solution - Two S-K circuits can be used in cascade to realize the specified $H(s)$. Both poles are on the unit circle, and we have the following required design parameters:

$$\begin{aligned}
\omega_{01}^2 &= 1 = \omega_{02}^2 \\
2\alpha_1 &= 1.848 \\
2\alpha_2 &= 0.765 \\
K &= \mu_1 \mu_2
\end{aligned} \tag{5.47}$$

The low-Q pole is assigned to the first stage because its gain will be less than the peak gain of the composite function. This conclusion is not obvious, so we can use μCAP to check the gain distribution for the two possible pole arrangements in the cascade topology.

There is an infinite number of solutions for the element values since there are six unknown element values for the two required constraints, α and ω_0, for each stage. For $\omega_0 = 1$, the following common solution is virtually by inspection. Let

$$R_1 = R_2 = 1 = C_1 = C_2$$

Then,

$$2\alpha = 3 - \mu = 2 - R_4/R_3$$

5.4. CASCADE METHODS

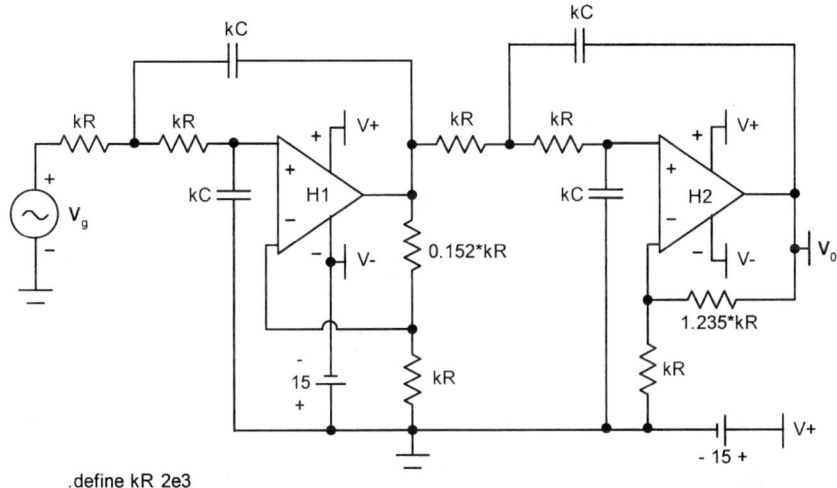

.define kR 2e3
.define kC 250e-9/pi
*** Wide bandwidth JFET i/p opamp
.MODEL LF351 OPA (LEVEL=2 TYPE=3 C=10P A=100K ROUTAC=50 ROUTDC=75
VOFF=5M IOFF=25P SRP=13MEG SRN=13MEG IBIAS=50P VEE=-15 VCC=15
VPS=14 VNS=-14 GBW=4MEG)

Figure 5.30. Prototype design for Example 4.

Next, with $R_3 = 1$, we get R_4 and μ:

$$R_4 = 2 - 2\alpha$$
$$\mu = 3 - 2\alpha \tag{5.48}$$

Elements $R_{1,2}$, $C_{1,2}$, and R_3 are the same for both stages. R_4 and μ have the following values for the required design:

$$\mu_1 = 3 - 2\alpha_1 = 3 - 1.848 = 1.152$$
$$R_{4,1} = 2 - 2\alpha_1 = 2 - 1.848 = 0.152 \ \Omega$$
$$\mu_2 = 3 - 2\alpha_2 = 3 - 0.765 = 2.235$$
$$R_{4,2} = 2 - 2\alpha_2 = 2 - 0.765 = 1.235 \ \Omega$$
$$K = \mu_1\mu_2 = (1.152)(2.235) = 2.575 \text{ (or 8.215 dB)}$$

The prototype design is given in Figure 5.30. The resulting topology is a neat solution since each stage only has one element value, R_4, that sets the location of the pole on the unit circle. Such pole arrangements are a property of Butterworth filters (discussed in Chapter 8), and their simplicity is one reason they are a popular choice. Simulated responses are shown in Figures 5.31 and 5.32 for the output of both Stage-1 and Stage-2 to check for the proper choice for positioning the low-Q pole. The results show that the low-Q pole should be positioned in the first stage

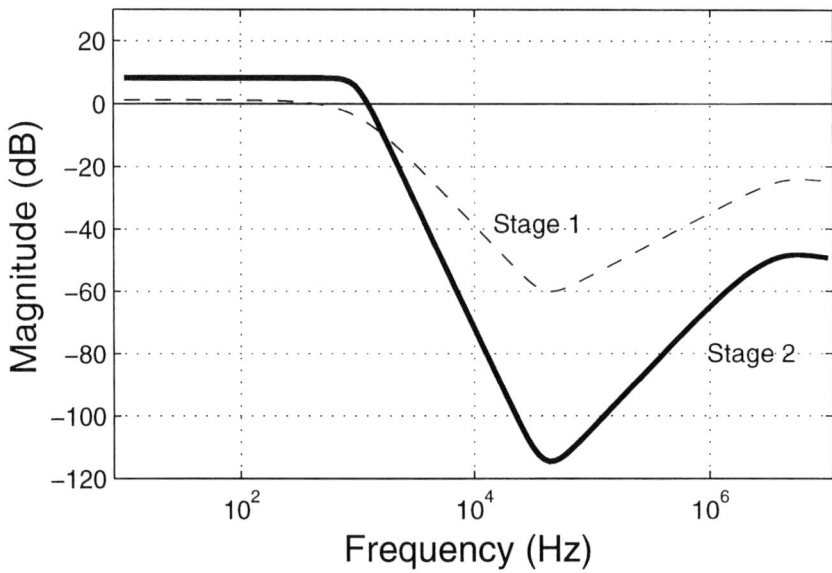

Figure 5.31. Frequency response for Example 4, low-Q pole at Stage-1.

to avoid the possibility of internal clipping, or nonlinear response. Note that the composite response is the same for both arrangements, but for either case, the ideal response (-80 dB/dec) only extends to about 30 kHz. At that point, the falling opamp gain begins to provide nonideal behavior, and the desired frequency response deteriorates appreciably. The stopband for this design can only meet a -56 dB relative attenuation specification for frequencies above 5 MHz. This is far from the ideal four-pole Butterworth function behavior and is a typical problem encountered in the active filter design problem. It should be noted that passive filter design, for all-pole functions, has a similar problem due to winding capacitance contributing to nonideal inductor behavior. The solution is to usually choose different components, topologies, or objective functions for the specified response.

5.4.3 The Twin-T Circuit

Just as it is desirable to have a complex pole circuit for the cascade design approach, it is sometimes useful to have a circuit to generate j-axis transmission zeros. Such a circuit is shown in Figure 5.33. The terminating resistors are included so that the network will not have a short circuit input impedance at high frequencies. The network function is found most directly by means of admittance parameters for the parallel networks shown in Figure 5.34. The use of general analysis is quite tedious and lengthy; the use of properties of two-port parameters simplifies the analysis for this circuit. First, note that the network is symmetrical so that Ports-1 and -2 are indistinguishable (i.e., $y_{11} = y_{22}$ and $y_{12} = y_{21}$). Next, recall that when networks

5.4. CASCADE METHODS

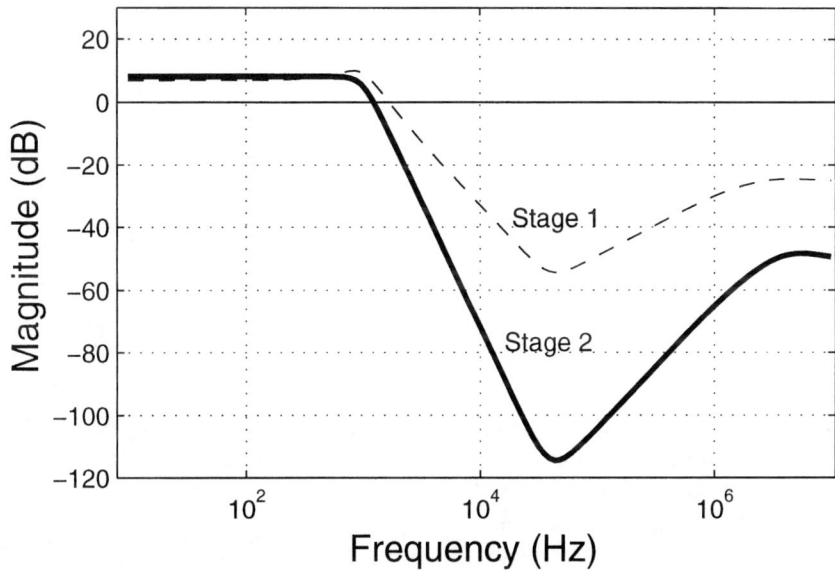

Figure 5.32. Frequency response for Example 4, high-Q pole at Stage-1.

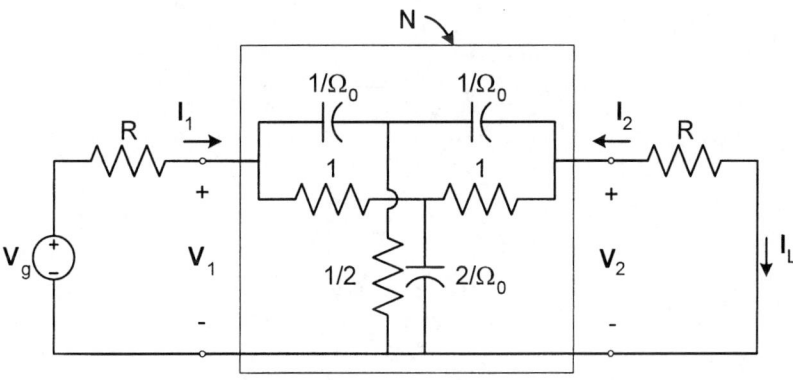

Figure 5.33. Terminated Twin-T prototype circuit used to generate j-axis transmission zeros.

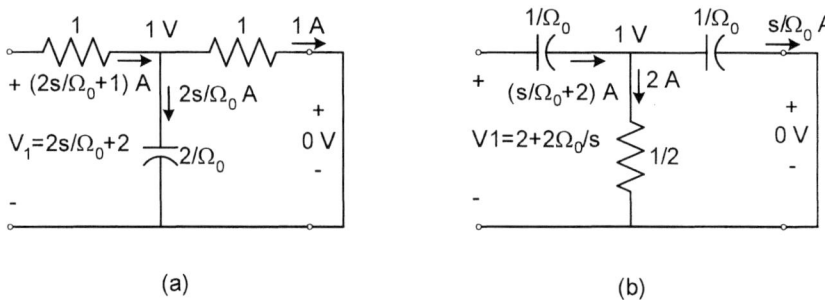

Figure 5.34. Parallel subnetworks used to analyze the Twin-T circuit: (a) series-R shunt-C, and (b) series-C shunt-R.

are connected in parallel, their y-parameters add to form the y-parameters for the combined networks. Letting y_{11} and y_{21} represent the y-parameters for the two parallel T circuits enclosed in the box, N, in Figure 5.33, we get the following network function for the terminated network:

At Port-1:

$$\begin{aligned} \mathbf{I}_1 &= G(\mathbf{V}_g - \mathbf{V}_1) = y_{11}\mathbf{V}_1 + y_{21}\mathbf{V}_2 \\ G\mathbf{V}_g &= (y_{11} + G)\mathbf{V}_1 + y_{21}\mathbf{V}_2 \text{ and} \\ \mathbf{V}_1 &= (G\mathbf{V}_g - y_{21}\mathbf{V}_2)/(y_{11} + G) \end{aligned} \quad (5.49)$$

At Port-2:

$$\begin{aligned} \mathbf{V}_2 &= -R\mathbf{I}_2 = -R(y_{21}\mathbf{V}_1 + y_{11}\mathbf{V}_2) \text{ so} \\ (1 + Ry_{11})\mathbf{V}_2 &= -Ry_{21}\mathbf{V}_1 \text{ and} \\ \mathbf{V}_2 &= -Ry_{21}\mathbf{V}_1/(1 + Ry_{11}) \end{aligned} \quad (5.50)$$

Equations (5.49) and (5.50) are combined to get the desired network function in terms of R and the subnetwork y-parameters:

$$\begin{aligned} \frac{\mathbf{V}_2}{\mathbf{V}_g} = \frac{R\mathbf{I}_L}{\mathbf{V}_g} &= \frac{-Gy_{21}}{(y_{11} + G)^2 - y_{21}^2} \\ &= \frac{-Gy_{21}}{(y_{11} + G + y_{21})(y_{11} + G - y_{21})} \end{aligned} \quad (5.51)$$

Now, the y-parameters for each subnetwork can be obtained by inspection of Figure 5.34 for the case where each output has been short-circuited ($\mathbf{V}_2 = 0$). We get the following results using ladder analysis (assume an output current and work to the input):

$$y_{12}^a = \left. \frac{\mathbf{I}_2}{\mathbf{V}_1} \right|_{\mathbf{V}_2=0} = \frac{-1}{(2 + 2s/\Omega_0)} = \frac{-\Omega_0}{2(s + \Omega_0)}$$

5.4. CASCADE METHODS

$$y_{11}^a = \left|\frac{I_1}{V_1}\right|_{V_2=0} = \frac{(2s+\Omega_0)/\Omega_0}{2(s+\Omega_0)/\Omega_0} = \frac{(2s+\Omega_0)}{2(s+\Omega_0)}$$

$$y_{12}^b = \left|\frac{I_2}{V_1}\right|_{V_2=0} = \frac{-s/\Omega_0}{2+2\Omega_0/s} = \frac{-s^2}{2\Omega_0(s+\Omega_0)}$$

$$y_{11}^b = \left|\frac{I_1}{V_1}\right|_{V_2=0} = \frac{(s+2\Omega_0)/\Omega_0}{2(s+\Omega_0)/s} = \frac{s(s+2\Omega_0)}{2\Omega_0(s+\Omega_0)} \quad (5.52)$$

Next, the y-parameters for the parallel connected networks are obtained by adding (i.e., $y_{jk} = y_{jk}^a + y_{jk}^b$):

$$\begin{aligned}
y_{12} &= y_{12}^a + y_{12}^b = -\frac{\Omega_0}{2(s+\Omega_0)} - \frac{s^2}{2\Omega_0(s+\Omega_0)} \\
&= \frac{-(s^2+\Omega_0^2)}{2\Omega_0(s+\Omega_0)} \quad \text{and}
\end{aligned} \quad (5.53)$$

$$\begin{aligned}
y_{11} &= y_{11}^a + y_{11}^b = \frac{(2s+\Omega_0)}{2(s+\Omega_0)} + \frac{s(s+2\Omega_0)}{2\Omega_0(s+\Omega_0)} \\
&= \frac{(s^2+4\Omega_0 s+\Omega_0^2)}{2\Omega_0(s+\Omega_0)}
\end{aligned} \quad (5.54)$$

Finally, we can substitute the composite y-parameters into (5.51) to get the desired network function:

$$\begin{aligned}
\frac{V_2}{V_g} &= \frac{-G[-(s^2+\Omega_0^2)/(2\Omega_0(s+\Omega_0))]}{\left(G+\frac{4\Omega_0 s}{2\Omega_0(s+\Omega_0)}\right)\left(G+\frac{2s^2+4\Omega_0 s+2\Omega_0^2}{2\Omega_0(s+\Omega_0)}\right)} \\
&= \frac{+G(s^2+\Omega_0^2)}{[4\Omega_0 s+2\Omega_0 G(s+\Omega_0)][G+(s+\Omega_0)/\Omega_0]} \\
&\equiv \frac{G(s^2+\Omega_0^2)}{2(2+G)(s+a)(s+b)} \quad \text{where} \\
a &= (G+1)\Omega_0 > \Omega_0 \\
b &= G\Omega_0/(2+G) < \Omega_0
\end{aligned} \quad (5.55)$$

Example 5 - Use the Twin-T circuit in a cascade approach to synthesize the following network function. The network function can have a $\pm$ sign on the multiplicative constant, which is also required to have unity magnitude. There should be no undesired poles or zeros in the result.

$$H(s) = \frac{\pm(s^2+9)}{(s+1)^2} \quad (5.56)$$

Solution - Inspection of the Twin-T network function, (5.55), shows that in addition to the desired j-axis transmission zeros, the network generates two real poles at prescribed locations, one greater and one less than Ω_0. The required $H(s)$ has

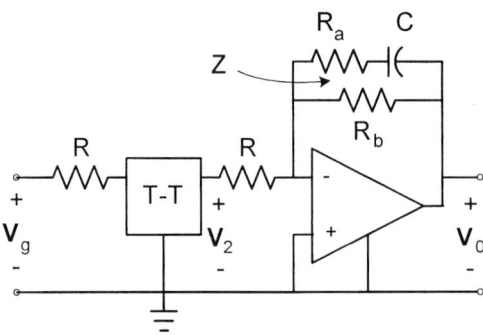

Figure 5.35. Proposed Twin-T (t-T) design for Example 5.

two real poles at $s = -1$ with $\Omega_0 = 3$. It is not possible to get two equal real poles from the given Twin-T circuit. However, it is possible to set the pole at $s = -b$ to -1 and then to use a cascade stage with a zero to cancel the undesired pole at $s = -a$ and to insert another pole at $s = -1$. The pole cancellation and shifting by insertion of another pole is a form of compensation. A proposed design topology is shown in Figure 5.35, where the Twin-T circuit is followed by an opamp stage, which provides the following transfer function:

$$H_A = -\frac{Z}{R} = \frac{-K(s+a)}{(s+1)} \tag{5.57}$$

where

$$Z = \frac{1}{(G_a + G_b)} \cdot \frac{(s + 1/R_a C)}{(s + 1/(R_a + R_b)C)}$$

The following results are obtained when the network function for the proposed design is equated to the desired network function:

$$\frac{G(s^2 + \Omega_0^2)}{2(2+G)(s+a)(s+b)} \cdot \frac{-G(s + 1/R_a C)}{(G_a + G_b)(s + 1/(R_a + R_b)C)} = \frac{-(s^2 + 9)}{(s+1)^2}$$

$$\therefore \Omega_0 = 3$$

$$G\Omega_0/(2+G) = b = 1 \text{ yields } G = 1$$

then with $(G+1)\Omega_0 = a$, $\quad a = 6$ is obtained.

Three independent equations for the three unknown parameters in the opamp feedback circuit are then obtained to complete the design:

$$1/(R_a C) = a = 6$$
$$1/((R_a + R_b)C) = 1$$
$$\frac{G^2}{2(2+G)(G_a + G_b)} = 1$$

These last conditions combine to yield $R_a = 7.2 \ \Omega$; $R_b = 36 \ \Omega$; and $C = 5/216$ F.

5.4. CASCADE METHODS

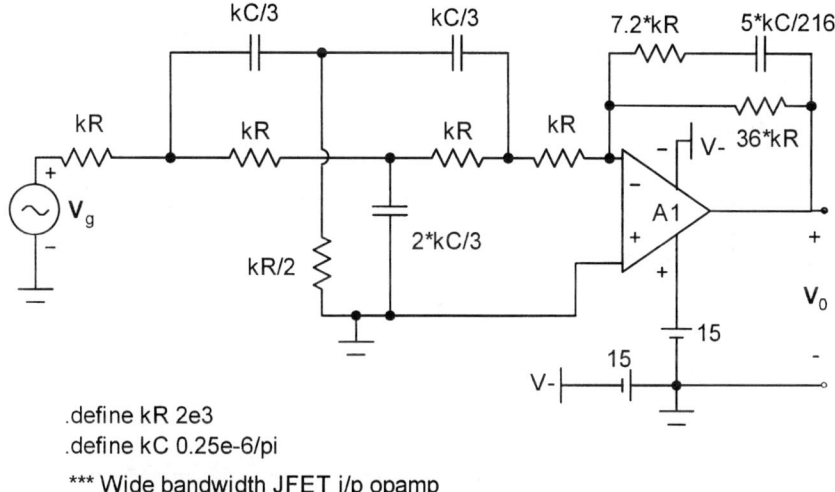

.define kR 2e3
.define kC 0.25e-6/pi
*** Wide bandwidth JFET i/p opamp
.MODEL LF351 OPA (LEVEL=2 TYPE=3 C=10P A=100K ROUTAC=50
ROUTDC=75 VOFF=1M IOFF=5P SRP=13MEG SRN=13MEG
IBIAS=10P VEE=-15 VCC=15 VPS=14 VNS=-14 GBW=4MEG)

Figure 5.36. Completed prototype design using a Twin-T j-axis zero.

There is no choice in the sequence of the stages. Since the Twin-T stage does not have a voltage source output, it needs to drive a high impedance amplifier or deliver its load current as input to the opamp feedback circuit, as shown in Figure 5.35.

The final prototype circuit is shown in Figure 5.36. Each component is labeled with its design value times its impedance scale factor. The resulting frequency response is shown in Figure 5.37 for the prototype scaled by $k_R = 2$ kΩ and $k_\omega = 2\pi 10^3$. The design fits the theoretical response very well all the way to about 1 MHz, where the circuit response drops away from the desired theoretical response. This divergence, beyond 100 kHz, is due to the fact that the opamp open-loop gain is cutting off. Use of a wider bandwidth amplifier would extend the range for achieving the desired response. A straightforward solution, requiring several opamps, could be obtained using the state variable method. See Problems 5.33 and 5.34.

5.4.4 The Biquad Circuit

A four opamp circuit, called a biquad (short for biquadratic), is a standard component used to generate cascade network functions of the form shown in (5.58):

$$H(s) = \frac{\mathbf{V}}{\mathbf{V}_g} = \frac{\pm(cs^2 + ds + e)}{(s^2 + as + b)} \qquad (5.58)$$

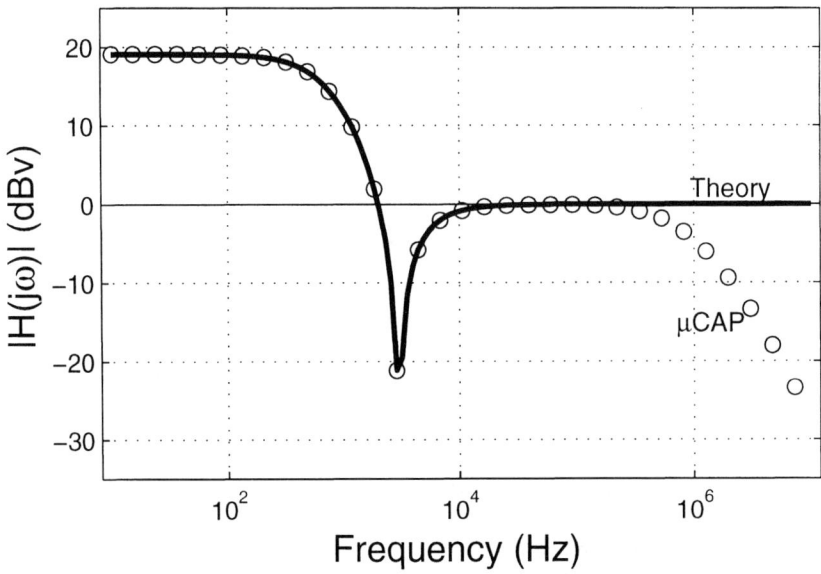

Figure 5.37. μCAP simulated response for the lowpass Example 5.

The circuit has the unique feature that parameters a and b in $H(s)$ are each controlled by a single resistor. Normally, as in a state variable circuit, two or more elements are required to adjust a single network parameter without affecting other parameters in the network function.

The design of the biquad is started with an all-pole function, as for the state variable approach, but the realization is accomplished by means of a different algebraic separation to obtain the all-pole response as follows:

$$\mathbf{V}_a = \frac{-K\mathbf{V}_g}{(s^2 + as + b)} \tag{5.59}$$

A gain parameter is included so as to provide explicit control of that parameter. The $(-)$ sign is chosen so as to use the simpler inverting multiplier summing circuit. The desired circuit form is accomplished by dividing both sides of the cleared form of (5.59) by the factor, $(s+a)$:

$$\frac{(s^2 + as + b)}{(s+a)}\mathbf{V}_a = \frac{-K\mathbf{V}_g}{(s+a)} \text{ or}$$

$$\left[s + \frac{b}{(s+a)}\right]\mathbf{V}_a = \frac{-K\mathbf{V}_g}{(s+a)} \text{ so}$$

$$s\mathbf{V}_a = \frac{-(K\mathbf{V}_g + b\mathbf{V}_a)}{(s+a)} \tag{5.60}$$

It can be recognized that when the factor, $z = 1/(s+a)$, represents a parallel RC network used as feedback around a summing opamp, then K and b are conductances

5.4. CASCADE METHODS

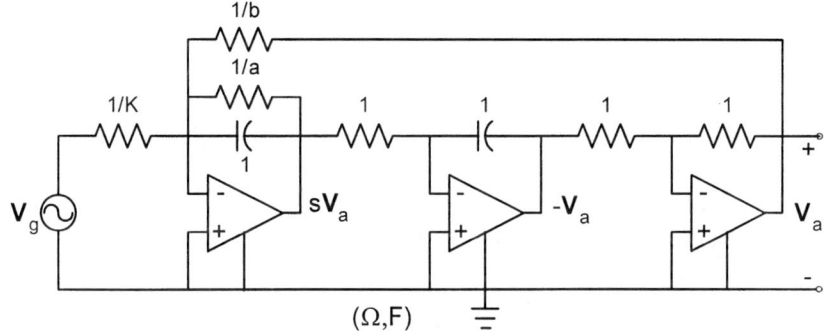

Figure 5.38. Circuit used to generate the biquad all-pole function, (5.59).

at the opamp input from the voltage sources, $\mathbf{V}_g$ and $\mathbf{V}_a$, respectively. An implementation of the all-pole function is shown in Figure 5.38. Note that in this circuit each parameter, a, b, and K, is controlled by an independent resistor value. The four-opamp version, shown in Figure 5.39, includes a summing output amplifier to form an arbitrary biquadratic ratio as follows:

$$\begin{aligned} \mathbf{V}_0 &= -(c_0 \mathbf{V}_g + c_1 s \mathbf{V}_a + c_2 \mathbf{V}_a) \text{ or} \\ \frac{-\mathbf{V}_0}{\mathbf{V}_g} &= (c_1 s + c_2)\frac{\mathbf{V}_a}{\mathbf{V}_g} + c_0 \end{aligned} \qquad (5.61)$$

Substitution for $\mathbf{V}_a/\mathbf{V}_g$ yields a final result as given by (5.62):

$$\begin{aligned} \frac{-\mathbf{V}_0}{\mathbf{V}_g} &= c_0 + \frac{(c_1 s + c_2)(-K)}{(s^2 + as + b)} \text{ and} \\ \frac{\mathbf{V}_0}{\mathbf{V}_g} &= \frac{-[c_0 s^2 + (ac_0 - c_1 K)s + (bc_0 - c_2 K)]}{(s^2 + as + b)} \end{aligned} \qquad (5.62)$$

An association of (5.62) to the desired $H(s)$ of (5.58) yields the following identifications for parameters, c, d, and e, where $H(s)$ is realized with a $(-)$ sign multiplier.

$$\begin{aligned} c &= c_0 \\ d &= ac_0 - c_1 K \\ e &= bc_0 - c_2 K \end{aligned} \qquad (5.63)$$

The conductances, $c_{0,1,2}$, can be set to different values to get a variety of network functions.

i - Bandpass:

$$H_{BP} = \frac{+as}{(s^2 + as + b)} \qquad (5.64)$$

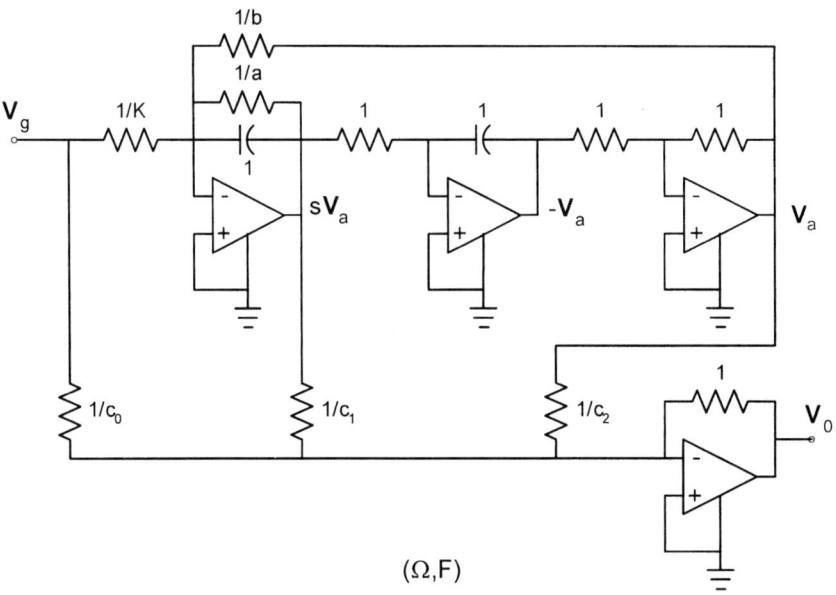

Figure 5.39. Four-opamp biquad circuit.

The output summing conductances (to yield H_{BP}) are obtained by associating the required form with the general function in (5.62), on a term-by-term basis, to obtain the following results: $-c_0 = 0$; $-ac_0 + c_1 K = +a$, so $c_1 = a/K$; and $-bc_0 + c_2 K = 0$, so $c_2 = 0$.
Note: A zero conductance is an open circuit, or no connection between nodes.

ii - Lowpass:

$$H_{LP} = \frac{b}{(s^2 + as + b)} \tag{5.65}$$

Term-by-term association yields the following conductances for H_{LP}: $-c_0 = 0$; $-ac_0 + c_1 K = 0$, so $c_1 = 0$; and $-bc_0 + c_2 K = +b$, so $c_2 = b/K$.

iii - Highpass:

$$H_{HP} = \frac{-s^2}{(s^2 + as + b)} \tag{5.66}$$

Term-by-term association yields the following conductances for H_{HP}: $+c_0 = +1$; $-ac_0 + c_1 K = 0$, so $c_1 = a/K$; and $-bc_0 + c_2 K = 0$, which yields $c_2 = b/K$.

iv - Bandstop:

$$H_{BS} = \frac{-(cs^2 + e)}{(s^2 + as + b)} \tag{5.67}$$

5.4. CASCADE METHODS

Term-by-term association yields the following conductances for H_{BS}: $+c_0 = +c$; $-ac_0 + c_1 K = 0$, so $c_1 = ac/K$; and $-bc_0 + c_2 K = e$, which yields $c_2 = (e + bc)/K$.

v - *Allpass*:

$$H_{AP} = \frac{-(s^2 - as + b)}{(s^2 + as + b)} \tag{5.68}$$

Term-by-term association yields the following conductances for H_{AP}: $+c_0 = +1$; $-ac_0 + c_1 K = +a$, so $c_1 = 2a/K$; and $-bc_0 + c_2 K = -b$, which yields $c_2 = 0$.

The functions, H_{BP} through H_{AP}, are advertised as filters when it is clear that they are all derived from the one general network function represented by (5.62). They are all simple functions, but it should be noted that the conductances are interdependent for the different functions. In addition, there can be serious problems related to gain distribution in the biquad circuit. Interior amplifiers, with respect to $\mathbf{V}_0$ in the circuit of Figure 5.39, may have considerably higher peak gain than $\mathbf{V}_0$, especially for high-Q poles.

Example 6 - Use the biquad circuit to design the network function given by (5.69):

$$H(s) = \frac{\pm K(s^2 + 10)}{(s^2 + s + 1)(s + 1)} \tag{5.69}$$

Specify K for maximum output so that the peak signal occurs on the output amplifier. Note that the specified network function is the same as that used for Example 5.

Solution - With $a = 1 = b = K$, the prototype biquad takes the form shown in Figure 5.40, and the network function, (5.62), specializes to (5.70):

$$\frac{\mathbf{V}_0}{\mathbf{V}_g} = \frac{-K_0[c_0 s^2 + (c_0 - c_1)s + (c_0 - c_2)]}{(s^2 + s + 1)} \tag{5.70}$$

Matching (5.70) to (5.69) requires that $c_0 = 1$, $c_1 = 1$, $c_2 = -9$, and $K_0 = K/(s+1)$. Since negative conductance is not possible, the sign change is accomplished by taking the output (for c_2) from $-\mathbf{V}_a$ instead of $+\mathbf{V}_a$. The change in connection is indicated with the dashed line in Figure 5.40. The requirement for K_0 to provide the real pole is easily met by using a parallel RC network with $R = K$ and $C = 1/K$, as shown in Figure 5.41.

In order to check the gain requirement, the circuit is simulated in μCAP using a circuit scaled by $k_R = 2000$ and $k_\omega = 2\pi 10^3$. The μCAP circuit is shown in Figure 5.42, and the simulated response is compared to theory, for $K = 1$, in Figure 5.43. Note the output stage provides the real pole to complete the design for this example. In the μCAP simulation, it was observed that the peak gain is obtained on the output amplifier, $H4$, for any K value greater than $1/8$. The plot in Figure 5.43 uses $K = 1$ in order to compare to the network function used for the previous example.

This design provides a j-axis zero with as good a null depth as was obtained for the passive Twin-T used in Example 5. The frequency response fits theory over

196   CHAPTER 5. OPERATIONAL DESIGN OF ACTIVE FILTERS

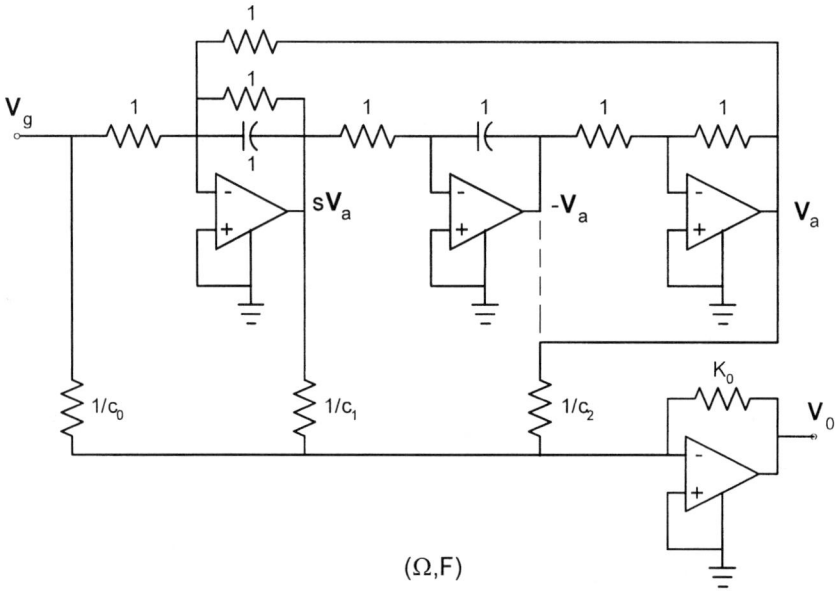

Figure 5.40. Prototype circuit for Example 6.

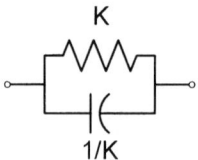

Figure 5.41. Realization of K_0 as a frequency-dependent factor.

5.4. CASCADE METHODS

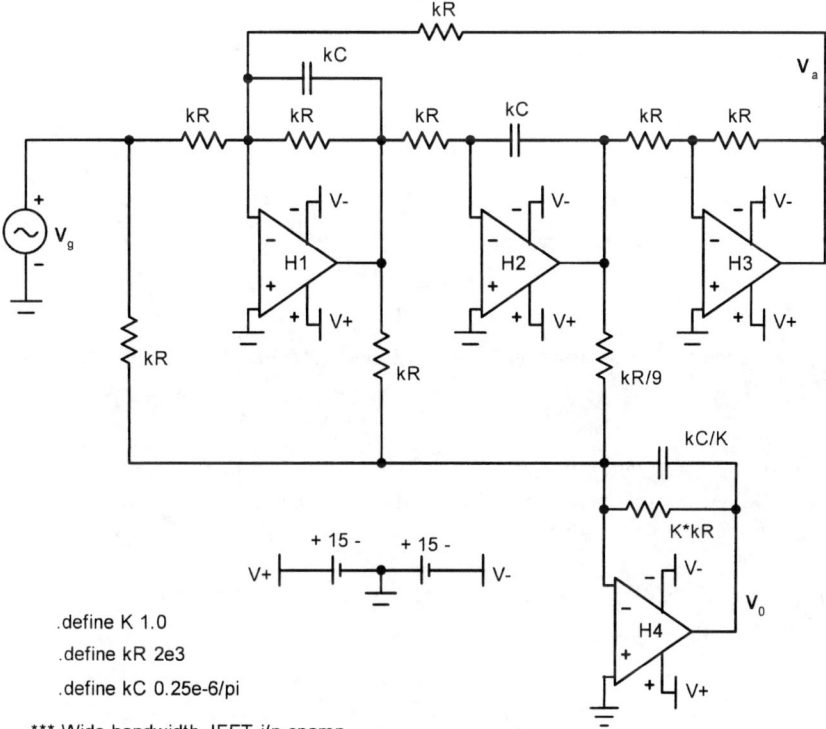

.define K 1.0
.define kR 2e3
.define kC 0.25e-6/pi

*** Wide bandwidth JFET i/p opamp
.MODEL LF351 OPA (LEVEL=2 TYPE=3 C=10P A=100K ROUTAC=50 ROUTDC=75 VOFF=5M IOFF=25P SRP=13MEG SRN=13MEG IBIAS=50P VEE=-15 VCC=15 VPS=14 VNS=-14 GBW=4MEG)

Figure 5.42. Final design for Example 6, $K = 1$.

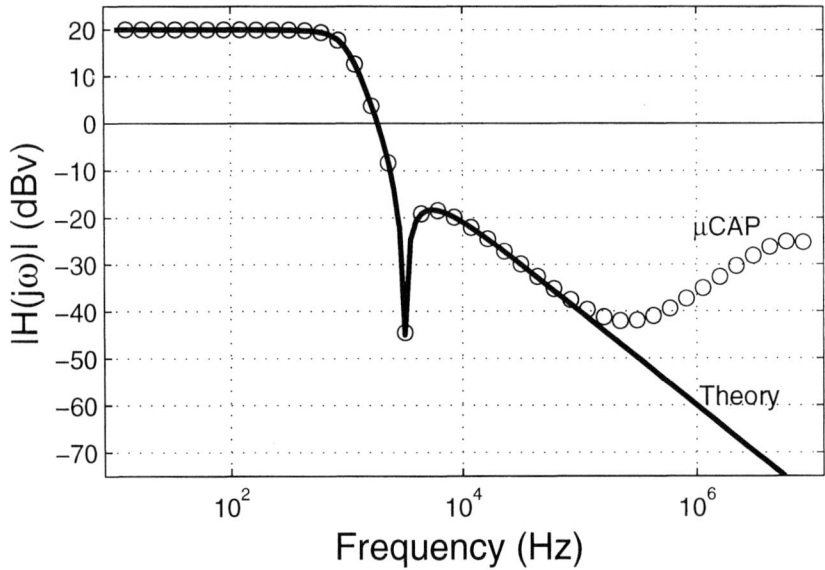

Figure 5.43. Comparison to theoretical response for Example 6.

about the same frequency range, but the spurious response in the stopband is everywhere below the peak theoretical gain allowed in the stopband. Thus, the circuit is useful over a wide frequency range.

Example 7 - We conclude this chapter by designing a state variable filter to provide the all-pole response specified in (5.59):

$$\frac{\mathbf{V}_a}{\mathbf{V}_g} = \frac{-K}{(s^2 + as + b)}$$

It will then be shown that the state variable form of the biquad is actually easier to use to provide the various biquadratic filter functions i through v in the list developed for the biquad circuit.

Solution - Start by forming $s^2 \mathbf{V}_a$ from the given all-pole function, and then realize the sum using an inverting summing amplifier:

$$s^2 \mathbf{V}_a = -K\mathbf{V}_g - (as + b)\mathbf{V}_a$$

Following the state variable procedure, two integrators are used to obtain $s\mathbf{V}_a$ and $\mathbf{V}_a$ for the realization of the all-pole network function. The resulting prototype is shown in Figure 5.44. Compared to the traditional biquad, shown in Figure 5.39, this solution requires an additional amplifier to provide the full biquadratic function. However, note the simplicity for obtaining desired numerator coefficients in

5.4. CASCADE METHODS

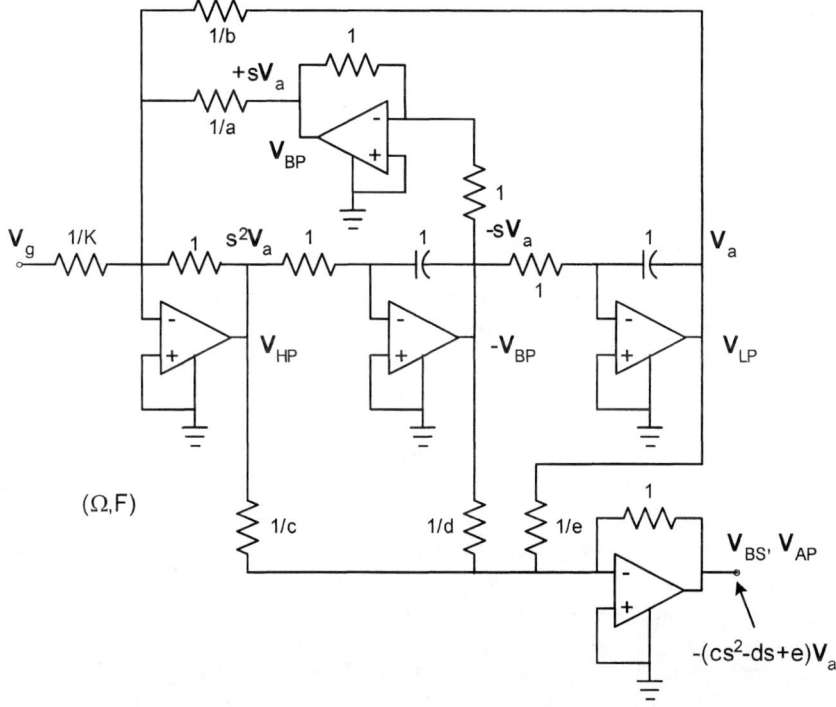

Figure 5.44. The state variable design for biquadratic filter functions.

this solution versus that in Figure 5.39. If it is desired to control each parameter of the biquad function with a single resistor value, the circuit in Figure 5.44 is superior to that in Figure 5.39. The five filter functions are provided as follows (at the node voltages shown with corresponding subscripts). Each output is taken off of an amplifier output and is thus off of a voltage source.

i-Bandpass:

$$H_{BP} = \frac{s\mathbf{V}_a}{\mathbf{V}_g} = \frac{-Ks}{s^2 + as + b}$$

ii-Lowpass:

$$H_{LP} = \frac{\mathbf{V}_a}{\mathbf{V}_g} = \frac{-Ks}{s^2 + as + b}$$

iii-Highpass:

$$H_{HP} = \frac{s^2\mathbf{V}_a}{\mathbf{V}_g} = \frac{-Ks^2}{s^2 + as + b}$$

iv-Bandstop ($c = 1, d = 0$):

$$H_{BS} = \frac{-(s^2 + e)\mathbf{V}_a}{\mathbf{V}_g} = \frac{+K(s^2 + e)}{s^2 + as + b}$$

v-All-pass ($c = 1, d = a, e = b$):

$$H_{AP} = \frac{-(s^2 - as + b)\mathbf{V}_a}{\mathbf{V}_g} = \frac{+K(s^2 - as + b)}{s^2 + as + b}$$

It should be clear that this form of the biquad is straightforward to use in order to obtain a biquadratic network function of any type. Note that the circuit in Figure 5.44 is configured for $-d$ coefficients of s in the general numerator (zeros) form. Whenever $+d$ coefficients are taken, the conductance, d, is connected to V_{BP} rather than at $-V_{BP}$, as shown in the schematic.

In this chapter we have looked at designing active filter prototype networks from specified network functions. We looked at two basic ways to accomplish the design. The first method is the state variable approach, which uses analog operations to build the poles from integrator outputs, then uses these responses to form the desired zeros. The method realizes the network function as one complete entity featuring negative feedback loops. The second method uses a cascade approach, where the desired network function is obtained as a product of smaller network functions. Thus, the critical frequencies are generated independently of each other by using separate networks. The cascade approach generally allows fewer amplifiers to be used in the synthesis of a given network function. The S-K and biquads were examples of circuits used to generate complex-pole functions.

In the next chapter, we will take a look at network function sensitivity to element value errors and the effect of temperature, and we will develop a specific architecture that exhibits superior insensitivity to parameter variations. The architecture, referred to as the leapfrog circuit, has much in common with the state variable circuit.

5.5 Problems

5.1 The differential amplifier, shown in Figure 5.45, is sometimes referred to as a transconductance amplifier/integrator. Using the approximation, $i_0 \approx I_k v_i/(2\gamma V_T) = g_i v_i$, show how $\mathbf{V}_0(s)$ is related to $\mathbf{V}_i(s)$ so as to obtain an integrator relationship and show how this unit could be used as a building block to design state variable filters.

5.2 Realize a state variable filter for the following network function using an inverting rather than the conventional noninverting summing amplifier. Specify the inputs for the circuit of Figure 5.46, $\mathbf{V}_{1,2,3,4}$, with appropriate s^k factors and signs, so as to properly complete the sum for $s^3\mathbf{V}_a$ (e.g., $\mathbf{V}_2 = (+)$ or $(-)2s^2\mathbf{V}_a$).

$$\mathbf{V}_a = \frac{\mathbf{V}_g}{s^3 + 2s^2 + 2s + 1}$$

5.5. PROBLEMS

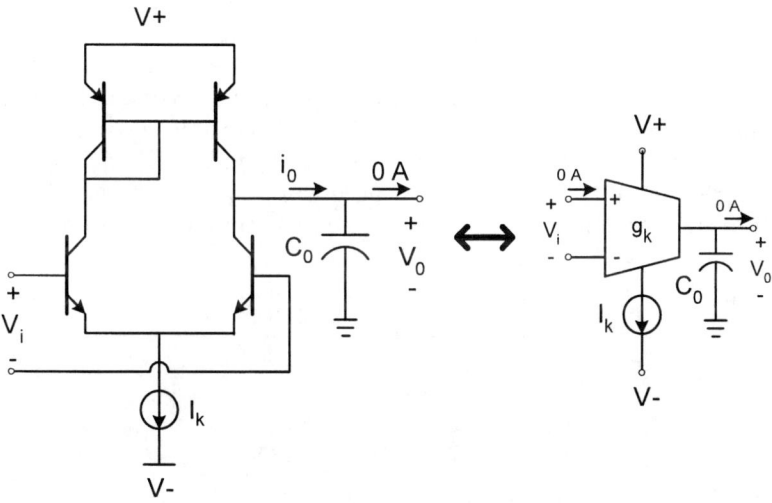

Figure 5.45. Circuit for Problem 5.1.

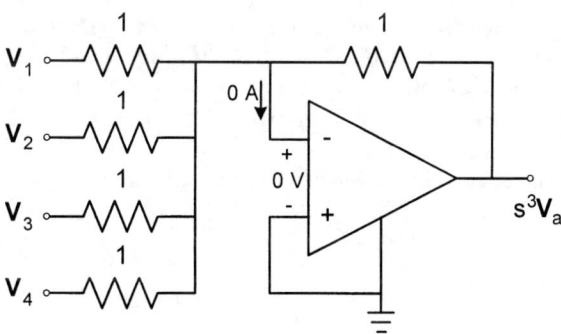

Figure 5.46. Circuit for Problem 5.2.

5.3 The following transfer function, using a prototype frequency scale, is required for an instrumentation project. Submit a design incorporating opamps and RC networks to realize the required transfer function. The final design should be impedance scaled to 1 kΩ, and the prototype frequency of 1 rad/sec should be scaled to 1 kHz. The customer is worried about cost and would like all capacitors to be equal if possible. Your job is to achieve this as much as possible, while satisfying the requirement to equalize all amplifier peak gains to within ± 3 dB. Note that the overall gain requirement is specified in the given transfer function, so you may need an extra amplifier to meet both the specification and a distributed gain performance.

$$T(s) = \frac{0.1(s^2 + 0.25)(s^2 + 4)}{(s^2 + 0.1s + 0.64)(s^2 + 0.15625s + 1.5625)}$$

5.4 Realize the following network function using the state variable method and an inverting summing multiplier to realize $s^2 \mathbf{V}_a$ instead of the noninverting amplifier topology used in the text.

$$\frac{\mathbf{V}_a}{\mathbf{V}_g}(s) = \frac{-20}{s^2 + 4s + 20}$$

5.5 Design a state variable active filter for the following network function using an inverting, in place of a noninverting, summing multiplier at the input to the filter.

$$H(s) = \frac{\mathbf{V}}{\mathbf{V}_g} = \frac{(\pm)5s^2}{s^4 + s^3 + 2.5s^2 + s + 1}$$

5.6 In the state variable circuit in Figure 5.47, the peak gain at Amplifier 4 needs to be lowered by a factor of 4 (12 dB). Identify which resistors need to be changed and specify their new values in order to meet this requirement without changing the remaining amplifier output network functions.

5.7 The circuit in Figure 5.48 has been submitted as a design for the following network function for $\mathbf{V}_2/\mathbf{V}_1$. What is wrong with the network, and how would you realize the required function?

$$\frac{\mathbf{V}_2}{\mathbf{V}_1} = T(s) = \frac{10(s + 6000)}{(s + 1000)^3}$$

The following LP network function is given for Problems 5.8 and 9:

$$T(s) = \frac{1}{5.06s^5 + 6.26s^4 + 10.2s^3 + 7.15s^2 + 4.10s + 1.00} \quad (5.71)$$

5.5. PROBLEMS

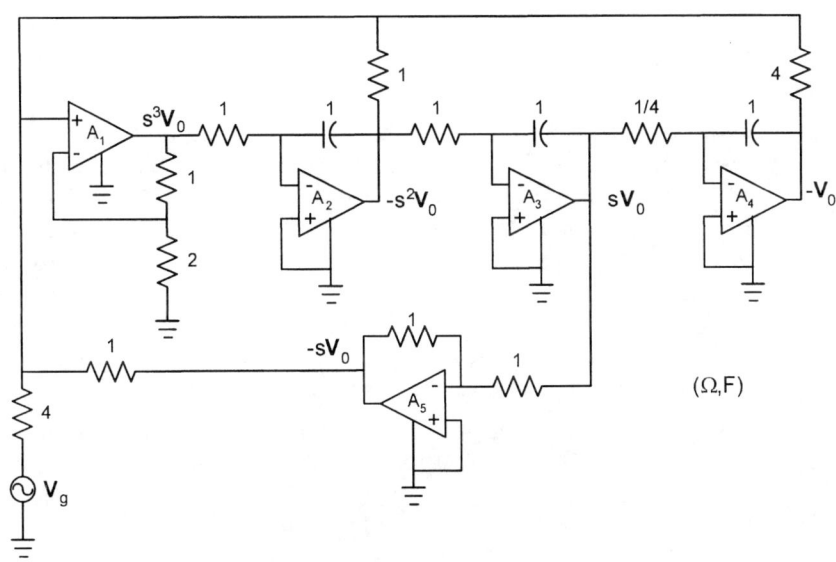

Figure 5.47. Circuit for Problem 5.6.

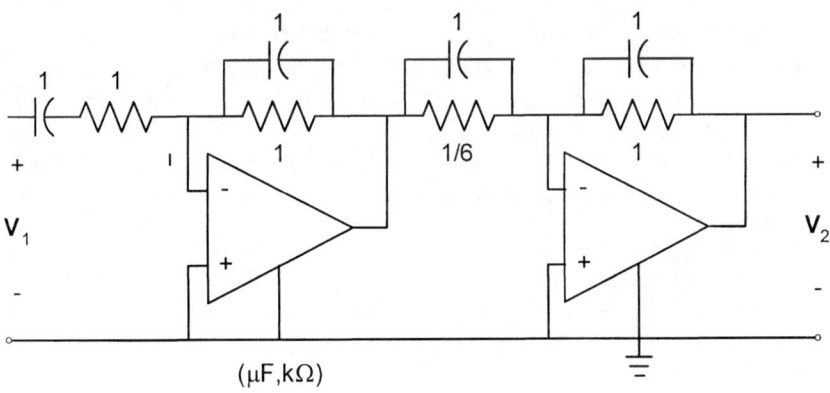

Figure 5.48. Circuit for Problem 5.7.

5.8 Apply a BP transform to the LP network function, (5.71), so as to obtain a filter Q of 6 at $\omega_0 = 1$ (i.e., a prototype BP function, in s'). Design a state variable filter for your answer, and scale it to an impedance level of 2 kΩ and a center frequency of 12 kHz. Use μCAP and Matlab to check the frequency response and accuracy of your design.

5.9 This problem illustrates a different way to get a state variable active BP filter with one-half the number of integrators as used in Problem 5.8. Design a 1 Ω – 1 rad/sec state variable filter for the LP function given in (5.71). Frequency-scale this result using the scale factor, $k_\omega = 1/A$. Next, perform an LP-to-BP frequency transform, $s = s' + 1/s'$, to get a 1 Ω –1 rad/sec BP prototype. Scale this last result by the same factors used in Problem 5.8, and check this filter response with μCAP. Note that the capacitor integrators have been converted to parallel LC (bandpass) integrators. Compare the performance of this design with the one obtained for Problem 5.8, and discuss the results.

The following LP network function is given for Problems 5.10 and 11.

$$T(s) = \frac{(s^2 + 6.25)}{10(s + 0.5)(s^2 + 0.4s + 1.25)} \tag{5.72}$$

5.10 Apply a BP transform to the LP network function, (5.72), so as to obtain a filter Q of 5 at $\omega_0 = 1$ (i.e., a prototype BP function, in s'). Design a state variable filter for your answer, and scale it to an impedance level of 5 kΩ and a center frequency of 10 kHz. Use μCAP and Matlab to check the frequency response and accuracy of your design.

5.11 This problem illustrates a different way to get a state variable active BP filter with one-half the number of integrators as used in Problem 5.10. Design a 1 Ω –1 rad/sec state variable filter for the LP function given in (5.72). Frequency-scale this result using the scale factor, $k_\omega = 1/Q$. Next, perform an LP-to-BP frequency transform, $s = s' + 1/s'$, to get a 1 Ω –1 rad/sec BP prototype. Scale this last result by the same factors used in Problem 5.10 and check this filter response with μCAP. Note that the capacitor integrators have been converted to parallel LC (bandpass) integrators. Compare the performance of this design with the one obtained for Problem 5.10 and discuss the results.

5.12 Use cascade methods with S-K stages to design the following network function. Choose the order of the stages so that the output amplifier has the maximum gain. Check the gain response on μCAP using $k_R = 5$ kΩ and $k_\omega = 2\pi 10^3$. Specify K.

$$H(s) = \frac{K}{(s^2 + s + 0.25)(s^2 + 0.5s + 1)}$$

5.13 Use cascade methods and the Twin-T circuit to design the following network function. Check the response on μCAP by scaling the prototype with $k_R = 5$ kΩ and 1 rad/sec to 1 kHz. Specify K, and order stages so that the output

5.5. PROBLEMS

amplifier has maximum gain. Specify the range of frequency where your design performance is closest to theoretical response. The design should be doable with just two stages.

$$H(s) = \frac{K(s^2 + 4)}{(s+1)(s+2)}$$

5.14 Repeat Problem 5.13 using a state variable design. Compare the resulting frequency response performance to the theoretical magnitude response function versus frequency (on the same graph).

5.15 The purpose of this problem is to obtain a gain-corrected state variable design for the following network function:

$$\frac{V_1}{V_g} = \frac{s^2 + 16}{4(s+1)(s^2 + s + 4)}$$

(a) Design a first-cut prototype solution for the required network function, and use Matlab to predict frequency responses for each opamp output. Specify suitable gain-reduction factors to use at each opamp so that each peak gain is brought to within 0 to -4 dB of the peak desired output gain. Plot theoretical gains versus frequency for a (correctly) scaled second-cut design.

(b) Obtain a circuit design for the second-cut solution to accomplish the desired gain redistribution. Scale the circuit to check your design on μCAP by using $k_R = 5$ kΩ and $k_\omega = 2\pi 10^3$ (i.e., 1 Ω to 5 kΩ and 1 rad/sec to 1 kHz). Use μCAP to obtain frequency responses for each opamp output. Over what frequency range does the circuit perform according to theoretical requirements?

5.16 Repeat Problem 5.15 for the following network function:

$$\frac{V_2}{V_g} = \frac{(s^2 + 6.25)(s^2 + 25)}{(s^2 + s + 9.25)(s^2 + s + 16.25)}$$

5.17 Design a biquad circuit to provide the following prototype network function. Specify all element values. What type of filter (e.g., LP, BP) does this function represent? Explain your answer.

$$T(s) = \frac{\pm(s^2 + 1)}{(s^2 + 2s + 5)}$$

5.18 (a) Find the network function between the output of Amplifier 2 and the source voltage for the circuit in Figure 5.49.

(b) Show how you would modify the circuit to lower peak gain by 12 dB at the outputs of Amplifiers 3 and 4 without changing the network function and the gains at the remaining amplifier outputs.

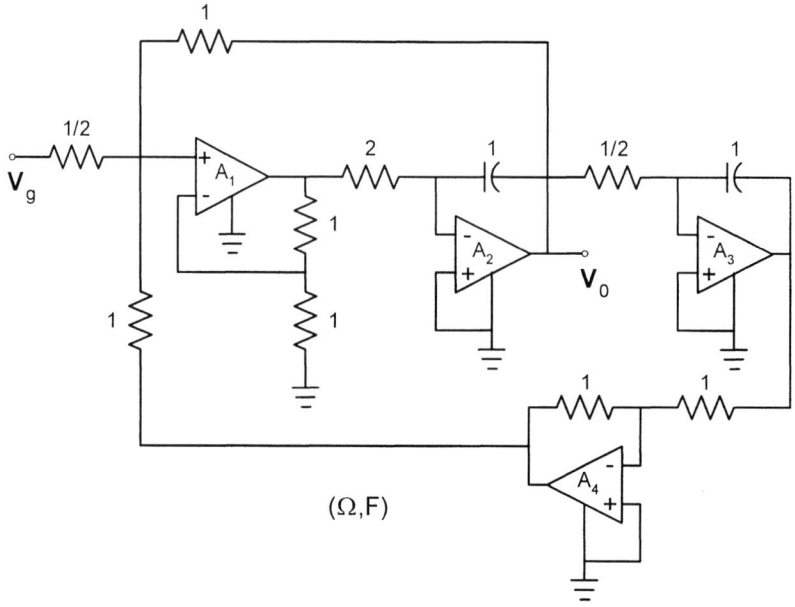

Figure 5.49. Circuit for Problem 5.18.

5.19 Use cascade circuit design procedures to design a prototype circuit for the following network function. Determine the best order for the sections in the cascade, while providing a buffered input and a voltage source at the output.

$$H(s) = \frac{\pm 10(s^2 + 9)}{(s+2)(s^2 + 2s + 5)}$$

5.20 Design a prototype state variable opamp circuit to realize the following network function:

$$\frac{V_0}{V_g} = \frac{\pm(s^2 + 9)}{(s+1)(s^2 + s + 1)}$$

(a) Distribute loop gains so that the output amplifier has the maximum gain and all other amplifier peak gains fall within 0 and -6 dB of this value. The polarity of the network function is not considered a critical specification, so either sign ($+$ or $-$) is appropriate.

(b) Use μCAP to check your design by using LF351 opamps and a scaled version of the design obtained in part a. Choose a scale frequency factor so the transmission null occurs at 12 kHz, and the 1-Ω impedance is scaled to 5 kΩ. Check the peak gain response for each opamp. Run a response out to at least 10 MHz to check the input-to-output result for your design.

5.5. PROBLEMS

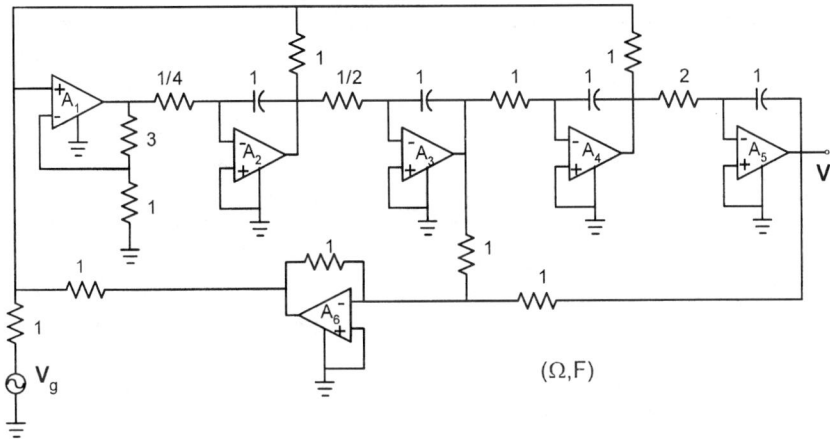

Figure 5.50. Circuit for Problem 5.21.

(c) Frequency-scale the desired function in Matlab, and compare the result to the μCAP simulation out to a bandwidth of at least 50 kHz.

(d) Discuss all results.

5.21 In the circuit in Figure 5.50, the peak values at Amplifiers 2, 3, 4, and 6 each need to be reduced by 6 dB, without affecting the gains at Amplifiers 1 and 5. Show how you would modify the circuit to meet this requirement.

5.22 A prototype filter function is given as follows:

$$\frac{V}{V_g}(s) = \frac{K(s^2 + 9)}{(s^2 + 2s + 2)(s^2 + 2s + 5)}$$

(a) Use the state variable method to design an equal capacitor valued active filter to realize the given function. Redistribute gain so that the peak gains are all within 0 and -3 dB of the output amplifier peak gain.

(b) What are limits on the value of K so as to maintain fully linear operation of the filter within the gain tolerance given in part a? This is a thought question. K may be partly absorbed in the initial design and may also be used at the input to adjust the overall function gain. Use the condition that no opamp has more than 20 dB of gain versus frequency.

(c) Scale the the prototype circuit so that the minimum load resistance on any opamp is greater than 2 kΩ and the 1 rad/sec prototype frequency is shifted to 1 kHz. Use μCAP to simulate the scaled circuit and compare the amplifier output frequency responses to the part a gain analyses.

5.23 A high-Q network function is given as follows:

$$\frac{V_0}{V_g} = \frac{\pm 2}{s^2 + 0.02s + 1}$$

(a) Design a state variable filter to realize the function. Scale the resulting design to provide peak response at 100 Hz along with a 10-kΩ impedance scale factor for the circuit. Use μCAP to check performance as follows:

  i. First, use $\pm 1\%$ element (Lot) values on all RC components, and obtain frequency plots of magnitude and phase over $90 < f < 110$ Hz for a run of 10 cases at $T = +25°$C. Use Monte Carlo techniques for the run so that all 10 responses are shown on the same graph.

  ii. Assign temperature coefficients of $+350$ ppm/$°$C to each resistor and -200 ppm/$°$C to each capacitor. Obtain frequency plots of magnitude and phase for the network function over the given frequency range for $0 \leq T \leq 75°$C in steps of $25°$C for the single case of nominal element values. Use a 10-dB range for the magnitude plot. *Note*: You will need to modify your library opamp model so that it has both Lot tolerances and temperature coefficients for both the low-frequency gain, A, and the unit-gain bandwidth, GBW, parameters.

(b) Repeat part a using the S-K circuit to realize the given network function.

5.24 Use cascade methods and simple RC impedance networks, as discussed in the text, to design the following required servo prototype compensator network function. The gain factor, K, should be a variable in your design. Use μCAP and Matlab to check your design on a scaled circuit such that the 1 rad/sec prototype frequency goes to 1 kHz, a practical impedance level is used, and $K = 10$. Note that two opamps should be sufficient, and the cascade order should be chosen so that peak amplifier gain occurs at the output.

$$H(s) = \frac{V_{out}}{V_{in}} = \frac{K(s+5)(s+1/5)(s+8)(s+1/8)}{(s+5/4)(s+4/5)(s+2)(s+1/2)}$$

5.25 Design a prototype active filter for the following network function using a minimum number of opamps. Either sign (+ or −) is allowed, but a Twin-T circuit is not desired. It is not necessary to redistribute peak gain for this problem since doing so would not change the number of opamps required in the design.

$$T(s) = \frac{\pm(s^2 + 16)}{(s^2 + s + 1)(s^2 + 2s + 5)}$$
$$= \frac{\pm(s^2 + 16)}{s^4 + 3s^3 + 8s^2 + 7s + 5}$$

5.26 Use a biquad active filter to realize the following delay compensation (allpass) network function. A -1 multiplier, times $T(s)$, is not desirable.

$$T(s) = \frac{s^2 - s + 1}{s^2 + s + 1}$$

5.5. PROBLEMS

5.27 Realize each of the following compensator Bode plot representations using cascade methods. Use as few opamps as possible and arrange the order of the pole-zero factors so as to optimize the dynamic range performance of the resulting structure (i.e., ensure that the output amplifier will control the peak-to-peak input signal limit). Use μCAP to check each design for its fit to the specified asymptotic response, and confirm that your designs are correct. $x = log_{10}(f)$ for both problems, and f is in hertz.

(a) $|H|_{dB} = 10 - 20(x - 2.5)u(x - 2.5) + 20(x - 3)u(x - 3)$
$\qquad -20(x - 4)u(x - 4) + 20(x - 4.5)u(x - 4.5)$ dB

(b) $|H|_{dB} = 20(x - 1.5)u(x - 1.5) - 20(x - 2)u(x - 2)$
$\qquad -20(x - 3)u(x - 3) + 20(x - 4)u(x - 4)$ dB

5.28 Design a prototype cascade circuit to obtain the following transfer function using as few opamps as possible. Scale the prototype design so that the transmission zero will occur at 10 kHz and 1 Ω goes to 1 kΩ. Use μCAP to check the design performance versus frequency.

$$H(s) = \frac{s^2 + 10}{(s + 5)(s^2 + s + 2)}$$

5.29 Use the state variable method and design a prototype active filter for the following transfer function:

$$H(s) = \frac{V}{V_g} = \frac{s^2 + 25}{(s^2 + 2s + 2)(s^2 + 2s + 10)}$$

(a) Check the peak gain of each opamp versus ω, and distribute the loop gain so that all internal opamps have less gain than the output opamp. The peak gain spread should not be greater than 6 dB.

(b) Use Matlab (or MathCad) to obtain a nominal gain plot, and compare the result to a μCAP simulation by scaling the prototype circuit from 1 rad/sec to 1 kHz and 1 Ω to 1 kΩ. The simulated response should be observed out to 100 MHz to check the effect of various opamp models on the design performance.

5.30 Modify the circuit in Figure 5.51 to redistribute the gain so that the output of Amplifiers 2, 3, and 4 will all increase by a factor of 4 (12 dB) while keeping the same gain and output function of s on Amplifier 1. Show all changes required to modify the circuit appropriately.

5.31 The circuit shown in Figure 5.52 is one stage from a prototype cascade filter design. Amplifier A_1 does not perform well due to the high-frequency short circuit provided by its load. Redesign the load stage so that the same transmission function is obtained while getting rid of the capacitive short circuit on A_1. Specify all element values.

210   CHAPTER 5. OPERATIONAL DESIGN OF ACTIVE FILTERS

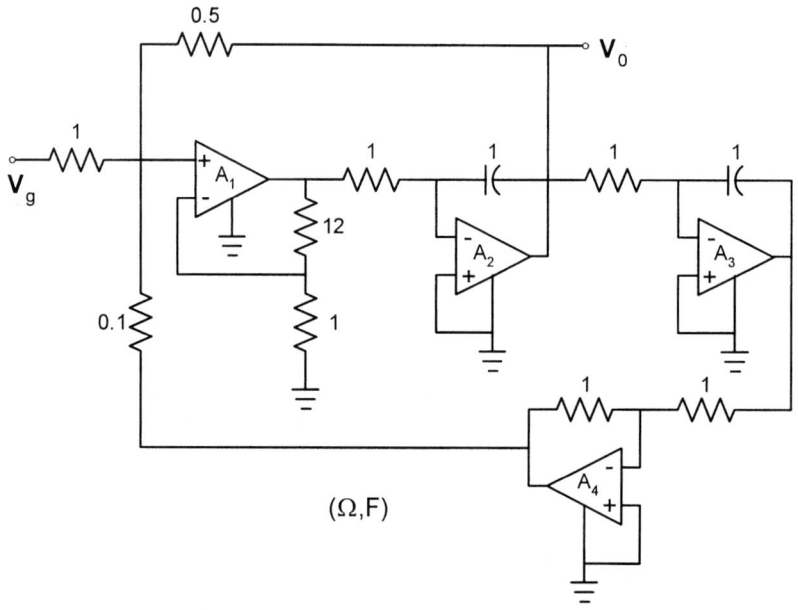

Figure 5.51. Circuit for Problem 5.30.

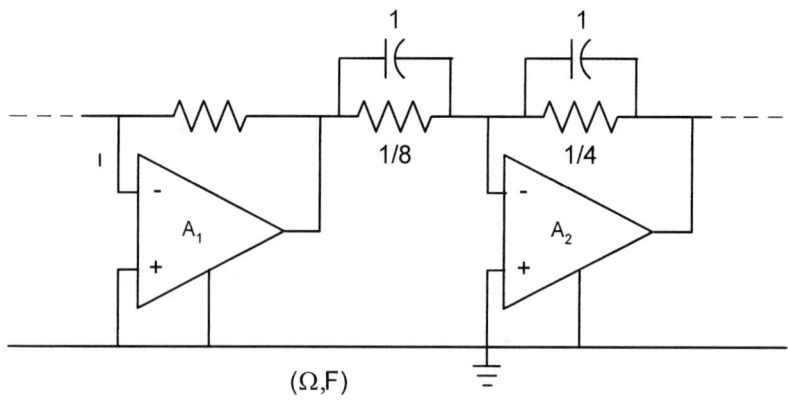

Figure 5.52. Circuit for Problem 5.31.

5.5. PROBLEMS

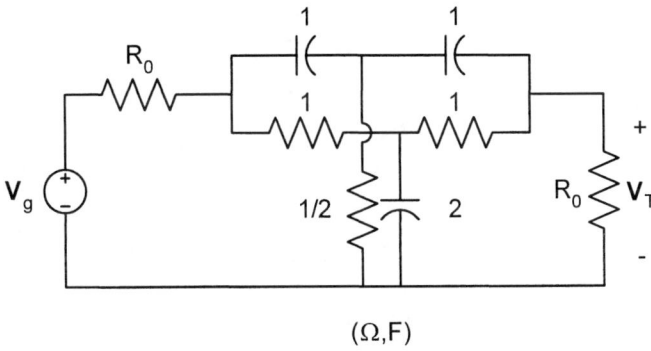

Figure 5.53. Circuit for Problem 5.32.

5.32 The following network function applies to the Twin-T circuit shown in Figure 5.53:
$$\frac{V_T}{V_g} = \frac{(s^2+1)}{2(s+1+G_0)((1+2R_0)s+1)}$$

(a) Frequency-scale the given circuit so that the j-axis zeros occur at $s' = \pm j4$ rad/sec. Specify the transformed network function for the scaled network, and draw a new circuit diagram for the circuit. Specify all element values.

(b) Use the result from part a to obtain a prototype S-K cascade design for the following desired network function. *Note*: It is necessary to choose a value for R_0, so use the choice to your advantage.
$$T(s) = \frac{V_{out}}{V_{in}} = \frac{s^2+16}{8(s+1)(s^2+2s+2)}$$

5.33 Use the state variable method to design an active filter for the network function used in Example 5.

5.34 Use the state variable method to design an active filter for the following transfer function:
$$\frac{V}{V_g} = \frac{K(s^2+10)}{(s^2+s+1)(s+1)}$$

(a) Obtain a design, and specify the maximum value of K for the function so as to obtain the maximum output signal with fully linear operation of the filter. Plot the required transfer function magnitude in decibels versus frequency, and confirm that your design produces the desired response. Scale to 1 kΩ and 1 kHz, and use μCAP to check the design.

(b) Redo part a using cascade methods with just two opamps. Select the order of the stages to ensure that the output amplifier is the first to clip.

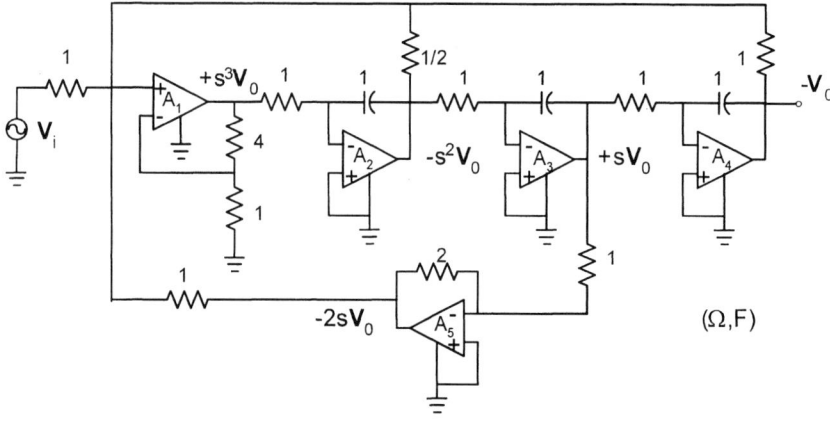

Figure 5.54. Circuit for Problem 5.35.

5.35 The circuit given in Figure 5.54 realizes a specified LP prototype network function, $T(s) = -\mathbf{V}_0/\mathbf{V}_i$. Show how to modify the circuit without adding any opamps so that it is possible to obtain a gain of 10 times $T(s)$.

5.36 The filter shown in Figure 5.55 is a state variable prototype for a specified network function, $T(s) = \mathbf{V}_2/\mathbf{V}_1$. Show how you would obtain the function $T_2 = (s^2 + 2s + 10)T(s)$ from the given filter.

5.37 Use the state variable method to design an active filter for the following transfer function:
$$\frac{\mathbf{V}}{\mathbf{V}_g}(s) = \frac{K(s^2 + 10)}{(s+2)(s^2 + s + 4.25)}$$

(a) Use as few opamps as possible, and specify the maximum value of K for the function so as to obtain the maximum output signal with fully linear operation of the filter. Plot the theoretical magnitude response (in decibels) versus frequency and confirm that your design produces the desired response. Use μCAP with the network scaled to 1 kΩ and 1 kHz to check the design performance. Use a wideband opamp for the design to get best frequency-range performance for the circuit.

(b) Redo part a using cascade methods with just two opamps. Select the stage order so that the output amplifier will be the first to clip for maximum input signal. Use the same procedure to check the design with μCAP.

5.38 Obtain an active filter design for each of the Bode compensator asymptotic frequency response curves, (a) and (b), shown in Figure 5.56. Use as few opamps as possible, and arrange the sequence of pole-zero pairs so as to optimize the dynamic range performance for each design.

5.5. PROBLEMS

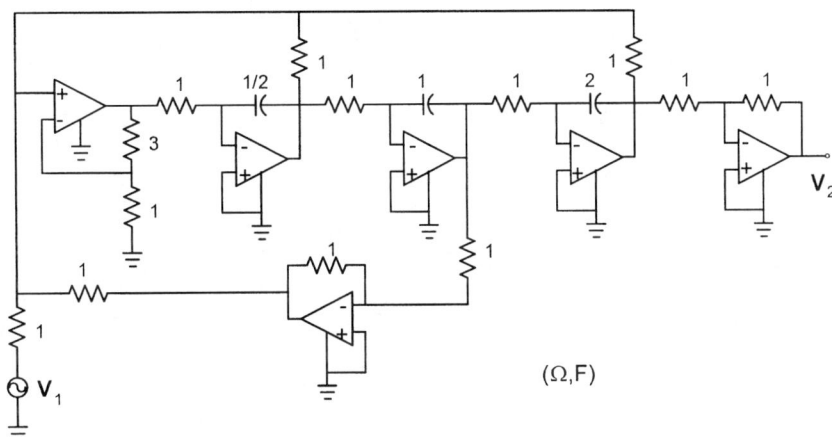

Figure 5.55. Circuit for Problem 5.36.

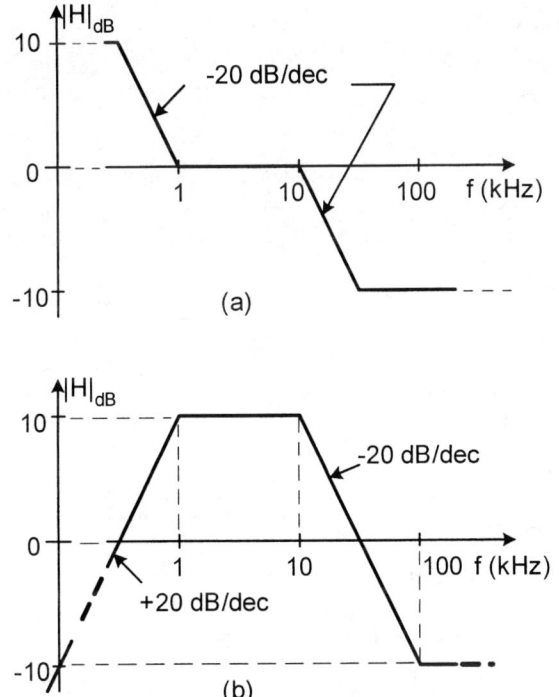

Figure 5.56. Response curves, (a) and (b), for Problem 5.38.

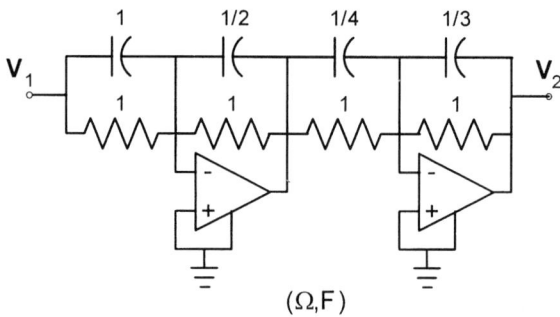

Figure 5.57. Circuit for Problem 5.39.

5.39 The prototype circuit shown in Figure 5.57 was designed to provide the transmission function, $\mathbf{V}_2/\mathbf{V}_1$:

$$\frac{\mathbf{V}_2}{\mathbf{V}_1} = \pm \frac{(1+s/1)(1+s/4)}{(1+s/2)(1+s/3)}$$

(a) Explain why this is a poor design.

(b) Redesign the prototype to provide the same transfer function (times + or −) using a single opamp and only two capacitors (plus resistors of course!).

5.40 Design a prototype circuit for the following equalizer transfer function, $T(s)$. Parameters a and b must be adjustable by means of a single resistor for each parameter, and the adjustment must not otherwise affect $T(s)$.

$$T(s) = \pm \frac{s^2 - bs + 1}{s^2 + as + 1}$$

5.41 Use a biquad cascade procedure to design a prototype for the following network function. Specify all element values.

$$T(s) = \frac{\pm 1.6(s^2 + 25)}{(s+2)(s^2 + 4s + 20)}$$

5.42 An alternate complex-pole generating circuit is shown in Figure 5.58. As compared to the S-K circuit, it does not employ positive feedback in its design. Analyze the circuit to obtain its network function, $\mathbf{V}_2/\mathbf{V}_1$, in terms of circuit parameters, $R_{1,2,3}$ and $C_{1,2}$. Define the quadratic factor parameters, 2α and ω_0^2, from your answer, and discuss a tuning procedure for setting 2α and ω_0 in production.

5.43 Design a state variable circuit to obtain the network function, (5.29), of Example 2. Scale the circuit in the same way as was done in the example, and

5.5. PROBLEMS

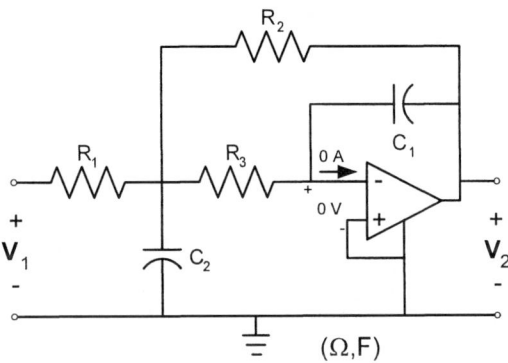

Figure 5.58. Circuit for Problem 5.42.

compare its response to that obtained for Design c, the single amplifier result. Discuss pros and cons for each design.

5.44 Obtain a state variable solution for the network function of Example 5 (as follows). Use μCAP, and compare the frequency response to that obtained for the cascade circuit using the Twin-T circuit with j-axis zeros. Discuss the pros and cons of each design.

$$H = \frac{\pm(s^2 + 10)}{(s^2 + s + 1)(s + 1)}$$

Reference

[1] Johnson, D. E., and Hilburn, J. L., *Rapid Practical Designs of Active Filters*, New York: Wiley-Interscience, 1975.

Chapter 6

Network Sensitivity and Leapfrog Filters

In the last chapter, we learned about state variable and other architectures used to synthesize linear network functions directly from a specified ratio of polynomials in the complex frequency variable, s. So far, we have not considered what is referred to as network sensitivity to element value variations and environmental factors. In this chapter, we look at a formal definition for the sensitivity factor and then at a particular active circuit architecture, called leapfrog, that is used to emulate low-sensitivity performance of passive ladder networks. Once again, operational ideas are used to realize a particular set of equations and to achieve a desired network function.

6.1 A Filter Sensitivity Definition

Filter sensitivity to specific element value variation is visualized as shown in the circuit in Figure 6.1. The network function, $\mathbf{V}_2/\mathbf{V}_g$, has an implicit dependence upon the kth element, E_k, of the network. Usually, we are interested in some parameter, P, of $T(s)$, such as the 3-dB bandwidth, peak gain, or anything that depends upon some other parameter, x. The other parameter can be a circuit element value or any environmental variable, such as temperature. The sensitivity of parameter P with respect to x is represented by the function, S_x^P. It is defined as in (6.1):

$$S_x^P \doteq \frac{\% \text{ change in } P}{\% \text{ change in } x} = \frac{\Delta P/P}{\Delta x/x} = \frac{x}{P}\left(\frac{\Delta P}{\Delta x}\right) \tag{6.1}$$

A particular sensitivity factor reflects the change obtained in the parameter, P, due to changes in x only. Thus, the derivative, identified in (6.1), is really a partial derivative as Δx goes to the limit of a differential quantity. The true sensitivity factor is then given by (6.2):

$$S_x^P \equiv \frac{x}{P}\left(\frac{\partial P}{\partial x}\right) \tag{6.2}$$

CHAPTER 6. NETWORK SENSITIVITY AND LEAPFROG FILTERS

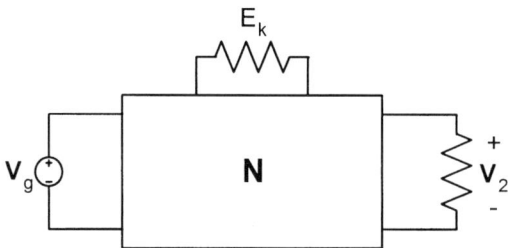

Figure 6.1. V_2 has a dependence upon the value of the kth element, E_k.

An equivalent form, frequently encountered in quality control, is given by (6.3), where the differential of the natural logarithm of P is used, $\partial(\ln P) = \partial P/P$:

$$S_x^P = \frac{\partial(\ln(P))}{\partial(\ln(x))} \tag{6.3}$$

Normally, the sensitivity factors are determined empirically by means of statistical records taken on designs that are in production. In some cases, a theoretical function is worked out for critical cases. In any event, no matter how the S_x^P are determined, they can be used to obtain estimates for the total parameter variation by reversing its definition:

$$\frac{\partial P}{\partial x} = \frac{P}{x} S_x^P \tag{6.4}$$

The partial derivative(s) given by (6.4) then combine to yield the total differential for P as shown in (6.5):

$$\begin{aligned} dP &= \left(\frac{\partial P}{\partial x_1}\right) dx_1 + \left(\frac{\partial P}{\partial x_2}\right) dx_2 + \ldots \text{ or} \\ dP &= \sum_{k=1}^{K} \left(\frac{\partial P}{\partial x_k}\right) dx_k \end{aligned} \tag{6.5}$$

For a zero-mean process, we have the expectation (average) that dP is zero. This requires that the production process be adjusted so that each element variation is also set at zero:

$$\begin{aligned} E(dP) &= <dP> \equiv \sum_{k=1}^{K} \left(\frac{\partial P}{\partial x_k}\right) <dx_k> = 0 \\ \therefore <dx_k> &= 0 \text{ for all } k \end{aligned} \tag{6.6}$$

The variance for dP is obtained by squaring (6.5) and averaging:

$$<dP^2> = \sum_{k,\ell=1}^{K} \left(\frac{\partial P}{\partial x_k}\right)\left(\frac{\partial P}{\partial x_\ell}\right) <dx_k dx_\ell> \equiv \sigma_P^2 \tag{6.7}$$

6.1. A FILTER SENSITIVITY DEFINITION

For parameters that are independent of each other (the usual case), we have the expectation that $dx_k dx_\ell$ is zero for all $\ell \neq k$. Thus, (6.7) reduces to (6.8):

$$\sigma_P^2 = \sum_{k=1}^{K} \left(\frac{\partial P}{\partial x_k}\right)^2 <dx_k^2>$$

$$\equiv \sum_{k=1}^{K} \left(\frac{P}{x_k} S_{x_k}^P\right)^2 \sigma_{x_k}^2 \qquad (6.8)$$

This last result shows how the sensitivity factors are related to the total parameter variance for a zero-mean independent variable process.

A key sensitivity relationship is that for the parameter dependence of a given sinusoidal steady-state network function upon a parameter x. The result is obtained as follows. With $T(j\omega) = |T(j\omega)|e^{j\theta(\omega)}$ and dependent upon some element, x, we follow the definition for S_x^T to its conclusion in (6.9):

$$S_x^T = \frac{x}{T}\frac{\partial T}{\partial x} = \frac{x}{|T|e^{j\theta}}\left[\frac{\partial}{\partial x}|T|e^{j\theta}\right]$$

$$= \frac{x}{|T|e^{j\theta}}\left[\frac{\partial |T|}{\partial x}e^{j\theta} + j\frac{\partial \theta}{\partial x}|T|e^{j\theta}\right]$$

$$\frac{x}{|T|}\frac{\partial |T|}{\partial x} + jx\frac{\partial \theta}{\partial x} \equiv \Re e[S_x^T] + j\Im m[S_x^T]$$

$$= S_x^{|T|} + j\theta S_x^\theta$$

$$\therefore S_x^{|T|} = \Re e[S_x^T]$$

$$\theta S_x^\theta = \Im m[S_x^T] \qquad (6.9)$$

Equation (6.9) shows how to find a network function magnitude or phase sensitivity function from the basic sensitivity definition.

Example 1 - Find the magnitude and phase sensitivity factors for the following network function in terms of their dependence upon parameter $2a$.

$$T(j\omega) = \frac{2aj\omega}{(-\omega^2 + 2aj\omega + \omega_0^2)}$$

Solution - We have the following:

$$S_{2a}^T = \frac{2a}{T}\frac{\partial T}{\partial(2a)} \quad \text{and}$$

$$\frac{\partial T}{\partial(2a)} = j\omega\frac{1\cdot(-\omega^2 + 2aj\omega + \omega_0^2) - 2a(j\omega)}{(-\omega^2 + 2aj\omega + \omega_0^2)^2} \quad \text{so that}$$

$$S_{2a}^T = \frac{(-\omega^2+\omega_0^2)(-\omega^2+\omega_0^2-2aj\omega)}{(-\omega^2+\omega_0^2)^2 + 4a^2\omega^2} \quad \text{but}$$

$$S_{2a}^T \equiv S_{2a}^{|T|} + j\theta S_{2a}^\theta \quad \text{so that}$$

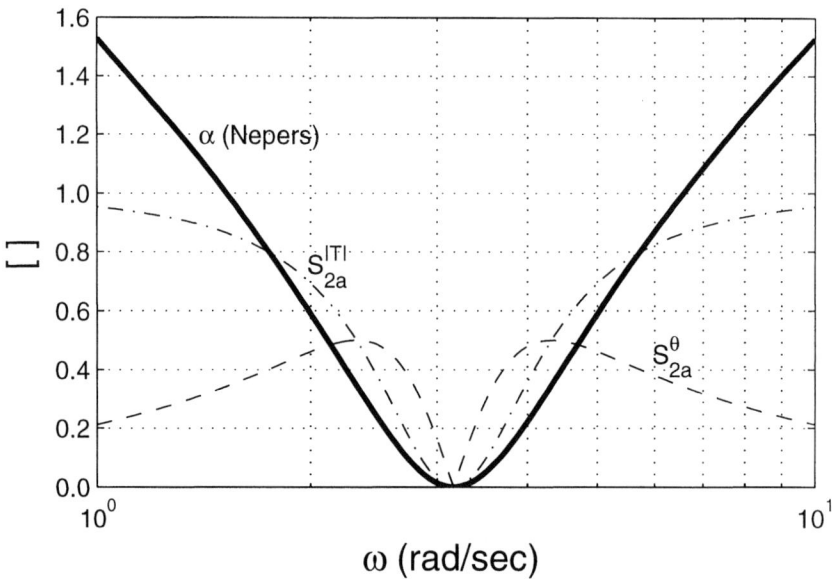

Figure 6.2. Example 1, sensitivity and loss functions.

$$S_{2a}^{|T|} = \frac{(-\omega^2 + \omega_0^2)^2}{(-\omega^2 + \omega_0^2)^2 + 4a^2\omega^2} \quad \Leftarrow$$

$$\theta S_{2a}^{\theta} = \frac{-2a\omega(-\omega^2 + \omega_0^2)}{(-\omega^2 + \omega_0^2)^2 + 4a^2\omega^2} \quad \Leftarrow$$

The answers found for the sensitivity functions are plotted in Figure 6.2 along with the network loss (attenuation) function. An interesting observation is that the minimum of the magnitude of the sensitivity function occurs where the loss function, $\alpha = -20\ln|T|$, has its minimum.

Sensitivity analysis can be used as a design criterion to choose designs that have lower sensitivity to element value variations for critical performance measures. The S-K complex-pole generating circuit, shown in Figure 6.3, provides a good example. This circuit was introduced in the previous chapter but is redrawn here for convenience. The network function and design parameters are as follows:

$$\frac{\mathbf{V}_2}{\mathbf{V}_1} = T(s) = \frac{\mu\omega_0^2}{s^2 + 2\alpha s + \omega_0^2} = \frac{\mu}{1 + \frac{2\alpha}{\omega_0^2}s + \frac{1}{\omega_0^2}s^2}$$

$$\text{where} \quad \mu = 1 + G_3 R_4$$

$$\omega_0^2 = \frac{G_1 G_2}{C_1 C_2}$$

$$2\alpha = \frac{(1-\mu)G_2}{C_2} + \frac{G_1 + G_2}{C_1}$$

6.1. A FILTER SENSITIVITY DEFINITION

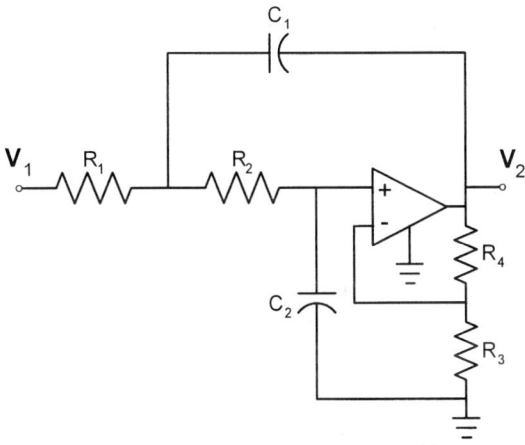

Figure 6.3. S-K circuit and component definitions.

The parameter, μ, is the amplifier closed-loop gain and was referred to as K in a previous discussion. The coefficient, 2α, is the 3-dB bandwidth for the network function, $T(s)$. Letting $2\alpha = B$, $\omega_0 = 1$, and $\mu = 2$, we get the following design constraints for the circuit:

$$\omega_0^2 = 1 = \frac{G_1 G_2}{C_1 C_2}$$
$$2\alpha = B = -\frac{G_2}{C_2} + \frac{G_1 + G_2}{C_1} \quad (6.10)$$

Equation (6.10) provides two equations for four unknowns, and so, we are free to choose two parameter values and solve for the remaining two. A common choice is $C_1 = C_2 = 1$, from which we get $G_1 = B$ and $G_2 = 1/B$. Note that the pole-$Q = \omega_0/2\alpha = 1/B$ for the given parameters.

Now, consider that we are interested in the sensitivity of the 3-dB bandwidth, 2α, to the gain parameter, μ. Then, we get the following result:

$$S_\mu^{2\alpha} = \frac{\mu}{2\alpha}\frac{\partial(2\alpha)}{\partial\mu} = \frac{\mu}{2\alpha}\left(\frac{-G_2}{C_2}\right)$$
$$= \frac{2}{B}\left(\frac{-1}{B\cdot 1}\right) = -2Q^2 \quad \Leftarrow$$

Thus, the 3-dB sensitivity is proportional to the square of the pole-Q. This can be quite large for moderate- to high-Q values.

A different choice than C_1 and C_2 leads to an entirely different result for the bandwidth gain-sensitivity factor for this circuit. An alternate choice is $G_1 = 1 = G_2$. We then have the following for C_1 and C_2:

$$C_1 C_2 = 1$$

$$B = \frac{1}{Q} = \frac{-1}{C_2} + \frac{2}{C_1} = -C_1 + 2/C_1 \text{ so that}$$
$$C_1^2 + C_1/Q - 2 = 0 \text{ or}$$
$$C_1 = \frac{1}{2}\left(-\frac{1}{Q} + \sqrt{\frac{1}{Q^2} + 8}\right) = \frac{1}{C_2} \quad \Leftarrow$$

The sensitivity factor then evaluates to the following result:

$$S_\mu^{2\alpha} = \frac{\mu}{2\alpha}\left(\frac{-G_2}{C_2}\right) = \frac{2}{B}(-1 \cdot C_1) = 2Q\left(\frac{-1}{2}\left(\frac{-1}{Q} + \sqrt{\frac{1}{Q^2} + 8}\right)\right)$$
$$= +1 - Q\sqrt{8 + 1/Q^2} \quad \Leftarrow$$

For this choice of parameter values, the bandwidth sensitivity is only proportional to the pole-Q to the first power. This is an appreciable reduction in gain sensitivity.

Although it is tempting to equalize capacitor values in a circuit, it is not always the best solution.

6.1.1 A Sensitivity Property for Terminated Passive Filters

Sensitivity ideas are applied to the insertion loss function of a doubly terminated passive filter to demonstrate that the passive filter architecture has zero sensitivity to all elements at particular passband frequencies. At those frequencies where the network function, $H(s) = \mathbf{V}_L/\mathbf{V}_g$, is at a maximum (in the filter passband), the attenuation function is at a minimum. When the insertion loss function, IL, is used (see Chapter 1 definition), this minimum can never be negative for a passive network. The passive network can only transfer power by matching the load to the source. It cannot amplify power; thus, the insertion loss is physically always greater than or equal to zero for a passive network. This conclusion is not true for an active filter.

The above conclusion can be formalized through the following discussion of the relation of a minimum insertion loss value dependence upon the circuit's nominal element values. Figure 6.4 depicts insertion loss dependence at a passband frequency, ω_p, where the filter loss is at a minimum when all elements are at their nominal design value. When an element value changes, the insertion loss can only go more positive since it cannot be negative for a passive network. So, at $\omega = \omega_p$ we have the conditions given in (6.11):

$$\left[\frac{\partial \alpha}{\partial L_k}\right]_{\omega=\omega_p} = 0 \quad \text{and} \quad \left[\frac{\partial \alpha}{\partial C_k}\right]_{\omega=\omega_p} = 0 \quad (6.11)$$

where the attenuation function, $\alpha(\omega)$, is at a minimum for $\omega = \omega_p$. Likewise, since $|T| = 10^{-\alpha/20}$, the transfer function is at a maximum whenever α is at a minimum. Then, (6.12) is also true:

$$\left[\frac{\partial |T|}{\partial L_k}\right]_{\omega=\omega_p} = 0 \quad \text{and} \quad \left[\frac{\partial |T|}{\partial C_k}\right]_{\omega=\omega_p} = 0 \quad (6.12)$$

6.2. THE LEAPFROG FILTER ARCHITECTURE

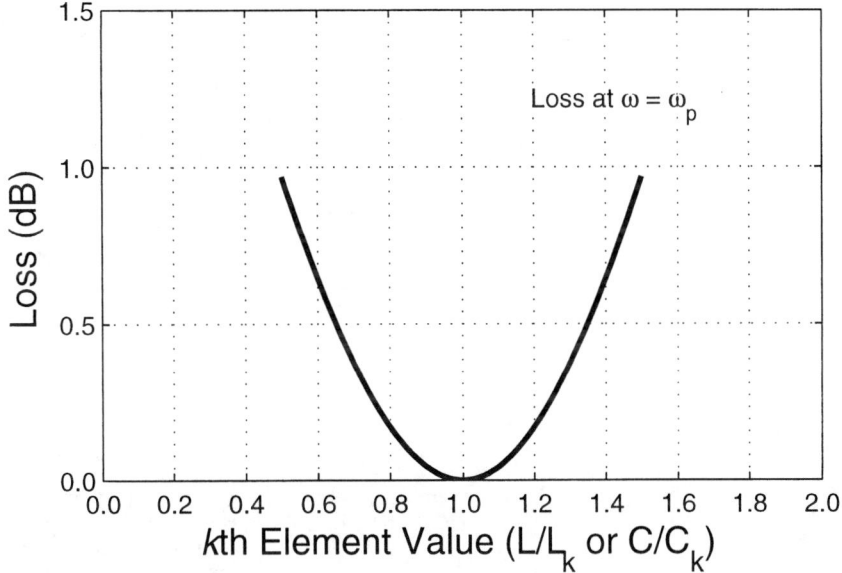

Figure 6.4. Minimum loss coincides with all circuit elements at nominal value at a particular passband frequency, ω_p.

These conclusions are based upon lossless LC networks that are connected between resistive load and source impedances. Finally, the sensitivity definition yields the following result for passive networks operated at those frequencies, ω_p, that yield minimum insertion loss values:

$$\begin{aligned} S_{L_k}^{|T|} &= \left[\frac{L_k}{|T|}\frac{\partial |T|}{\partial L_k}\right]_{\omega=\omega_p} = 0 \text{ and} \\ S_{C_k}^{|T|} &= \left[\frac{C_k}{|T|}\frac{\partial |T|}{\partial C_k}\right]_{\omega=\omega_p} = 0 \end{aligned} \qquad (6.13)$$

6.2 The Leapfrog Filter Architecture

The implicit zero sensitivity conditions for the passive network, with respect to all element values, provide motivation to duplicate the interaction of the passive elements in forming a network function into an active circuit architecture [1]. The procedure leads to what are referred to as leapfrog filters. The reason for the name will become clear very soon.

Figure 6.5 illustrates a ladder network structure and terminology, which is characteristic of a typical prototype filter. Admittances are used to describe the series branches, while impedances are used to describe shunt branches. Only six branches are used because they are sufficient to illustrate the general procedure. The branch

224 CHAPTER 6. NETWORK SENSITIVITY AND LEAPFROG FILTERS

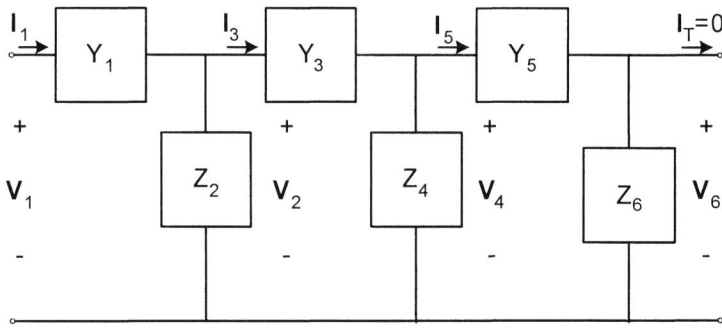

Figure 6.5. A ladder network topology and branch variable definitions.

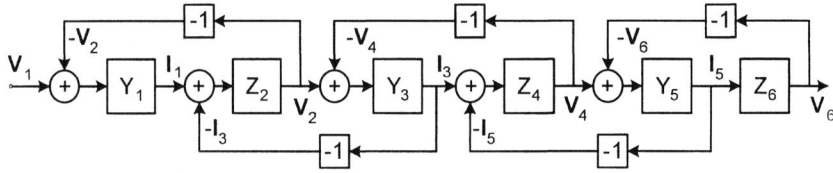

Figure 6.6. The flow of equation operations for a ladder network.

equations for the network are as follows:

$$\begin{aligned}
\mathbf{I}_1 &= Y_1(\mathbf{V}_1 - \mathbf{V}_2) \\
\mathbf{V}_2 &= Z_2(\mathbf{I}_1 - \mathbf{I}_3) \\
\mathbf{I}_3 &= Y_3(\mathbf{V}_2 - \mathbf{V}_4) \\
\mathbf{V}_4 &= Z_4(\mathbf{I}_3 - \mathbf{I}_5) \\
\mathbf{I}_5 &= Y_5(\mathbf{V}_4 - \mathbf{V}_6) \\
\mathbf{V}_6 &= Z_6(\mathbf{I}_5 - \mathbf{I}_T) = Z_6 \mathbf{I}_5
\end{aligned} \quad (6.14)$$

Note that Y_1 and Z_6 are the terminations for the passive filter consisting of elements Z_2 through Y_5.

The set of equations in (6.14) can be visualized in terms of a set of operations as shown by the signal flow graph in Figure 6.6. In the graph, each block represents a transfer function between an entering and leaving signal. It is assumed that values can be picked off a node without affecting the variable at the node.

The signal flow graph illustrates the source of the name leapfrog from the topology, which shows the jump back over a summation to provide the input for the calculation. For example, $\mathbf{V}_2$ is fed back through a sign change to form $(\mathbf{V}_1 - \mathbf{V}_2)$ in the determination of $\mathbf{I}_1 = Y_1(\mathbf{V}_1 - \mathbf{V}_2)$. In the flow graph it looks like $\mathbf{V}_2$ has jumped back (over some operations) to form $\mathbf{I}_1$. Likewise, $\mathbf{I}_3$ on the bottom path is fed back to form the difference, $\mathbf{I}_1 - \mathbf{I}_3$, to determine $\mathbf{V}_2$, and so forth.

6.2. THE LEAPFROG FILTER ARCHITECTURE

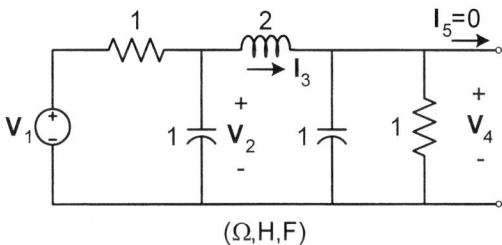

(Ω,H,F)

Figure 6.7. Third-order Butterworth LP filter prototype with variable definitions.

Now, with active filters, the currents, $\mathbf{I}_k$, are not readily realized, so they are simulated with voltage variables as in constructing an analog circuit model. The proportionalities, Y_k and Z_k, are simulated via transfer functions as follows. We let $\mathbf{I}_k \to \mathbf{V}_{I_k}$ and $Y_k \to T_{Y_k}$.

The equations in (6.14) are thus converted to those shown in (6.15) for an analog circuit:

$$\begin{aligned}
\mathbf{V}_{I_1} &= T_{Y_1}(\mathbf{V}_1 - \mathbf{V}_2) \\
\mathbf{V}_2 &= T_{Z_2}(\mathbf{V}_{I_1} - \mathbf{V}_{I_3}) \\
\mathbf{V}_{I_3} &= T_{Y_3}(\mathbf{V}_2 - \mathbf{V}_4) \\
\mathbf{V}_4 &= T_{Z_4}(\mathbf{V}_{I_3} - \mathbf{V}_{I_5}) \\
\mathbf{V}_{I_5} &= T_{Y_5}(\mathbf{V}_4 - \mathbf{V}_6) \\
\mathbf{V}_6 &= T_{Z_6}(\mathbf{V}_{I_5} - \mathbf{V}_{I_T}) = T_{Z_6}\mathbf{V}_{I_5}
\end{aligned} \quad (6.15)$$

Note that in the analog circuit, prototype branch voltages are represented using the same variable name (i.e., $\mathbf{V}_k \to \mathbf{V}_k$). A signal flow graph for (6.15) looks identical to that in Figure 6.6 with appropriate changes in the variable and transfer function names. We will now work an example.

Example 2 - Design a leapfrog circuit realization for the double terminated LP prototype shown in Figure 6.7. Scale the resulting design to 2 kΩ and 1 kHz, and use μCAP to check the leapfrog filter frequency response. Also, measure the response of each amplifier to check the gain distribution for the active filter realization.

Solution - The analog functions for the given prototype are as follows:

$$\begin{aligned}
\mathbf{V}_{I_1} &= 1 \cdot (\mathbf{V}_1 - \mathbf{V}_2) \\
\mathbf{V}_2 &= \frac{1}{s} \cdot (\mathbf{V}_{I_1} - \mathbf{V}_{I_3}) \\
\mathbf{V}_{I_3} &= \frac{1}{2s} \cdot (\mathbf{V}_2 - \mathbf{V}_4)
\end{aligned}$$

226   CHAPTER 6. NETWORK SENSITIVITY AND LEAPFROG FILTERS

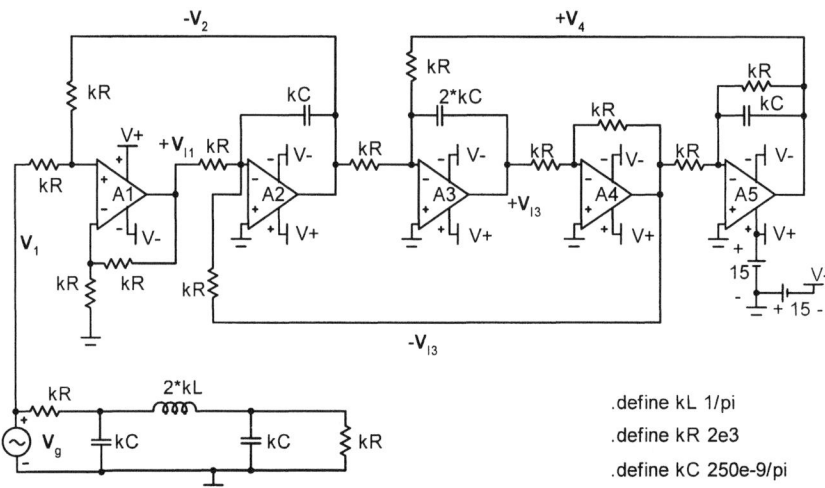

Figure 6.8. A given passive filter and a leapfrog active implementation.

$$\mathbf{V}_4 = \frac{1}{(s+1)} \cdot (\mathbf{V}_{I_3} - 0) \qquad (6.16)$$

These functions are implemented with the impedance- and frequency-scaled circuit shown in Figure 6.8. The circuit builds the equations given in (6.16) by using the topology outlined in the flow graph of Figure 6.6. For example, the output of Amplifier $A1$ gives $\mathbf{V}_{I_1}$, the voltage analog for the current $\mathbf{I}_1$, as required:

$$\mathbf{V}_{I_1} = (1 + k_R/k_R)\frac{(\mathbf{V}_1/k_R - \mathbf{V}_2/k_R)}{(1/k_R + 1/k_R)} \equiv 1 \cdot (\mathbf{V}_1 - \mathbf{V}_2)$$

Then, $A2$ just provides $-\mathbf{V}_2$, rather than $+\mathbf{V}_2$, so as to save a sign change, since $-\mathbf{V}_2$ is required to form $\mathbf{V}_{I_1}$. Amplifier $A3$ forms $+\mathbf{V}_{I_3}$ since the input is $(-\mathbf{V}_2 + \mathbf{V}_4)$, which gets a sign change from the inverting integration. A sign change is required in order to feed back $-\mathbf{V}_{I_3}$ to A_2 to build $-\mathbf{V}_2$, and so forth. The required variables are developed from a logical flow similar to that shown in Figure 6.6.

Note in the scaled circuit that $k_R k_C = 1/k_\omega$, so $k_R k_C s = s/k_\omega$ represents the frequency-scaling operation. The output of Amplifier $A4$ is thus as follows:

$$\begin{aligned}\mathbf{V}_4 &= \frac{-Z_2}{Z_1}(-\mathbf{V}_{I_3}) = +\frac{1}{k_R}\frac{k_R}{(k_R k_C s + 1)}\mathbf{V}_{I_3} \\ &\equiv \frac{1}{(s/k_\omega + 1)} \cdot \mathbf{V}_{I_3}\end{aligned}$$

The result for $\mathbf{V}_4$ is the frequency-scaled version required by (6.16).

Also noteworthy is the fact that each energy storage element in the given passive prototype appears as a capacitor in the leapfrog analog circuit design. The leapfrog

6.2. THE LEAPFROG FILTER ARCHITECTURE

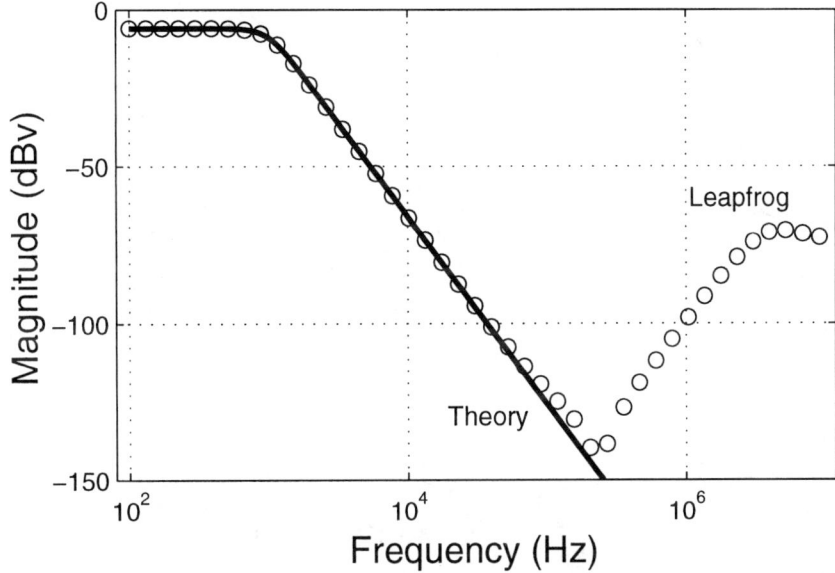

Figure 6.9. Frequency response comparison for the leapfrog analog circuit.

implements the inductor network function in the form of a capacitor analog of the inductor function. It is easy to make a one-to-one relationship between the given prototype value and the leapfrog capacitor values.

The scaled frequency response for the above design is shown in Figure 6.9. The active circuit follows the desired passive response for almost two full decades beyond the filter passband cutoff frequency to a gain magnitude of −110 dB. However, for higher frequencies, as the opamp frequency response comes into consideration, the active filter response falls off, degrading to a peak stopband loss of 70 dBv in the vicinity of 50 MHz. For most applications, the nonideal stopband response is acceptable for losses greater than some minimum, usually 60 dB or more. So, this design represents a pretty good implementation of the desired passive filter response.[1]

It should be observed that, in the response shown in Figure 6.9, the analog filter also provides the flat loss (6 dB) of the passive filter. This is a consequence of duplicating all the network branch constraints for the passive filter. When we look at the gain distribution curves for the leapfrog circuit, shown in Figure 6.10, we see a 6-dB discrepancy between the output amplifier, $A5$, and the peak gain, which occurs on Amplifier $A1$. So, in order to equalize the gain distribution, it is desirable to raise the output by a factor of 2 (6 dB) so as to avoid distortion on amplifiers internal to the filter. A revised circuit is shown in Figure 6.11 where $2 \cdot \mathbf{V}_4$

---

[1] Note if the passive filter modeled winding capacitance in the inductor, the passive filter would exhibit a similar stopband degradation. See Problem 6.6 for an example of this effect.

228   CHAPTER 6. NETWORK SENSITIVITY AND LEAPFROG FILTERS

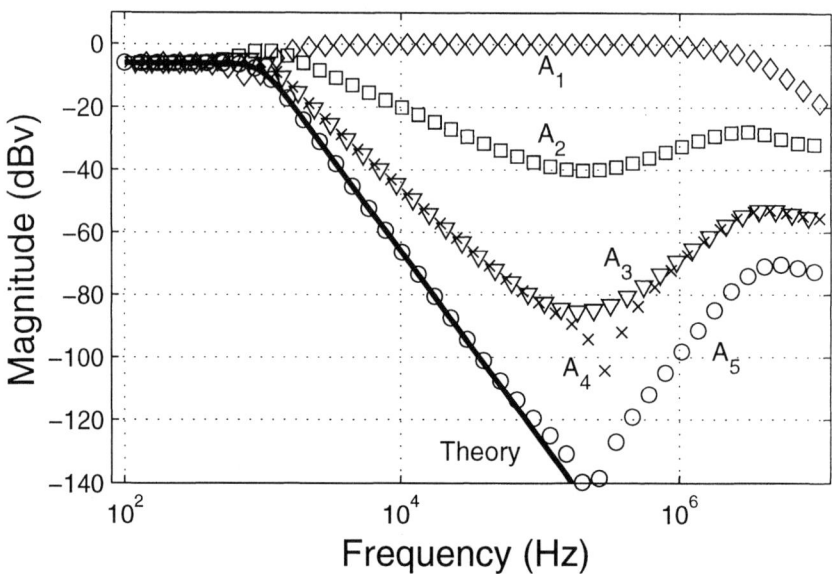

Figure 6.10. Amplifier output response curves for the leapfrog circuit in Figure 6.8.

is implemented at $A5$ simply by changing its input resistor, k_R, to $k_R/2$. In order to preserve the integrity of the desired branch equations, it is then necessary to go back in the circuit and modify those branches that depend on using the variable $1 \cdot \mathbf{V}_4$ rather than $2 \cdot \mathbf{V}_4$.

The required modification at $A3$, an inverting summing integrator, has the following branch relation for the feedback capacitor current for the two voltage levels:

$$\mathbf{I}_\Sigma = -\mathbf{V}_2/k_R + 1 \cdot \mathbf{V}_4/k_R \equiv -\mathbf{V}_2/k_R + 2 \cdot \mathbf{V}_4/(2 \cdot k_R)$$

When the gain-scaled output is fed back to a summing inverter, the scale factor only affects the input resistor that weights the scaled voltage. It is a more complicated change whenever a scaled voltage is fed back to a noninverting sum circuit such as the input to $A1$ in Figure 6.8 or 6.11. For this type of circuit, both the input conductance and the inverting gain are changed to cancel modified input voltage scale factors, which cause a change in the ΣG input conductance sum. The change in ΣG requires a change in the closed-loop gain factor, $K = \Sigma G$, so that the output of the summing amplifier will not see the change in scaled voltage fed back to the circuit. This issue will be encountered in subsequent examples.

The amplifier gain responses for the gain-redistributed circuit in Figure 6.11 are shown in Figure 6.12. The output, $A5$, is shown compared to twice (+6 dB) the ideal passive filter response. The curves for each of the other four amplifiers are clearly unaffected by the gain redistribution, as inspection of Figure 6.10 reveals. The only change is in the gain of $A5$, and that response matches the peak gain of

6.2. THE LEAPFROG FILTER ARCHITECTURE

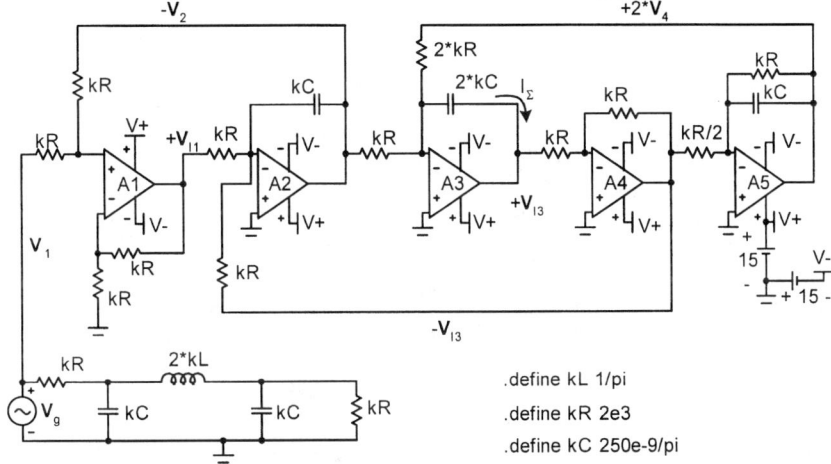

Figure 6.11. The circuit in Figure 6.8 with gain redistribution.

$A1$, so the output distorts at the same input level as it would for $A1$. An interesting result is that the stopband loss on the output, $A5$, is still at 70 dB, so that the relative stopband loss has improved by 6 dB. The conclusion is that the nonideal frequency response is not proportional to the filter signal levels. It is a nonlinear effect.

Before concluding this rather lengthy example, it is convenient to point out a useful operation in implementing an analog circuit for the response of the passive filter shown in Figure 6.7. An amplifier can be eliminated by performing a source transformation (Norton and Thévenin's equivalent) on the circuit. The modified prototype is shown in Figure 6.13. The analog functions that need to be implemented are given in (6.17):

$$\begin{align}
\mathbf{V}_{I_1} &= 1 \cdot \mathbf{V}_1 \\
\mathbf{V}_2 &= \frac{1}{(s+1)} \cdot (\mathbf{V}_{I_1} - \mathbf{V}_{I_3}) \\
\mathbf{V}_{I_3} &= \frac{1}{2s} \cdot (\mathbf{V}_2 - \mathbf{V}_4) \\
\mathbf{V}_4 &= \frac{1}{(s+1)} \cdot (\mathbf{V}_{I_3} - 0)
\end{align} \tag{6.17}$$

Note that the $\mathbf{I}_1$ current for this circuit is different from that for the previous circuit. It is a source variable in this circuit, while it was a resistor current in the first example. The equations in (6.17) are easily implemented in a leapfrog topology as shown in the four-amplifier design in Figure 6.14. The amplifier response for this design is shown in Figure 6.15. The output follows the ideal response (for a $2 \cdot \mathbf{V}_4$ implementation) through the pass and transition bands, but the recovery in the stopband is worse (by about 8 dB) than for other designs. So again, we

230   CHAPTER 6. NETWORK SENSITIVITY AND LEAPFROG FILTERS

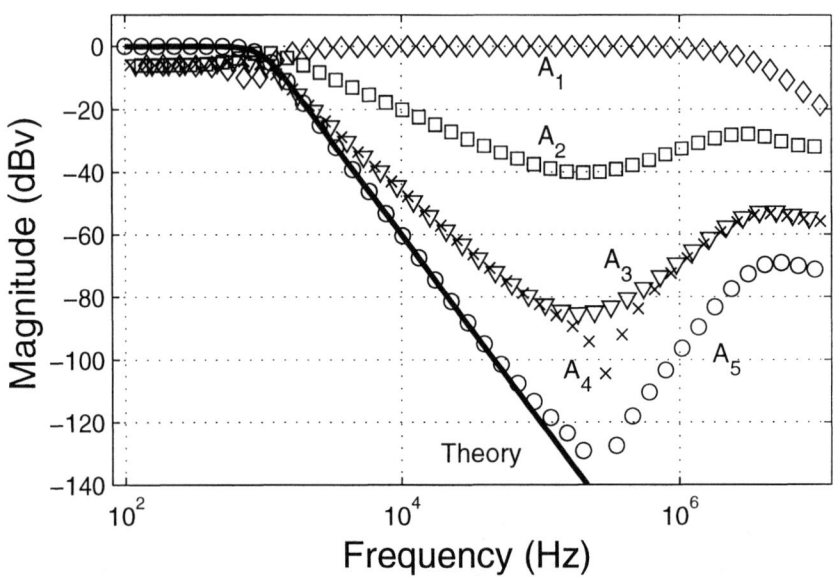

Figure 6.12. Amplifier response for the gain-redistributed circuit in Figure 6.11.

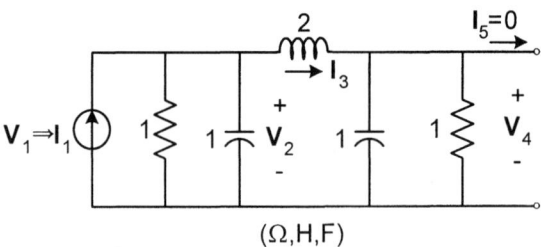

Figure 6.13. The given prototype modified by a source transformation.

6.2. THE LEAPFROG FILTER ARCHITECTURE

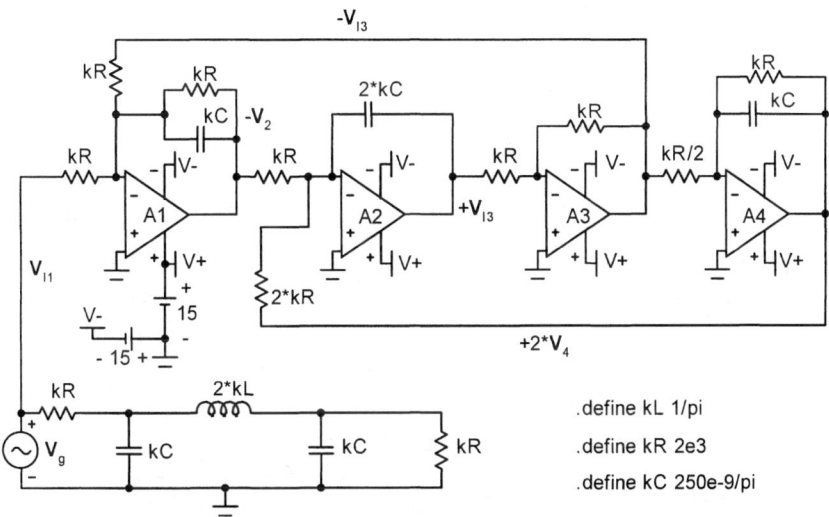

Figure 6.14. The source-transformed prototype having one less opamp.

see a design trade-off. The simpler (lower-cost) circuit requires a larger minimum loss specification in the stopband than does the more complicated design. The fact that this will happen is not obvious or easy to explain and highlights the value of simulating a proposed design.

6.2.1 Leapfrog Example of Sensitivity Performance

Before moving on to the next subject, it is instructive to look at typical sensitivity performance for a leapfrog analog circuit realization for a passive filter. The circuit shown in Figure 6.14 implements a three-pole all-pole lowpass filter and serves as a good example for this discussion.

Recall the introductory discussion that led to the choice of the leapfrog architecture, namely, that sensitivity factors are zero where the insertion loss is minimum (or the transmission gain is maximum). In this example LP filter, the gain is maximum in the passband, which lies between dc and 1 kHz.

Figure 6.16 illustrates gain errors, measured as a dB difference between the output of the active filter $(2 \cdot V_4)$ and twice the output of the passive (ideal) filter shown in Figure 6.14. The curve in Figure 6.16(a) is a temperature test where all elements have zero temperature dependence except for the inductor $(2 \cdot kL)$ in the passive filter and its corresponding analog capacitor $(2 \cdot kC)$ in the active filter. Both elements were modeled to have a +350 ppm/°C temperature coefficient. The test was run over a 50°C range in three steps, as shown. At 25°C, the nominal operating temperature, the circuits exhibit virtually identical thermal dependence through the full passband and well into the stopband (see Figure 6.15 for different band ranges). For other temperatures, however, the circuits do not track exactly,

232   CHAPTER 6. NETWORK SENSITIVITY AND LEAPFROG FILTERS

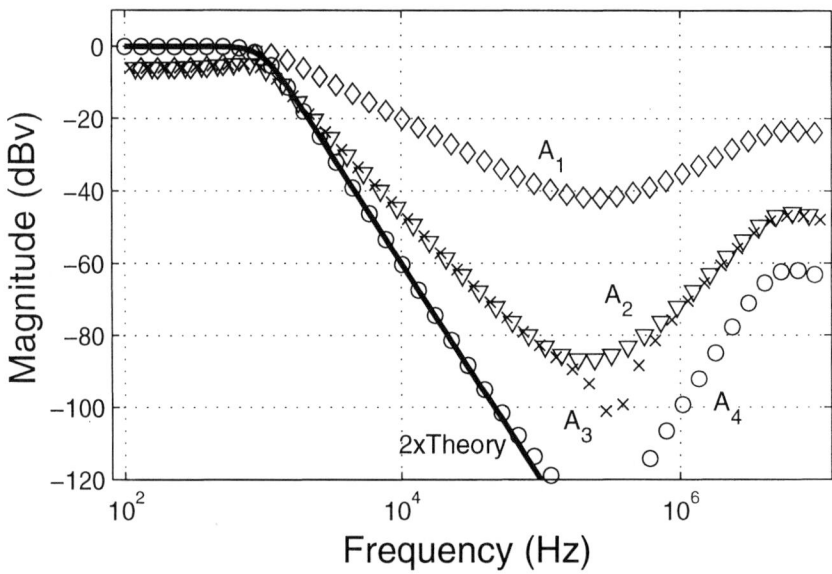

Figure 6.15. Amplifier output responses for the source-transformed design.

but the error goes to zero in the passband for low frequencies where the transmission gain is maximum.

Figure 6.16(b) shows gain sensitivity at the fixed nominal operating temperature of 25°C for the active filter in response to element value perturbation of a single resistor. The resistor chosen for the test was the series input resistor to Amplifier $A3$. The variation amounted to a ±20% range, which is extreme for a linear filter application. The results show a significant gain dependence (±2 dB) upon this resistor for frequencies outside the passband, but again, the sensitivity goes to zero in the passband of the filter. Similar results are obtained when the other resistors are allowed to vary. Some have more effect than others, but the general result is that the network is not as sensitive to element value variations for passband frequencies as it is for out-of-band frequencies.

Finally, Figure 6.16(c) shows gain sensitivity, at a nominal operating temperature, to the open-loop dc gain parameter of Amplifier $A1$. The gain was allowed to change by ±50%, and the result, shown on a very sensitive scale, shows virtually no sensitivity to this parameter over the full frequency range tested. Similar results are obtained for the other amplifiers. This result shows the remarkable effect of negative feedback and the fact that the opamp gain only has to be large, not reach a specific number.

With these results obtained for the leapfrog filter, it is interesting to compare how well a state variable implementation performs. In order to do this, we need the

6.2. THE LEAPFROG FILTER ARCHITECTURE

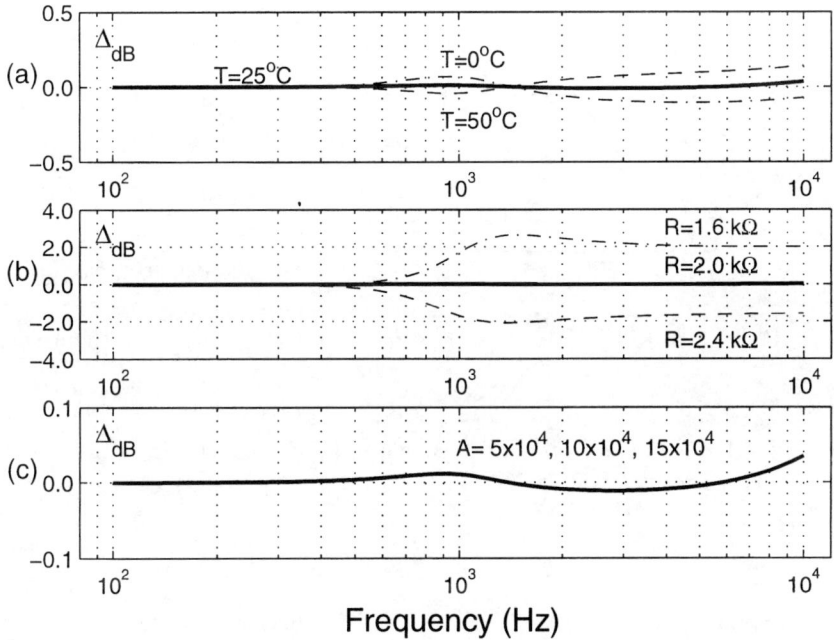

Figure 6.16. Gain-error variations for leapfrog sensitivity comparisons: (a) inductor variation, (b) resistor variation, and (c) amplifier open-loop gain variation.

234   CHAPTER 6. NETWORK SENSITIVITY AND LEAPFROG FILTERS

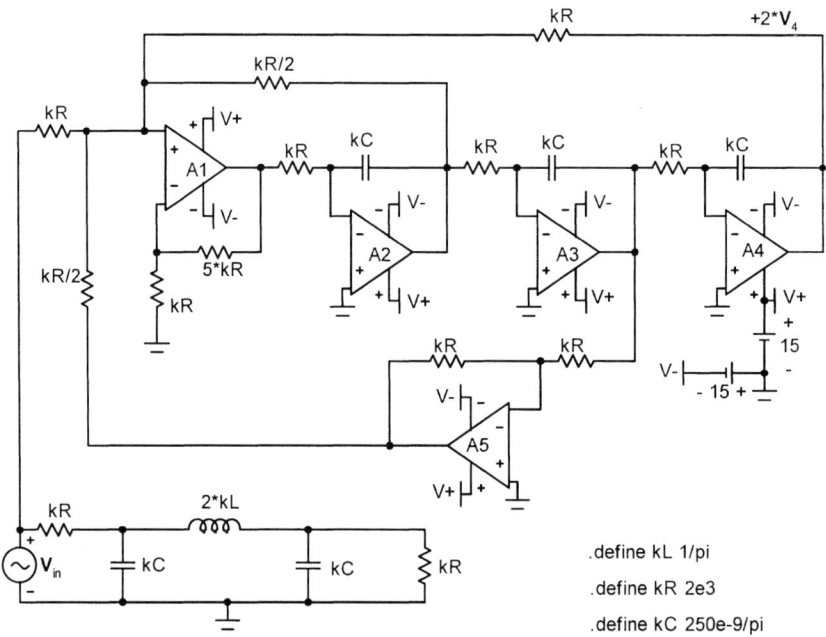

Figure 6.17. The state variable realization for the leapfrog circuit in Figure 6.11.

network function for the example filter shown in Figure 6.7. Ladder analysis of the circuit yields the result given in (6.18):

$$\frac{2\mathbf{V}_4}{\mathbf{V}_1} = \frac{1}{s^3 + 2s^2 + 2s + 1} \tag{6.18}$$

A state variable design for (6.18) is shown in Figure 6.17, where the amplifier gains have been adjusted to provide the peak gain on the output of $A4$. The network function for this filter fits the desired function extremely well through the transition band (out to $\approx$50 kHz), but the stopband performance is poor compared to the leapfrog circuit. The minimum relative stopband attenuation for this filter is $\approx$37 dB, which is a very small number. Apparently, the leapfrog topology gets more mileage out of the opamp bandwidth than does the corresponding state variable topology. So, that is a plus for the leapfrog circuit.

The sensitivity of this circuit is checked in the same fashion as was done for the leapfrog circuit. The results are shown in Figure 6.18. For the temperature range, the passive circuit inductor is given a +350 ppm/°C temperature coefficient, while in the state variable circuit, the feedback capacitor around Amplifier $A3$ is given the same temperature coefficient. All other components are held constant. The results are substantially the same as those obtained for the leapfrog circuit. The peak-to-peak variation is the same, and the difference approaches zero at the peak passband gain (at dc).

6.2. THE LEAPFROG FILTER ARCHITECTURE 235

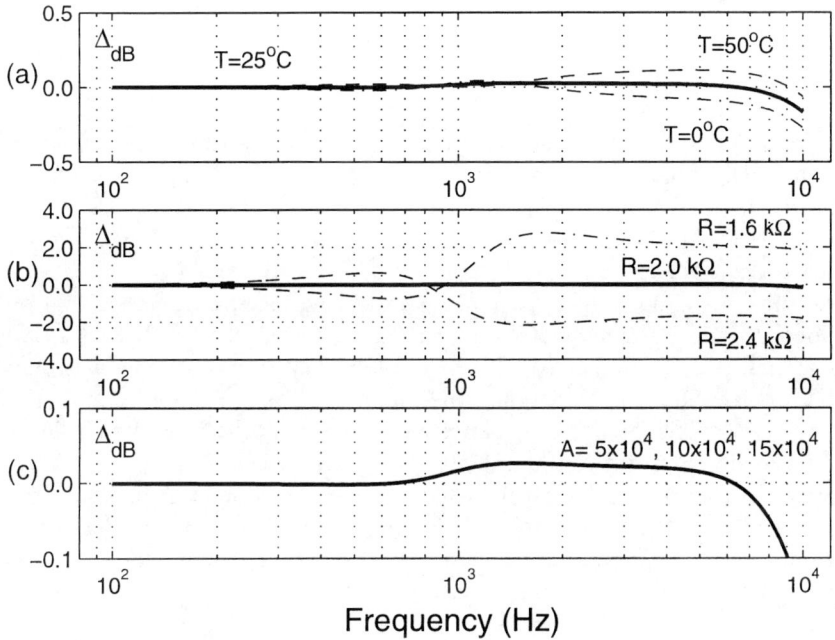

Figure 6.18. Gain-error variations for the state variable realization in Figure 6.17: (a) inductor variation, (b) resistor variation, and (c) amplifier gain variation.

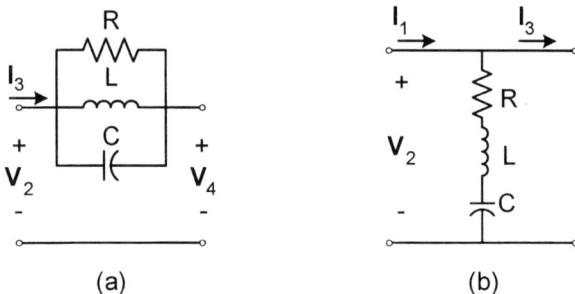

Figure 6.19. Type I resonant branches used in passive filters: (a) series, and (b) shunt.

In Figure 6.18(b), ±20% perturbations on Resistor $R3$ (input to Amplifier $A3$) were made, and the response curves were compared to the unperturbed ideal response of the passive filter. Again, the result is substantially the same as that obtained earlier; however, the variation in the passband is larger than it was for the leapfrog realization.

The third test, shown in Figure 6.18(c), involved perturbations (±50%) on the open-loop dc gain parameter of Amplifier $A1$. The resulting frequency responses were compared to the unperturbed passive filter ideal response. The variation has only a nominal effect and is virtually the same as the previous result. It is hard to tell which circuit has the better sensitivity behavior, and these results challenge the basis for the development of the leapfrog design approach.

Every filter has different sensitivity issues, depending upon its nominal frequency response and the architecture selected for implementation. Before making or accepting filter specifications for gain error performance, it is essential to perform sensitivity tests to determine acceptable parameter limits for production tolerances.

6.2.2 The Elliptic Filter LP Topology

The previous discussion has illustrated a procedure to implement active filter analog circuits for tabulated all-pole LP prototype filters. The procedure yields circuits referred to as leapfrog filters due to the feedback nature of the signal flow graph implementation. The leapfrog procedure thus far developed needs to be expanded to handle other popular LP prototype designs, such as the elliptic filter and other types that use resonant branches in their design. These filters are not all-pole filters and are characterized by their producing finite (usually j-axis) transmission zeros.

Figure 6.19 illustrates two commonly occurring branches for filters that produce finite transmission zeros. The branch shown in Figure 6.19(a) is a parallel resonant circuit that is used as a series branch in an LP filter design. Figure 6.19(b) is a series resonant circuit that is used as a shunt branch in an LP filter design. When the parallel circuit resonates, it presents a large impedance (R) which in turn reduces the current that can get from the source to the load. This large impedance creates

6.2. THE LEAPFROG FILTER ARCHITECTURE

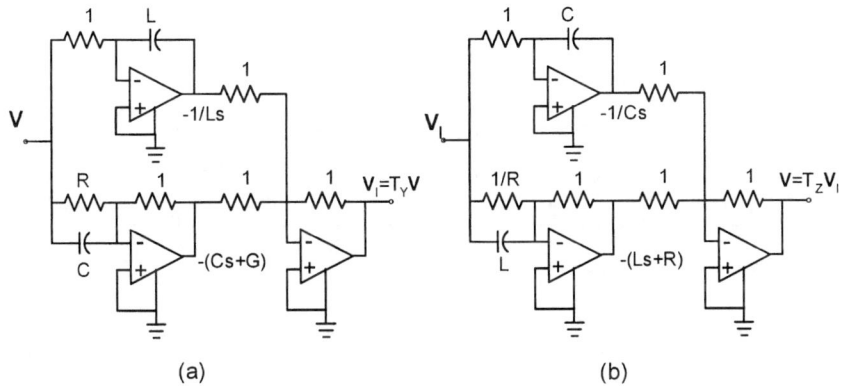

Figure 6.20. Active circuit realizations for the Type I resonant branches: (a) series, and (b) shunt.

a small (or zero if $R \to \infty$) output, or transmission minimum. Likewise, when the series branch resonates, it provides a small resistance (R) to the circuit and limits the voltage that can be developed across the branch. This small resistance also tends to short out the source from connecting to the load, and so, a minimum transmission gain is developed. The branch equations are as follows:

a - For the series branch [Figure 6.19(a)]:

$$\begin{aligned} \mathbf{I}_3 &= Y(\mathbf{V}_2 - \mathbf{V}_4) \\ \text{where} \quad Y &= Cs + G + 1/Ls \end{aligned} \quad (6.19)$$

b - For the parallel branch [Figure 6.19(b)]:

$$\begin{aligned} \mathbf{V}_2 &= Z(\mathbf{I}_1 - \mathbf{I}_3) \\ \text{where} \quad Z &= Ls + R + 1/Cs \end{aligned} \quad (6.20)$$

Note that both Y and Z have the same form and can be realized in their operational form by using the same circuit topology.

Feedforward realizations for (6.19) and (6.20) are shown in Figure 6.20. The operational circuit formation for the analog results given in (6.21) are straightforward to obtain from the circuits given in Figure 6.20:

$$\begin{aligned} \mathbf{V}_I &= T_Y \mathbf{V} \text{ for the parallel resonator} \\ \text{and } \mathbf{V} &= T_Z \mathbf{V}_I \text{ for the series resonator} \\ \text{where } T_Y &= Cs + G + 1/Ls \\ \text{and } T_Z &= Ls + R + 1/Cs \end{aligned} \quad (6.21)$$

238 CHAPTER 6. NETWORK SENSITIVITY AND LEAPFROG FILTERS

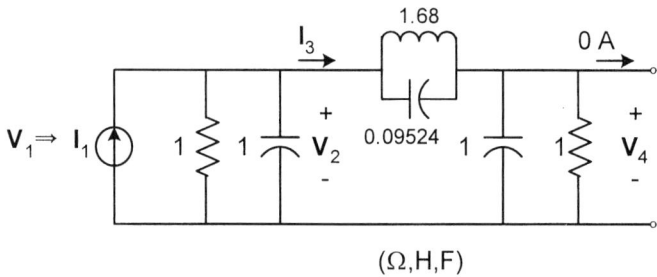

Figure 6.21. Third-order LP prototype with finite transmission zeros.

Example 3 - The use of the parallel resonator analog circuit is demonstrated on the LP prototype circuit given in Figure 6.21. This filter has transmission zeros at $s = \pm j2.25$ and has a topology similar to the elliptic filter class. You may recall that it was used extensively in Chapter 2 for several examples. Its branch equations are as follows:

$$\begin{align}
\mathbf{V}_{I_1} &= \mathbf{V}_1 \\
\mathbf{V}_2 &= \frac{1}{(s+1)}(\mathbf{V}_{I_1} - \mathbf{V}_{I_3}) \\
\mathbf{V}_{I_3} &= (0.095s + 1/1.68s)(\mathbf{V}_2 - \mathbf{V}_4) \\
\mathbf{V}_4 &= \frac{1}{(s+1)}(\mathbf{V}_{I_3} - 0)
\end{align} \tag{6.22}$$

Equation (6.22) is realized by the circuit shown in Figure 6.22 where a transmission network, Figure 6.20(a), is used to realize the parallel LC branch behavior. Note that it is a lossless resonator, so $R \to \infty$ for this branch. When the circuit realizes $\mathbf{V}_4$ directly, the output amplifier, $A6$, does not have the peak gain within a 4 dB discrepancy, and so, the output is scaled to $2 \cdot \mathbf{V}_4$, as was done in the previous example. Note also that the transmission factor, $T_Y = Cs + 1/Ls$, is realized with a sign change by using a noninverting amplifier on the output.

The response for the circuit of Figure 6.22 is shown in Figure 6.23 compared to the desired response for the ideal LP prototype. The response fits the desired magnitude function well into the stopband where a spurious peak is encountered at ≈ 200 kHz. This peak exceeds the minimum stopband attenuation of the function and is therefore unacceptable in a final design. We can deduce that the peak is caused by the ac short circuit of the $0.095 \cdot kC$ capacitor input to Amplifier $A4$. One way to subvert this effect is to insert a small resistance in series with the capacitor to provide a minimum load impedance for the source amplifier, $A2$. How small should the resistance be? The reactance required, in the prototype design, for the 0.095-F capacitor at the transmission zero, $\omega = 2.25$, is 4.7 Ω. Therefore, using a 0.1-Ω resistor provides a resistance of about 2% of the required reactance. For smaller frequencies the percentage is also smaller. The modified design response, shown in Figure 6.23, is for the case where a $0.1 \cdot kR$ resistor was inserted in series

6.2. THE LEAPFROG FILTER ARCHITECTURE

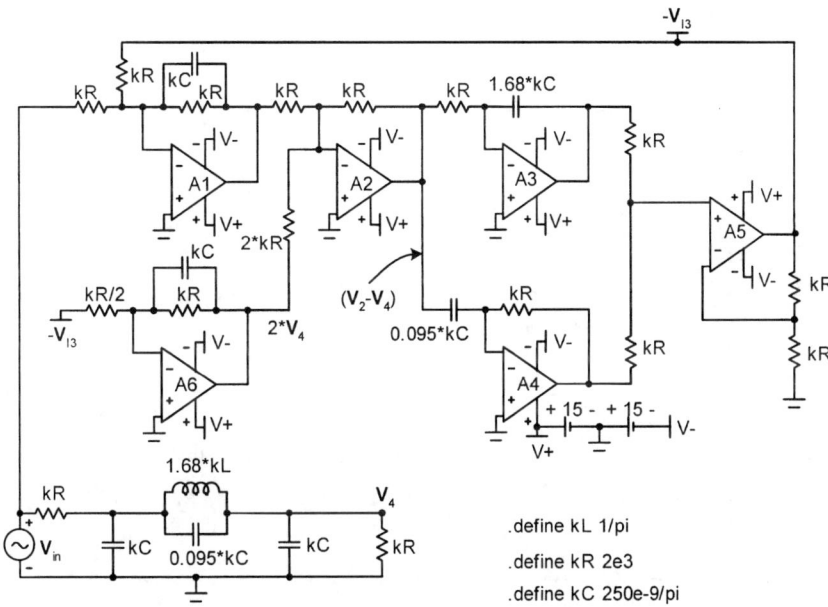

Figure 6.22. Leapfrog realization for the LP prototype in Figure 6.21.

with the capacitor input to $A4$. The resulting response shows that the spurious peak is completely damped by this small loss, with the result that the filter provides an acceptable response throughout the stopband shown for this example.

6.2.3 The All-Pole BP Filter Topology

Now that we have developed a working capability for LP filters using the leapfrog design, it is reasonable to look at BP filter requirements. Again, we start with the all-pole structure and finish with a look at filters that produce finite transmission zeros.

Two branches from a bandpass all-pole network are shown in Figure 6.24. These are two commonly occurring branches in the bandpass filter for an all-pole LP prototype. Note that usually R, in the series branch, and G, in the shunt branch, are zero. At resonance in the series branch, the small value of R connects the input and output nodes of the branch. Likewise, at resonance in the shunt branch, the circuit goes to a high impedance, and the input current is allowed to pass to the output. Thus, these branches are connected so as to pass energy from input to output when they are at resonance. The branch equations are as follows:

a - For the Series Branch [Figure 6.24(a)]:

$$\mathbf{I}_3 = Y(\mathbf{V}_2 - \mathbf{V}_4)$$

240   CHAPTER 6. NETWORK SENSITIVITY AND LEAPFROG FILTERS

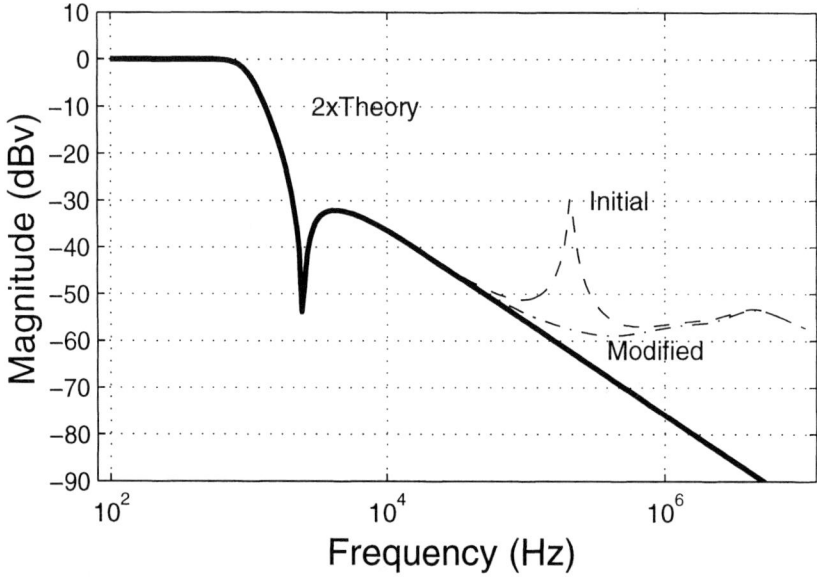

Figure 6.23. Response for the circuit in Figure 6.22. A small resistor is inserted in series with the capacitor input to $A4$ for the modified design response.

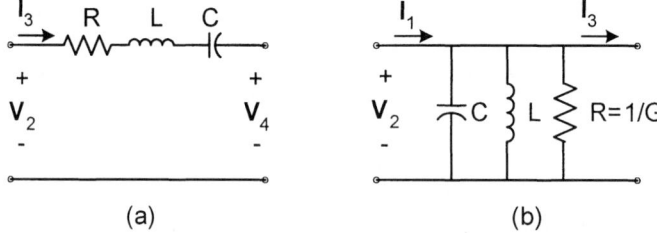

Figure 6.24. Type II resonant branches used in filters for bandpass all-pole filters: (a) series, and (b) shunt.

6.2. THE LEAPFROG FILTER ARCHITECTURE

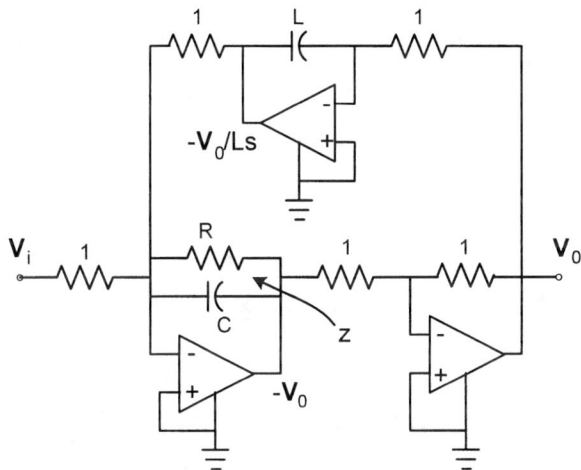

Figure 6.25. Biquad circuit, Type II, realizing the branch impedance, (6.24).

$$\text{where} \quad Y = \frac{1}{(Ls + R + 1/Cs)} = \frac{s/L}{(s^2 + Rs/L + 1/LC)} \quad (6.23)$$

b - For the Shunt Branch [Figure 6.24(b)]:

$$\mathbf{V}_2 = Z(\mathbf{I}_1 - \mathbf{I}_3)$$
$$\text{where} \quad Z = \frac{1}{(Cs + G + 1/Ls)} = \frac{s/C}{(s^2 + Gs/C + 1/LC)} \quad (6.24)$$

Both branch relationships, (6.23) and (6.24), are of the same form and can be realized using a biquad circuit as shown in Figure 6.25. This circuit realizes the network function, $\pm \mathbf{V}_0/\mathbf{V}_i = T_Z \doteq Z$, for the shunt branch where the source voltage, $\mathbf{V}_i$, is taken from an opamp output or any low-impedance source. The circuit function is obtained as follows (*Note:* $z = 1/(Cs + G)$ and the input opamp yields $-\mathbf{V}_0$):

$$-\mathbf{V}_0 = -z(\mathbf{V}_i \cdot 1 - \mathbf{V}_0/Ls) \text{ so}$$
$$(Cs + G)\mathbf{V}_0 = \mathbf{V}_i - \mathbf{V}_0/Ls \text{ or}$$
$$(Cs + G + 1/Ls)\mathbf{V}_0 = \mathbf{V}_i \text{ and}$$
$$\mathbf{V}_0 = T_Z \mathbf{V}_i$$
$$\text{where} \quad T_Z = \frac{s}{C(s^2 + Gs/C + 1/LC)} \quad (6.25)$$

The circuit in Figure 6.25 can also be used to obtain the transmission function, T_Y, for the series branch by changing $C \to L$, $R \to G = 1/R$, and $L \to C$ so that we get (6.26):

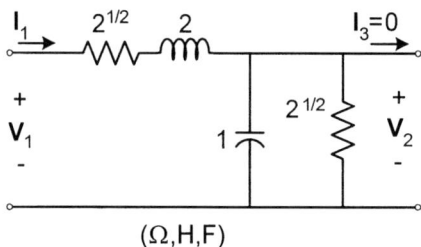

Figure 6.26. Second-order all-pole LP prototype circuit for Example 3.

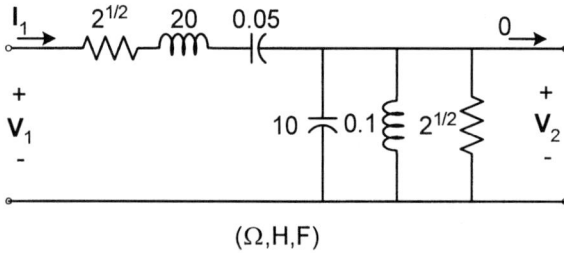

Figure 6.27. BP transform for the circuit in Figure 6.26 using a filter Q of 10.

$$T_Y = \frac{s}{L(s^2 + Rs/L + 1/LC)} \qquad (6.26)$$

Example 4 - Transform the second-order LP filter given in Figure 6.26 to a BP filter that has a filter Q of 10. Use a leapfrog architecture to realize the resulting circuit.

Solution - The BP prototype is given in Figure 6.27 following procedures developed in Chapter 2. The leapfrog design equations are as follows:

$$\mathbf{V}_{I_1} = T_Y(\mathbf{V}_1 - \mathbf{V}_2) = \frac{1}{(20s + \sqrt{2} + 20/s)}(\mathbf{V}_1 - \mathbf{V}_2)$$

$$\mathbf{V}_2 = T_Z(\mathbf{V}_{I_1} - 0) = \frac{1}{(10s + 1/\sqrt{2} + 10/s)}\mathbf{V}_{I_1} \qquad (6.27)$$

The active leapfrog circuit in Figure 6.28 is implemented directly from the prototype in Figure 6.27 using the Type II transmission network and its complement for the two resonant branches. Note the use of R and G in the two sections used to emulate the series RLC and the parallel GCL, respectively.

When the circuit in Figure 6.28 is checked for gain distribution, the peak gain for Amplifier $A1$ has the maximum global gain exceeding the output amplifier, $A6$,

6.2. THE LEAPFROG FILTER ARCHITECTURE

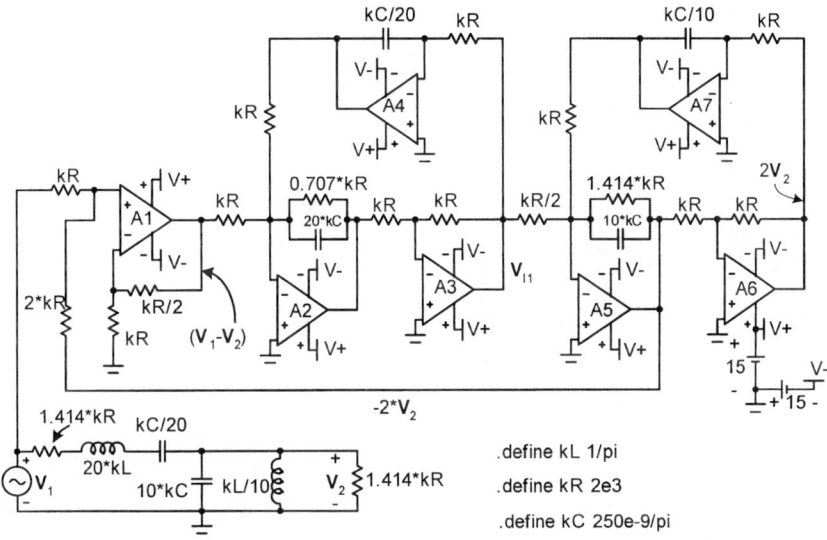

Figure 6.28. Leapfrog realization, with gain scaling, for the BP filter in Figure 6.27.

by ≈ 6.9 dB. The circuit is thus scaled by a factor of 2 (+6 dB) to provide $2 \cdot \mathbf{V_2}$ rather than $1 \cdot \mathbf{V_2}$. Note that this is done by halving the series input to the second network, T_Z, to provide $-2V2$ on $A5$ and $+2V2$ on $A6$. When this voltage is fed back to the input summing amplifier, $A1$, to provide $(\mathbf{V_1} - \mathbf{V_2})$, the factor of 2 is removed by the input scaling resistor coming from $-2V2$. This makes the input prototype ΣG go from 2.0 to 1.5, forcing a change in the feedback resistor around $A1$. This complication is always encountered when a noninverting summing amplifier is required to process gain-scaled node voltages.

The frequency response for the leapfrog design is shown in Figure 6.29, where it is compared to ideal response over two decades of frequency range. The comparison is quite good down to -75 dB, where the leapfrog circuit begins to deviate from ideal behavior. The stopband response for this circuit is actually rather poor for larger frequencies as it comes up to about 40-dB minimum attenuation around 6.5 MHz. Whether or not this design is a good solution depends upon the specific application and the need for meeting a minimum stopband attenuation specification. If the filter is only required to filter signals in the 0.10- to 10-kHz range, then the circuit in Figure 6.28 is a good solution.

A disadvantage of the leapfrog design procedure that we can notice at this point is that it is difficult to analyze the gain distribution using a closed-form network function in Matlab. The leapfrog circuit is more complex in the design process than is the state variable circuit. Nevertheless, gain distribution is important and should be considered for every active filter design procedure.

244 CHAPTER 6. NETWORK SENSITIVITY AND LEAPFROG FILTERS

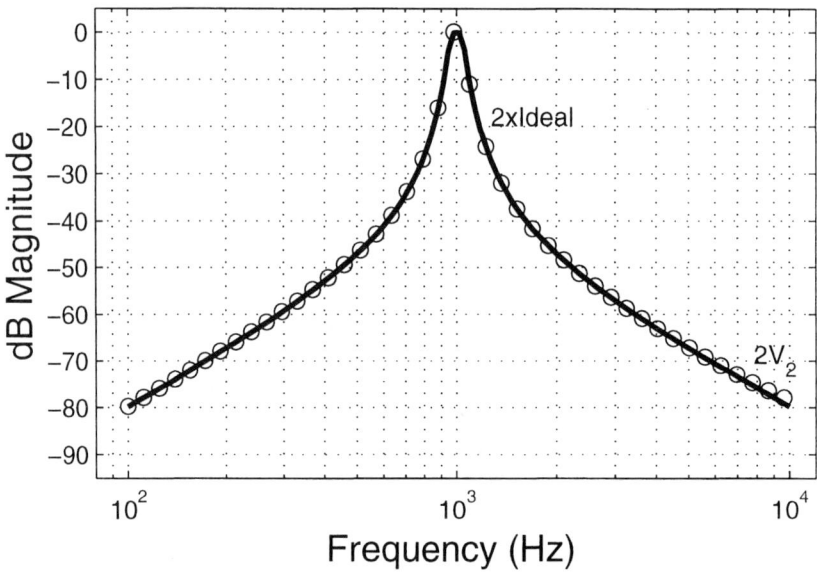

Figure 6.29. Frequency response for the circuits in Figure 6.28.

6.2.4 Topology for BP Filters with Finite Transmission Zeros

This chapter concludes with an example for the BP realization of a non-all-pole network function. The LP prototype shown in Figure 6.21 is such a filter. Its BP realization was discussed in Chapter 2, and after transformation of the series branches to equivalent forms, the BP topology consists entirely of parallel resonant LC branches. A leapfrog realization would thus use impedance realizations of Type I(a) for the series branches and Type II(b) for the shunt branches. The leapfrog procedure is modified slightly to account for the two series branches, as illustrated by the following example.

Example 5 - Use leapfrog design procedures to realize the BP filter shown in Figure 6.30. This is the BP form for the circuit in Figure 6.21, and this equivalent form was developed in Chapter 2 (see Figure 2.23) for a filter Q of 5. The circuit is reprinted in Figure 6.30 to define design variables for the branch currents and voltages.

Solution - The leapfrog design procedure builds an analog circuit to replicate a passive prototype filter branch current and voltage requirements. The circuit shown in Figure 6.30 has two branches in series for the current, $\mathbf{I}_3$. The required leapfrog equations are thus as follows:

$$\mathbf{V}_{I_1} = \mathbf{V}_1$$

6.2. THE LEAPFROG FILTER ARCHITECTURE

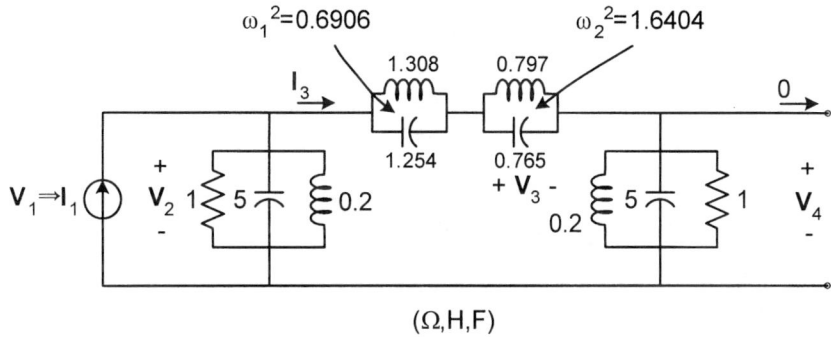

Figure 6.30. BP prototype for the circuit in Figure 6.31, $Q_F = 5$.

$$\mathbf{V}_2 = \frac{1}{(5s + 1 + 5/s)}(\mathbf{V}_{I_1} - \mathbf{V}_{I_3})$$
$$\mathbf{V}_{I_3} = (1.254s + 1/1.308s)(\mathbf{V}_2 - \mathbf{V}_3 - \mathbf{V}_4)$$
$$\mathbf{V}_3 = \frac{1}{(0.765s + 1/0.797s)}\mathbf{V}_{I_3}$$
$$\mathbf{V}_4 = \frac{1}{(5s + 1 + 5/s)}(\mathbf{V}_{I_3} - 0) \tag{6.28}$$

The desired output is $\mathbf{V}_4$. Note that the series branch current, $\mathbf{I}_3$, depends upon an additional drop, $\mathbf{V}_3$, for the added series branch as compared to previous examples of this procedure. It should be clear from this example how the procedure can be implemented for higher-order circuits that have more branches.

The equations in (6.28) are implemented[2] by the leapfrog circuit shown in Figure 6.31. Thirteen opamps are required to realize (6.28) along with eight capacitors for the LC elements of the prototype. By contrast, a state variable filter would only require 10 opamps and 6 capacitors in order to realize this 6-pole network function (6 integrators, 2 sign changers, and 2 summation amplifiers). However, it would be necessary to know the network function for the filter. It is usually straightforward to determine a network function for most circuits, especially the ladder-type circuit.

Figure 6.32 shows the response of the leapfrog filter compared to the ideal passive response over two decades of operating frequency. The active filter is a perfect match to the passive filter over this range. A check of stopband performance out to 10 MHz reveals that the attenuation never comes back up to the -38-dB maximum stopband gain level of the ideal function. This design provides a good replica of the passive filter ideal response.

A check of the peak gain response, however, yields some bad news. The peak output gain of -6 dB on opamps $A11$ and $A13$ ($\pm\mathbf{V}_4$, the desired output) is about

---

[2]The opamps are shown with zero bias connections due to exceeding the allowed number of components for this student μCAP simulation package. This does not affect the calculation for ac response.

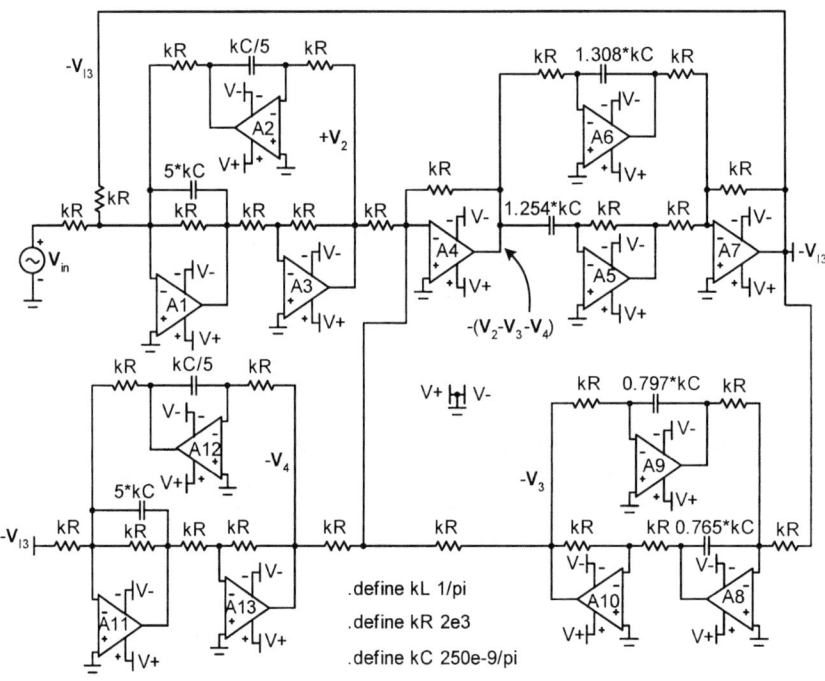

Figure 6.31. Leapfrog active realization for the BP prototype shown in Figure 6.30.

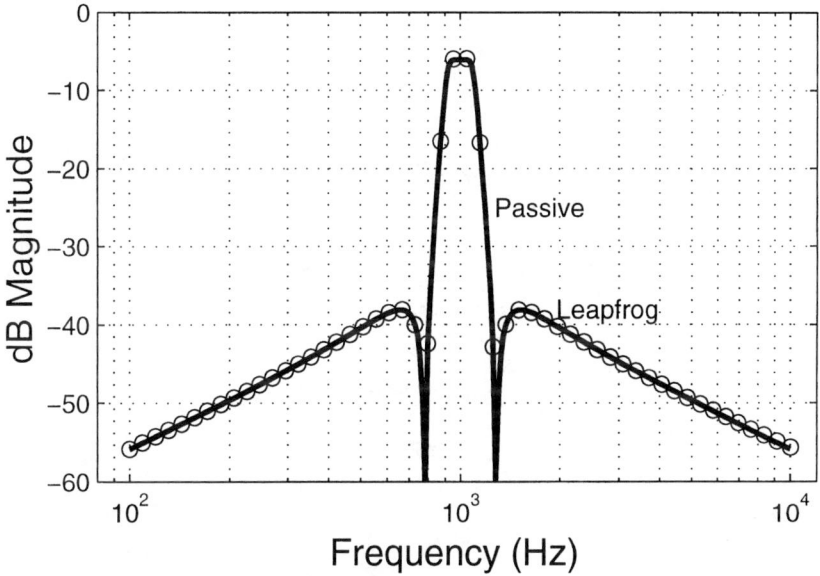

Figure 6.32. Comparison of leapfrog and passive filter responses for Example 5.

the lowest peak of all 13 opamps! Therefore, if one seriously wanted to use this design, the gain (at $A13$) would have to be appropriately scaled to be the largest value in the circuit. The other amplifiers have high peak gains because they are replicating high-Q resonant LC branch behavior. Some opamp peak gains were found to be $+12.4$ dB on $A2$, $+4.8$ dB on $A6$, $+6.3$ dB on $A9$, $+5.1$ dB on $A10$, and $+8.4$ dB on $A12$. Since the highest gain is $+12.4$ compared to the peak output of -6 dB, the output amplifier should be boosted by $+18$ dB, which is equivalent to a factor of 7.94 ($= 10^{18/20}$). After gain redistribution is achieved by changing all the resistors connected to the output of $A13$ from $1 \cdot kR$ to $8 \cdot kR$, this modification raises the $A13$ output from $-\mathbf{V}_4$ to $-8 \cdot \mathbf{V}_4$, while keeping all other amplifier outputs the same. With this gain redistribution, the leapfrog circuit provides a good active realization for the network function of the given BP prototype ($+18$ dB).

6.3 Summary

The above examples conclude our introduction to leapfrog design procedures. The method provides a good basis for replicating the performance of tabulated passive filters. An advantage is that the network function is not required in order to obtain a good circuit design. The basis for the design procedure was predicted to provide circuits with low sensitivity factors; however, the theory does not account for amplifier errors. Thus, the design does not appear to perform much better than a state variable circuit.

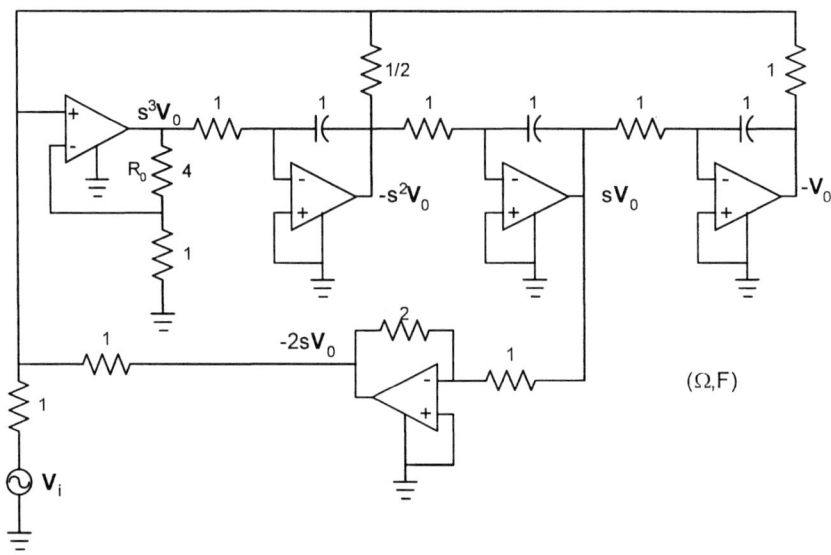

Figure 6.33. Circuit for Problem 6.1.

Disadvantages of the leapfrog design include the requirement for more opamps than the state variable circuit and the fact that the determination of peak gain is awkward when compared to the state variable filter. Nevertheless, the leapfrog is quite often the basis for circuits that are implemented using the switched capacitor topology, hence the need to learn about their design procedure. The switched capacitor topology substitutes switches and capacitors to model charge transfer through resistors, thus saving appreciable area in integrated circuit implementation of active filters. The next chapter provides an introduction to switched capacitor active filter design procedures.

6.4 Problems

6.1 In the circuit in Figure 6.33, the transmission function may be written as follows. Determine the sensitivity of the a coefficient to the resistor, R_0 (i.e., find $S_{R_0}^a$).

$$T(s) = \frac{-\mathbf{V}_0}{\mathbf{V}_i} = \frac{K}{s^3 + as^2 + bs + c}$$

6.2 The purpose of this problem is to observe stability, or the lack of it, between the S-K circuit and any state variable circuit that you use to make the same function (including gain) as the S-K circuit.

(a) Design prototype circuits for the following network resonator function. Specify K, and use resistors that have temperature coefficients of $+500$

6.4. PROBLEMS

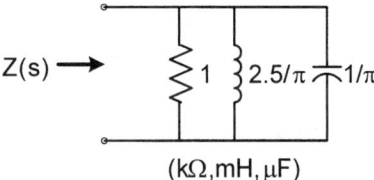

Figure 6.34. Circuit for Problems 6.4 and 6.5.

ppm/°C and capacitors with +300 ppm/°C. Scale $\omega = 1$ rad/sec to 100 Hz and $R = 1\ \Omega$ to 5 kΩ. Use μCAP to check gain versus frequency over the -20-dB bandwidth (relative to peak gain) for temperatures spanning 0°C to 75°C. Use stepped response to plot the curves on the same plot for different temperature values.

$$H(s) = \frac{+K}{s^2 + 0.1s + 1}$$

(b) Analyze the circuits designed in part a so as to obtain bandwidth sensitivity factors with respect to the capacitors (i.e., $S_{C_1}^{2\alpha}$ and $S_{C_2}^{2\alpha}$). Note that you will need to find the parameter 2α as a function of all circuit element values.

6.3 An active filter, complex-pole, squared magnitude depends upon circuit elements as follows. If all resistors have a temperature coefficient of +500 ppm/°C, what temperature coefficient is required for all capacitors so that the temperature coefficient for ω will be zero?

$$\omega_0^2 = \frac{1}{R_1 R_2 C_1 C_2}\left(1 + \frac{R_4}{R_3}\right)$$

6.4 In the tuned circuit of Figure 6.34, the nominal design elements are given for an operating temperature of 25°C. The circuit is built in production using elements whose values are uniformly distributed across a range equal to the nominal value ±5%. Find and plot limits in the s-plane for the location of the second-quadrant complex pole for the impedance function of the tuned circuit. Note that this can be done either by Monte Carlo simulation or in Matlab by analysis of the equation of the expected pole location as a function of element value extremes. Specify and plot the limits with respect to the nominal pole value.

6.5 Repeat Problem 6.4 for the case where the nominal element values have zero error at $T = 25$°C, but the operating temperature ranges from -50°C to $+100$°C. The element temperature coefficients are as follows: $TC_R = +500$ ppm/°C; $TC_C = -200$ ppm/°C; and $TC_L = +400$ ppm/°C.

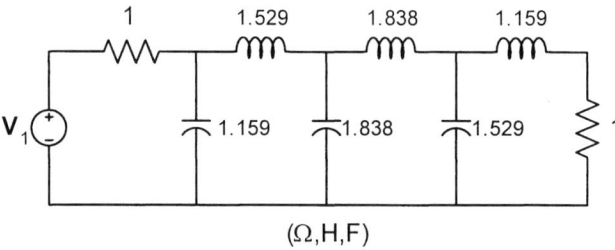

Figure 6.35. Circuit for Problem 6.7.

6.6 Modify the inductor, in the passive circuit in Figure 6.8 to have a parasitic winding capacitance in parallel with the inductance so that the branch resonates at 1 MHz. Repeat the simulation for the passive filter over the same frequency range as was used in Example 2 and observe degradation to the stopband response. Discuss your results. Does the value of the resonating capacitance qualify as a parasitic value?

6.7 The purpose of this problem is to check whether lowpass leapfrog circuits are actually less sensitive than state variable circuits that yield the same network function. The test is to be made against element value variations by using Lot values of 5% on both circuits.

   (a) Design a leapfrog filter for the LP filter given in Figure 6.35.

   (b) Find the network function for the LP filter shown in Figure 6.35, and use this function to design a state variable circuit.

   (c) Redistribute the gain at the output amplifier for designs (a) and (b) to overcome the flat loss of 6 dB caused by the equal terminations; and then, scale both networks to 4 kΩ and 10 kHz. Collect 10 Monte Carlo runs for element value effects and compare results for both designs. In particular, compare passband sensitivities over the range 100 Hz $< f <$ 10 kHz, using an output magnitude scale with as few dB peak-to-peak as possible.

6.8 Design a prototype active leapfrog circuit for the passive LP filter in Figure 6.36. Scale both passive and active circuits by 10 kHz and 10 kΩ. Run μCAP to compare the responses of both circuits using 2% Lot variations for all capacitors and resistors. Run 10 cases, and show frequency response plots for each filter under the Monte Carlo routine to observe the leapfrog circuit's sensitivity to element values as compared to its generating passive filter.

6.9 Design a corresponding leapfrog circuit for the passive filter in Figure 6.37. Scale the circuit using the same factors as in Problem 6.8 and use μCAP to obtain the nominal frequency response for your design. A Monte Carlo run is not necessary for this problem. Look at frequencies out to at least 10 MHz to

6.4. PROBLEMS

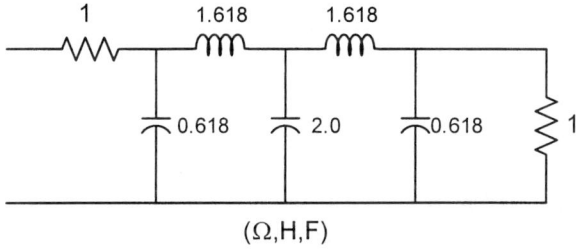

Figure 6.36. Circuit for Problem 6.8.

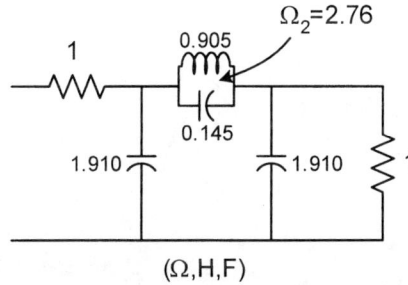

Figure 6.37. Circuit for Problem 6.9.

determine how well the active filter provides the desired filter response. Does the apparent high-frequency short in the analog for the series tank (parallel LC) circuit cause any problems with the filter response? How many opamps are required in order to design this filter using a state variable approach?

6.10 Find the sensitivity factor, S_a^T, for $T(s)$ with respect to parameter a for the given lowpass prototype function. What is the behavior of the sensitivity factor for small ω values?

$$T(s) = \frac{0.16s^2 + 1}{s^3 + as^2 + 1.84s + 1}$$

6.11 Use the variables defined in the passive BP filter in Figure 6.38 to show that the equation flow graph, given in Figure 6.39, correctly represents the first circuit. Convert the equations to a leapfrog prototype realization for the passive filter.

6.12 Find the network function, $\mathbf{V}_2/\mathbf{V}_g$, for the RLC resonator shown in Figure 6.40, and show that 2α and ω_0^2 have the following dependence upon the circuit element values. Use $G_1 = 1 = G_2$, $C = 2Q$, and $L = 1/(2Q)$ for nominal values.

$$2\alpha = (G_1 + G_2)/C \text{ and } \omega_0^2 = 1/(LC)$$

252 CHAPTER 6. NETWORK SENSITIVITY AND LEAPFROG FILTERS

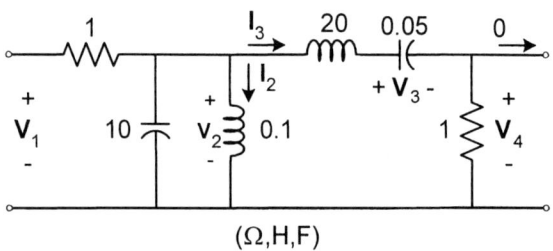

Figure 6.38. Circuit for Problem 6.11.

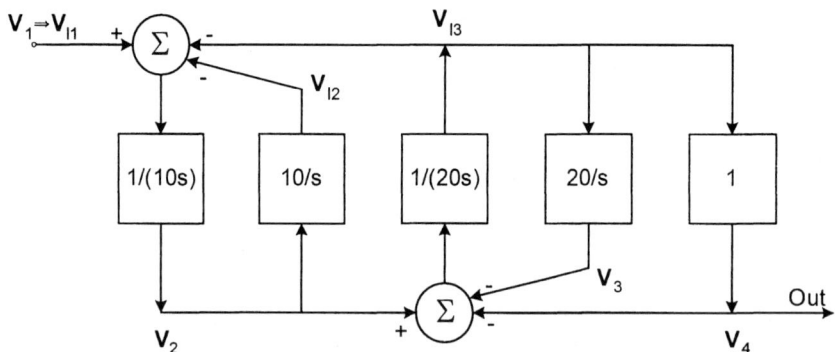

Figure 6.39. Equation flowgraph for Problem 6.11.

6.4. PROBLEMS

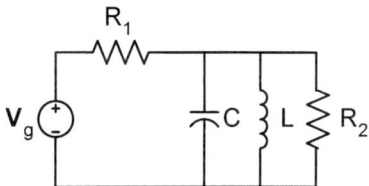

Figure 6.40. Circuit for Problem 6.12.

(a) Evaluate the total derivatives for both 2α and ω_0^2 in terms of the nominal circuit parameters and their stability factors (i.e., dG/G, dC/C, and dL/L). *Note*: Do not substitute nominal values until you have determined the complete equation for $d(2\alpha)$ and $d(\omega_0^2)$. Is it possible to choose temperature coefficients for R, L, and C so that both differentials can be zero? Explain your answer.

(b) Determine the total variance for 2α in terms of the four element variations and the element nominal values.

6.13 This problem wants to compare the S-K active resonator with the passive resonator of Problem 6.12 using the S-K circuit parameters, C_1, C_2, G_1, and G_2. The parameter μ is treated as a constant since it is the ratio of two resistors of the same type.

(a) Find both $d(2\alpha)$ and $d(\omega_0^2)$ in terms of the four parameters, and evaluate them for the following two sets of possibilities:

(1) $\mu = 1$, $G_1 = 1 = G_2$, $C_1 = 2Q$, and $C_2 = 0.5/Q$
(2) $\mu = 2$, $C_1 = 1 = C_2$, $G_1 = 1/Q$, and $G_2 = Q$

Is it possible to choose temperature coefficients so that both differentials can be zero? Explain your answer.

(b) Determine the total variance for 2α in terms of the four element variations and the element nominal values for both cases. Is one set better than the other? Explain your answer.

6.14 Using no more than five opamps, design an ordinary prototype leapfrog circuit for the circuit given in Figure 6.41. Insert gain (without adding extra opamps) so that the filter will have a passband gain of 0 dB.

6.15 Design a leapfrog active filter for the passive filter given in Figure 6.42.

6.16 Design a leapfrog filter for the RLC prototype network given in Figure 6.43. What type of filter is provided by this circuit? Explain your answer.

6.17 Design a leapfrog circuit for the prototype shown in Figure 6.44. Specify all element values and use as few opamps as possible to achieve the design.

254   CHAPTER 6. NETWORK SENSITIVITY AND LEAPFROG FILTERS

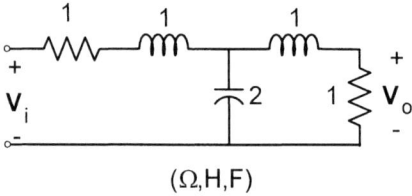

Figure 6.41. Circuit for Problem 6.14.

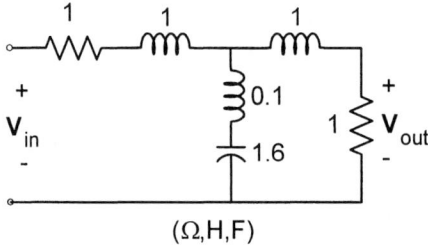

Figure 6.42. Circuit for Problem 6.15.

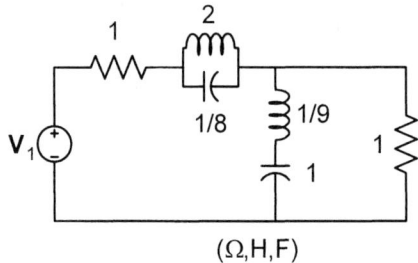

Figure 6.43. Circuit for Problem 6.16.

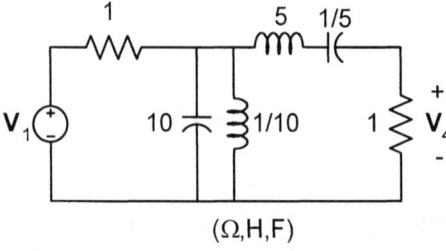

Figure 6.44. Circuit for Problem 6.17.

6.4. PROBLEMS

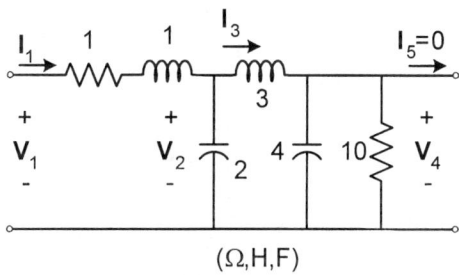

Figure 6.45. Circuit for Problem 6.18.

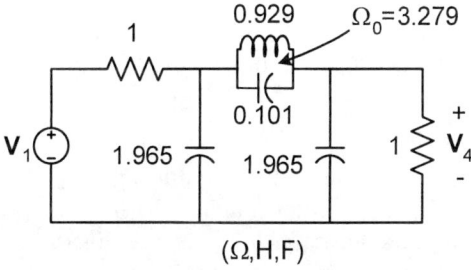

Figure 6.46. Circuit for Problem 6.19.

6.18 Design a prototype leapfrog active circuit to realize the lowpass circuit given in Figure 6.45.

6.19 Design a leapfrog filter for a bandpass equivalent of the circuit in Figure 6.46. A filter Q of 10 is desired.

(a) Find the required passive BP prototype with center frequency at 1 rad/sec, and convert the resulting BP series branch LC tank or trap to the series connection of two LC tanks as developed in Chapter 2. Specify all element values for this prototype.

(b) Design a corresponding leapfrog circuit for the answer to part a. Scale your answer by 1 kΩ and 1 kHz, and use μCAP to check the correctness of design for both the active and passive BP filters.

6.20 Design a leapfrog filter for the 1-dB third-order elliptic LP filter given in Figure 6.47. Scale the active design from 1 Ω to 5 kΩ and 1 rad/sec to 1 kHz, and check the design using a μCAP simulation. You can compare the simulated response to theory by running the passive version (appropriately scaled) of the same simulation off of the same input source and plotting output curves on the same graph, or plotting error (differences) on a separate graph.

256   CHAPTER 6. NETWORK SENSITIVITY AND LEAPFROG FILTERS

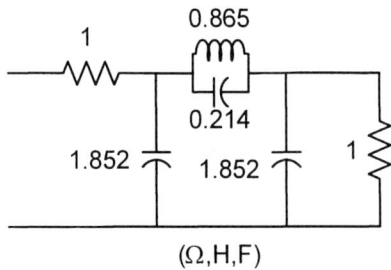

Figure 6.47. Circuit for Problems 6.20 and 6.21.

6.21 Use the filter in Problem 6.20 to design a bandpass filter having a Q of 10, and obtain a leapfrog design for the bandpass filter. Use the same impedance and frequency scale factors as in Problem 6.20, and compare the response of the leapfrog filter to the theoretically expected response.

6.22 Use simulation to check the amplifier gain distribution for nominal element values at 25°C at each opamp output for the filter in Problem R-20. Discuss your results, and show how the leapfrog architecture can be appropriately modified, where necessary, to obtain suitable gain distribution for the circuit. Do you think sensitivity to element values would be affected by gain redistribution in the leapfrog circuit?

6.23 (a) Use the LP passive prototype circuit in Figure 6.47 to design a leapfrog half-octave bandpass filter. The filter should have Q of 1.414 centered at 1 kHz. Use a 10-kΩ impedance factor, and provide a peak gain of +3 dB in the passband of the filter. Check your design to see that it provides the correct response.

  (b) Use +150 ppm/°C temperature coefficients for just the capacitors in the design for the half-octave BP filter, and check the sensitivity of the filter responses to a +50°C temperature change.

6.24 Design a leapfrog prototype circuit for the coupled coil circuit shown in Figure 6.48. Explain why it is desirable for the squared mutual inductance, M^2, to be less than the product of $L_1 L_2$. Note that the coupled coils can be represented with the following set of equations (models):

$$\text{from} \quad \mathbf{V}_2 = L_1 s \mathbf{I}_2 - M s \mathbf{I}_3 \text{ and}$$
$$\mathbf{V}_4 = M s \mathbf{I}_2 - L_2 s \mathbf{I}_3 \text{ rearrangement yields}$$
$$\mathbf{I}_2 = \frac{1}{L_1 s} \mathbf{V}_2 + \frac{M}{L_1} \mathbf{I}_3$$
$$\mathbf{I}_3 = \frac{-1}{L_2 s} \mathbf{V}_4 + \frac{M}{L_2} \mathbf{I}_2$$

6.4. PROBLEMS

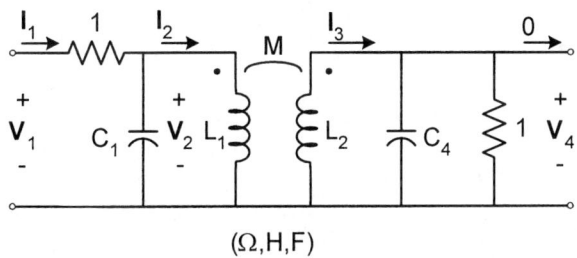

Figure 6.48. Circuit for Problem 6.24.

Reference

[1] Orchard, H. J., "Inductance Filters," *Electronic Letters*, Vol. 2, Sept. 1966.

Chapter 7

Switched Capacitor Concepts

The CMOS IC-fabrication technology has evolved as a practical choice for the implementation of low-power, relatively fast, signal-processing devices. The process is efficient, reliable, and low cost, while providing a high density of transistor-like components. A drawback to the process is that the fabrication of resistors uses large amounts of area, and so, the switched capacitor concept was developed to virtually eliminate the requirement for resistors in the fabrication of active filter components.

The purpose of this chapter is to introduce the switched capacitor equivalent of resistance concept and to show how this technology can be used to implement complex filter designs on a chip. This chapter does not go into the theory of the design or operation of CMOS devices. The student is referred to their basic electronics course textbooks or notes. A review of CMOS design concepts is presented in [1].

7.1 The CMOS Switch or Transmission Gate

The circuit shown in Figure 7.1 illustrates the conceptual design for a CMOS switch. It is often referred to as a transmission gate, and it allows current to flow in either direction. This action is modeled with a simpler schematic as shown in Figure 7.2. This latter schematic just indicates the voltage controlled action of the switch; however, inspection of Figure 7.1 can provide some limitations on the range and values of the gate voltage, $\mathbf{V}_{gate}$, in order to obtain proper switch action.

Let $\mathbf{V}_G = \mathbf{V}_{gate}$ for the following discussion. In Figure 7.1, we note that $\mathbf{V}_G$ is applied directly to the gate terminal of the NMOS device, while an inverted version of $\mathbf{V}_G$ is applied to the gate terminal of the PMOS device. This arrangement allows both MOS devices to be in a conducting state, or ON, at the same time, thus allowing load current to flow in either direction as required by the load/source combination. The source-to-drain resistance for this ON state is on the order of a few kilohms (e.g., $R_{sw} = 10$ kΩ for the ON state).

Likewise, when the gate voltage is less than the gate-to-source threshold voltage, both MOS devices are cut off and are in the OFF state. The source-to-drain resistance for the OFF state is on the order of tens of megohms (e.g., $R_{sw} = 100$ MΩ for

260   CHAPTER 7. SWITCHED CAPACITOR CONCEPTS

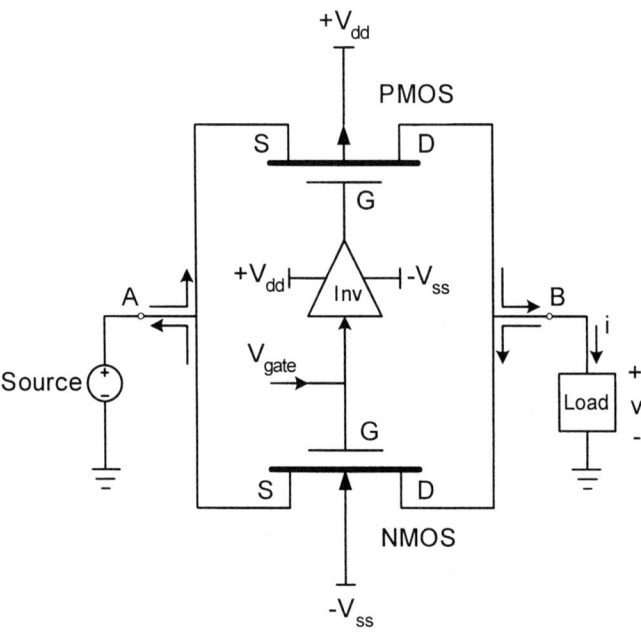

Figure 7.1. CMOS gate switch concept that allows bipolar current flow.

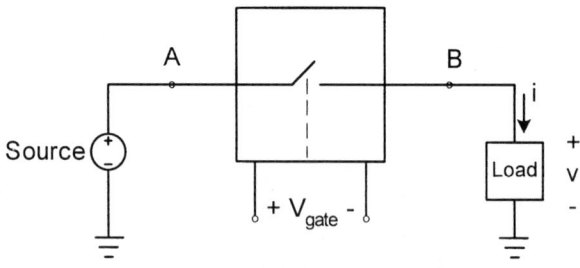

Figure 7.2. Simplified μCAP (or SPICE) model for the CMOS switch in Figure 7.1.

7.1. THE CMOS SWITCH OR TRANSMISSION GATE

the OFF state). The ratio of OFF-to-ON resistance is thus $10^8/10^4 = 10^4/1$, or 80 dB. An 80-dB discrimination ratio makes for a good electronic switch; however, one has to keep in mind that the ON resistance is a long way from being a short circuit. As we will see, the switch's ON resistance is an important design consideration.

Limitations on the range of the gate voltage are determined through the following considerations.

In order for $\mathbf{V}_G$ to turn on the NMOS device, shown in Figure 7.1, (7.1) must be true:

$$\begin{aligned} \mathbf{V}_{GS} &= \mathbf{V}_G - \mathbf{V}_S > \mathbf{V}_{TH} \text{ for ON} \\ \mathbf{V}_{GS} &= \mathbf{V}_G - \mathbf{V}_S < \mathbf{V}_{TH} \text{ for OFF} \end{aligned} \qquad (7.1)$$

$\mathbf{V}_{TH}$ is the threshold voltage for the device.

For bipolar operation, we have the input range given by (7.2):

$$\begin{aligned} -\mathbf{V}_m &\leq \mathbf{V}_S \leq +\mathbf{V}_m \\ \text{or} \quad |\mathbf{V}_S| &\leq \mathbf{V}_m \end{aligned} \qquad (7.2)$$

Substitution for the range of the source voltage into (7.1) yields an associated range for the gate voltage as follows:

$$\begin{aligned} \mathbf{V}_G &> \mathbf{V}_{TH} + \mathbf{V}_m \text{ for ON} \\ \mathbf{V}_G &< \mathbf{V}_{TH} - \mathbf{V}_m \text{ for OFF} \end{aligned} \qquad (7.3)$$

Normally, the gate control and the inverter output are operated at CMOS logic levels, and both $\mathbf{V}_G$ and $\mathbf{V}_{in}$ lie between the limits of $\mathbf{V}_L$ for the logic low level and $\mathbf{V}_H$ for the logic high level. In addition, the threshold voltage is negative for the PMOS device. We thus get the following conditions on both devices for proper control:

ON:

$$\begin{aligned} \mathbf{V}_H &> \mathbf{V}_{TH} + \mathbf{V}_m \text{ for NMOS} \\ \mathbf{V}_L &< -\mathbf{V}_{TH} - \mathbf{V}_m \text{ for PMOS, or} \\ -\mathbf{V}_L &> \mathbf{V}_{TH} + \mathbf{V}_m \end{aligned}$$

OFF:

$$\begin{aligned} \mathbf{V}_L &< \mathbf{V}_{TH} - \mathbf{V}_m \text{ for NMOS, or} \\ -\mathbf{V}_L &> -\mathbf{V}_{TH} + \mathbf{V}_m \\ \mathbf{V}_H &> -\mathbf{V}_{TH} + \mathbf{V}_m \text{ for PMOS} \end{aligned} \qquad (7.4)$$

Clearly, $\mathbf{V}_H$ and $-\mathbf{V}_L$ have the same constraint set by the sum, $\mathbf{V}_{TH} + \mathbf{V}_m$, providing the worst case. Thus, it is proper to set $\mathbf{V}_H = +\mathbf{V}_0$ and $\mathbf{V}_L = -\mathbf{V}_0$, which provides a symmetrical bipolar logic family for the CMOS components. Typically,

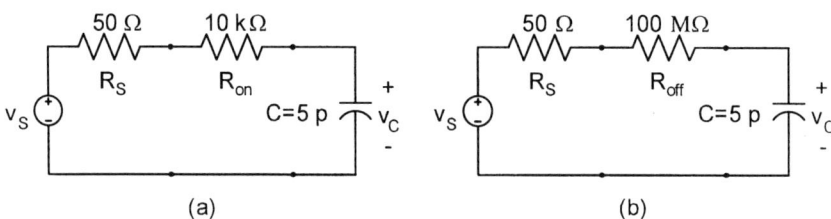

Figure 7.3. A switched capacitor alternately connected to a source: (a) ON, and (b) OFF.

for $V_{TH} = 1.5$ V and $V_m = 1.0$ V, $V_0 > 2.5$ V is required to properly control the switch state.

Note that the above discussion also applies to the gate-to-drain voltage in order to guarantee full bipolar current flow for the switch. However, since the drain voltage is usually less than or equal to the source voltage in this configuration, similar results are obtained for the limits on V_H and V_L for the logic control signals. We will not go through a similar discussion for the gate-to-drain voltage since CMOS transistors are symmetrical devices.

7.1.1 The Switch Clock Rate and Sampling

The switched capacitor concept involves alternate charging and discharging to effect charge transfer. This is visualized as the value that transfers the same charge per cycle as does the capacitor. Thus, the equivalence is based on a measure of total charge per cycle. The implementation strongly represents sampling ideas, especially in making decisions about an appropriate clock rate at which to switch the capacitor charge.

Figure 7.3 illustrates the two states of a capacitor connected to a source through a CMOS transmission gate. In Figure 7.3(a), the switch is ON, and so, the gate resistance is R_{ON}, which yields the time constant, $\tau_{ON} = 50$ nsec, for the given values.

$$\tau_{ON} = R_{eq}C = (R_S + R_{ON})C \approx 50 \text{ nsec} \tag{7.5}$$

When the switch is OFF, as in Figure 7.3(b), the gate resistance is R_{OFF}, which yields the time constant, $\tau_{OFF} = 500$ μsec, for the given values.

$$\tau_{OFF} = R_{eq}C = (R_S + R_{OFF})C = 500 \text{ μsec} \tag{7.6}$$

The ratio between τ_{OFF} and τ_{ON} is essentially the switch discrimination factor, R_{OFF}/R_{ON}, which allows the switched capacitor to acquire a charge (sample) and then hold the charge for a relatively long time (compared to the acquisition time), as shown in Figure 7.4. In order to acquire the signal with good accuracy during half of the clock period, the half-period should be greater than five charging time constants.

7.1. THE CMOS SWITCH OR TRANSMISSION GATE

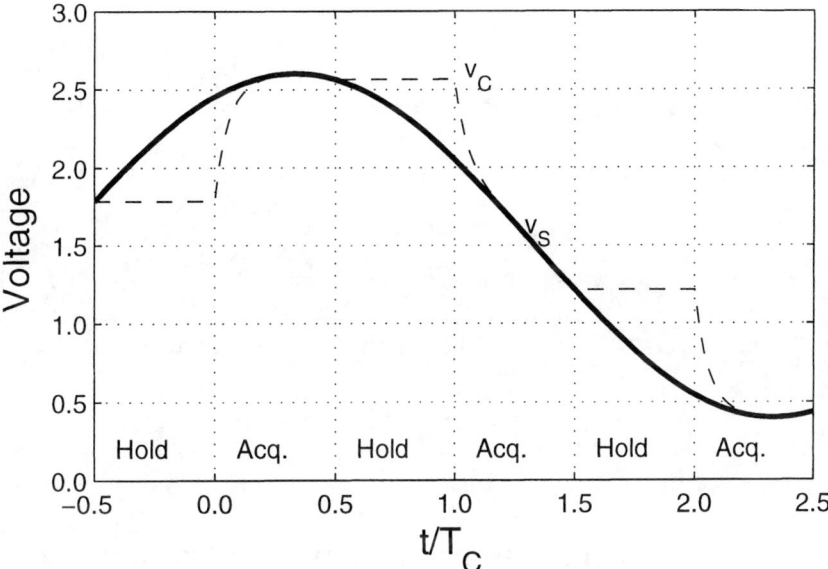

Figure 7.4. The switched capacitor waveform acquiring the source voltage in one half-cycle (switch ON) and holding the acquired value for one half-cycle (switch OFF).

$$0.5T_C > 5\tau_{ON} = 5(R_S + R_{ON})C$$

For $R_{ON} \gg R_S$, the following condition is required for the switching clock rate:

$$T_C > 10R_{ON}C \tag{7.7}$$

For the given typical values, $T_C > 500$ nsec, or $f_C < 2$ MHz, is required to accurately acquire the source voltage in one half-period.

Once the signal is acquired, it is usually desirable to hold it for the second half-cycle with a minimal change in the held value. The change in voltage on a capacitor in a source-free circuit for small changes is linear with time so that (7.8) represents the approximate change in half a clock period for an initial value of V_{max}:

$$\Delta V_m \simeq \frac{V_{max}}{\tau_{OFF}} \frac{T_C}{2} < V_{max}\rho \tag{7.8}$$

ρ is a desired fractional change allowed for the discharge during the OFF time of the switch.

When (7.7) and (7.8) are combined, we get the following limits on the switching clock period:

$$10R_{ON}C < T_C < 2\rho R_{OFF}C \tag{7.9}$$

In this result, the source resistance is neglected in comparison to both R_{ON} and R_{OFF}. This last result tells us that the specification for the allowable fractional discharge per cycle, ρ, is limited by the switch discrimination factor, R_{OFF}/R_{ON}.

$$\frac{\Delta V_m}{V_{max}} \leq \rho > \frac{5R_{ON}}{R_{OFF}} \tag{7.10}$$

ρ cannot be arbitrarily specified without consideration of the switch resistances.

7.1.2 Switch Configurations and Parasitic Capacitance

The CMOS transmission gate is used as a building block, or basic unit, to obtain different switching forms, such as single-pole-double-throw (SPDT) and double-pole-double-throw (DPDT). By itself, the transmission gate is a single-pole-single-throw (SPST) switch. These different switching architectures require the use of multiple phased clocks, depending upon the number of states (throws) per clock cycle. Double-throw switches require two-phase clocks, as shown in Figure 7.5 for a non-overlapping set of gate signals, $v_{g1} = \phi_1$ and $v_{g2} = \phi_2$. The nonoverlapping of the transmission intervals, from OFF-to-ON or ON-to-OFF, is important to establish "break-before-make" (bbm) in the capacitor switching cycle. "Make-before-break" (mbb) requires overlapping transitions for the two waveforms. It is important that bbm switching be used on capacitors so as to obtain controlled charge transfer from one node to another.

An SPDT switching circuit and simplified schematic are shown in Figure 7.6 in two forms. The schematic shown in Figure 7.6(a) shows the two gates and the clocks explicitly, whereas the schematic in Figure 7.6(b) adopts a symbolic representation. In the latter figure, the conventional SPDT switch is shown with

7.1. THE CMOS SWITCH OR TRANSMISSION GATE 265

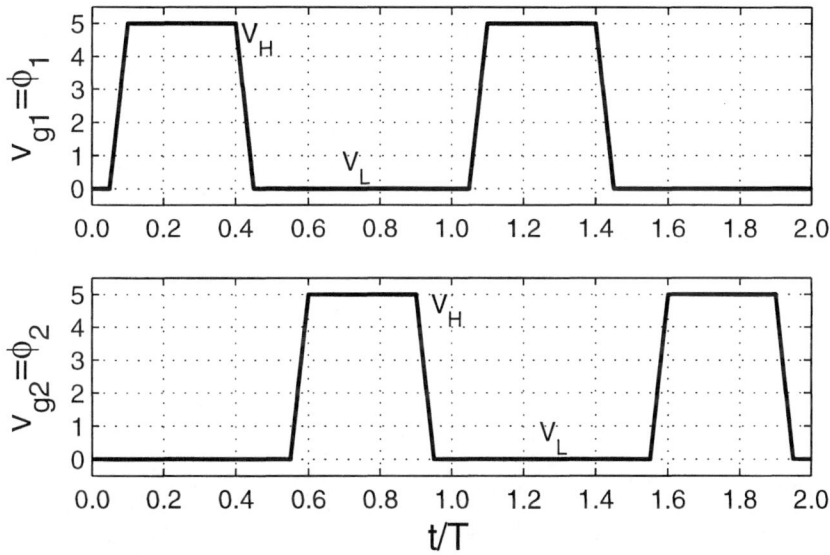

Figure 7.5. A nonoverlapping two-phase clock.

an attached arrow to show the direction of closure for the first half clock cycle. The opposite state occurs for the second half-cycle. The symbolic circuit is much simpler than the one showing each switch plus all clock sources and their connections. The reduced complexity is essential for developing schematics that require tens of switched capacitors in their implementation.

Two SPDT switches are used to create a DPDT switch, as shown in Figure 7.7. The switch arrows indicate the transition direction for the first half-period of the clock cycle. The opposite direction is enabled in the second half-cycle.

A practical application for DPDT switching is illustrated in Figure 7.8. During the first half-cycle the capacitor acquires the charge to sustain $v_1(kT)$, and then, in the second half-cycle, the capacitor is inverted so that $v_2(kT) = -v_1(kT)$ is obtained. Recall that the leapfrog architecture requires an inverter for each branch variable (leap), and so, it appears possible that the inverting amplifier can be replaced with a DPDT switched capacitor.

DPDT switching is also used to deal with capacitor parasitic capacitance in what is referred to as symmetrical switching. A capacitor implemented in MOS technology has parasitic capacitance associated with both the top and bottom plates [2, pp. 64–66]. Figure 7.9 illustrates a typical CMOS model for a floating capacitor, C. C_T is a parasitic capacitance associated with the top plate (layer) of the capacitor, and its value is on the order of 2% of the nominal capacitance, C. C_B is the parasitic capacitance associated with the bottom plate (layer) of the capacitor, and its value is on the order of 10% of C. The actual values of C_T and C_B depend upon the process, but C_B can vary over wide ranges from run to run. C_T depends more on

266				CHAPTER 7. SWITCHED CAPACITOR CONCEPTS

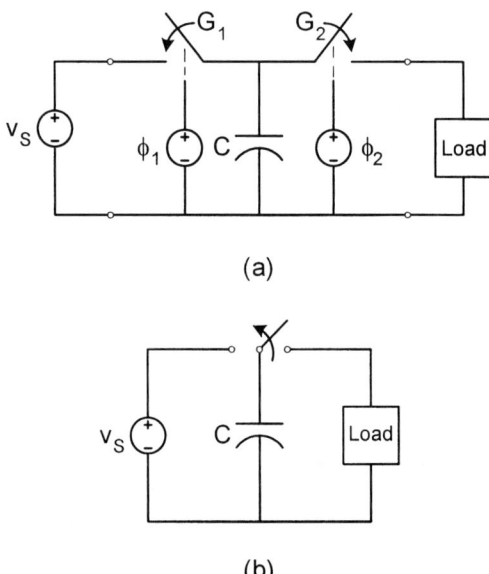

Figure 7.6. An SPDT circuit example: (a) schematic for SPDT switching, and (b) simplified schematic for SPDT switching.

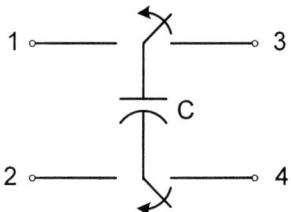

Figure 7.7. A DPDT configuration using two SPDT switches.

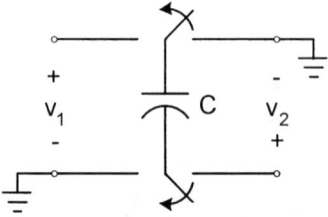

Figure 7.8. A DPDT used to obtain an inverter operation.

7.1. THE CMOS SWITCH OR TRANSMISSION GATE

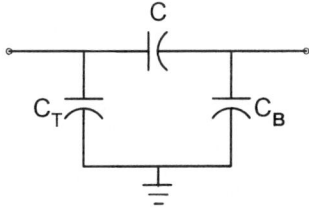

Figure 7.9. Circuit model for a CMOS capacitor.

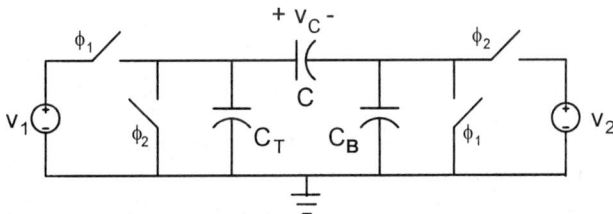

Figure 7.10. Symmetrically switched capacitor with parasitic capacitances.

the layout and has a smaller variance from run to run. A goal of good circuit design is to make the charges delivered to C_T and C_B independent of the charge flow on C.

A symmetrically switched capacitor is shown in Figure 7.10, where four SPST switches are used for the operation. The clock phases are included to provide clarity to the operation of the circuit. When ϕ_1 is ON, ϕ_2 is OFF, and so, C and C_T are connected in parallel (in the dc steady state) across the source, v_1, while C_B is connected to ground, and v_2 is isolated from C. Thus, the capacitor C obtains the exact charge to support voltage v_1 (as does C_T), since it is connected to a voltage source. When ϕ_2 is ON and ϕ_1 is OFF, the situation is reversed. C_T is discharged through its parallel switch to ground, and C and C_B are connected in parallel across the source, v_2. Again C obtains the exact charge to sustain the voltage, v_2 (as does C_B), and its charge is unaffected by the presence of C_B. The key to the entire performance is that the capacitor is switched onto voltage sources, so the parasitics are charged in parallel with the desired nominal capacitance.

We can also account for parasitic effects by using more complex differential signal circuit architectures, but such circuits will not be discussed in this text.

7.1.3 The Equivalent Resistance Concept

With all of the above background we now turn our attention to a development of the equivalent resistance concept for a switched capacitor. A typical arrangement is shown in Figure 7.11, where a capacitor, C_R, is alternately connected to two different

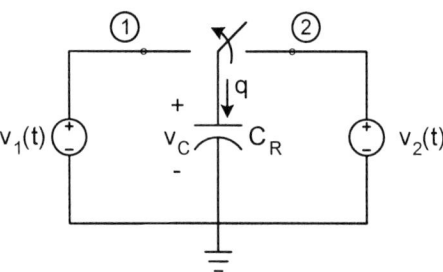

Figure 7.11. A capacitor switched between two voltages.

nodes where specific voltages are impressed, or maintained. For the purposes of this development, it is assumed that both voltages are virtually constant across a full switching period, T_C. This is a first-order approximation to the switching behavior, but it is a standard approach in sampling theory.

The capacitor is first connected to Node 1 where it charges to the value of $v_1(kT_C)$ (on the kth switching cycle). Then, on the second half of the cycle, it is connected to Node 2, where it charges to the value of $v_2(kT_C)$. This process is represented by the graph given in Figure 7.12 for the kth switching period on the capacitor. As discussed previously, the half-period, $T_C/2$, must be greater than or equal to five times the charging time constant $(R_{ON}C_R)$. Now, the net charge flowing from Node 1 to Node 2 over the period T_C is obtained as follows:

$$\begin{aligned} \Delta Q &= C_R(v_1(kT_C) - v_2(kT_C)) \text{ and} \\ \Delta T &= T_C \end{aligned} \qquad (7.11)$$

Using the definition of current, $i = \Delta Q/\Delta T$, and equating to the current obtained by placing a resistor, R_C, between Nodes 1 and 2, we arrive at the switched capacitor equivalent resistance:

$$\begin{aligned} i(kT_C) &= \frac{\Delta Q}{\Delta T} = \frac{C_R(v_1(kT_C) - v_2(kT_C))}{T_C} \\ &\equiv \frac{v_1(kT_C) - v_2(kT_C)}{R_C} \\ \therefore R_C &= \frac{T_C}{C_R} = \frac{1}{f_C C_R} \ \Omega \end{aligned} \qquad (7.12)$$

Note that this equivalent resistance does not carry the usual 2π that is present in the normal capacitor reactance relationship. Equation (7.12) illustrates the importance of using the value of v_2 at the same time as v_1; otherwise, it is not possible to write the current based on Ohm's law for a resistor. That is a key assumption for this result, but as we will see, it is an assumption that provides a useful and practical result. Based on the above assumptions, we then have the result that the switched capacitor in Figure 7.11 delivers incremental charge per switching period that is equivalent to the resistor arrangement given in Figure 7.13.

7.1. THE CMOS SWITCH OR TRANSMISSION GATE

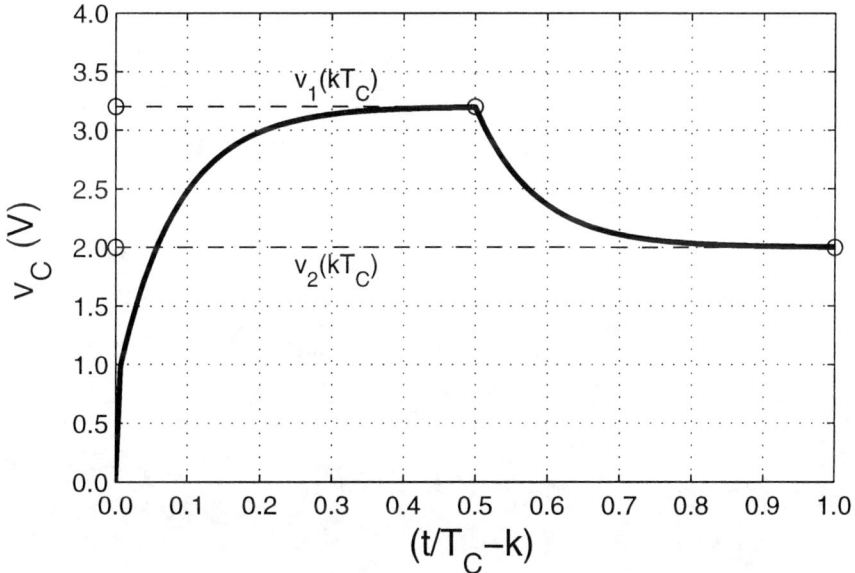

Figure 7.12. Approximate switched capacitor voltage over one full switch period.

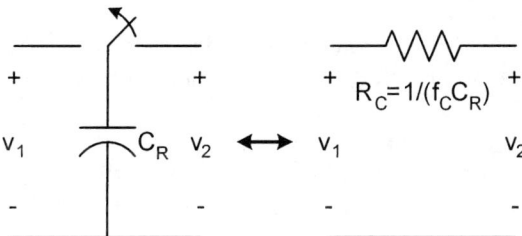

Figure 7.13. Per-period basis resistance equivalent for a switched capacitor.

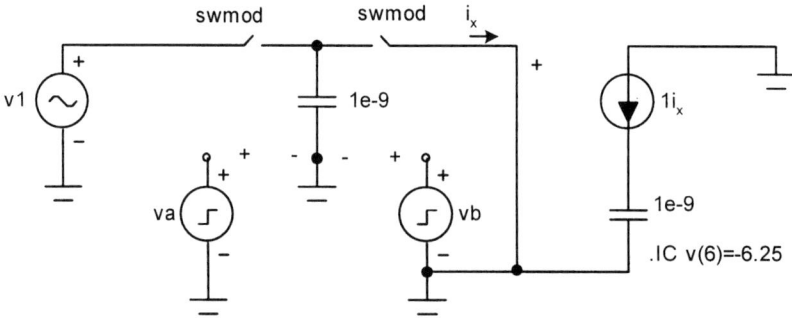

.MODEL swmod VSWITCH (RON=1e3 ROFF=1e7 VON=1 VOFF=-1)
.MODEL V1 SIN (F=1000 A=2 DC=0 PH=0 RS=1M RP=0 TAU=0)
.MODEL VA PUL (VZERO=-5 VONE=5 P1=0.1u P2=0.2u
P3=24.8u P4=24.9u P5=50u)
.MODEL VB PUL (VZERO=-5 VONE=5 P1=25.1u P2=25.2u
P3=49.8u P4=49.9u P5=50u)

Figure 7.14. Circuit used to simulate equivalent resistance effect.

Example 1 - The circuit shown in Figure 7.14 is used to simulate (in μCAP) the equivalent resistance effect for a switched capacitor. The circuit parameters are all defined in the figure. The clocks, v_A and v_B, are nonoverlapping and have a frequency of $f_C = 20$ kHz. v_A turns on and connects the capacitor at Node 2 to the sine wave source, v_1, for approximately one half-period. v_B turns on next and connects the capacitor to ground for the second half-cycle. Note that both the source, v_1, and the connection to ground represent voltage sources for the capacitor connection over the full switching period. When the capacitor is grounded, the discharge current, between the sample of v_1 and ground, is coupled by a unit-gain ideal current-controlled current source to the capacitor connected at Node 6. This output capacitor integrates the current delivered to it as a consequence of its branch $i - v$ characteristic. The switches are both modeled to have identical ON resistance of 1 kΩ and OFF resistance of 10 MΩ. The source voltage is a 1-kHz 4-V_{pp} sine wave.

The results of the simulation are shown in Figure 7.15, where v_1 and v_2 are compared on one graph, and v_6 is on the second graph. The output approximates a 12.8 V_{pp} sine wave (for an initial condition of -6.25 V), which represents the integral of the net current delivered between v_1 and ground. (The waveform is shifted by 90°, as it should be for the integral relationship.) In terms of peak-to-peak amplitudes and phasor current voltage quantities, we have the following results:

$$\mathbf{I}_{1pp} = \mathbf{V}_{1pp}/R_{eq} = 4/R_{eq}$$

7.1. THE CMOS SWITCH OR TRANSMISSION GATE

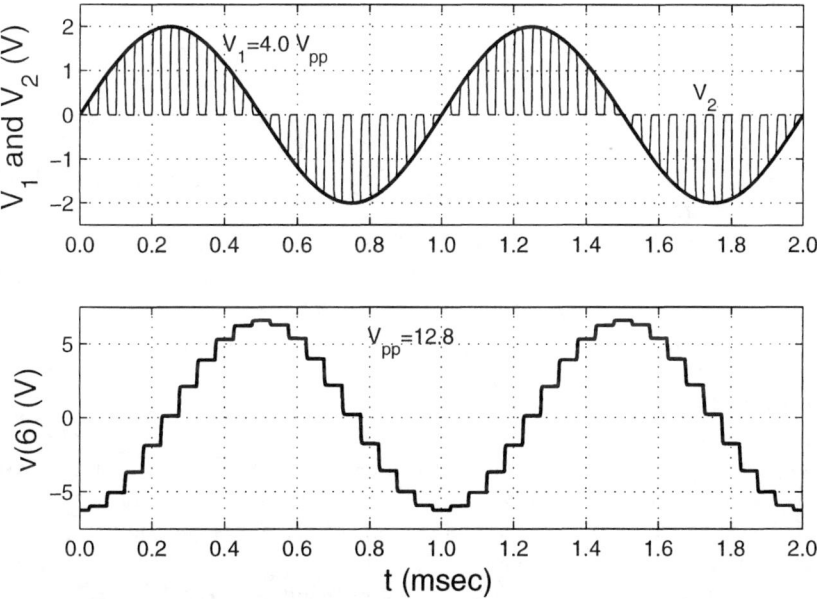

Figure 7.15. Waveforms for the switched capacitor simulation.

$$\mathbf{V}_{6pp} = 12.8 = \frac{\mathbf{I}_{6pp}}{\omega C} \text{ or}$$
$$\mathbf{I}_{6pp} = \omega C \mathbf{V}_{6pp}$$

By the control source constraint, it is required that $\mathbf{I}_{1pp} = \mathbf{I}_{6pp}$. From the simulation data, we get $\mathbf{I}_{6pp} = 2\pi 10^3 \cdot 10^{-9} \cdot 12.8 = 80.4$ μA. Using the result, (7.12), for R_{eq}, we get $\mathbf{I}_{1pp} = \mathbf{V}_{1pp} f_C C_R = 4 \cdot 20 \cdot 10^3 \cdot 10^{-9} = 80$ μA. The difference of 0.4 μA in the two estimates represents only 0.5% error in the test, and it is quite acceptable for validation of the equivalent resistance concept, or model.

The quality of the result is affected by the clock rise and fall times and the switch ON and OFF values. Slow rise times affect switching efficiency, and large ON resistances introduce switching losses. The values chosen for this example provide good performance for the test to illustrate the validity of the procedure being developed.

7.1.4 Typical Resistance and Clock Frequencies

The equivalent resistance for switched capacitance is based on a per-period condition. Both node voltages, v_1 and v_2, are referenced to the same time, kT_C, and they do not change significantly over one clock period, T_C. The capacitors are switched at a frequency, f_C, which is larger in magnitude than the complex frequency, s, that is used to represent signals in linear network analysis. Under the complex frequency concept, s appears in the argument of complex exponentials, as in (7.13):

$$v(t) = \mathbf{V}(s)e^{st} \text{ so}$$

$$v_1(kT_C) = \mathbf{V}_1 e^{skT_C} \text{ and}$$
$$v_2(kT_C) = \mathbf{V}_2 e^{skT_C} \qquad (7.13)$$

The assumption is that we agree to look at responses only at specific times, $t = kT_C$, so then the resistance based on charge transfer is independent of the source frequency, s. Thus, loosely, we can use R_{eq} and its associated current for a fictitious resistor connecting Node 1 to Node 2.

For regular active filters, we have $1/RCs$ on the order of $10^4/s$ for most practical applications (i.e., typically $RC = 10^{-4}$ sec). With $C = 10$ nF, $R = 10$ kΩ is a typical circuit resistance value. For switched capacitors, with $C = 10$ pF, we get $R = 10$ MΩ as a typical value for an audio bandwidth application. The element values for switched capacitors require smaller capacitance and larger resistor values by two to three orders of magnitude.

From your digital-signal-processing courses, you should recall the sampling requirement that if a signal contains frequencies out to some f_{max}, then in order to reconstruct the sampled signal without aliasing, the original signal has to be sampled at a clock rate greater than $2f_{max}$. This fact, based on the sampled signal, provides a lower limit for the clock frequency.[1] Previous discussion, based on a settling time of five time-constants ($5\tau_{ON}$), provides an upper limit. This upper limit is strongly tied to the fabrication process and may range from 2 to 20 MHz or so. Equation (7.14) spells out the range for the choice of clock frequency, f_C:

$$2f_{max} \leq f_C \leq \frac{1}{5R_{ON}C_R} \text{ or}$$
$$f_C \leq f_{PROCESS} \qquad (7.14)$$

The highest choice depends upon which is smaller, $5\tau_{ON}$ or the $f_{PROCESS}$ limit. For example, FM radio has a maximum signal bandwidth on the order of $f_{max} = 100$ kHz, and with $1/5\tau_{ON} \approx 2$ MHz, we get the following available range for f_C:

$$200 \text{ kHz} < f_C <\approx 2 \text{ MHz}$$

Clearly, the limit allows for a broad range of applications. When we look at different circuit architectures, such as bandpass filters, we will find further restrictions on the choice of the clock frequency in order to control the range of element values required to fabricate a specific filter design.

Before we move on to some switched capacitor examples, it is helpful to indicate at least one method for obtaining a nonoverlapping two-phase clock. Note that simply inverting a logic signal will not work since the inverted signal is delayed from the clock input to the inverter. A typical generator is shown in Figure 7.16, where an inverter is used on the input to one side of a pair of NOR gates that are arranged to form a flip-flop circuit. Waveforms for the generator circuit are given in Figure 7.17. A stable clock, ϕ_A, serves as an input to the two-phase generator to control the clock frequency. The input q corresponds to ϕ_A, and the input p is a delayed and inverted version of ϕ_A. The delay is one gate-propagation delay.

---

[1] See [3, p. 36]. The frequency where aliasing occurs is called the *Nyquist* frequency.

7.1. THE CMOS SWITCH OR TRANSMISSION GATE

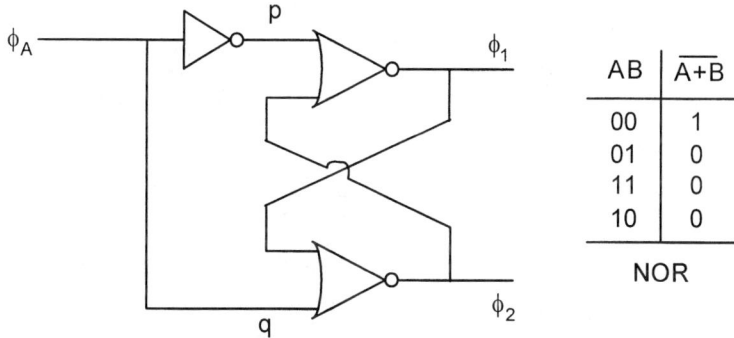

Figure 7.16. Nonoverlapping two-phase clock generator.

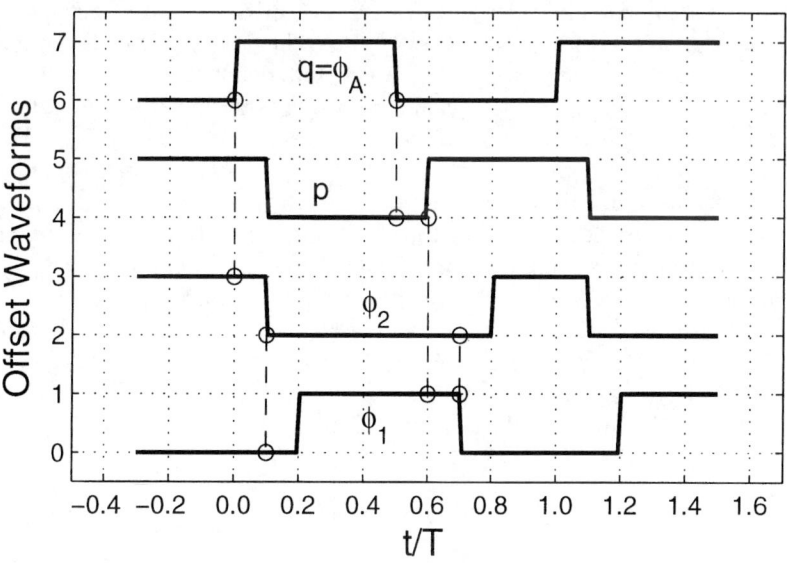

Figure 7.17. Two-phase clock generator waveforms.

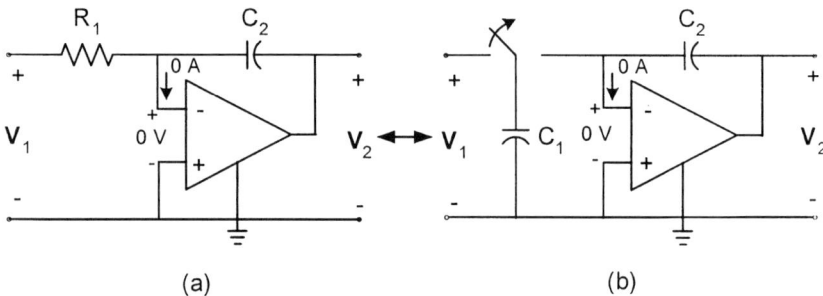

Figure 7.18. An inverting integrator: (a) analog, and (b) switched capacitor.

Note that p and q do not form a nonoverlapping set of waveforms, as mentioned above. In order to talk about the flip-flop state changes, it is helpful to note the logic combinational dependence for each output. $\phi_1 = \overline{(p + \phi_2)}$, and $\phi_2 = \overline{(q + \phi_1)}$, which means ϕ_1 is high whenever p and ϕ_2 are both low, and ϕ_2 is high whenever q and ϕ_1 are both low. They are low for any other combination. To explain the waveform in Figure 7.17, it is assumed that it takes one gate-propagation delay for a change at the input to change the NOR gate output. In the beginning, assume that with p high and ϕ_2 high, ϕ_1 has to be low and, with both q and ϕ_1 low, ϕ_2 has to be high. The assumed state is correct. The first transition is in q going low to high, and so, both p and ϕ_2 change one delay after q. Then, with p and ϕ_2 both low, ϕ_1 must go high, and this happens one additional gate delay after the p-ϕ_2 transition. The generator is now stable until q changes from high back to the low state. When q goes low with ϕ_1 high, they are different, so there is no change in ϕ_2. The next change is when p goes high at a gate delay after q goes low. With p going high, ϕ_1 changes its state from high to low one gate delay after the p transition. The change in ϕ_1 to a low state with q also low then drives ϕ_2 high, thus completing one full period. Note that the ϕ_1 high state is inside the ϕ_2 low state by a gate delay on each end of the state. Thus, the waveforms are nonoverlapping two-phase signals, as required. ϕ_1 has the same duty cycle as the input, ϕ_A, and lags ϕ_A by two logic gate delays. ϕ_2 has an altered duty cycle where its high (ON) state is reduced by two gate delays so that the duty cycle is $ON/T = 0.5 - 2\tau_g/T$. With $\tau_g << T$, the duty cycle reduction in ϕ_2 is not usually a problem.

7.2 Switched Capacitors and Analog Operations

Before we work examples, it is useful to catalog some basic analog circuit operations in their switched capacitor equivalent forms.

a. Inverting Integrator
The switched capacitor version of the inverting integrator is shown in Figure 7.18. We have the following network functions for the two circuits:

7.2. SWITCHED CAPACITORS AND ANALOG OPERATIONS

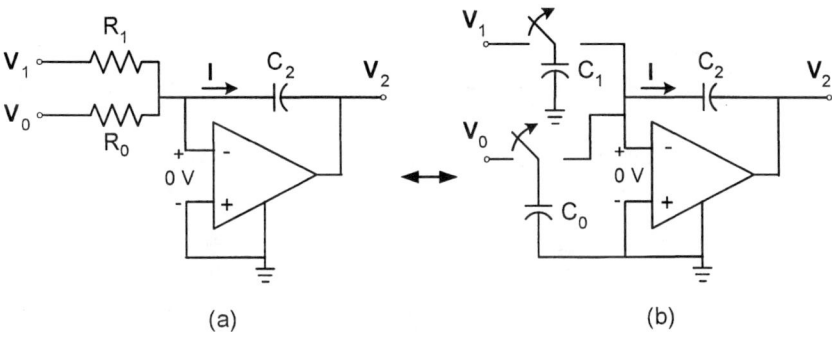

Figure 7.19. Summing and integrating circuit: (a) analog, and (b) switched capacitor.

Analog:
$$\frac{V_2}{V_1} = \frac{-1}{R_1 C_2 s}$$

Switched Capacitor:
$$V_2 = \frac{-I}{C_2 s} = \frac{-1}{C_2 s}\frac{V_1}{R_{C_1}}$$
$$\frac{V_2}{V_1} = \frac{-f_C C_1}{s C_2} \tag{7.15}$$

f_C is the switch clock frequency, and s is the signal, $v_1(t)$, complex frequency. The switched capacitor circuit is considered in terms of the equivalent resistance model for the switched capacitor.

The integrator circuit can be modified, as in Figure 7.19, to handle two (or more) inputs as a summing integrator.

b. Sum and Integrate

The network functions for the circuits in Figure 7.19 are as follows. The added input just injects more current (charge) into the feedback capacitor. Note that the switches need to be phased to transfer charge to the opamp input at the same time.

Analog:
$$V_2 = \frac{-1}{sC_2}(G_0 V_0 + G_1 V_1)$$

Switched Capacitor:
$$V_2 = \frac{-I}{sC_2} \text{ yields}$$
$$V_2 = \frac{-f_C}{sC_2}(C_0 V_0 + C_1 V_1) \tag{7.16}$$

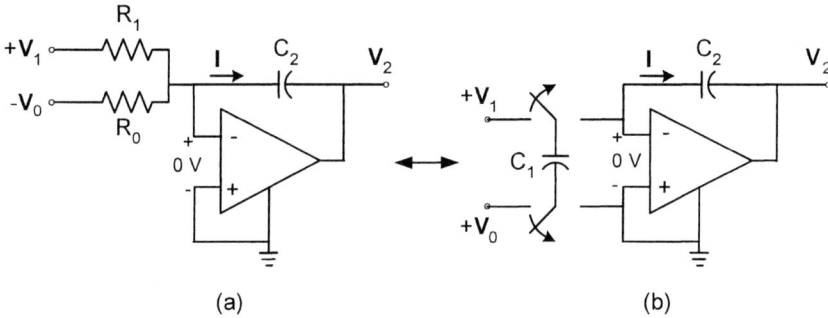

Figure 7.20. Subtracting and integrating circuits: (a) analog, and (b) switched capacitor.

With $C_0 = C_1 = C$, we get a straight summation of two signals for the switched capacitor circuit:

$$\mathbf{V}_2 = \frac{-f_C C}{s C_2}(\mathbf{V}_0 + \mathbf{V}_1) \tag{7.17}$$

c. *Subtract and Integrate*

A circuit for this operation is given in Figure 7.20. Note that the analog circuit requires one input to be inverted (sign changed) in order to create the difference, but the switched capacitor does not require the inverter in order to accomplish the difference.

Analog:

$$\mathbf{V}_2 = \frac{-1}{sC_2}(-G_0 \mathbf{V}_0 + G_1 \mathbf{V}_1)$$
$$\mathbf{V}_2 = \frac{G}{sC_2}(\mathbf{V}_0 - \mathbf{V}_1) \quad \text{for} \quad G_0 = G_1 = G$$

Switched Capacitor:

$$\mathbf{V}_2 = \frac{-f_C C_1}{s C_2}(\mathbf{V}_1 - \mathbf{V}_0) \quad \text{or}$$
$$= \frac{f_C C_1}{s C_2}(\mathbf{V}_0 - \mathbf{V}_1) \tag{7.18}$$

By switching the capacitor between two voltages, the difference can be obtained through the use of a DPDT switch rather than an inverting amplifier to change the sign of a signal. This feature is a decided advantage for switched capacitor circuits.

7.2. SWITCHED CAPACITORS AND ANALOG OPERATIONS

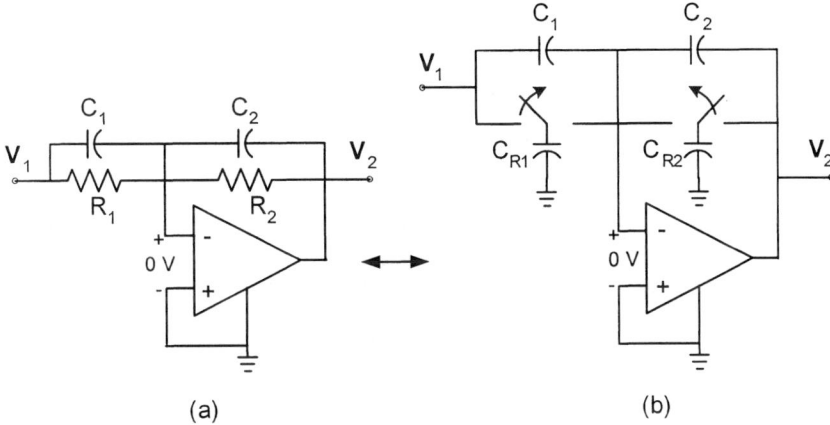

Figure 7.21. A lead-lag compensation network: (a) analog, and (b) switched capacitor.

d. Lead-Lag Compensation Networks
The type of circuit shown in Figure 7.21 allows the cascade of real poles and zeros. It is typically used for stabilizing feedback networks and as equalizers in sound systems and is a commonly occurring circuit.

Analog:
$$\frac{\mathbf{V}_2}{\mathbf{V}_1} = \frac{-(C_1 s + G_1)}{(C_2 s + G_2)}$$

Switched Capacitor:
$$\frac{\mathbf{V}_2}{\mathbf{V}_1} = \frac{-(C_1 s + f_C C_{R_1})}{(C_2 s + f_C C_{R_2})} \tag{7.19}$$

In all of the above examples, the given network functions are based on the phasor complex frequency network response functions and are presented loosely as functions of s. Strictly speaking, the switched circuits are based on discrete sample sequences and have z-transform representation. As long as the clock frequency is chosen properly, so as to prevent aliasing of the sampled signals, the given functions are all useful for circuit design purposes. In summary, the functions are given in the signal s-domain where from $v_1(kT_C) = \mathcal{R}e[\mathbf{V}_1(s)e^{skT_C}]$ and $v_2(kT_C) = \mathcal{R}e[\mathbf{V}_2(s)e^{skT_C}]$, we are working with the time-independent s-domain variables, $\mathbf{V}_1(s)$ and $\mathbf{V}_2(s)$. This domain is commonly used in most filter design procedures (e.g., the state variable filter).

Example 2 - The purpose of this example is to design a switched capacitor circuit for the single-pole analog circuit shown in Figure 7.22. The circuit is required to have its pole at $\alpha = 1000\pi$ Nep/sec (500 Hz), with 500-kΩ equivalent resistance and with switching at 40 kHz. In addition, the circuit should be tested (via simulation) to verify conformance to theory.

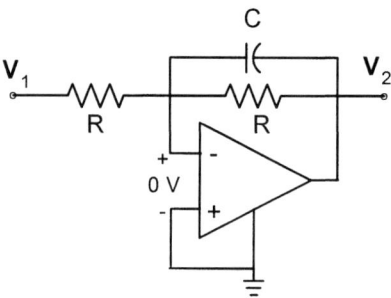

Figure 7.22. A single-pole integrator circuit.

Solution - The network function for the circuit in Figure 7.22 is obtained by inspection.

$$\frac{V_2}{V_1} = H(s) = \frac{-1}{R(Cs+G)} = \frac{-1}{1+RCs}$$

$$\equiv \frac{-1}{1+s/\alpha} \qquad (7.20)$$

where $\alpha = 1/RC$. The resistors are specified at 500 kΩ each, and so, we need the value for C so that α meets the specified pole value.

$$\alpha = 10^3\pi = \frac{1}{RC} = \frac{2 \times 10^{-6}}{C}$$

$$\therefore C = \frac{2 \times 10^{-9}}{\pi} \text{ F}$$

A switched capacitor implementation for the single-pole network function is shown in Figure 7.23. The switched capacitor required for the equivalent resistance of 500 kΩ is obtained as follows:

$$R_C = 500 \text{ k}\Omega = \frac{1}{f_C C_R} \text{ or}$$

$$C_R = \frac{1}{f_C R_C} = \frac{1}{40 \times 10^3 \times 500 \times 10^3} = \frac{10^{-9}}{20}$$

$$= 50 \text{ pF}$$

The two-phase clocks, v_a and v_b, are nonoverlapping with 25-μsec periods as required for the 40-kHz switching specification. Note that the capacitors are phased so that they both deliver charge to the opamp input at the same time.

Now, we want to test the performance of the switched capacitor circuit. Simulation of mixed signal (digital and analog) circuits is very tricky, and one must proceed carefully. Proper simulation requires transient analysis, not ac steady state, in order to account for fast rise and fall times on the switch signals, even though the

7.2. SWITCHED CAPACITORS AND ANALOG OPERATIONS

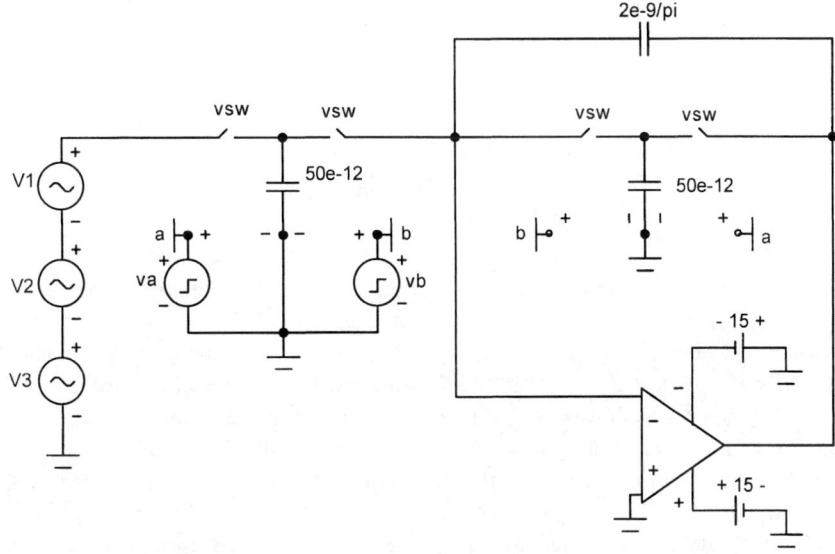

```
.IC v(5)=0 v(8)=0 v(9)=0
.MODEL VA PUL (VZERO=-5 VONE=5 P1=0.1u P2=0.2u P3=12.3u P4=12.4u P5=25u)
.MODEL VB PUL (VZERO=-5 VONE=5 P1=12.6u P2=12.7u P3=24.8u P4=24.9u P5=25u)
.MODEL V1 SIN (F=200 A=2 DC=0 PH=0 RS=1M RP=0 TAU=0)
.MODEL V2 SIN (F=2040 A=2 DC=0 PH=0 RS=1M RP=0 TAU=0)
.MODEL V3 SIN (F=10040 A=2 DC=0 PH=0 RS=1M RP=0 TAU=0)
```

Figure 7.23. A switched capacitor single-pole circuit design.

actual signal response of interest is usually low frequency in nature. Thus, one way to test for the desired frequency response is to use multiple sine waves and compute the discrete Fourier transform (DFT) for the samples collected by the simulation. When a fast Fourier transform (FFT) is used to compute the DFT spectral measures for a sine wave test, it is important that the test frequencies be chosen in an exact manner to obtain the best result. (See your digital signal processing notes.) In particular, the DFT only represents a discrete set of frequencies, $f_k = (0, 1, 2, \ldots k, \ldots N-1)F_s/N$, and when quantization is present, it is customary to select the kth test frequency to be an odd number, or relatively prime to N.

Thus, for this simulation, we arbitrarily choose N to be 1000 samples and choose three test frequencies on odd integer values of the FFT set. The resolution is $\Delta F = F_s/N = 40 \cdot 10^3/1000 = 40$ Hz, and so, with test frequency k-values of 5, 51, and 251, we get the following three test frequencies:

$$\begin{aligned} f_1 &= 5 \times 40 = 200 \text{ Hz} \\ f_2 &= 51 \times 40 = 2040 \text{ Hz} \\ f_3 &= 251 \times 40 = 10040 \text{ Hz} \end{aligned}$$

This choice puts one frequency in the passband, one in the transition band, and one into the stopband.

Now, when the simulation is run on μCAP, it is necessary to specify the time duration for the test and the number of samples to save to a file, as well as a limit on the time step size allowed for the test. Since we want 1000 samples from the simulation over the duration, $T = (N-1)/F_s = 999/40,000 = 24.975$ msec, if we only specify 1000 samples for the run, the simulator will try to use time steps that are too large. Thus, we specify a larger number of samples for the simulation to maintain integrity in the transient analysis and then down-sample the file that the simulator saves. In conclusion, for this test we specify a run time of 24.975 msec to collect 4996 samples at a maximum step size of 1 μsec. After the data are collected, the file is edited to remove the μCAP headers, and then the file is down-sampled to get the desired number of samples. The following Matlab file shows how the analysis is performed on the resulting data. Note that the resulting set of samples is windowed to reduce the effect of truncation of the sample set [4].

```
load dat7_24.dat-ascii
%3 col., Time, Vin, Vout for opamp One pole low pass

N=1e3;
Fs=40e3;
df=Fs/N;
f=df*[0:N/2-1]';
V0=2 ./(1+j*f/500); % 2 Vpk input signal ampl/sine wave
Vdb=db(abs(V0));

k=[0:N-1]';
% Sin cubed window
```

7.2. SWITCHED CAPACITORS AND ANALOG OPERATIONS

```
wk=0.75*pi*sin(pi*k/N).^3;
% rectangular window
% wk=ones(size(k));

vin=dat7_24(1:5:4996,2);
vout=dat7_24(1:5:4996,3);
VIN=2*fft(vin.*wk)/N;
VOUT=2*fft(vout.*wk)/N;

figure(1)
subplot(211)
plot(f*1e-3,db(VIN(1:N/2)),...
    f*1e-3,6.02*ones(size(f)),'r--'),grid
axis([Fs*1e-3*[0 0.5] -50 10])
ylabel('Input (dBv)','Fontsize',14)
subplot (212)
plot (f*1e-3,db(VOUT(1:N/2)),...
    f*1e-3,Vdb,'r--'),grid
axis([Fs*1e-3*[0 0.5] -110 10])
ylabel('Output (dBv)','Fontsize',14)
xlabel('Frequency (kHz)','Fontsize',14)

t=k*1e3/Fs;   sample time in msec
figure (2)
subplot (211)
plot(t,vin),grid
axis([0 25 -7.5 7.5])
ylabel('Input (V)','Fontsize',14)
subplot (212)
plot(t,vout),grid
axis([0 25 -2.5 2.5])
ylabel('Output (V)','Fontsize',14)
xlabel('Time (msec)','Fontsize',14)
```

The signal waveforms for input and output are shown in Figure 7.24. The three 2 V amplitude sine waves combine to yield a noisy input waveform that spans a ±6 V range. The output is basically the 200-Hz sine wave that is passed by the filter. The 2040-Hz component is quite evident (10 periods to 1 of the 200-Hz signal), while the 10,040-Hz signal is not very visible in this time-domain graph. Clearly, the circuit is acting like the lowpass filter it is supposed to be. Note that the time scale of the graph is large compared to the switching period of 25 μsec. If the output were viewed on a higher-resolution time scale, it would show a succession of steps after each switching period. This is normally the case where the capacitor switching frequency is large compared to the critical frequencies (poles and zeros) of a desired frequency response.

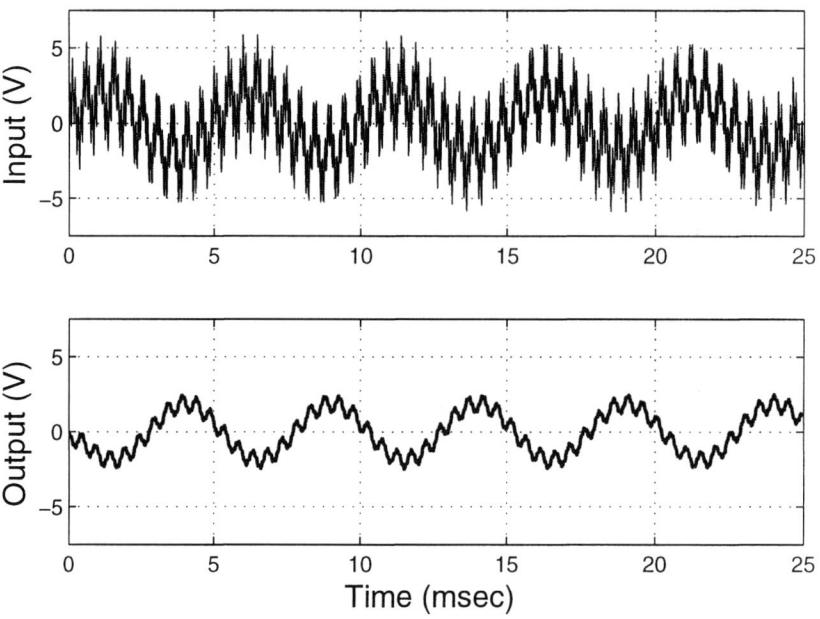

Figure 7.24. Input-output waveforms for the single-pole switched capacitor example.

7.2. SWITCHED CAPACITORS AND ANALOG OPERATIONS

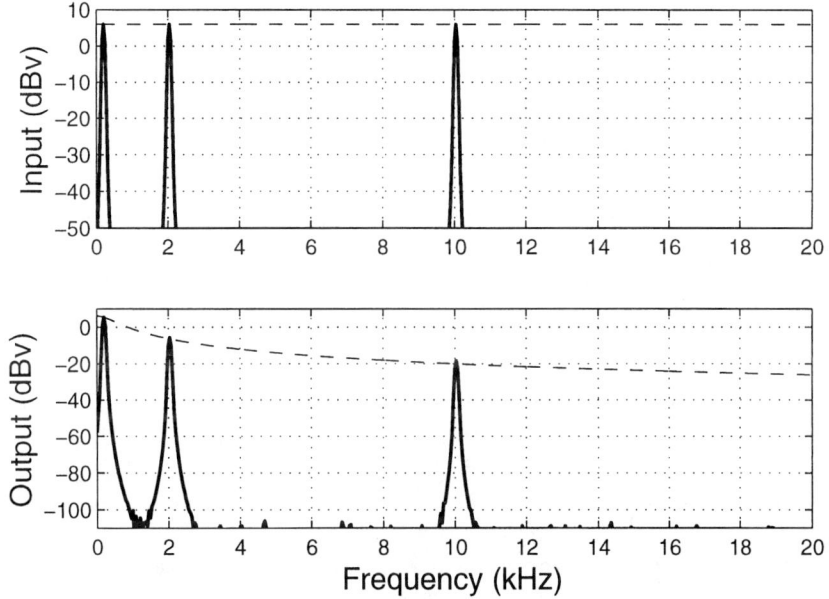

Figure 7.25. Spectral response for the single-pole switched capacitor example.

The filtering action is illustrated spectrally as shown in the graph of Figure 7.25. The DFT components are spikes for pure sine waves, and these are indicated by the peaks in the graph for both the input and output spectra.

The input consists of three sine waves all at 2-V amplitude. This spectrum is, therefore, compared to the dBv value for each phasor component:

$$\mathbf{V}_{in} = 2\angle 0° \text{ or}$$
$$|\mathbf{V}_{in}|_{dB} = +6.0$$

The dashed line on the input graph is thus set at +6 dBv for reference purposes. It shows that the sample equation and all the factors used to calculate the FFT are correct. The output spectrum is also compared to expected phasor magnitudes (in dBv) for the desired single-pole (analog) responses as an ideal measure. Here we have the following:

$$|\mathbf{V}_{out}| = |\mathbf{V}_{in}|/|1 + jf/500| = |2/(1 + jf/500)|$$

The magnitude is plotted in dBv units. The data show that the results are excellent for this filter. It provides response virtually identical to the desired single-pole analog filter.

If the cutoff frequency is increased closer to the switching frequency, the results will degenerate as there is less bandwidth available ($F_s > 2f_{max}$) to obtain a meaningful stopband for the filter. Our next example shows the design for a filter

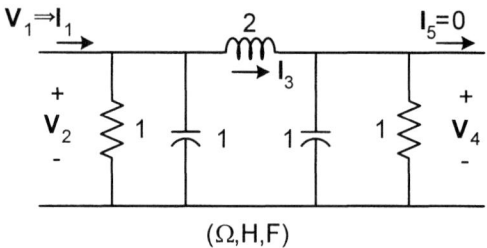

Figure 7.26. A three-pole all-pole prototype filter.

that provides a shorter transition band and better filtering versus the switching frequency.

Example 3 - Design and simulate the response for a switched capacitor implementation for the all-pole three-pole lowpass filter shown in Figure 7.26. The circuit should be frequency-scaled from 1 rad/sec to 500 Hz. In addition, the realization should use the leapfrog topology and provide a gain of 2 so as to overcome the doubly terminated loss (i.e., the switched capacitor circuit should realize $2CV_4$ in its implementation).

Solution - The leapfrog branch equations that must be duplicated are as follows:

$$\mathbf{V}_2 = \frac{1}{(s+1)}(\mathbf{V}_{I_1} - \mathbf{V}_{I_3})$$

$$\mathbf{V}_{I_3} = \frac{1}{2s}(\mathbf{V}_2 - \mathbf{V}_4)$$

$$\mathbf{V}_4 = \frac{1}{(s+1)}(\mathbf{V}_{I_3} - 0) \qquad (7.21)$$

$\mathbf{V}_2$ and $\mathbf{V}_4$ require the single-pole circuit in Figure 7.23, and $\mathbf{V}_{I_3}$ requires a switched capacitor integrator like that in Figure 7.18.

Equation (7.21) needs to be modified to provide $2\mathbf{V}_4$ instead of $1\mathbf{V}_4$. The procedure is simplified by allowing the circuit also to develop $2\mathbf{V}_2$, since $\mathbf{V}_2$ is only used to form a difference with $\mathbf{V}_4$ in the feedback for $\mathbf{V}_{I_3}$. The modified (but equivalent) branch equations are thus as follows:

$$2\mathbf{V}_2 = \frac{2}{(s+1)}(\mathbf{V}_{I_1} - \mathbf{V}_{I_3})$$

$$\mathbf{V}_{I_3} = \frac{1}{2(2s)}(2\mathbf{V}_2 - 2\mathbf{V}_4)$$

$$2\mathbf{V}_4 = \frac{2}{(s+1)}(\mathbf{V}_{I_3} - 0) \qquad (7.22)$$

Next, we need circuit scale factors for the switched capacitor realization. As in the previous example, a switching frequency, $f_C = 40$ kHz, is used along with the

7.2. SWITCHED CAPACITORS AND ANALOG OPERATIONS

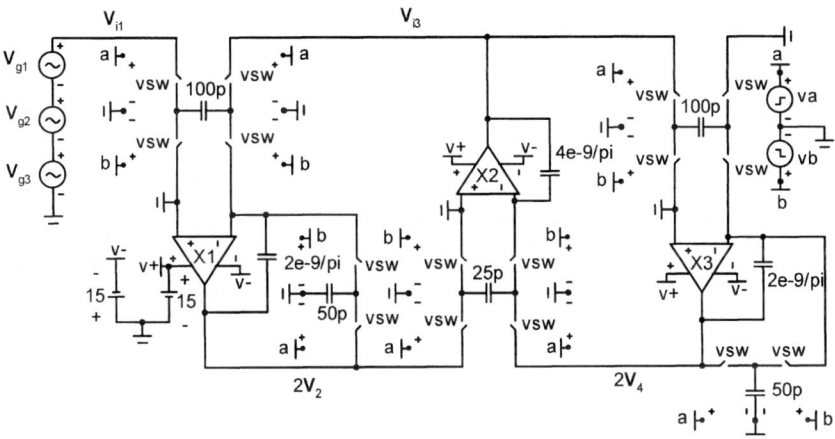

Figure 7.27. A switched capacitor implementation for the all-pole filter in Figure 7.26.

impedance scale factor, $k_R = 500$ kΩ. The switched capacitor value, C_1, to emulate a 1 Ω prototype resistor is thus 50 pF:

$$C_1 = \frac{1}{f_C k_R} = \frac{1}{40 \times 10^3 \times 10^6/2} = 50 \text{ pF}$$

The frequency scale factor is specified to scale 1 rad/sec to 500 Hz so that $k_\omega = 2\pi 500 = 1\pi 10^3$. Therefore, the scale factor for other analog capacitors is given by k_C:

$$k_C = \frac{1}{k_R k_\omega} = \frac{1}{\frac{10^6}{2}\pi 10^3} = \frac{2}{\pi} \text{ nF}$$

The equations in (7.22) are then realized (with scaled values) by the circuit shown in Figure 7.27. The current variables, $\mathbf{V}_{I_1}$ and $\mathbf{V}_{I_3}$, have nodes across the top of the circuit, and the voltages, $2\mathbf{V}_2$ and $2\mathbf{V}_4$, have nodes across the bottom of the circuit. $2\mathbf{V}_2$ is realized using the single-pole circuit with the prototype input resistance equal to 0.5 Ω and a parallel RC impedance used in the feedback circuit. (The prototype components are shown with their scaled values in Figure 7.27.) Note that switching is used to form the difference voltages, $\mathbf{V}_{I_1} - \mathbf{V}_{I_3}$ and $2\mathbf{V}_2 - 2\mathbf{V}_4$, in the switched capacitor circuit. Next $\mathbf{V}_{I_3}$ is realized using a switched capacitor multiplying integrator circuit with its prototype equivalent input resistance set to 2 Ω to cancel the factor of 2 in the difference, $2\mathbf{V}_2 - 2\mathbf{V}_4$. The prototype feedback capacitor is 2 F so as to provide the transfer, $1/2s$, for $\mathbf{V}_{I_3}$ in (7.22). Finally, $2\mathbf{V}_4$ is realized with a circuit identical to that for $2\mathbf{V}_2$ since it requires the same transfer function.

Observe the symmetry in Figure 7.27 for the implementation of the leapfrog all-pole circuit. The topology easily generalizes for nth-order all-pole circuits and finds extensive use in IC fabrication of tabulated passive filter designs.

286   CHAPTER 7. SWITCHED CAPACITOR CONCEPTS

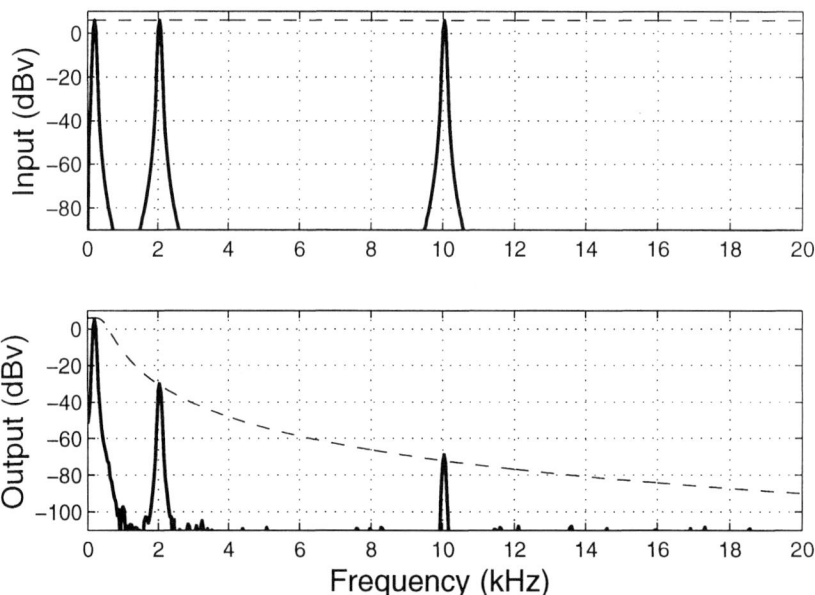

Figure 7.28. Measured spectral response for the switched capacitor design of a three-pole lowpass filter.

The final design is tested in exactly the same manner as that used in the previous example. Three sine waves are input simultaneously to test the passband, transition band, and stopband of the nominal filter design. The same frequencies and number of samples are taken in μCAP to obtain the results given in Figures 7.28 and 7.29. Figure 7.28 provides the frequency spectrum measure compared to the theoretically expected response for each sinusoid. The expected response, $\mathbf{V}_4/\mathbf{V}_1$, is easily obtained by circuit analysis to obtain the following result:

$$\frac{\mathbf{V}_4}{\mathbf{V}_1} = \frac{1}{2(s^3 + 2s^2 + 2s + 1)} = \frac{1}{2(s+1)(s^2+s+1)} \text{ and}$$

$$\frac{2\mathbf{V}_4}{\mathbf{V}_1} = \frac{1}{(s+1)(s^2+s+1)}$$

The Matlab code used for the previous example is slightly modified by changing the expression, V0, for the expected output response for each $2\mathbf{V}_4$ sine wave (and changing the data file name).

```
...
f=df*[0:N/2-1];
s=j*f/500;
V0=2 ./(1+s)./(s.^2+s+1);
VdB=db(abs(V0));
...
```

7.2. SWITCHED CAPACITORS AND ANALOG OPERATIONS

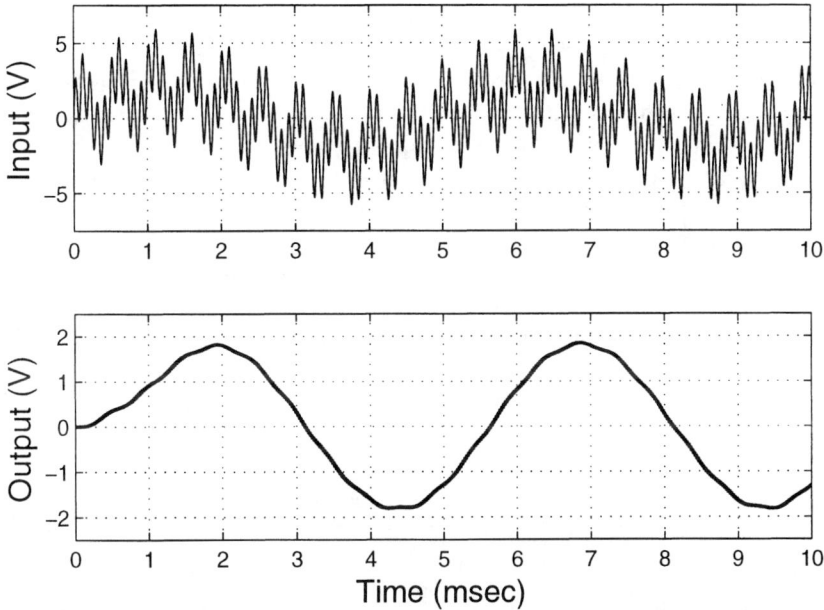

Figure 7.29. Measured time-domain response for the three-pole example.

The resulting spectral measures compare favorably to theory, as the graphs in Figure 7.28 indicate. The resolution exceeds a 100 dB dynamic range as shown in the graph for the output data. The measures conform closely to expected values into the stopband for this design. It is noted that the 10 kHz component is about 3 dB too large, but this is most likely due to aliasing. This design procedure simulates a better inductance than could be obtained in practice for this (audio) frequency range. Consequently, the switched capacitor architecture finds much use in practice due to its simplicity of design and performance.

Time-domain data are shown in Figure 7.29 for the three-frequency test signal. The data are plotted for the first 10 msec in the 5-μsec steps used to collect the data from μCAP. (Recall that the 4996 μCAP samples were down-sampled to obtain 1000 samples at $\Delta T = 25$ μsec for the Fourier analysis.) The graph shows the transient decay into steady state. The three-pole filter has done a superior job of filtering out both the 2-kHz and 10-kHz tones as compared to the previous single-pole example. The circuit clearly demonstrates the improved selectivity that can be obtained from a higher-order filter with a smaller transition bandwidth and more stopband attenuation.

7.2.1 Switched Capacitor s-Plane Distortion

While the previous examples have served to introduce switched capacitor design ideas, the examples have not illustrated a serious deficiency of the method. By implementing the leapfrog topology in order to simulate branch equations for a specific prototype design, we obtain a desired network function by an indirect procedure. There is no exact control over the placement of poles and zeros, as in the state variable procedure. The concept is that by implementing branch equations for a suitable prototype, we obtain a suitable network function implicitly. Since the switched capacitor only approximates an analog resistor by transferring a charge from one node to another, per clock period, the circuit response is actually operating on a discrete-time basis rather than on a continuous-time basis. This fact affects the performance of the leapfrog integrators, which are assumed to provide the transfer function, $1/s$, of continuous time circuits.

Consider the analog and switched capacitor integrators shown in Figure 7.18(a, b). Let the voltage on the capacitor be $\mathbf{V}_C = -\mathbf{V}_2$ (+ polarity on opamp input). The capacitor voltage for the analog circuit is the standard result, as given by (7.23):

$$\mathbf{V}_C(s) = \frac{1}{R_1 C_2 s} \mathbf{V}_1 \text{ or}$$

$$\frac{\mathbf{V}_C(s)}{\mathbf{V}_1(s)} = \frac{C_1}{T C_2 s} \quad (7.23)$$

In (7.23), the switched capacitor equivalent resistance per switch period T has been inserted for R_1 to show the $1/s$ transfer function commonly used to implement leapfrog branch equations.

Analysis of Figure 7.18(b), using the switch capacitor assumption that $v_1(t)$ does not change significantly over the switching interval, T, yields the following:

$$v_C(kT+T) = \frac{1}{C_2} \int_{kT}^{kT+T} dt \frac{\Delta Q}{\Delta T} \approx \frac{1}{C_2} \int_{kT}^{kT+T} dt \frac{[C_1 v_1(kT) - 0]}{T} + v_C(kT)$$

$$\approx \frac{C_1}{C_2} \frac{v_1(kT)}{T} \cdot T + v_C(kT) \text{ and so}$$

$$v_C(kT+T) - v_C(kT) \approx \frac{C_1}{C_2} v_1(kT) \quad (7.24)$$

Equation (7.24) shows the switched capacitor integrator response in a difference equation form for its response at the discrete switching times, $t = \ldots 0, T, \ldots kT, \ldots$. As known from digital-signal-processing techniques, the difference equation for discrete-time sequences can be written in its z-transform form. Thus, (7.24) is changed to the following, where $z = e^{sT}$:

$$(z-1)\mathbf{V}_C(z) = \frac{C_1}{C_2} \mathbf{V}_1(z) \text{ or}$$

$$\frac{\mathbf{V}_C(z)}{\mathbf{V}_1(z)} = \frac{C_1}{C_2} \frac{1}{(z-1)} \quad (7.25)$$

Comparison of the two transfer functions, (7.23) and (7.25), shows the following assumed correspondence for the two representations:

$$sT \leftrightarrow z - 1 = e^{sT} - 1 \quad (7.26)$$

7.2. SWITCHED CAPACITORS AND ANALOG OPERATIONS

The result tells us that we are assuming s corresponds to $(e^{sT} - 1)/T$ in order to use $1/s$ for the integrator response in the design procedure. A condition for this assumption to be true is derived by substituting the power series form for e^{sT}.

$$\begin{aligned}\frac{e^{sT}-1}{T} &= \frac{1}{T}\left[1 + sT + \frac{(sT)^2}{2!} + \frac{(sT)^3}{3!} + \ldots - 1\right] \\ &= s + \frac{Ts^2}{2!} + \frac{T^2 s^3}{3!} + \ldots \end{aligned} \quad (7.27)$$

The desired correspondence is true only as $T \to 0$ (i.e., for arbitrarily large switching frequencies).

Example 4 - Design and simulate a switched capacitor circuit to provide the response for the LP filter given in Figure 6.21. This circuit has finite j-axis transmission zeros and provides the following network function.

$$\frac{\mathbf{V}_4}{\mathbf{V}_1} = \frac{(0.16s^2 + 1)}{2(s^3 + 1.84s^2 + 1.84s + 1)} \quad (7.28)$$

Scale the 1 rad/sec design frequency to 800 Hz, along with an impedance factor of 0.5 MΩ for the circuit, and use a switching frequency of 40 kHz. Determine the expected pole-zero distortions due to the discrete-time implementation procedure and test the resulting design for its passband, transmission zero (null), and stopband performance by comparing simulation results to theory.

Solution - First the all-pole filter design procedure needs to be modified to provide a finite j-axis transmission zero. This is accomplished by the following branch equation analysis. Note, in Figure 6.21, let the current, $\mathbf{I}_3$, flow through the inductor, $L_3 = 1.68$ H. The branch relations are then as follows:

$$\begin{aligned} \mathbf{V}_{I_1} &= \mathbf{V}_1 \\ \mathbf{V}_2 &= \frac{1}{(C_2 s + 1)}[\mathbf{V}_{I_1} - \mathbf{V}_{I_3} - C_3 s(\mathbf{V}_2 - \mathbf{V}_4)] \text{ or} \\ (1 + (C_2 + C_3)s)\mathbf{V}_2 &= \mathbf{V}_{I_1} - \mathbf{V}_{I_3} + C_3 s \mathbf{V}_4 \text{ yields} \\ \mathbf{V}_2 &= \frac{1}{(1 + (C_2 + C_3)s)}[\mathbf{V}_{I_1} - \mathbf{V}_{I_3} + C_3 s \mathbf{V}_4] \\ \mathbf{V}_{I_3} &= \frac{1}{L_3 s}(\mathbf{V}_2 - \mathbf{V}_4) \\ \mathbf{V}_4 &= \frac{1}{(C_4 s + 1)}[\mathbf{V}_{I_3} - 0 - C_3 s(\mathbf{V}_2 - \mathbf{V}_4)] \text{ or} \\ (1 + (C_3 + C_4)s)\mathbf{V}_4 &= \mathbf{V}_{I_3} + C_3 s \mathbf{V}_2 \quad \text{yields} \\ \mathbf{V}_4 &= \frac{1}{(1 + (C_3 + C_4)s)}[\mathbf{V}_{I_3} + C_3 s \mathbf{V}_2] \end{aligned} \quad (7.29)$$

where $C_2 = 1 = C_4$, $C_3 = 0.09524$ (all in F), and $L_3 = 1.68$ H. The equations in (7.29) show that the node voltages are modified from the all-pole form by providing

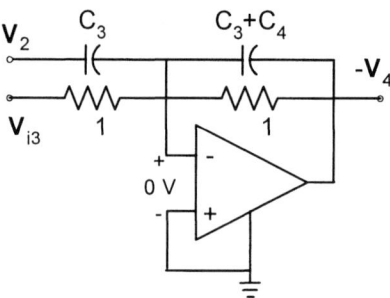

Figure 7.30. The addition of capacitive coupling from V_2 to the formation of node voltage, $-V_4$.

coupling to each other. The circuit in Figure 7.30 shows how this coupling can be implemented on a summing integrator to obtain the desired form for V_4 in (7.29). Note that the result shown in Figure 7.30 is for an inverting integrator.

With the desired branch equations thus determined, the switched capacitor circuit element values can then be obtained:

$$\begin{aligned}
k_R &= 0.5 \text{ M}\Omega \\
k_\omega &= 2\pi 800 = 1.6\pi 10^3 \\
k_C &= \frac{1}{k_R k_\omega} = \frac{1.25}{\pi} 10^{-9} \text{ and with} \\
f_C &= 40 \text{ kHz, the 1 } \Omega \text{ switched capacitor values are} \\
C_1 &= \frac{1}{k_R f_C} = 50 \text{ pF and for the analog filter values,} \\
C_{L3} &= 1.68 \to 1.68 k_C = 688.5 \text{ pF} \\
C_{23} &= C_2 + C_3 = 1.09524 \to 1.09524 k_C = 435.8 \text{ pF} \\
C_3 &= 0.09524 \to 0.09524 k_C = 37.9 \text{ pF}
\end{aligned}$$

The switched capacitor implementation is shown in Figure 7.31. Note that the node voltage variable, V_2, is implemented using coupling from $-V_4$ through the inverting integrator. $-V_4$ is then realized using coupling from $+V_2$. So, it is convenient, in this type of circuit, to implement the node voltage variables so as to have alternating signs. Extra switches are required, however, in order to change the sign of V_4 for the construction of $+V_{I_3}$. Clearly, the architecture can be generalized for higher-order filters.

As in the previous two examples, the circuit is tested by collecting 4996 samples at 5-μsec intervals to down-sample to 1000 samples on the switching period of 25 μsec. The DFT resolution is thus $\Delta F = 40$ kHz/10^3 samples= 40 Hz/sample bin. The three test frequencies are placed on odd bin numbers at $F_1 = 280$ Hz (bin 7), $F_2 = 2040$ Hz (bin 51), and $F_3 = 6040$ Hz (bin 151). The expected null frequency is at $\Omega_0 k_\omega/2\pi = 2.5(800) = 2000$ Hz for the j-axis zero of the network.

7.2. SWITCHED CAPACITORS AND ANALOG OPERATIONS

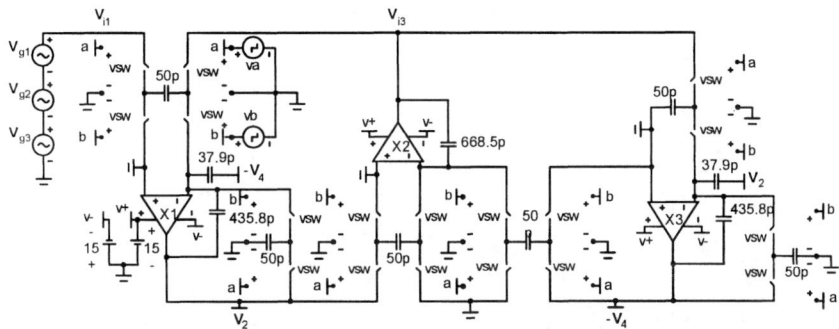

Figure 7.31. A switched capacitor design for obtaining finite j-axis zeros.

The following Matlab file is used to determine the pole-zero locations expected for the design based upon the discrete sampling distortion predicted by (7.26):

```
load dat7_32.dat -ascii
% 3 col., Time, Vin, Vout for opamp One pole low pass

N=1e3;
Fs=40e3;
df=Fs/N;
f=df*[0:N/2-1]';
s=j*f/800;
k=[0:N-1]';
% Sin cubed window
wk=0.75*pi*sin(pi*k/N).^3;
vin=dat7_32(1:5:4996,2);
vout=dat7_32(1:5:4996,3);
V1=2*fft(vin.*wk)/N;
V2=2*fft(vout.*wk)/N;

%get pole-zero frequencies due to sampling distortion
Ps=0.16*[1 0 6.25];
Qs=[1 1.84 1.84 1];
zros=roots(Ps);
poles=roots(Qs);
kw=2*pi*800;
T=1/Fs;
zr=zros*kw;
po=poles*kw;
zprime=(exp(zr*T)-1)/T;
pprime=(exp(po*T)-1)/T;
Pz=0.16*poly(zprime/kw);
```

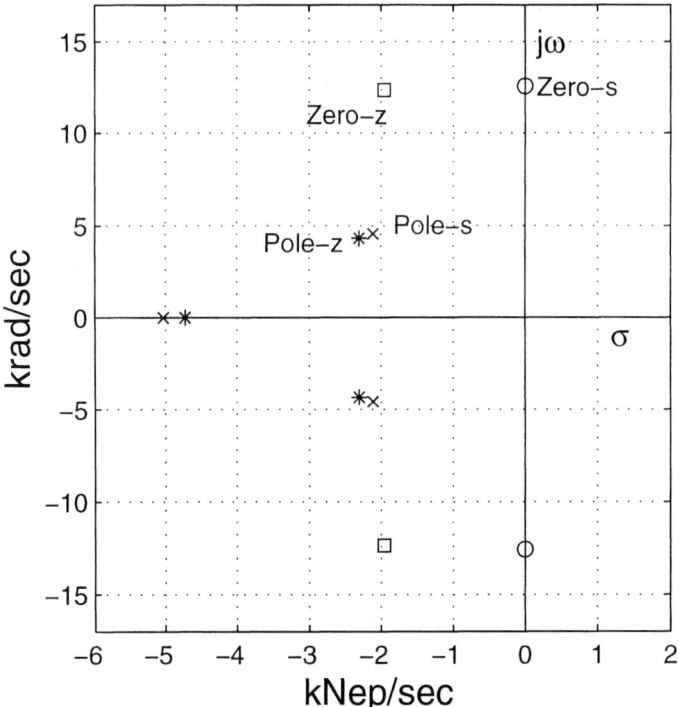

Figure 7.32. Pole-zero shifts obtained from the discrete-time distortion of a switched capacitor filter.

```
Qz=poly(pprime/kw);
Hs=polyval(Ps,s)./polyval(Qs,s);
Hz=polyval(Pz,s)./polyval(Qz,s);
```

The resulting pole-zero locations are shown in Figure 7.32. The nominal design poles are shown with crosses and the zeros are shown with circles. The actual poles and zeros provided by the discrete-time z-transform result are shown using asterisks for poles and squares for zeros. Note that the pole distortion is small with the real-part shift nearly cancelling for this example. The finite j-axis zeros, however, show a huge shift as the desired roots are shifted about 2 kNep/sec into the left half-plane. Such a shift in the transmission zeros eliminates the desired j-axis null and has a significant effect upon the expected sinusoidal steady-state response for this filter.

Simulated response to three sinusoids is shown in Figure 7.33 for the switched capacitor circuit in Figure 7.31. The upper curve shows the DFT magnitude (in dBv) for the input signal, and the lower curve shows the output at Node-V4. The three tones simultaneously test the passband, the null at the end of the transition band, and a point in the stopband. The passband and stopband compare well with the prototype s-plane design objective, but clearly the j-axis transmission

7.2. SWITCHED CAPACITORS AND ANALOG OPERATIONS

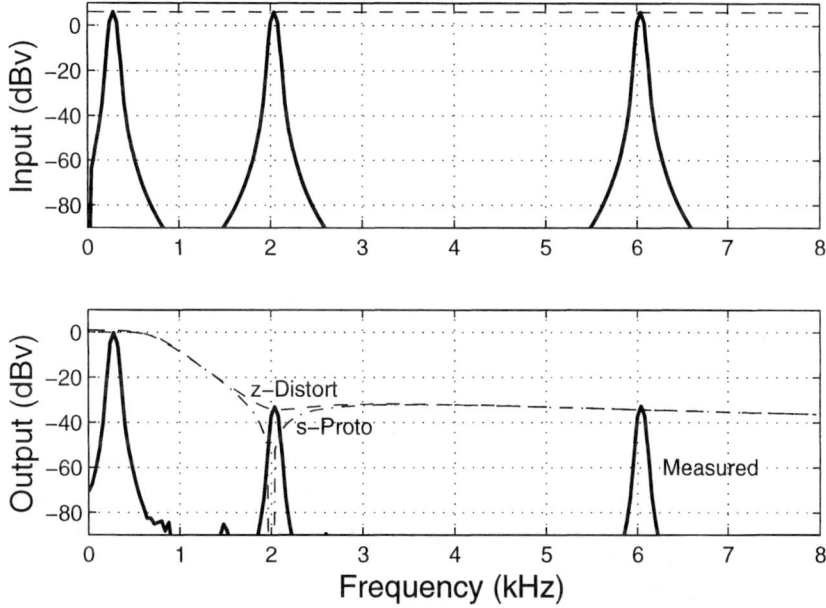

Figure 7.33. Measured versus expected responses for the filter in Figure 7.31.

zero has been lost. There is a transmission gain minimum near the design null, and this provides slightly more loss than the minimum stopband attenuation. Such a result could be acceptable for most applications, but if the design had been chosen specifically to null an undesired frequency, then this design would not be satisfactory.

Recall that z-transform distortion is reduced as the switching period goes to zero. Thus, the null depth for this example can be increased by going to a higher switching frequency (e.g., 200 kHz instead of 40 kHz). A serious limit on the switching frequency is the opamp used to implement the filters. In order to effect the proper charge transfer, it is important that the opamp maintain a good approximation to its virtual ground at its input, and this condition is required to be maintained at the selected switching frequency for filter design.

The shift in the j-axis zero is not surprising when one studies the discrete-time frequency distortion, (7.27), a little more carefully. Let $s = j\omega_0$, a j-axis frequency, and then (7.27) yields the following:

$$(e^{j\omega_0 T} - 1)/T = j\omega_0 - T\omega_0^2/2 - jT^2\omega_0^3/6 + \ldots \qquad (7.30)$$

For the previous example, $T = 25$ μsec, and $\omega_0 = 1600\pi$, so that the real-part frequency shift is approximately $-T\omega_0^2/2 = -200\pi^2 = -1974$ Nep/sec. This result shows, to a first-order approximation, that the shift shown in Figure 7.32 is expected for this design. Reducing the clock period has a proportionate effect on reducing the shift. Such frequencies are shifted into the left half-plane and toward the origin. The previous example verifies this effect. See Problem 7.16 to check this conclusion.

Finally, it should be noted that lowering the impedance scale factor, k_R, does not affect the result obtained for this example. Lowering k_R improves the switch resolution for a fixed OFF resistance relative to the equivalent resistance value determined from a 1-Ω prototype value. In this example the OFF resistance is modeled at 10 MΩ, and the 1-Ω prototype resistance is scaled to 0.5 MΩ for a resolution of 20:1 (26 dB). Reducing the scale factor to 50 kΩ provides a resolution of 200:1, but the resulting response is unchanged from that shown in Figure 7.33. See Problem 7.18 to check this conclusion. However, the point is that the lack of a well-defined null for this circuit is due to distortion of the network pole-zeros as a consequence of the discrete-time integration failure to model the $1/s$ branch equation requirement of the design procedure properly.

7.2.2 Precompensated Network Functions

The previous example showed that the switched capacitor filter implementation provides severe distortion for j-axis critical frequencies for a given filter design. The undesired shift is caused by the discrete-time operation of the integrators so that the desired $1/s$ integrator function is distorted, as given in (7.26). In some cases, it would be useful to compensate for the discrete-time integrator pole-zero shifts, and this section illustrates how such compensation can be accomplished.

Since the distortion is affected by the value of the switching clock period, T, and the actual (scaled) complex frequency, $s' = k_\omega s$, (7.26) gives us the following complex function for the integrator prototype network function:

$$\frac{1}{w} = \frac{k_\omega T}{(z-1)} \approx \frac{1}{s}$$

where $z = e^{s'T} = e^{sk_\omega T}$. With this variable definition, the switched capacitor network function, (7.25), is changed to the following:

$$\frac{-\mathbf{V}_2}{\mathbf{V}_1}(z) = \frac{C_1}{C_2(z-1)} \approx \frac{C_1}{C_2 s k_\omega T} \text{ but}$$

$$C_1 = \frac{T}{k_R R} \text{ and } C_2 = \frac{C_{proto}}{k_\omega k_R} \text{ so}$$

$$\frac{-\mathbf{V}_2}{\mathbf{V}_1}(z) \approx \frac{T}{k_R R} \cdot \frac{k_\omega k_R}{C_{proto}} \cdot \frac{1}{s k_\omega T} = \frac{1}{RC_{proto}} \cdot \frac{1}{s} \quad (7.31)$$

Equation (7.31) just shows that the actual network function approximates the desired prototype integrator network function.

In Example 4, it was desired to produce a network function to obtain a quadratic numerator over a cubic denominator (i.e., two j-axis zeros and three poles). If we wish to precompensate the poles and zeros for the discrete-time integrator distortion, then usually it is not straightforward to find a prototype circuit to realize the precorrection. Consequently, the use of leapfrog circuit implementation procedures does not help to realize precorrected pole-zero patterns in the switched capacitor

7.2. SWITCHED CAPACITORS AND ANALOG OPERATIONS

design procedure. However, the pole-zero patterns are known, and it is thus feasible to look at implementing the state variable design procedure using switched capacitor design techniques.

The signal-flow block diagram shown in Figure 7.34 illustrates a configuration for the discrete-time integrator using the $1/w$ variable to approximate a state variable circuit that provides a quadratic-over-cubic network function. Analysis of the block diagram proceeds as follows:

By integration:

$$\begin{aligned} \mathbf{V}_1 &= w\mathbf{V}_a \\ \mathbf{V}_2 &= w\mathbf{V}_1 = w^2\mathbf{V}_a \\ \mathbf{V}_3 &= w\mathbf{V}_2 = w^3\mathbf{V}_a \end{aligned} \qquad (7.32)$$

At the input summation:

$$\begin{aligned} \mathbf{V}_3 &= \pm\mathbf{V}_g - (b_2\mathbf{V}_2 + b_1\mathbf{V}_1 + b_0\mathbf{V}_a) \text{ or} \\ w^3\mathbf{V}_a &= \pm\mathbf{V}_g - (b_2w^2 + b_1w + b_0)\mathbf{V}_a \text{ so} \\ \frac{\mathbf{V}_a}{\mathbf{V}_g} &= \frac{\pm 1}{(w^3 + b_2w^2 + b_1w + b_0)} \end{aligned} \qquad (7.33)$$

The output summation yields $\mathbf{V}_0$:

$$\begin{aligned} \mathbf{V}_0 &= a_0\mathbf{V}_a + a_1\mathbf{V}_1 + a_2\mathbf{V}_2 \text{ or} \\ &= (a_2w^2 + a_1w + a_0)\mathbf{V}_a \text{ so} \\ \frac{\mathbf{V}_0}{\mathbf{V}_g} &= \frac{\pm(a_2w^2 + a_1w + a_0)}{(w^3 + b_2w^2 + b_1w + b_0)} \end{aligned} \qquad (7.34)$$

The network function, (7.34), is the desired form of quadratic over cubic, where w approximates s.

Since the integrator variable, w, in (7.34), depends upon the specific frequency scale factor, k_ω, and the switching period, T, the design parameters, a_k and b_k, have to be determined for each application. Using the same design parameters that were used for Example 4, we get $T = 25$ μsec$= 1/(40$ kHz$)$ and $k_\omega = 1600\pi$ with pole frequencies $p_{1,2,3} = roots([1\ 1.84\ 1.84\ 1]) = -1.0, -0.4200 \pm j0.9075$, and zero frequencies $z_{1,2} = roots([1\ 0\ 6.25]) = \pm j2.500$.

The b_k design parameters of the state variable circuit are then found by requiring the denominator of (7.34) to be zero for those values of ω that correspond to p_k, while the a_k values are found by requiring the numerator to be zero for those values of ω that correspond to z_k. The specified roots provide exactly enough independent conditions to solve for all of the b_k and two of the a_k parameters. The parameter a_2 is specified by the asymptotic behavior of the network function for large s values. For the given example, $a_2 = 0.16$ is specified so that $H(s) \to 0.16/s$ as $s \to \infty$. Thus, let w_{pk} equal the value of ω for $s = p_k$, $k = 1, 2, 3$, and w_{zk} equal the value

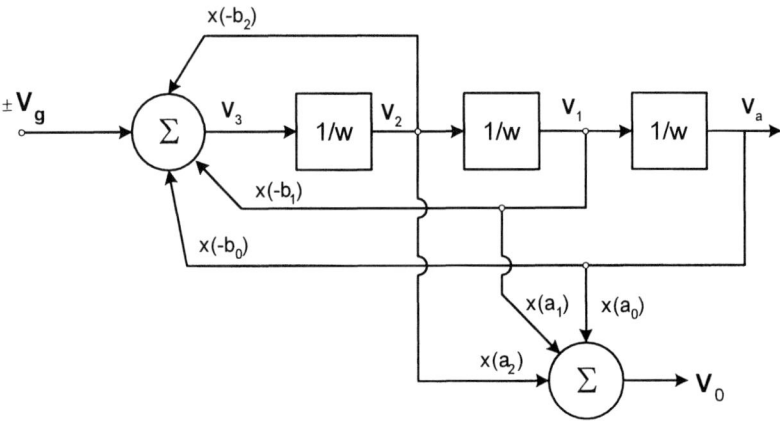

Figure 7.34. A signal-flow schematic for a discrete-time state variable circuit.

of ω for $s = z_k$, $k = 1, 2$. We then get the following formulation by point-matching the data:

$$[W_p][\mathbf{b}] = \begin{bmatrix} w_{p1}^2 & w_{p1} & 1 \\ w_{p2}^2 & w_{p2} & 1 \\ w_{p3}^2 & w_{p3} & 1 \end{bmatrix} \begin{bmatrix} b_2 \\ b_1 \\ b_0 \end{bmatrix} = - \begin{bmatrix} w_{p1}^3 \\ w_{p2}^3 \\ w_{p3}^3 \end{bmatrix} = -[w_p^3] \quad (7.35)$$

and

$$[W_z][\mathbf{a}] = \begin{bmatrix} w_{z1} & 1 \\ w_{z2} & 1 \end{bmatrix} \begin{bmatrix} a_1 \\ a_0 \end{bmatrix} = -a_2 \begin{bmatrix} w_{z1}^2 \\ w_{z2}^2 \end{bmatrix} = -a_2[w_z^2] \quad (7.36)$$

Applying the given design specifics, k_ω, T, a_2, p_k, and z_k, to (7.35) and (7.36) yields the following state variable design coefficients for the precorrected network function: $a_2 = 0.16$, $a_1 = 0.1246$, $a_0 = 0.9918$; $b_2 = 1.8560$, $b_1 = 1.8088$, and $b_0 = 0.8906$. These answers were obtained from the following Matlab file:

```
T=1/40e3;
kw=1600*pi;
a2=0.16;
pk=roots([1 1.84 1.84 1]);
zk=roots([1 0 6.25]);
wp=(exp(pk*kw*T)-1)/kw/T;
wz=(exp(zk*kw*T)-1)/kw/T;
Wp=[wp.^2 wp ones(3,1)];
Wz=[wz ones(2,1)];
b=-inv(Wp)*wp.^3;
a=-a2*inv(Wz)*wz.^2;
```

Now we are ready to implement a switched capacitor circuit to provide the state variable network functions in (7.33) and (7.34). One such circuit is shown in

7.2. SWITCHED CAPACITORS AND ANALOG OPERATIONS

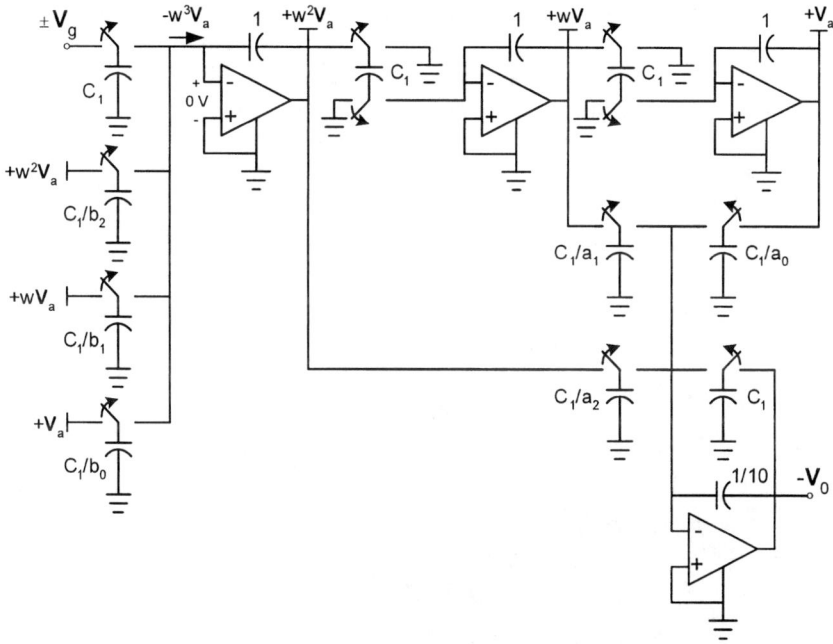

Figure 7.35. Switched capacitor state variable circuit.

Figure 7.35. Three integrators are used to obtain the three poles required, and the desired output, $-\mathbf{V_0}$, is obtained via a single-pole lowpass filter summing circuit. Note that the bandwidth of this output summation circuit is set at 10 times the bandwidth of the desired filter response. We are forced to use this circuit to sum the output since the opamp should not be run open-loop, as would be the case if we tried to implement a switched capacitor version of a conventional resistive summing opamp circuit. This is also why the variable $-w^3\mathbf{V}_a$ is only implemented as a charge transfer, since the switched capacitor equivalent resistance is not practical to use as the only feedback component for a high-gain opamp circuit. The variable $-w^3\mathbf{V}_a$ is integrated and inverted to provide $+w^2\mathbf{V}_a$. This in turn is inverted on the switched capacitor and integrated to provide $+w\mathbf{V}_a$, and repeated again, to reach $+\mathbf{V}_a$ for the all-pole response. The schematic in Figure 7.35 assumes positive real coefficients for all the a_k and b_k parameters; however, if one is required to be negative, a charge inverting capacitor can always be used (e.g., in front of the integrators for $w\mathbf{V}_a$ and $\mathbf{V}_a$).

Technically, a proper design should consider the opamp gain distribution problem in the same manner as for a regular state variable circuit. This procedure was discussed earlier and will be omitted for this example discussion.

Note that the state variable implementation relies entirely on the implementation of the integrator transfer function. That is why we have taken steps to precorrect the transfer function to correspond to the resulting discrete-time frequency mapping given by (7.30). In addition, we should also be aware that in order actually to obtain or observe a good transmission null frequency, all components need to be accurate to a high resolution, typically one part per 10^3 for a 60-dB null. Therefore, for this example, we lower the impedance scale factor, k_R, by a factor of 10 to a value of 50 kΩ. This is a departure from the previous examples but is necessary since we are looking for a null frequency response in our test. Thus, the application capacitor values are obtained as follows for the circuit in Figure 7.35:

$$k_R = 50 \text{ k}\Omega$$
$$f_C = 40 \text{ kHz}$$
$$k_\omega = 1600\pi \text{ rad/sec}$$
$$C_1 = \frac{1}{f_C k_R \cdot 1} = \frac{1}{40 \times 10^3 \times 50 \times 10^3} = 500 \text{ pF}$$
$$k_C = \frac{1}{k_R k_\omega} = \frac{1}{50 \times 10^3 \times 1600\pi} = \frac{12.5}{\pi} \text{ nF/F}$$
$$C_{1/b_2} = b_2 C_1 = 928.0 \text{ pF}$$
$$C_{1/b_1} = b_1 C_1 = 904.4 \text{ pF}$$
$$C_{1/b_0} = b_0 C_1 = 453.3 \text{ pF}$$
$$C_{1/a_2} = a_2 C_1 = 80.0 \text{ pF}$$
$$C_{1/a_1} = a_1 C_1 = 62.3 \text{ pF}$$
$$C_{1/a_0} = a_0 C_1 = 459.9 \text{ pF}$$

The final design is shown in Figure 7.36. The design implements (7.28) times a factor of 2. The previous example used leapfrog procedures to implement branch equations for the circuit in Figure 6.21, and it did not remove the loss factor of 2 caused by the equal terminations on the passive filter. The simulated measured response for this design is shown in Figure 7.37. The results are excellent, as a null of 62 dB is observed at the desired frequency, and the other test frequencies coincide with expected values. Thus, the circuit validates the discussion about the switched capacitor s-plane distortion to the leapfrog design procedure and confirms that precorrection can be used to provide exact j-axis zeros. Deeper nulls can be observed by using better ON-OFF values for the switches, but this procedure is left as a homework exercise to observe the switch effect on the j-axis zero. So far, the same switch parameters have been used for all examples to provide a consistent progression of ideas on switched capacitor filter design procedures. See Problem 7.19 for a look at these issues.

7.2.3 Scale Factors and Bandpass Filter Considerations

As we have seen, choosing design scale factors is not a trivial issue. It is not always possible to set the switching frequency, f_C, to be independent of the impedance factor, k_R, for several reasons. These two parameters are inversely proportional to the switched capacitor value, C_1, that provides the correspondence to the prototype resistance of 1 Ω used as an input to transfer charge to the opamp feedback

7.2. SWITCHED CAPACITORS AND ANALOG OPERATIONS

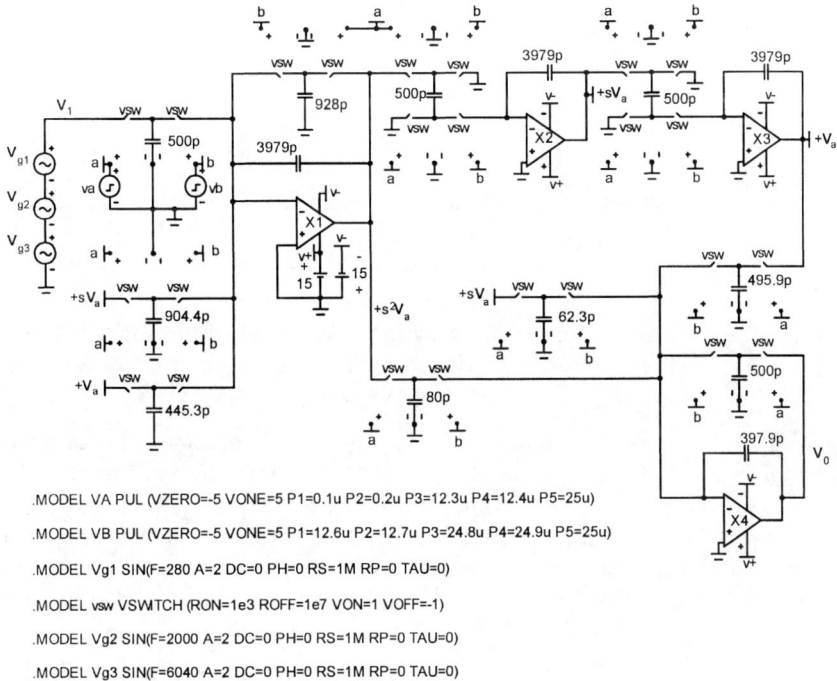

Figure 7.36. Example switched capacitor implementation for a state variable design.

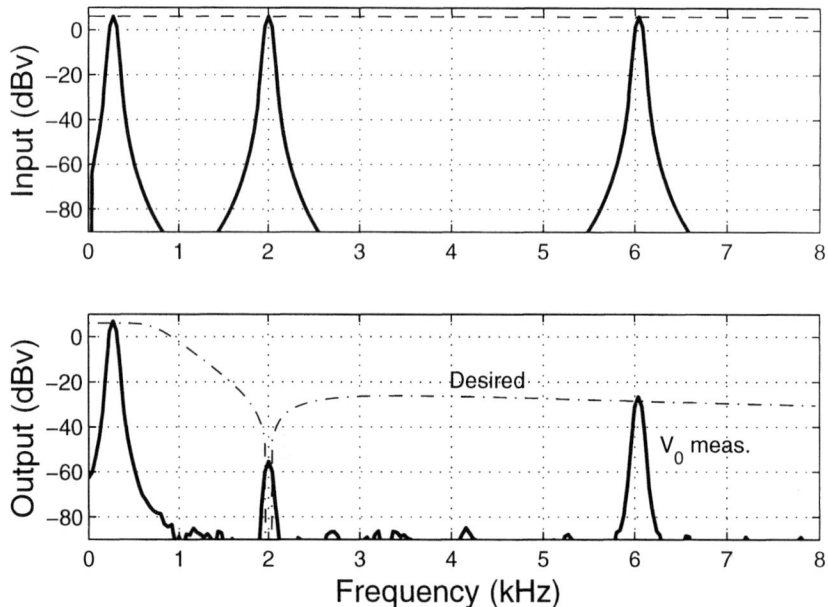

Figure 7.37. Spectral response for the switched capacitor state variable circuit.

capacitance. The switching frequency has to be less than some limit provided by the dynamics of the amplifier, as well as larger than twice the maximum frequency required to process information through the filter (from the sampling theorem). The impedance scale factor is limited by the achievable OFF resistance of the switches and by practical capacitor values that the fabrication process can easily support. However, there are constraints that come about through the design element value ratios that are required versus the element value ratio that can actually be supported by the fabrication process. k_R and f_C limits are dependent upon these element value ratios in a not-so-obvious manner. This section explores some of these considerations.

Assume that the fabrication process provides a minimum least-capacitance value of C_L and an upper maximum value of C_U, both within good tolerances for analog filter design purposes. Typical values might be 0.1 to 100 pF for one process (a 10^3:1 ratio) or 20 fF to 20 pF (again 10^3:1) in a different process. In addition, if the design procedure is implementing a leapfrog procedure, then the element value ratio, C_{min} to C_{max}, of the prototype circuit is also a consideration for the proper selection of scale factors. Appropriate constraints are determined as follows:

Suppose the prototype element value range (for the switched capacitor design, e.g., Figure 7.31) is C_{min} to C_{max} for analog values, and all switched capacitors are for 1 Ω. f_C and k_R are then constrained as shown in (7.37):

$$C_L \leq C_1 = \frac{1}{f_C k_R \cdot 1} \leq C_U \tag{7.37}$$

7.2. SWITCHED CAPACITORS AND ANALOG OPERATIONS

In addition, $k_C = 1/(k_\omega k_R)$ for the analog capacitors, and this fact yields (7.38):

$$\frac{C_L}{C_{min}} \leq k_C = \frac{1}{k_\omega k_R} \leq \frac{C_U}{C_{max}} \quad (7.38)$$

Equation (7.38) is only valid for $C_{max}/C_{min} < C_U/C_L$, from which it provides upper and lower constraints for k_R:

$$\frac{k_\omega C_L}{C_{min}} \leq \frac{1}{k_R} \leq \frac{k_\omega C_U}{C_{max}} \text{ or}$$

$$\frac{C_{max}}{k_\omega C_U} \leq k_R \leq \frac{C_{min}}{k_\omega C_L} \quad (7.39)$$

Once k_R is chosen, the switching frequency can then be selected from the limits given in (7.37):

$$k_R C_L \leq \frac{1}{f_C} \leq k_R C_U \text{ or}$$

$$\frac{1}{k_R C_U} \leq f_C \leq \frac{1}{k_R C_L} \quad (7.40)$$

For example, if k_R is selected to equal its minimum value in (7.39), then the limits for f_C appear as follows:

$$\frac{k_\omega}{C_{max}} \leq f_C \leq \frac{C_U}{C_L} \frac{k_\omega}{C_{max}} \quad (7.41)$$

This result shows that the choice of f_C is dependent on the circuit design parameters. Using a minimum value for k_R allows for the selection of the largest possible switching frequency as implied by the upper constraint of (7.37).

Considering that the switching frequency needs also to be greater than twice the desired signal bandwidth, Bw, and less than the amplifier unit gain bandwidth, GBW, (7.40) needs to be modified to the following constraint:

$$max\left\{2Bw, \frac{1}{k_R C_U}\right\} \leq f_C \leq min\left\{GBW, \frac{1}{k_R C_L}\right\} \quad (7.42)$$

The last result is useful in choosing, or setting, design parameters for switched capacitor filters. For example, consider the previous design for the scaling of the precorrected state variable filter. Assume $C_L = 5$ pF, $C_U = 5$ nF, and the given parameters $k_\omega = 1600\pi$, $Bw = 10$ kHz, $GBW = 4$ MHz, $C_{max} = 1.0$, and $C_{min} = 0.1$. Equation (7.39) yields the following range for k_R:

$$\frac{1.0}{1600\pi \times 5 \times 10^{-9}} \leq k_R \leq \frac{0.1}{1600\pi \times 5 \times 10^{-12}} \text{ so}$$

$$39.8 \text{ k}\Omega \leq k_R \leq 3.98 \text{ M}\Omega$$

$k_R = 50$ kΩ was used for the design given in Figure 7.36, and so, a suitable range for the selection of the switching frequency is obtained from (7.40):

$$max\left\{20 \cdot 10^3, \frac{1}{50 \cdot 10^3 \cdot 5 \cdot 10^{-9}}\right\} \leq f_C$$

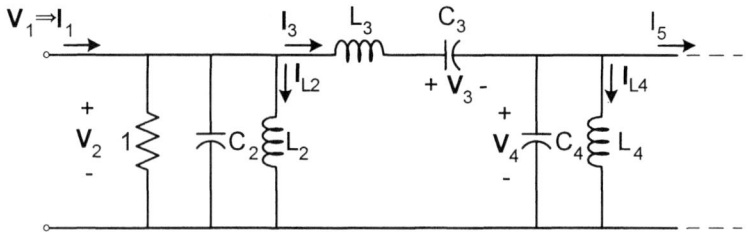

Figure 7.38. Prototype all-pole bandpass filter architecture.

$$f_C \leq min\left\{4 \cdot 10^6, \frac{1}{50 \cdot 10^3 \cdot 5 \cdot 10^{-12}}\right\}$$

so
$$max\{20 \cdot 10^3, 4000\} \leq f_C \leq min\{4 \cdot 10^6, 4 \cdot 10^6\}$$
$$20 \text{ kHz} \leq f_C \leq 4 \text{ MHz}$$

It is seen that the use of 40 kHz lies within the allowable design range for f_C in the state variable filter design.

In addition, note for this example that the use of the upper limit for k_R (4 MΩ) would greatly reduce the range of f_C. In addition, the upper limit for k_R [from (7.39)] is also affected by a consideration that it should be at least an order of magnitude less than the switch OFF resistance and less than the amplifier input impedance. Both of these parameters are process dependent and will differ for different designs.

A particular design that stresses scale factor selection is the all-pole bandpass filter architecture. A typical prototype is shown in Figure 7.38, where $L_2 = 1/C_2, C_3 = 1/L_3$, and so forth for a 1 Ω–1 rad/sec prototype (see Chapter 2). The branch equations for a leapfrog implementation proceed as follows:

$$\begin{aligned}
\mathbf{V}_{I_1} &= \mathbf{V}_1 \\
\mathbf{V}_2 &= \frac{1}{(C_2 s + 1)}(\mathbf{V}_{I_1} - \mathbf{V}_{I_3} - \mathbf{V}_{I_{L_2}}) \\
\mathbf{V}_{I_{L_2}} &= \frac{1}{L_2 s}\mathbf{V}_2 \\
\mathbf{V}_{I_3} &= \frac{1}{L_3 s}(\mathbf{V}_2 - \mathbf{V}_3 - \mathbf{V}_4) \\
\mathbf{V}_3 &= \frac{1}{C_3 s}\mathbf{V}_{I_3} \\
\mathbf{V}_4 &= \frac{1}{C_4 s}(\mathbf{V}_{I_3} - \mathbf{V}_{I_5} - \mathbf{V}_{I_{L_4}}) \\
\mathbf{V}_{I_{L_4}} &= \frac{1}{L_4 s}\mathbf{V}_4 \\
&\ldots
\end{aligned} \quad (7.43)$$

7.2. SWITCHED CAPACITORS AND ANALOG OPERATIONS

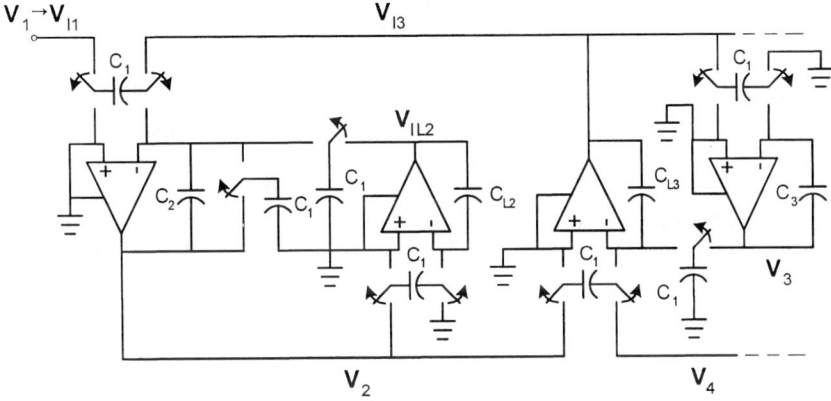

Figure 7.39. Prototype iterative structure for the switched capacitor bandpass filter.

New variables are introduced for the shunt inductors and series capacitors. A switched capacitor topology to implement (7.43) is shown in Figure 7.39. The circuit gives a clear picture of how resonating elements interact with their lowpass originating element (e.g., C_2 for L_2).

Note that the switched capacitor circuit capacitor values are numerically equal to the prototype capacitor and inductor values. Thus, C_{max} and C_{min} for the switched capacitor filter includes L and C values from the passive filter prototype. In addition, since the bandpass prototype is derived from a lowpass prototype, the values for C_2, L_3, and so forth, are each proportional to the transformation factor, $A = Q$, while resonating elements L_2, C_3, and so forth, are inversely proportional to $A = Q$. Normally, $Q > 1$.

Now, let P_{max} equal the maximum LP prototype L or C value and P_{min} equal the minimum value. Then, in the BP prototype, the maximum element value will be QP_{max}, while the minimum element value will be the lesser of $1/QP_{max}$ or QP_{min}:

$$\frac{1}{QP_{max}} < \frac{1}{QP_{min}} \text{ or } QP_{min}$$

Letting $C_{max} = QP_{max}$ and $C_{min} = 1/QP_{max}$ (the usual case) for a switched capacitor bandpass filter, the limits for k_R in (7.39) are modified as in (7.44):

$$\frac{QP_{max}}{k_\omega C_U} \leq k_R \leq \frac{1}{k_\omega C_L Q P_{max}} \tag{7.44}$$

This last result can be solved to determine the value of Q that forces the upper limit to equal the lower limit:

$$Q^2 \leq \frac{1}{P_{max}^2} \frac{C_U}{C_L} \text{ or}$$

$$Q \leq \frac{1}{P_{max}} \sqrt{\frac{C_U}{C_L}} \tag{7.45}$$

Equation (7.45) is a useful result to tell us the maximum Q that can be obtained in the all-pole bandpass filter architecture for a switched capacitor implementation. For example, let $C_U/C_L = 10^3$ and $P_{max} = 2$; then, from (7.45) we get $Q_{max} = 15.8$ as the maximum bandpass Q that can be supported by the design process. For most fabrication processes, it appears that very high-Q circuits are not likely to be achieved.

The problem is not radically changed by using bandpass filters with j-axis zeros, since even if the branch transformations are used, as discussed in Chapter 2, the shunt elements and their resonators are going to yield a similar result as given by (7.45). We can conclude that the limit in (7.45) is certainly in the ballpark for general decisions about design feasibility for a given bandpass filter requirement.

7.3 Problems

7.1 The purpose of this problem is to study the switched capacitor circuit of Figure 7.40, where ideal switches are used to accomplish charge transfer through a short circuit and a current controlled current source.

   (a) Using the switched capacitor model for equivalent resistance, determine the network function from v_1 to v_C as a function of source frequency, s. Determine if the circuit response fits the equivalent resistance model by changing the source frequency to different values. Beware of operating frequency limits.

   (b) Change the clock to run at 1 MHz. Rescale the network to give the same nominal transfer function as for the given circuit, and determine the apparent network function (magnitude) versus source frequency through repeated simulation runs on the circuit.

7.2 This problem uses a transmission gate, as shown in Figure 7.41, for bipolar switched current operation. It is desired to obtain a transfer function similar to that in Problem 7.1. Test the circuit, and see if it yields a desired response. You will have to select, or define, an appropriate model for the field effect transistor (FET) in this switching application. Note the impedance and time scaling of this circuit. Try running it for a larger equivalent switching resistance (increase k_R), modify k_C appropriately, and observe what happens. What causes the unexpected charging of the capacitor during the time when no charge should be delivered to the capacitor? A goal for this problem is to explain this switch behavior and to determine good parameters for switching FETs.

7.3 A standard biquad circuit is shown in Figure 7.42. Use switched capacitor concepts to replace resistors, and design a switched capacitor prototype for the biquad circuit. Take advantage of the fact that opamps are not required for inverters in the switched capacitor procedure. Specify all element values for your design.

7.3. PROBLEMS

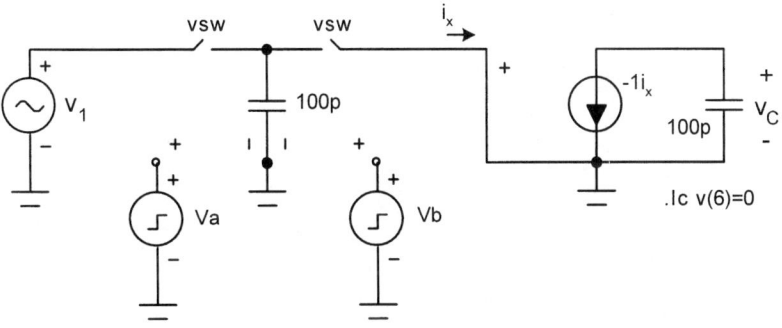

.MODEL VA PUL (VZERO=0 VONE=5 P1=1u P2=2u P3=49u P4=50u P5=100u)
.MODEL VB PUL (VZERO=5 VONE=0 P1=1u P2=2u P3=49u P4=50u P5=100u)
.MODEL VSW VSWITCH (RON=1e-3 ROFF=100MEG VON=4 VOFF=1)
.MODEL V1 SIN (F=1e3 A=1 DC=0 PH=0 RS=1M RP=0 TAU=0)

Figure 7.40. Circuit for Problem 7.1.

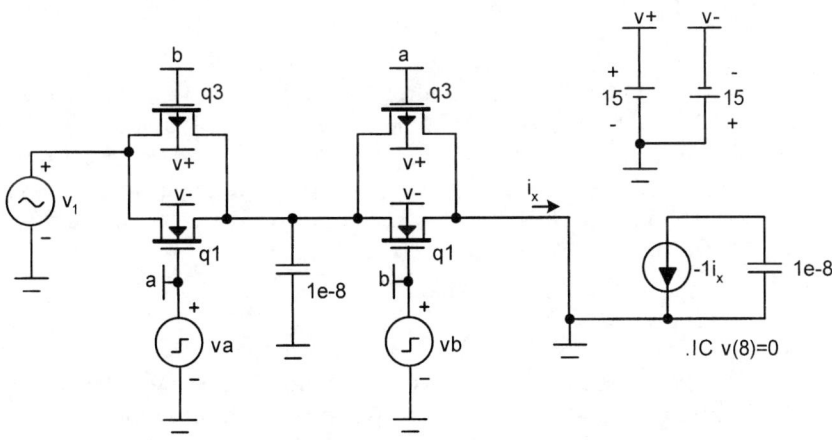

.MODEL VA PUL (VZERO=-10 VONE=10 P1=10n P2=20n P3=490n P4=500n P5=1000n)
.MODEL VB PUL (VZERO=10 VONE=-10 P1=10n P2=20n P3=490n P4=500n P5=1000n)

Figure 7.41. Circuit for Problem 7.2.

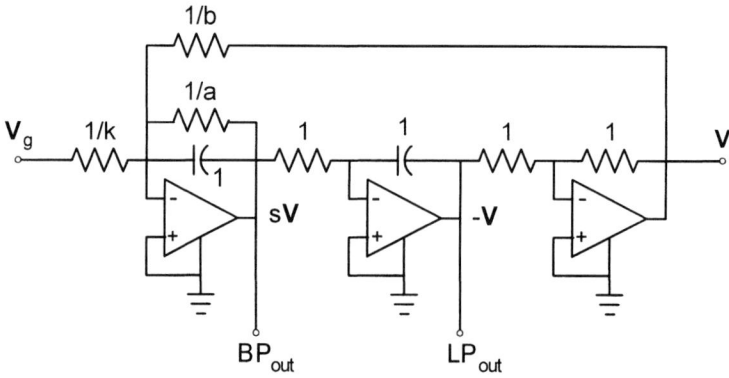

Figure 7.42. Circuit for Problem 7.3.

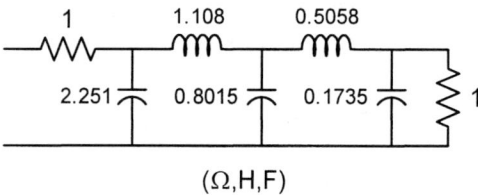

(Ω, H, F)

Figure 7.43. Circuit for Problem 7.5.

7.4 The following filter function is required for a servo control compensation system. It is desired to realize the function using a switched capacitor filter. Find a two opamp design for the given network function, and convert the design to a switched capacitor form. Choose an appropriate impedance scale factor, and specify a suitable clock frequency assuming that $C_{max} = 100$ pF.

$$\frac{\mathbf{V}_2}{\mathbf{V}_1} = \frac{+100(1 + s/100)}{(1 + s/2000)(1 + s/10^4)}$$

7.5 The fifth-order Gaussian filter shown in Figure 7.43 is required to operate with a 3 kHz passband in a switched capacitor realization. Obtain a switched capacitor design, and specify a suitable clock frequency and impedance scale factor, assuming that capacitance has to fit the range $0.1 < C < 100$ pF. Note that the prototype Gaussian filter 3-dB bandwidth is normalized to 1 rad/sec.

7.6 It is desired to use a switched capacitor circuit to realize a frequency-scaled version of the prototype integrator shown in Figure 7.44(a). The desired scale factor is from 1 rad/sec to 5 kHz.

7.3. PROBLEMS

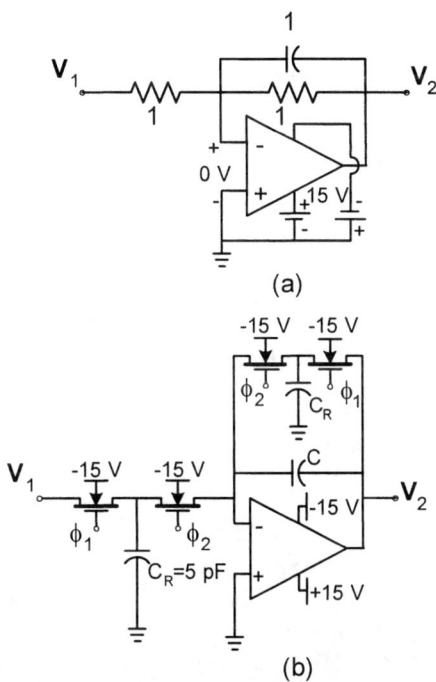

Figure 7.44. Circuits for Problem 7.6: (a) analog integrator, and (b) switched capacitor integrator.

(a) Use a 1-MHz switching rate and find the feedback capacitance, C, required for the circuit in Figure 7.44(b), when the switching capacitors each have the value $C_R = 5$ pF.

(b) Use an appropriate NMOS switch and transient analysis in μCAP to simulate the performance of the switched capacitor filter for signal frequencies at 1, 5, and 20 kHz. Use a 1-V peak amplitude for $\mathbf{V}_1$ in each case. Note that you will need to define appropriate pulse-switching functions for ϕ_1 and ϕ_2.

(c) Use sinusoidal steady-state analysis, find the expected peak voltage for $\mathbf{V}_2$ for each signal input frequency, and compare these values with those obtained from the simulations in part b.

(d) What happens, or what do we have to watch out for, if we try to put a gain factor, G_F, into the circuit by changing the input switched capacitor from 5 pF to $5 \times G_F$ pF?

7.7 It is desired to realize the following $T(s)$ using a state variable design approach in a switched capacitor architecture. It is proposed to realize the function with a short circuit current, as shown in Figure 7.45, which could subsequently be

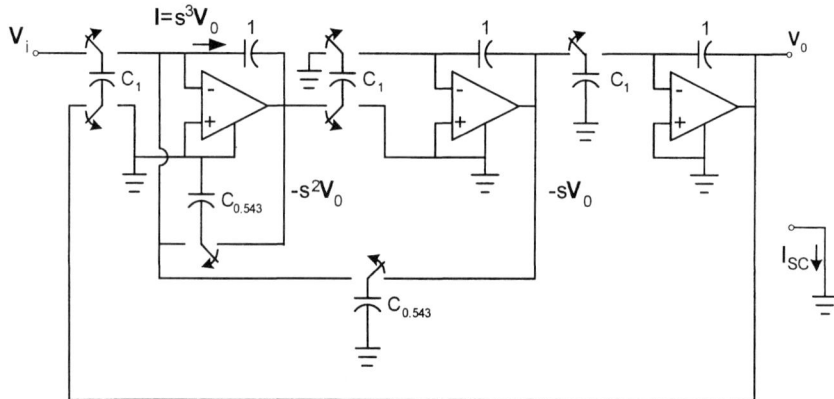

Figure 7.45. Circuit for Problem 7.7.

integrated to provide the desired transmission function. The circuit realizes the poles for the function in a fashion similar to that for a conventional state variable circuit. The problem is to show how the required short circuit current can be obtained from this structure using only components that are compatible with a switched capacitor circuit. Note that in the given circuit, the switched capacitor, C_x, is simulating a prototype resistor of value x.

$$T(s) = \frac{0.16(s^2 + 6.25)}{s^3 + 1.84s^2 + 1.84s + 1}$$

Show what components are required and how they are connected to the given circuit in order to form the desired short circuit current.

7.8 It is desired to use switched capacitors to build a bandpass all-pole filter. The filter is to have a center frequency at 1 kHz and as large a Q as possible. The analog prototype circuit capacitance range is $1/2Q < C < 2Q$, while the switched capacitors are all based on 1 Ω resistors. A practical design is desired, and so, it is required that the capacitors for the scaled circuit should all lie within the range $0.1 < C < 100$ pF. What is the maximum Q achievable under these conditions? What is the required impedance scale factor? What is the allowable range for the clock frequency?

7.9 In the circuit of Figure 7.46, the C_1 capacitors yield an equivalent resistance of 1 Ω in each clock period. The circuit is required to be scaled by 5 kHz, while it is desired to keep all capacitors in the range of 2 to 50 pF. Choose the smallest impedance scale factor that can be used and the largest clock frequency that can be used to design the practical (scaled) circuit.

7.10 What is the transfer function, $\mathbf{V_0}/\mathbf{V}_i$, for the prototype circuit given in Figure 7.47? C_1 is the switched capacitance required to yield a 1 Ω equivalent resistance in each clock period.

7.3. PROBLEMS

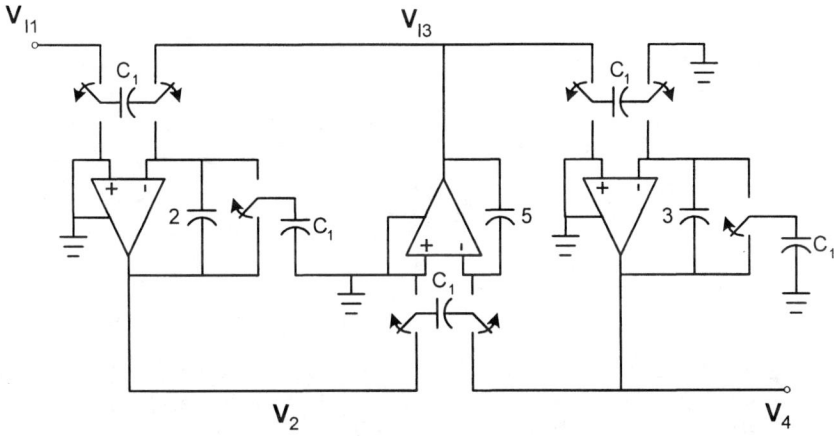

Figure 7.46. Circuit for Problem 7.9.

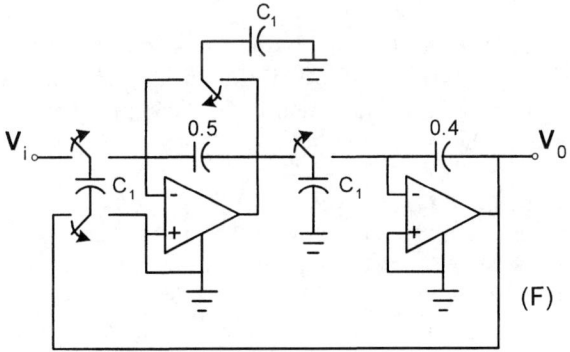

Figure 7.47. Circuit for Problem 7.10.

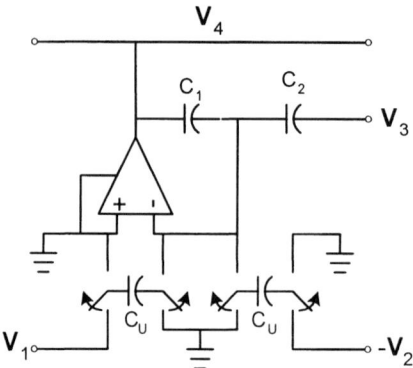

Figure 7.48. Circuit for Problem 7.12.

7.11 A passive RLC prototype filter is to be implemented with a switched capacitor circuit. The maximum and minimum prototype LC values are equal to 25 and 0.04, respectively. It is further required to frequency-scale the application from 1 rad/sec to 4 kHz. Specify a suitable impedance scale factor, k_R, and the limits for the switched capacitor clock frequency, f_C, so that all capacitors will fall in the range of $0.1 < C < 100$ pF.

7.12 Specify the analog function of s that is represented by the portion of a switched capacitor filter shown in Figure 7.48.

7.13 A certain active prototype circuit has analog capacitor and resistor values that range from 0.25 F to 8 F and 1 Ω to 4 Ω, respectively. It is desired to convert the active analog circuit to a switched capacitor circuit while frequency-scaling from $\omega = 1$ rad/sec to $f = 5$ kHz. Choose the largest possible switched capacitor clock frequency, and specify a suitable impedance scale factor so that all capacitors fall into the range $1 < C < 50$ pF. Are the Nyquist criteria satisfied for your choice of clock frequency?

7.14 It is desired to convert an RLC bandpass prototype to a switched capacitor filter. The maximum L or C value in the passive prototype is $1Q$ and the minimum is $0.2/Q$. The maximum and minimum prototype resistance values are 2 Ω and 0.5 Ω, respectively. The scaled capacitor range has to fit between the limits $0.1 < C < 100$ pF for the switched capacitor filter. Find min and max limits for the impedance scale factor, k_R, as a function of Q, and the frequency scale factor, k_ω, and thus determine the maximum Q that can be accommodated by the allowable capacitor range.

7.15 It is desired to implement a 10-kHz bandpass filter by applying switched capacitor technology on a leapfrog circuit architecture. The BP filter requires a 2-kHz bandwidth, and its lowpass prototype has an LC element range of

0.5 to 2.0. An available amplifier works best in an impedance environment of 200 kΩ or less.

(a) Using $k_R = 200$ kΩ, specify a suitable circuit capacitor value range to realize the circuit.

(b) Specify the maximum switching frequency that can be used with your specified capacitor range.

7.16 As suggested in the discussion for Example 4, the null depth should increase when a larger value is used for the capacitor switching frequency. Change the clock frequency for the design found for Example 4 from 40 kHz to 100 kHz, and determine the change in null depth, if any. Explain your results and check, via s-plane distortion estimates, agreement with a predicted null depth.

7.17 Repeat Problem 7.16 for a clock frequency of 200 kHz.

7.18 Confirm the discussion in Example 4 with regard to choosing the circuit scale factor, k_R, to be a lower value of 50 kΩ instead of 500 kΩ. Rescale the design found for the example, and use μCAP to obtain a response. Compare the response to theory as was done in the text and discuss your results.

7.19 As discussed for the design in Figure 7.36, this problem seeks to check the switch ON-OFF resistance effect upon the null depth (s-plane distortion) obtained for the circuit. Change the switch ON-OFF values to 10 Ω for ON and 100 MΩ for OFF in the μCAP simulation to provide a more nearly ideal switch resolution of $10^7 : 1$, or 140 dB, and determine the change in null depth, if any. Discuss your results.

References

[1] Baker, R. J., Li, H. W., and Boyce, D. E., *CMOS Circuit Design, Layout, and Simulation*, New York: IEEE Press, Inc., 1998.

[2] Allen, P. E., and Holberg, D. R., *CMOS Analog Circuit Design*, Ft. Worth, TX: Holt, Rinehart and Winston, 1987.

[3] Hamming, R. W., *Digital Filters*, 2nd ed., Upper Saddle River, NJ: Prentice Hall, 1983.

[4] Oppenheim, A. V., and Schafer, R. W., *Digital Signal Processing*, Upper Saddle River, NJ: Prentice Hall, 1975.

Chapter 8

The Approximation Problem

8.1 Traditional Methods

So far we have studied how to build filters to provide specified frequency response either by scaling or transforming given prototype designs or by designing from specified network functions. Most filter design is done this way; however, there are occasional instances where given prototypes don't quite provide the required response. When that happens, it is useful to be able to generate a desired prototype, or network function, to provide the required performance. Finding a suitable network function, as a ratio of polynomials in the complex frequency variable, s, is referred to as the approximation problem. This chapter provides an introduction to traditional methods that are used to find suitable network functions and concludes by presenting a general method to find a suitable polynomial ratio to fit specified loss functions.

8.1.1 Ideal LP Filter Characteristics

The sinusoidal frequency response magnitude function for an ideal filter is shown in Figure 8.1. It passes energy without loss for the frequency band, $-1 < \omega < +1$ rad/sec, and provides infinite loss for all frequencies, $|\omega| > 1$, that lie outside the passband. The transition bandwidth from passband to stopband is zero; thus, the filter has a shape factor,[1] $SF = (1+0)/1 = 1$. It is useful to determine the impulse function associated with this ideal function. Using $H(j\omega)$, described in (8.1), we get the following impulse response for the ideal characteristic from its inverse Fourier transform:

$$\begin{aligned} H(j\omega) &= 1e^{-j\omega\tau} \text{ for } |\omega| \leq 1 \\ &= 0 \text{ for } |\omega| > 1 \\ &= 1e^{-j\omega\tau}[u(\omega+1) - u(\omega-1)] \text{ for all } \omega \end{aligned} \qquad (8.1)$$

---

[1]Shape factor is the ratio of the frequency at the beginning of the stopband (or a specified loss in decibels) to the cutoff frequency (of the passband) for an LP filter. It is always ≥ 1. A small value, approaching 1, is difficult to realize in practice.

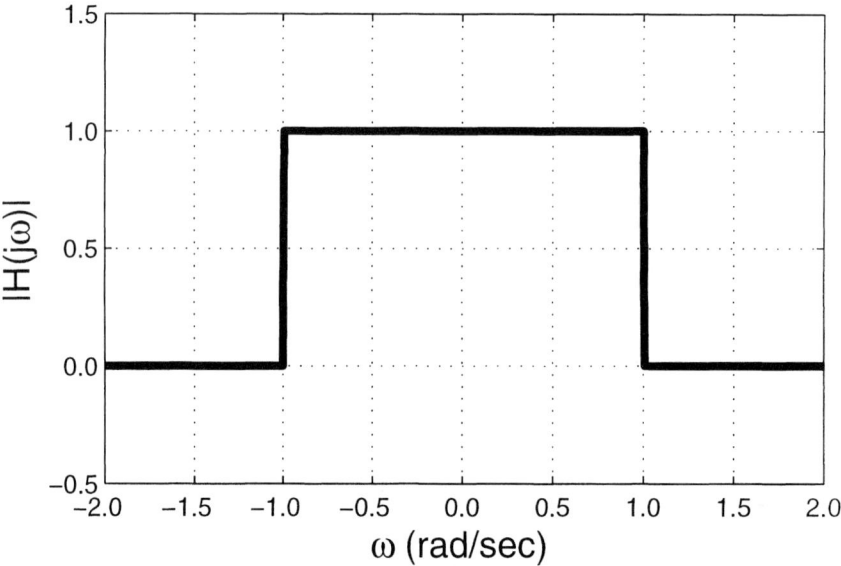

Figure 8.1. Frequency response magnitude for an ideal lowpass filter.

$$\begin{aligned} h(t) &= \frac{1}{2\pi} \int_{-\infty}^{\infty} d\omega e^{j\omega t} H(j\omega) \\ &= \frac{1}{2\pi} \int_{-1}^{1} d\omega 1 \cdot e^{j\omega(t-\tau)} \\ &= \frac{1}{\pi} \frac{\sin(t-\tau)}{(t-\tau)} \end{aligned} \qquad (8.2)$$

A sketch of the impulse response is shown in Figure 8.2, where it is seen to extend to $\pm\infty$ around the delay time, τ. An obvious conclusion is that a response should not start before the excitation (an impulse at $t = 0$), and so, this so-called ideal filter response does not represent a causal, or physically realizable, filter behavior. We can conclude, however, that insofar as the energy for all $t < 0$ is made to approach zero, for a sufficiently large delay factor, τ, the response approaches realizability. Thus, a feature of trying to design filters to have a very small transition band between their pass- and stopbands leads to filters with a large delay factor, τ.

Now consider a modified lowpass network function as shown in Figure 8.3. It is shown as a real function (zero phase), which is used for ease of computing without loss of generality, since treating the function as a magnitude with phase, $e^{-j\omega\tau}$, only delays the impulse response by τ seconds. Thus, let $H(j\omega)$ be broken into two parts, $H_2(j\omega) = a$ and $H_1(j\omega) = H(j\omega) - a$. The variable a represents a nonzero loss factor, while a nonzero transition band is given by $b = (\Omega_S - 1)$. These two nonzero parameters change the ideal response to a nonideal description. It is interesting to see the effect that the nonideal parameters have on the impulse response of the filter.

8.1. TRADITIONAL METHODS

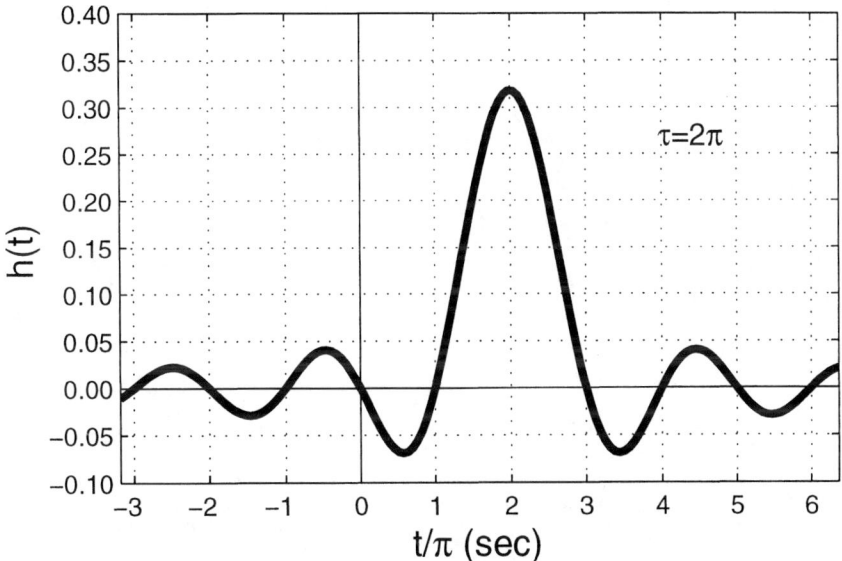

Figure 8.2. Impulse response for an ideal lowpass filter.

Note that $H_1(j\omega)$ has a trapezoidal form of connected lines and that the transition part is awkward to integrate into the inverse transform for $h(t)$. The problem is simplified by recalling the Fourier transform property given in (8.3):

$$\begin{aligned}
\text{For} \quad H(j\omega) &= H_1(j\omega) + H_2(j\omega) \\
H(j\omega) \leftrightarrow h(t) &= h_1(t) + h_2(t) \\
\text{and for} \quad H_1(j\omega) &\leftrightarrow h_1(t) \\
\frac{dH_1}{d\omega} &\leftrightarrow -jth_1(t)
\end{aligned} \quad (8.3)$$

The frequency derivative for $H_1(j\omega)$ is shown in Figure 8.4. The derivative of $H_1(j\omega)$ consists of the two transition bands where the slope of H_1 is constant and not zero. Then, $h_1(t)$ follows:

$$\begin{aligned}
-jth_1 &= \frac{1}{2\pi}\int_{-\infty}^{\infty} d\omega e^{j\omega t}\frac{dH_1}{d\omega} \\
&= \frac{1}{2\pi(\Omega_S - 1)}\left[\int_{-\Omega_S}^{-1} d\omega e^{j\omega t} \cdot 1 - \int_{1}^{\Omega_S} d\omega e^{j\omega t} \cdot 1\right] \\
&= \frac{1}{2\pi(\Omega_S - 1)}\frac{(e^{-jt} - e^{-j\Omega_S t}) - (e^{j\Omega_S t} - e^{jt})}{jt} \quad \text{so} \\
h_1(t) &= \frac{1}{\pi(\Omega_S - 1)}\frac{\cos(t) - \cos(\Omega_S t)}{+t^2}
\end{aligned}$$

316   CHAPTER 8. THE APPROXIMATION PROBLEM

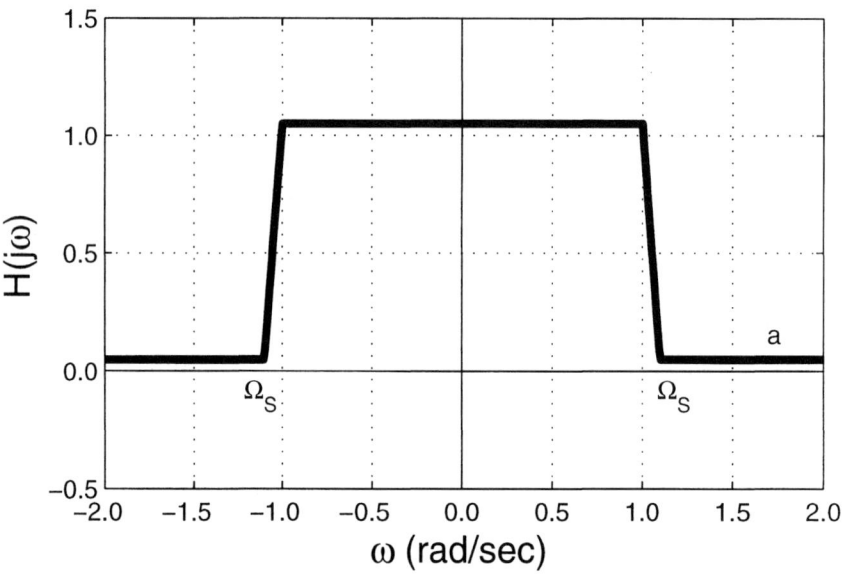

Figure 8.3. A modified lowpass function.

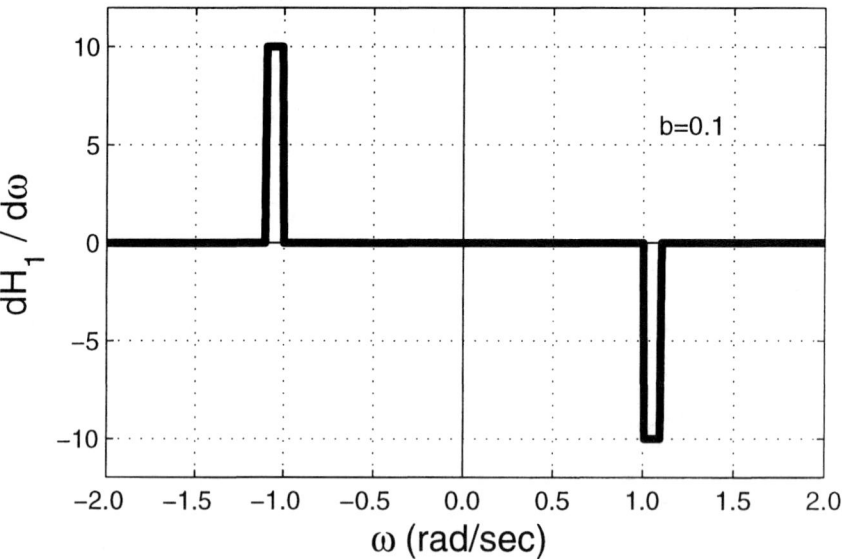

Figure 8.4. Frequency derivative for the H_1 part of $H(j\omega)$.

8.1. TRADITIONAL METHODS

$$\equiv \frac{2}{\pi(\Omega_S - 1)t^2} \sin\left((\Omega_S + 1)t/2\right) \sin\left((\Omega_S - 1)t/2\right)$$

$$h_1(t) = \frac{(\Omega_S + 1)}{2\pi} \frac{\sin\left((\Omega_S + 1)t/2\right)}{(\Omega_S + 1)t/2} \frac{\sin\left((\Omega_S - 1)t/2\right)}{(\Omega_S - 1)t/2} \quad (8.4)$$

The constant response, $H_2(j\omega) = a$, has the impulse response, $h_2(t) = a\delta(t)$, and so, the total impulse response for $H(j\omega)$ is as shown in (8.5). Note that the stopband frequency, $\Omega_S = 1 + b$, is the passband plus the transition bandwidth.

$$h(t) = h_1(t) + h_2(t) \text{ so that}$$

$$h(t) = a\delta(t) + \frac{(2+b)}{2\pi} \frac{\sin\left((2+b)t/2\right)}{(2+b)t/2} \frac{\sin(bt/2)}{bt/2} \quad (8.5)$$

We see that an arbitrarily small loss, a, over an infinite band feeds a portion of the input impulse straight through to the output. The added finite loss has nothing to do with terminating the impulse response for $t < 0$. The added transition band, b, has caused the impulse response, $h_1(t)$, to drop off as $1/t^2$. Thus, it is possible to find some value of τ to add delay to $H(j\omega)$ so that the total energy in the response for $t < 0$ will be less than a specified portion of the total energy. Note that as both $a \to 0$ and $b \to 0$, the response given in (8.5) goes to the response for the ideal filter given by (8.2). Figure 8.5 shows a comparison of the impulse responses, $h_1(t)$ of (8.4) and $h(t)$ of (8.2), for the transition band, $b = 0.25$ rad/sec. The impulse at $t = 0$ is omitted from the plot. The graph shows that adding a finite transition band, from passband to stopband, causes the impulse response to collapse towards the origin. As shown on the lower graph in Figure 8.5, there is a time beyond which the impulse response becomes negligible. Thus, a delay, τ, can be chosen so as to make the time response for the nonideal frequency response approach causal and realizable behavior for practical purposes.

In the procedures to be presented, it is possible to observe that as the frequency functions try to approach the ideal magnitude response of Figure 8.1, the filter slope of the phase function increases with respect to frequency so as to supply a larger delay through the filter. In conclusion, we note that filters approximating ideal LP response usually exhibit large delay.

8.1.2 Critical Frequencies and Steady-State Response

This section formulates a useful relationship between the poles and zeros of a given network function and its sinusoidal steady-state frequency response function. We start by letting $T(s) = \mathbf{V}_{OUT}(s)/\mathbf{V}_{IN}(s)$ be the transmission (voltage gain) function for any linear network.[2] For $s = j\omega$, we get the following relationships:

$$T(s) = \frac{\mathbf{V}_{OUT}}{\mathbf{V}_{IN}}(s) = \frac{N(s)}{D(s)}$$

$$|T(j\omega)| = |T(s)|_{s=j\omega} = \left|\frac{N(s)}{D(s)}\right|_{s=j\omega} \text{ and}$$

---
[2]Actually, $T(s)$ can be any type of response-to-excitation transfer function.

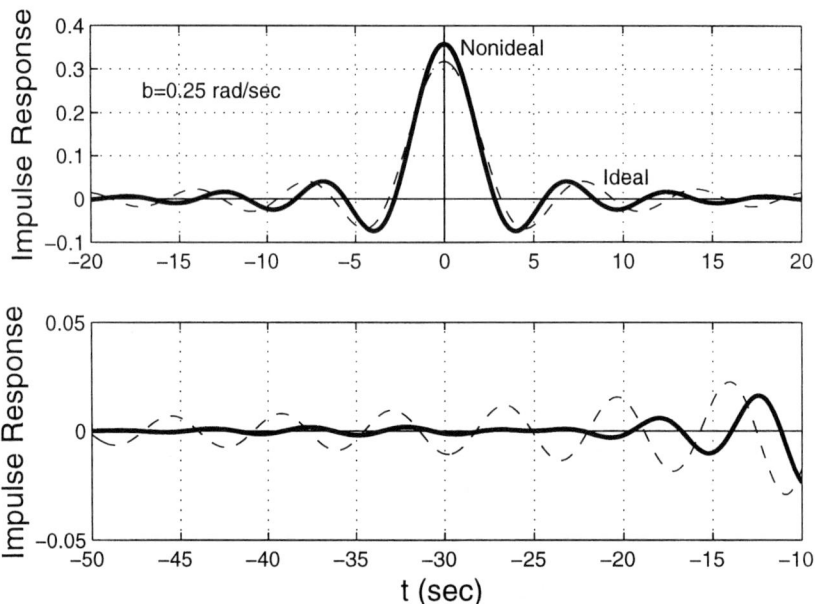

Figure 8.5. Impulse responses for the ideal and nonideal filter frequency functions.

$$\begin{aligned} |T(j\omega)|^2 &= T(j\omega)T^*(j\omega) \equiv T(s)T(-s)|_{s=j\omega} \\ &= \left| \frac{N(s)N(-s)}{D(s)D(-s)} \right|_{s=j\omega} \end{aligned} \tag{8.6}$$

The roots of N are the zeros of $T(s)$, and the poles are the roots of $D(s)$. In addition, the phase and delay relations are given in (8.7):

$$\begin{aligned} \angle T(j\omega) &= \text{phase lead} = \angle N(j\omega) - \angle D(j\omega) \text{ and} \\ \tau &= \text{group delay} = \frac{d\angle T(j\omega)}{d\omega} = \frac{d\angle N(j\omega)}{d\omega} - \frac{d\angle D(j\omega)}{d\omega} \end{aligned} \tag{8.7}$$

For a polynomial, $P(s)$, $P(-s)$ is its image in the s-plane with respect to the $j\omega$-axis. This property is deduced as follows:

$$\begin{aligned} P(s) &= K_P \prod_{r=1}^{m}(s - s_r) \text{ and} \\ P(-s) &= K_P \prod_{r=1}^{m}(-s - s_r) \equiv (-1)^m K_P \prod_{r=1}^{m}(s + s_r) \end{aligned} \tag{8.8}$$

The complex frequency, s_r, is a root (zero) of $P(s)$. Equation (8.8) shows that the roots of $P(-s)$ are the negative of the roots of $P(s)$.

8.1. TRADITIONAL METHODS

We can deduce by trial and error that the product of a polynomial, $P(s)$, with its $j\omega$-axis image, $P(-s)$, always yields a polynomial in s^2. The following are some examples:

1.
$$P(s) = s+1$$
$$P(s)P(-s) = (s+1)(1-s) = -s^2 + 1$$
2.
$$P(s) = s^2 + \sqrt{2}s + 1$$
$$P(s)P(-s) = (s^2 + \sqrt{2}s + 1)(1 - \sqrt{2}s + s^2) = s^4 + 0s^2 + 1$$
3.
$$P(s) = (s+1)(s^2 + s + 1) = s^3 + 2s^2 + 2s + 1$$
$$P(s)P(-s) = (s^3 + 2s^2 + 2s + 1)(1 - 2s + 2s^2 - s^3)$$
$$= -s^6 + 0s^4 - 0s^2 + 1$$
...

Clearly, these appear to be a special set of polynomials, but they all yield polynomials in s^2 when multiplied by their image. This is what makes $|T(j\omega)|^2$ a real function of $s = j\omega$.

Now we look at how (8.6) can be used to generate approximations to an ideal LP filter response.

8.1.3 The Butterworth Polynomials

Consider the network magnitude function given by (8.9):

$$|T(j\omega)|^2 = \frac{1}{(1 + \omega^{2n})} \qquad (8.9)$$

This is a first-order approximation to the design of an ideal LP prototype transmission function, as shown in Figure 8.6. The polynomials that yield the denominator in (8.9) are known as Butterworth polynomials. For $\omega \to 0$, $|T|^2 \to 1$ for all n, and for $\omega = 1$, $|T(j1)|^2 = 0.5$ for all n. The design parameter, n, is the only parameter that needs to be specified in order to design a Butterworth filter approximation. Now let's look at how the network polynomials, $N(s)$ and $D(s)$, are found from (8.9).

Equating (8.9) to the basic form defined in (8.6), we get the following result for $s = j\omega$ or $\omega = -js$:

$$|T(j\omega)|^2 = \left|\frac{N(s)N(-s)}{D(s)D(-s)}\right|_{s=j\omega} = \left|\frac{1}{1 + (-js)^{2n}}\right|_{s=j\omega} \equiv \frac{1}{1 + \omega^{2n}}$$

but $(-j)^{2n} = (-1)^n$ so we get

$$N(s)N(-s) = 1 \text{ and}$$
$$D(s)D(-s) = 1 + (-1)^n s^{2n} \qquad (8.10)$$

The left half-plane (lhp) roots in (8.10) are used to form $D(s)$, and $N(s) = 1$ is chosen. Let s_p be a root of $D(s)$. The roots are then as follows:

$$1 + (-1)^n s_p^{2n} = 0$$

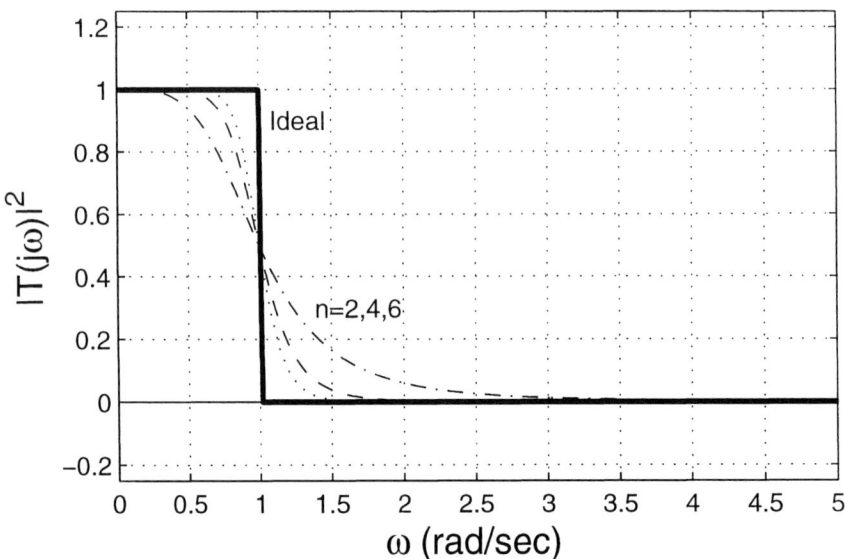

Figure 8.6. Butterworth approximations to an ideal LP response for $n = 2, 4, 6$.

$$\begin{aligned} s_p^{2n} &= \frac{-1}{(-1)^n} = -(-1)^n \\ &= +1 \text{ for } n \text{ odd} \\ &= -1 \text{ for } n \text{ even} \end{aligned} \tag{8.11}$$

There are $2n$ roots in $D(s) \cdot D(-s)$ for the nth-order Butterworth polynomial. The lhp roots are used to form $D(s)$ for the desired network function.

$$\begin{aligned} s_p &= +1,\ e^{j\pi/n},\ e^{j2\pi/n},\ \ldots,\ e^{j(2n-1)\pi/n} \text{ for } n \text{ odd} \\ &= e^{j\pi/2n},\ e^{j3\pi/2n},\ \ldots,\ e^{j\pi(4n-1)/2n} \text{ for } n \text{ even} \end{aligned} \tag{8.12}$$

The roots are distributed in the s-plane with angular spacing, $\phi = \pi/n$, around the unit circle, as shown in Figure 8.7 for $n = 5$ and 6. For both cases, even or odd, the lhp roots start at $\phi/2$ past the $j\omega$-axis. The Butterworth polynomials, with lhp roots, are written in factored form in (8.13):

$$\begin{aligned} \phi &= \pi/n \\ D(s) &= (s+1) \prod_{k=0}^{(n-3)/2} (s^2 + 2\sin((k+0.5)\phi) + 1) \text{ for } n \text{ odd} \\ \text{and } D(s) &= \prod_{k=0}^{(n-2)/2} (s^2 + 2\sin((k+0.5)\phi) + 1) \text{ for } n \text{ even} \end{aligned} \tag{8.13}$$

Equation (8.13) is useful to form the Butterworth polynomial for any value of n.

8.1. TRADITIONAL METHODS

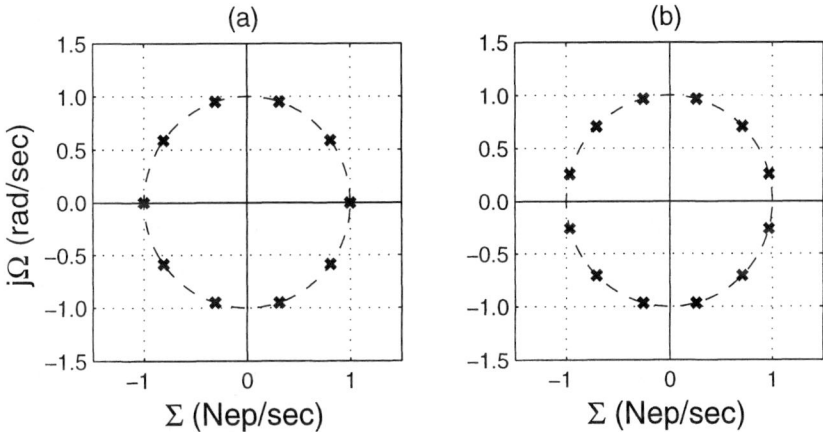

Figure 8.7. Example Butterworth polynomial roots for (a) $n = 5$, and (b) $n = 6$.

This is easily done in Matlab through the use of its conv function, which is used to convolve polynomial products (shown in Example 1).

As shown in Figure 8.6, the Butterworth polynomials used to form a network transfer function, $|T(j\omega)|^2$ in (8.10), provide an approximation to the LP filter response. The only design parameter is to choose a suitable value for n for a given requirement. We will work some examples to see how this is done; however, it is helpful first to consider the asymptotic behavior of the Butterworth network function. Let α be the loss in dB for $|T(j\omega)|$. We then get the following results:

$$\begin{aligned} \alpha &= -20\log_{10}|T(j\omega)| = -10\log_{10}|T(j\omega)|^2 \\ \therefore \alpha &= +10\log_{10}|1 + \omega^{2n}| \text{ and} \\ \text{for } \omega \to 0: \quad \alpha &\to 10\log_{10}(1) = 0 \text{ dB} \\ \text{at } \omega = 1: \quad \alpha &= 10\log_{10}(2) = 3.01 \text{ dB} \\ \text{for } |\omega| \gg 1: \quad \alpha &\to 20n\log_{10}|\omega| \end{aligned} \quad (8.14)$$

The asymptotic (Bode) response curves for the Butterworth transmission function are shown in Figure 8.8 for different values of n where they are plotted versus the logarithmic variable, $x = \log_{10}(\omega)$. Note that the cutoff frequency, $\omega = 1$, is the same for all n values and that at one decade from the passband cutoff point, $\omega = 10$, the loss increases by 20 dB for each integer value that n is increased. These properties help us choose an appropriate value of n for a specific requirement.

Example 1 - Determine a suitable value of n for a Butterworth polynomial to obtain the following LP filter performance. Also specify a suitable frequency scale factor to apply to the prototype design so as to obtain a maximum passband loss of 1 dB at $f_1 = 100$ Hz and a minimum stopband loss of 50 dB at $f_2 = 200$ Hz.

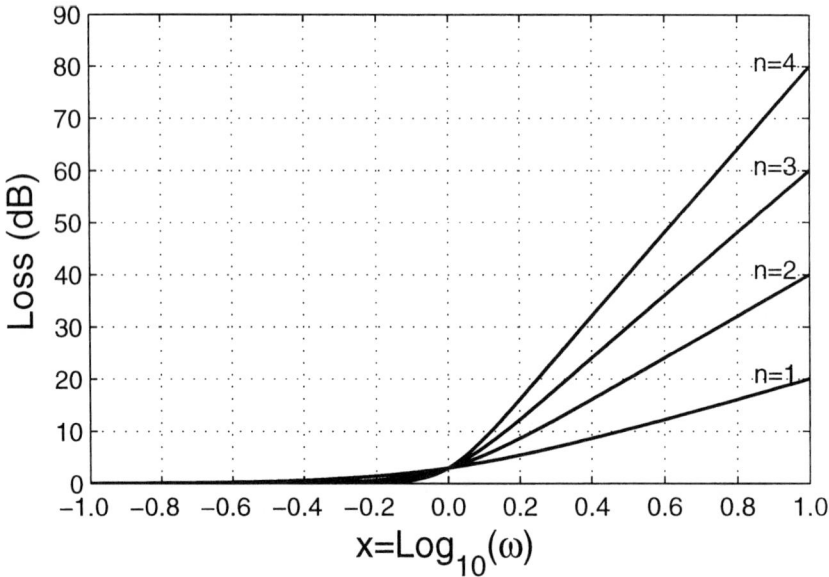

Figure 8.8. Asymptotic loss curves for Butterworth transmission functions.

Solution - The specified design requirements are sketched in Figure 8.9 where ω is a prototype frequency variable. A suitable value for n is chosen to meet both stop- and passband requirements by using the ratio $f_2/f_1 = 2$ as follows:

$$10\log_{10}(1+\omega_1^{2n}) \leq 1$$
$$10\log_{10}(1+\omega_2^{2n}) \geq 50 \qquad (8.15)$$

Equation (8.15) is then solved to obtain a minimum value for n.

$$(f_2/f_1)^{2n} = (\omega_2/\omega_1)^{2n} \geq \left(\frac{10^5-1}{10^{0.1}-1}\right) \text{ or}$$
$$(200/100)^{2n} = 4^n \geq \left(\frac{10^5-1}{10^{0.1}-1}\right)$$
$$\therefore n \geq \frac{1}{\log_e(4)}\log_e\left(\frac{10^5-1}{10^{0.1}-1}\right) = 9.28$$

Since n must be an integer, we choose $n = 10$ to meet the given loss requirement. This value of n exceeds the attenuation requirement slightly, and so, setting the prototype frequency scale factor allows for creating some margin on meeting the specifications with the design. For instance, solving (8.15) with equality for $n = 10$ yields the following results:

$$\omega_L = (10^{0.1}-1)^{1/20} = 0.9347 \geq 2\pi f_1/k_\omega$$

8.1. TRADITIONAL METHODS

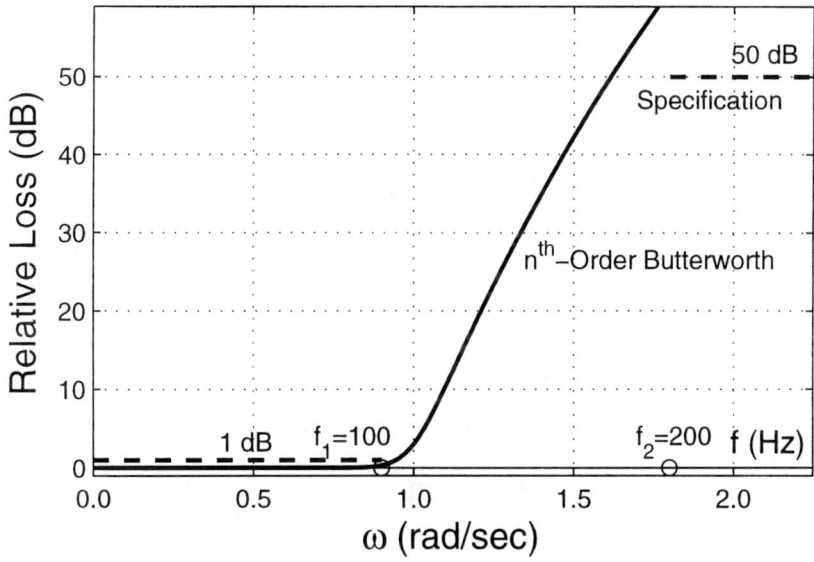

Figure 8.9. A Butterworth polynomial used to meet an LP filter requirement.

$$\omega_U = (10^5 - 1)^{1/20} = 1.7783 \leq 2\pi f_2/k_\omega$$
$$\omega_U/\omega_L = 1.9026 < f_2/f_1 = 2.0$$

The results tell us that the tenth-order Butterworth polynomial is capable of providing a 50-dB to 1-dB bandwidth ratio of 1.903, whereas we only require a ratio of 2.0 to meet the specification. One arbitrary choice would be to set $\omega_1 = 0.90$ and $\omega_2 = 2\omega_1 = 1.80$ for the prototype tenth-order filter. This yields a frequency scale factor of $0.9k_\omega = 200\pi$ or $k_\omega = 222.22\pi$ for the design. The polynomial coefficients are found by use of the following Matlab file:

```
function D=btrwrth(n)
phi=pi/n;

if(rem(n,2)==1
   D=[1 1]; % n is odd
   for k=0:(n-3)/2
      D=conv(D,[1 2*sin(k*phi+phi/2) 1]);
   end
else
   D=1; % n is even
   for k=0:(n-2)/2
      D=conv(D,[1 2*sin(k*phi+phi/2) 1]);
   end
end
```

For $n = 10$, we get the following polynomial data for the Butterworth polynomial from the above file and (8.13):

$$\begin{aligned}
D_{10}(s) &= (s^2 + 0.3129s + 1)(s^2 + 0.9080s + 1)(s^2 + 1.4142s + 1) \\
&\quad \cdot (s^2 + 1.7820s + 1)(s^2 + 1.9754s + 1) \\
&= s^{10} + 6.3925s^9 + 20.4317s^8 + 42.8021s^7 + 64.8824s^6 \\
&\quad + 74.2334s^5 + 64.8824s^4 + 42.8021s^3 + 20.4317s^2 \\
&\quad + 6.3925s + 1.0
\end{aligned} \qquad (8.16)$$

The circuitry required to implement the network function, $T(s) = 1/D_{10}(s)$, depends upon the architecture chosen for the design. The 10 poles could be realized using 5 S-K stages (5 opamps); 5 biquad stages (15 opamps); or an all-pole state variable circuit (10 integrators, 1 sign change, and 1 summation for a total of 12 opamps). A leapfrog implementation for a tabulated 10-pole Butterworth LP filter would have about the same number of opamps as the state variable form. The decision of which architecture to use is based upon the designer's preference and issues dealing with cost versus performance.

8.1.4 The Chebyshev Polynomials

In the previous section, we considered the Butterworth network function, which employed a single design parameter, n. It is about the simplest solution to the approximation problem that we can hope to achieve. A two-parameter formulation of the problem is achieved through the use of Chebyshev polynomials as in (8.17):[3]

$$|T(j\omega)|^2 = \frac{1}{[1 + \epsilon^2 V_n^2(\omega)]} \qquad (8.17)$$

$V_n(\omega)$ are nth-order Chebyshev polynomials that have the following definition:

$$\begin{aligned}
V_n(x) &= \cos(n\phi) \\
\text{where} \quad \phi &= \cos^{-1}(x)
\end{aligned} \qquad (8.18)$$

As shown in Figure 8.10, over the interval $-1 < x < +1$ for different values of n, the polynomials provide equiripple fits to the range ± 1. For values of $|x| > 1$, the defining function provides complex values for ϕ, which in turn provides real values greater than 1 for the polynomials. We will see how these properties are obtained through the following discussion. Note that the polynomials are odd functions of x for n odd and are even functions of x for n even.

The polynomials in x are generated by means of a recursive relationship that is derived using trigonometric identities applied to the defining (8.18).

$$\begin{aligned}
V_{n+1}(x) &= \cos((n+1)\phi) \equiv \cos(n\phi)\cos(\phi) - \sin(n\phi)\sin(\phi) \\
\text{and} \quad V_{n-1}(x) &= \cos((n-1)\phi) \equiv \cos(n\phi)\cos(\phi) + \sin(n\phi)\sin(\phi) \\
\text{so} \quad V_{n+1}(x) + V_{n-1}(x) &= 2\cos(\phi)\cos(n\phi) \equiv 2xV_n(x) \\
\therefore \quad V_{n+1}(x) &= 2xV_n(x) - V_{n-1}(x)
\end{aligned} \qquad (8.19)$$

---

[3]E. A. Guillemin preferred the spelling of Tschebyschef. Dr. Guillemin has been referred to as the father of modern network theory.

8.1. TRADITIONAL METHODS

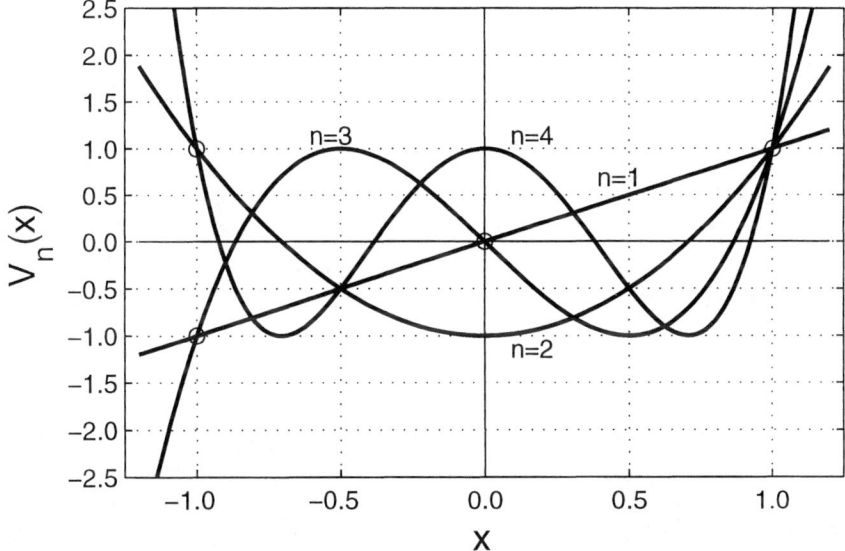

Figure 8.10. Chebyshev polynomial behavior for different values of n.

With $V_0(x) = 1$ and $V_1(x) = x$ to start the polynomials for $n = 0$ and 1, respectively, (8.19) is used to generate the polynomials listed in Table 8.1.

Observe the odd and even form of the polynomials for n odd and even. In addition, all the polynomials are equal to ± 1 at $x = \pm 1$, and the even polynomials are all equal to ± 1 at $x = 0$, while the odd polynomials equal zero at $x = 0$. It is interesting that even though the polynomials are defined trigonometrically with the cosine function, they nevertheless work out to be polynomials in x through elimination of the intermediate variable, ϕ. The cosine function is used solely to create the region where the function should have an equiripple magnitude equal to 1.

There is also an additional useful property that comes about through application of an identity for the cosine squared:

$$V_n^2 = \cos^2(n\phi) \equiv \frac{1}{2}(1 + \cos(2n\phi))$$
$$\therefore V_n^2(x) = \frac{1}{2}(1 + V_{2n}(x)) \tag{8.20}$$

For example, $V_3^2 = (4x^3 - 3x)^2 = 16x^6 - 24x^4 + 9x^2$ is clearly equal to $(1 + V_6)/2 = (32x^6 - 48x^4 + 18x^2 - 1 + 1)/2$.

This last result tells us that the Chebyshev network function, (8.17), is a function in ω^2 (V_{2n} is an even function of ω). Therefore, the same procedure of sorting right and left half-plane roots to create an all-pole function for $T(s)$ can also be applied to (8.17), as was done for the Butterworth polynomials.

Table 8.1. Generation Procedure for Chebyshev Polynomials

n	Recursion	V_n
0	—	1
1	—	x
2	$2x(x) - 1$	$2x^2 - 1$
3	$2x(2x^2 - 1) - x$	$4x^3 - 3x$
4	$2x(4x^3 - 3x) - (2x^2 - 1)$	$8x^4 - 8x^2 + 1$
5	$2x(8x^4 - 8x^2 + 1) - (4x^3 - 3x)$	$16x^5 - 20x^3 + 5x$
6	$2x(16x^5 - 20x^3 + 5x) - (8x^4 - 8x^2 + 1)$	$32x^6 - 48x^4 + 18x^2 - 1$
7	...	...

As noted earlier, this all-pole Chebyshev network function has two design parameters, ϵ and n. In order to understand how these are chosen, it is again helpful to look at the asymptotic behavior of the relative loss function for the Chebyshev network function. Again, let α be the loss in dB for the filter transmission function, $|T(j\omega)|$. In addition, let $x = \omega$ in order to apply the polynomials to the filter problem.

$$\alpha = -10\log_{10}|T(j\omega)|^2 = +10\log_{10}(1 + \epsilon^2 V_n^2(\omega)) \qquad (8.21)$$

$$\text{as } \omega \to 1, \quad V_n^2(1) = 1 \text{ and } \alpha \to 10\log_{10}(1 + \epsilon^2) \qquad (8.22)$$

This last result gives the maximum passband loss for this LP filter approximation function. Thus, the parameter, ϵ, is independent of when the stopband begins or of other specified performance for the desired filter response. The order, n, is determined by satisfying the stopband minimum attenuation requirement at the beginning of the stopband (or at the end of the transition band).

By induction, from Table 8.1 for increasing n, it is easy to determine the asymptotic loss as ω becomes much greater than 1:

$$\text{as } |x| \gg 1 \quad V_n \to 2^{n-1}x^n = 2^{n-1}\omega^n$$
$$\therefore \alpha = 10\log_{10}(1 + \epsilon^2 V_n^2) \to 20\log_{10}|\epsilon 2^{n-1}\omega^n| \text{ or}$$
$$\alpha \to 20\log_{10}(\epsilon 2^{n-1}) + 20n\log_{10}|\omega| \text{ for } |\omega| > 1 \qquad (8.23)$$

The asymptotic response for the Chebyshev filter has the same $\log_{10}|\omega|$ dependence as does the Butterworth filter, but in addition, it is offset from zero by the constant, $20\log_{10}(\epsilon 2^{n-1})$ dB. The Chebyshev filter, therefore, exhibits a sharper cutoff (smaller transition band) than the Butterworth function. The graph given in Figure 8.11 compares the two responses for $n = 6$ and $\epsilon = 0.5$. The offset on the Chebyshev asymptote is $20\log_{10}(0.5 \cdot 2^5) = 24.08$ dB. This offset gives the Chebyshev a lot of steepness at the edge of the passband of the filter. The Butterworth function is very smooth (and lazy appearing) as it transitions from its pass- to stopbands. In the passband, the Chebyshev filter function provides an equiripple response determined by the value of ϵ, as $\alpha_{max} = 10\log_{10}(1 + \epsilon^2)$.

8.1. TRADITIONAL METHODS

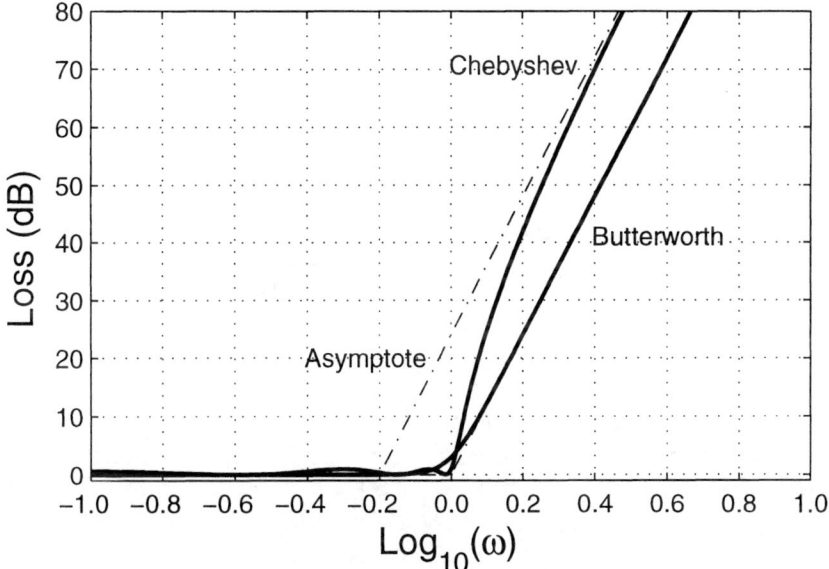

Figure 8.11. Chebyshev asymptotic response ($n = 6$, $\epsilon = 0.5$).

Example 2 - Find the Chebyshev parameters, n and ϵ, to satisfy the LP requirements of Example 1 and specify the frequency scale factor for the prototype design. In addition, find the poles for the prototype all-pole function and specify the network function, $T(s) = K/D(s)$.

Solution - We require a maximum loss of 1 dB for all frequencies less than $f_1 = 100$ Hz and a minimum loss for all frequencies greater than $f_2 = 200$ Hz. See Figure 8.12. With Chebyshev filters, the passband ends at $\omega = 1$, so this point gets scaled to $1 \cdot k_\omega = 2\pi f_1 = 200\pi$ and $\omega_1 = 1$. The maximum passband attenuation also occurs at $\omega = 1$ for the Chebyshev function, and this is used to determine ϵ.

$$\text{At} \quad \omega_1 = 1, \quad 10\log_{10}(1 + \epsilon^2 \cdot 1) \leq 1 \text{ dB}$$
$$\therefore \quad \epsilon \leq (10^{1/10} - 1)^{1/2} = 0.5088$$

Choosing $\epsilon = 0.50$ allows a small margin for the equiripple response in the passband. The minimum attenuation in the stopband is satisfied at $f_2 = 200$, which corresponds to $\omega_2 = 2$ on the prototype scale. Therefore, at $\omega_2 = 2$, we want

$$10\log_{10}(1 + \epsilon^2 V_n^2(2)) \geq 50$$
$$V_n(2) \geq (10^5 - 1)^{1/2}/\epsilon = 632.4524$$

Using the asymptote for $V_n(2)$ yields a choice for n:

$$2^{(n-1)}2^n \geq 632.4524 \text{ or}$$

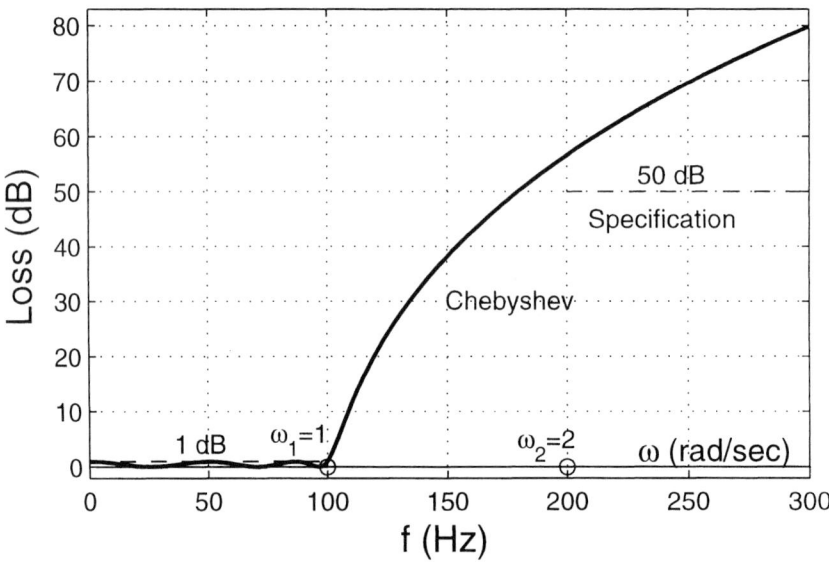

Figure 8.12. An nth-order Chebyshev polynomial that meets the LP filter requirement.

$$2^{2n} \geq 1264.9048$$
$$\therefore \quad n \geq \frac{\log_e(1264.9048)}{2\log_e(2)} = 5.1524$$

Since n is required to be an integer, we choose $n = 6$ for this design example. Recall that a tenth-order Butterworth polynomial was required to meet the same specification for Example 1.

The value for n was obtained by using the Chebyshev asymptote for $\omega > 1$, and so, the answer should be checked to ascertain that $n = 6$ meets the specification using the actual polynomial.

From the definition, (8.18), we get the following for $V_6(2)$:

$$V_6(2) = \cos(6\cos^{-1}(2)) = 1351.0$$

$V_6(2)$ is greater than 632.45, as required. The asymptote provided a safe estimate for n for this example. This is not always true as ω_2 is specified closer to 1 (a smaller transition band), so it is usually wise to check the actual polynomial behavior for the choice of n that is made for any given design. Now we can solve for the prototype loss function using $\epsilon = 0.5$ and $n = 6$, as follows:

$$|T(j\omega)|^2 = \frac{1}{1 + 0.5^2 V_6^2(\omega)} = \frac{1}{1 + 0.25 V_6^2(\omega)}$$

where
$$V_6^2(\omega) = (32\omega^6 - 48\omega^4 + 18\omega^2 - 1)^2$$
$$= 1024\omega^{12} - 3072\omega^{10} + 3456\omega^8 - 1792\omega^6 + 420\omega^4 - 36\omega^2 + 1$$

8.1. TRADITIONAL METHODS

Now we substitute $\omega^2 = -s^2$ for $\omega = -js$ to get $T(s)$:

$$T(s) \cdot T(-s) = \frac{1}{D(s) \cdot D(-s)}$$

where

$$\begin{aligned} D(s) \cdot D(-s) &= 1 + 0.25(1024s^{12} + 3072s^{10} + 3456s^8 \\ &\quad + 1792s^6 + 420s^4 + 36s^2 + 1) \\ &= 256s^{12} + 768s^{10} + 864s^8 + 448s^6 + 105s^4 + 9s^2 + 1.25 \end{aligned}$$

The lhp roots for $D(s)$ are at $-0.0629 \pm j0.9940$, $-0.1718 \pm j0.7277$, and $-0.2347 \pm j0.2663$. These roots then combine to yield the desired network function as shown in (8.24):

$$T(s) = \frac{1}{D(s)}$$
$$\begin{aligned} \text{where} \quad D(s) &= 16s^6 + 15.0180s^5 + 31.0481s^4 + 19.4953s^3 \\ &\quad + 15.1742s^2 + 4.9930s + 1.1180 \end{aligned} \quad (8.24)$$

The above operations were all completed with the help of the following Matlab .m-files:

```
% function P=cheby(n) Nov 6, 2002
% generates the nth order Chebyshev polynomial

function P=cheby(n)
v0=[0 1]; %V0(x)
v1=[1 0]; %V1(x)
for k=1:n-1
  v2=conv([2 0], v1)-[0 v0]; % recursive relation
  v0=[0 v1];
  v1=v2;
end
P=v2';

% Example 2 polynomial management for Chebyshev LP filters
e=0.5;
n=6;
Vn=cheby(n);
Vn2=conv(Vn,Vn);
DD=e^2*Vn2+[zeros(2*n,1);1]; %polynomial in w^2
% have to change coeff's of DD to those for w^2=-s^2
DDs=DD;
for k=1:2:2*n+1
  DDs(k)=DD(k)*(-1)^((k-1)/2);
end
```

```
zros=roots(DDs);
drts=zros(find(real(zros)<0));
D=poly(drts)*sqrt(DDs(1))
```

The result, $\epsilon = 0.5$, $n = 6$, with the network function of (8.24), was used to generate the polynomial curve shown in Figure 8.12 at the beginning of the discussion for this example. This answer has a good margin (6 dB) on meeting the minimum attenuation requirement, so it is possible to choose a smaller passband ripple (reduce ϵ) to improve the response with a smaller loss margin at f_2. Alternately, since $n = 5$ provides response very close to (but in excess of) the specified behavior, it is sometimes appropriate to renegotiate the specification in light of possible complexity or cost trade-offs. Specifications often have tenuous origin and can in fact be changed for good reasons.

Note that the .m-file listed above can be used with any value of ϵ and n to obtain a Chebyshev LP filter network function.

8.1.5 Inverted Chebyshev Polynomials

In the previous section, we learned about Chebyshev polynomials and their ability to provide equiripple fit in the passband of an LP filter. The result provided the same all-pole response as the Butterworth filter but with a sharper transition between passband and stopband.

Inverted Chebyshev filters are used to preserve the sharp transition between pass- and stopbands while preserving the smooth (maximally flat) passband response of the Butterworth filter.

The transmission function using the inverted Chebyshev polynomial is given in (8.25):

$$|T(j\omega)|^2 = \frac{1}{\left[1 + \frac{h^2}{V_n^2(1/\omega)}\right]} \tag{8.25}$$

Note not only that the reciprocal of the polynomial, V_n, is used, but also that the polynomial is a function of $1/\omega$ as well. The polynomial goes through a lowpass-to-highpass transform, and then its reciprocal is used to form an LP function in the denominator of $|T(jw)|^2$. Figure 8.13 illustrates the transformation achieved by these steps where an inverted fifth-order Chebyshev polynomial is compared to its normal form over the interval $0 < \omega < 3$. Note that both forms are greater than 1 for $\omega > 1$ and less than or equal to 1 for $\omega < 1$. Thus, both can be used to build LP filter approximants. $V_n^2(\omega)$ gets multiplied by ϵ^2 to control passband maximum attenuation, and $1/V_n^2(1/\omega)$ gets multiplied by h^2 to control minimum stopband attenuation. We can also observe that the stopband minimum attenuation begins at $\omega = 1$ for the inverted Chebyshev polynomial, whereas the passband ends at $\omega = 1$ for the normal Chebyshev polynomial.

Example 3 - Use the inverted Chebyshev form, (8.25), to design a prototype filter function for the specifications in Examples 1 and 2. Specify the prototype function.

8.1. TRADITIONAL METHODS

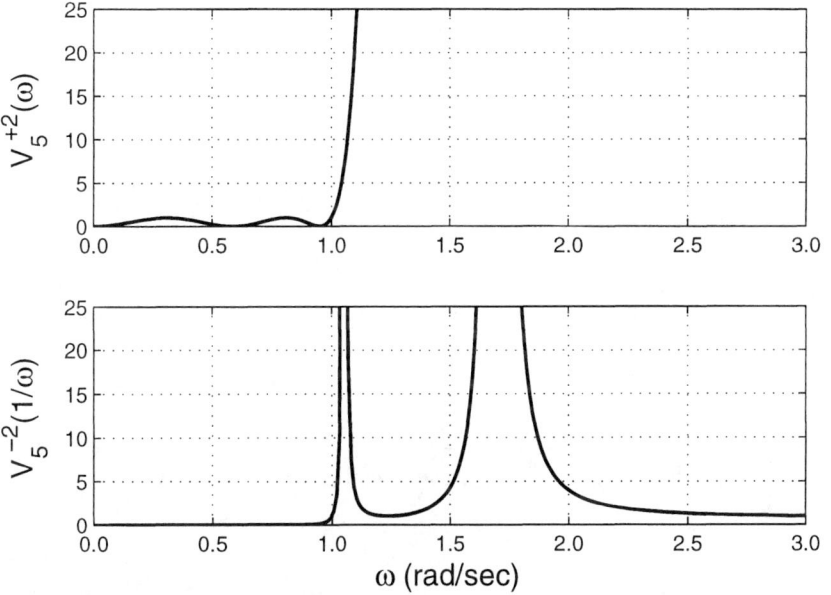

Figure 8.13. The inverted Chebyshev polynomial response for $n = 5$.

Solution - In this example, the stopband starts at $\omega = 1$ rad/sec in the prototype function. Therefore, this point is scaled to f_2 so that $1 \cdot k_\omega = 2\pi f_2 = 400\pi$. Then, the end of the passband is at $f_1 = 100$ Hz, or one-half of f_2. Thus, maximum passband attenuation occurs at $\omega = 0.5$ on the prototype scale. The loss function, α, is as follows:

$$\alpha = -10\log_{10}|T(j\omega)|^2 = 10\log_{10}\left(1 + \frac{h^2}{V_n^2(1/\omega)}\right) \quad (8.26)$$

So, at $\omega = 1$ (the beginning of the stopband), we get a specification for design parameter h:

$$\alpha = 50 \leq 10\log_{10}(1 + h^2/1) \text{ yields}$$
$$h \geq (10^5 - 1)^{1/2} = 316.2262$$

$h = 325$ is chosen to provide a slight margin and to round off the number. Now meeting the passband attenuation requirement provides a condition for the choice of parameter n.

Thus, at $\omega = 1/2$ (the end of the passband), we get the following condition:

$$\alpha = 1 \geq 10\log_{10}\left(1 + \frac{h^2}{V_n^2(2)}\right)$$
$$\therefore \quad \frac{h}{V_n(2)} \leq (10^{0.1} - 1)^{1/2} \text{ or}$$

$$V_n(2) \geq \frac{h}{(10^{0.1} - 1)^{1/2}} = 638.6987$$

Substitution of the asymptotic value for $V_n(2)$ yields the desired condition:

$$V_n(2) = 2^{n-1} \cdot (2)^n = 2^{2n-1} \geq h/(10^{0.1} - 1)^{1/2}$$
$$\therefore \quad n \geq \frac{1}{2} + \frac{\log_e[h/(10^{0.1} - 1)^{0.5}]}{2\log_e(2)} = 5.1595$$

With n being an integer, we choose $n = 6$ to meet the specification. This is the same order as was found for Example 2, but the response is considerably different. Also, again, the specification is close to requiring $n = 5$, so it would be worth looking at a very slight change in the specification for probably the minimum stopband attenuation.

With parameters $h = 325$ and $n = 6$ thus determined, we are ready to find the desired network function to provide the specified response. We get the following algebra:

$$V_6(\omega) = 32\omega^6 - 48\omega^4 + 18\omega^2 - 1$$
$$V_6(1/\omega) = 32/\omega^6 - 48/\omega^4 + 18/\omega^2 - 1$$
$$|T(j\omega)|^2 = \frac{1}{1 + \frac{h^2}{V_6^2(1/\omega)}} = \frac{1}{\left[1 + \frac{h^2}{(32/\omega^6 - 48/\omega^4 + 18/\omega^2 - 1)^2}\right]}$$
$$= \frac{1}{\left[1 + \frac{\omega^{12} h^2}{(-\omega^6 + 18\omega^4 - 48\omega^2 + 32)^2}\right]} \equiv T(s) \cdot T(-s)]_{s=j\omega}$$
$$T(s) \cdot T(-s) = \frac{(s^6 + 18s^4 + 48s^2 + 32)^2}{s^{12} h^2 + (s^6 + 18s^4 + 48s^2 + 32)^2} = \frac{N(s) \cdot N(-s)}{D(s) \cdot D(-s)}$$

Note that in the process of clearing the ω-terms in $V_6(1/\omega)$, the polynomial gets turned around end for end. A Matlab .m-file (see below) was written to perform the above algebra for this inverted Chebyshev network function.

The roots of $N(s)$, in theory, are supposed to all lie on the $j\omega$-axis, but the root finder will have very small error, δ, for the real part of one root. This δ is cancelled with a negative δ real part on its image root because the sum of all roots must equal zero. Thus, the real-part sorting appears to work for all examples tried to date; hence, the warning included in the code. The code yields the following network function for $h = 325$ and $n = 6$:

$$T(s) = \frac{(s^6 + 18s^4 + 48s^2 + 32)}{D(s)}$$
$$\text{where} \quad D(s) = (325.001s^6 + 819.972s^5 + 1034.44s^4 +$$
$$+831.688s^3 + 452.724s^2 + 160.942s + 32.0000)$$

The zeros are at $0 \pm j3.8637$, $0 \pm j1.4142$, and $0 \pm j1.0353$, and the poles are at $-0.7138 \pm j0.2412$, $-0.4194 \pm j0.5289$, and $-0.1282 \pm j0.60345$.

8.1. TRADITIONAL METHODS

% Example 3 polynomial management for
% inverted Chebyshev LP filters

```
h=325;
n=6;
Vn=cheby(n);
Vn2=conv(Vn,Vn);
NN=Vn2(end:-1:1);  % Turn polynomial end-for-end
DD=(h^2;zeros(2*n,1)]+NN; % polynomial in w^2
% change coeff's of NN and DD to those for w^2=-s^2
NNs=NN;
DDs=DD;
for k=1:2:2*n+1
   DDS(k)=DD(k)*(-1)^((k-1)/2);
   NNS(k)=NN(k)*(-1)^((k-1)/2);
end
zros=roots(NNS);
nrts=zros(find(real(zros)<0));  % The assumption that there is
% a real part may not always be true for these j-axis
% quadruplet root sets.  It pays to check the final polynomials
% for all real coefficients.  If they are not all real then a
% manual sort for the roots of N will have to be performed.  In
% theory the roots of NNs are all on the j-axis with 0 real part.
pols=roots(DDs);
drts=pols(find(real(pols)<0));
N=poly(nrts)*sqrt(NNs(1));
D=poly(drts)*sqrt(DDs(1));
```

The frequency response for this solution is shown in Figure 8.14. It is by far the best solution for the problem as it provides the flat (ripple-free) passband response of the Butterworth function, while providing a sharp transition (shape factor) to the stopband. The inverted Chebyshev polynomial solution to the LP approximation uses finite transmission zeros to facilitate the LP approximation.

8.1.6 A General Form for the LP Transmission Function

The three examples so far have illustrated progressively more complicated functional forms for the LP filter function. Equation (8.27) provides a general form for the network function where $F^2(\omega)$ is a function of ω^2, so that all critical frequencies lie in both the left and right halves of the s-plane.

$$|T(j\omega)|^2 = \frac{1}{1+F^2(\omega)} \equiv \left|\frac{N(s) \cdot N(-s)}{D(s) \cdot D(-s)}\right|_{s=j\omega} \qquad (8.27)$$

For the three examples, we had the following assignments for $F^2(\omega)$:

Example 1:
$$F^2(\omega) = \omega^{2n} \quad (n\text{th-order Butterworth})$$

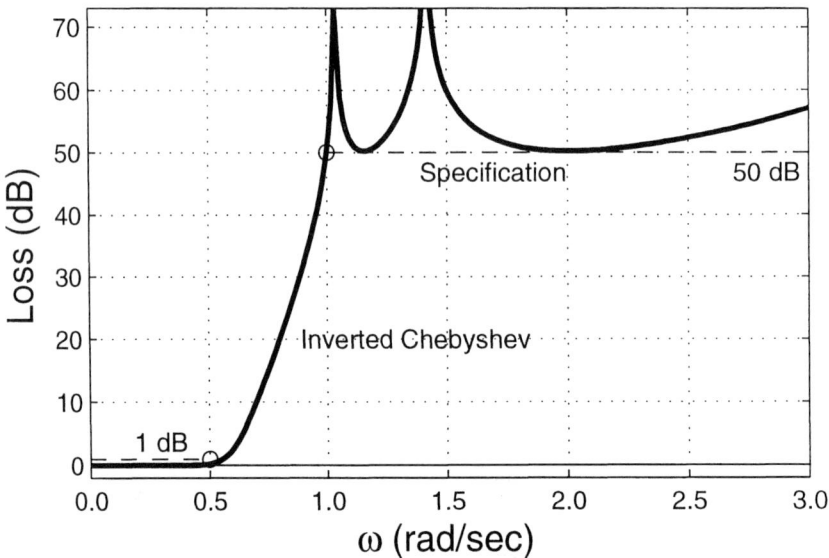

Figure 8.14. An inverted Chebyshev response for the example filter specification.

Example 2:
$$F^2(\omega) = \epsilon^2 V_n^2(\omega) \quad (n\text{th-order Chebyshev})$$

Example 3:
$$F^2(\omega) = h^2/V_n^2(1/\omega) \quad (n\text{th-order inverted Chebyshev})$$

Recall that $V_n^2(\omega) = (1+V_{2n}(\omega))/2$ is a polynomial in ω^2 with even power coefficients only. The Butterworth and Chebyshev functions provide all-pole responses while the inverted Chebyshev provides finite transmission zeros as well. As was pointed out in Chapter 2, it is traditional to design from an LP prototype, then do frequency transforms to obtain other filters, such as HP, BP, or BS.

Solution of the transmission function, (8.27), can be used either way. We can solve [find a suitable function, $F^2(\omega)$] for an LP equivalent, then frequency transform the result, or it can be solved directly for the desired filter type. This latter procedure can produce BP filters that do not have a derivable LP prototype form.

Many different forms can be proposed for $F^2(\omega)$. We will work an example to produce a BP filter directly by choosing a suitable function, and then conclude the chapter by developing a general method to solve the approximation problem for any filter type.

Example 4 - Design a prototype BP filter function referred to as an octave BP filter (i.e., the passband lies between $1/\sqrt{2} \leq \omega \leq \sqrt{2}$). The relative maximum passband attenuation should be less than 1 dB. The asymptotic response should

8.1. TRADITIONAL METHODS

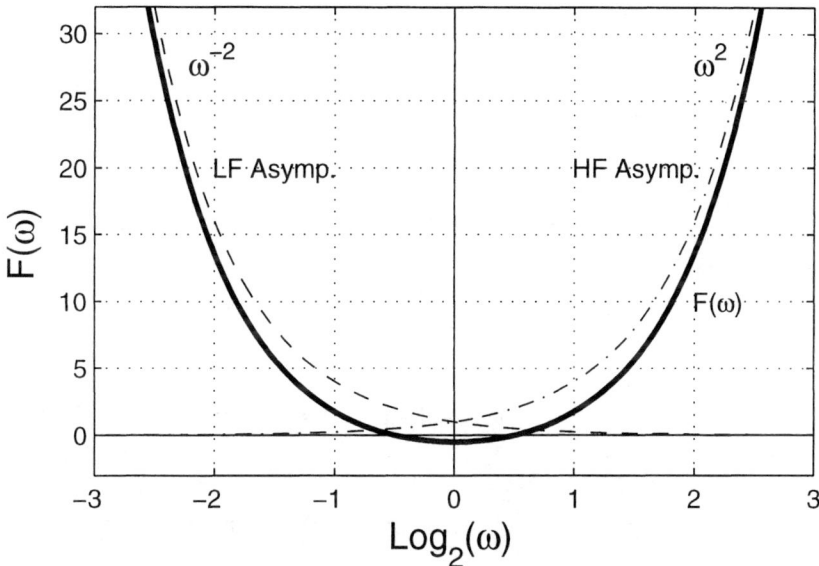

Figure 8.15. Proposed $F(\omega)$ for an octave band filter.

roll off at -40 dB/dec for both large and small ω-values (a second-order zero at both $\omega \to 0$ and $\omega \to \infty$).

Solution - We are free to create any $F^2(\omega)$ we like as long as we meet the given specifications. The key is that F^2 should be a small value in the passband and should become large for stopband frequencies. Such a function is given by (8.28) and sketched in Figure 8.15 for $\epsilon = 1$:

$$F(\omega) = \frac{\epsilon(1 - 2\omega^2)(1 - \omega^2/2)}{\omega^2} \qquad (8.28)$$

The maximum value of F for $1/\sqrt{2} \leq \omega \leq \sqrt{2}$ is controlled by zeros at $\omega_{1,2} = 1/\sqrt{2}$, $\sqrt{2}$, and ϵ. The asymptotic behavior for $\omega \to 0$ is proportional to $1/\omega^2$, so F gets large for small ω values, and it is proportional to ω^2 as ω gets large. So, this function appears to meet the specifications for a suitable value of ϵ.

To estimate a suitable value for ϵ we need to determine the maximum value of $F(\omega)$ in the passband and also insure that the loss is less than 1 dB. Inspection of Figure 8.15 indicates that this point occurs at $\log_2(\omega) = 0$, or $\omega = 1$. The fact that this is true can be confirmed by taking the derivative of F with respect to ω, setting it to zero and solving for ω. Thus, at $\omega = 1$, the attenuation specification yields the following:

$$\begin{aligned} \alpha &= 1 \text{ dB} \geq 10\log_{10}(1 + F^2(1)) \text{ or} \\ F^2(1) &\leq (10^{0.1} - 1) \end{aligned}$$

Substitution of (8.28) for F then yields a limit for ϵ:

$$F^2(1) = \frac{\epsilon^2(1-2)^2(1-1/2)^2}{1} = \epsilon^2(1)(1/4) \text{ so}$$
$$\epsilon^2 \leq 4(10^{0.1}-1) \text{ or}$$
$$\epsilon \leq 2(10^{0.1}-1)^{0.5} = 1.0177$$

It is reasonable to choose $\epsilon = 1.0$ for simplicity. The transmission function is then as follows:

$$|T(j\omega)|^2 = \frac{1}{1 + \frac{(1-2\omega^2)^2(1-\omega^2/2)^2}{\omega^4}} = \frac{\omega^4}{\omega^4 + (1-\omega^2)^2(1-\omega^2/2)^2}$$

Substituting $\omega^2 = -s^2$, we then get the desired network function from the lhp roots of $T(s) \cdot T(-s)$ as before:

$$T(s) \cdot T(-s) = \frac{s^4}{s^4 + (1+2s^2)^2(1+s^2/2)^2} = \frac{N(s) \cdot N(-s)}{D(s) \cdot D(-s)}$$

Factoring, sorting, and polynomial reconstruction yields the polynomials N and D, as in (8.29):

$$N(s) = s^2$$
$$D(s) = s^4 + 1.1118s^3 + 3.1180s^2 + 1.1118s + 1.0000$$
$$T(s) = \frac{N(s)}{D(s)} \qquad (8.29)$$

The poles of $T(s)$ are at $s_{1,2} = -0.1608 \pm j0.6174$ and at $s_{3,4} = -0.3951 \pm j1.5168$. The frequency response for $|T(j\omega)|$ is shown in Figure 8.16. As shown on the graph, the loss increases by 12 dB/oct (same as 40 dB/dec), and the passband loss is less than or equal to 1 dB, as required. The function chosen ends up with a denominator polynomial that has coefficients symmetrical about the center of the polynomial. This means that the roots are symmetrical to the unit circle in the s-plane, and so, an equivalent LP function does actually exist for $T(s)$. This was an accident and could be avoided by making different choices [e.g., by squaring one factor in $F(\omega)$ and not the other].

The resulting octave band response is not too sharp since the loss at the center of the next octave is only 6 dB. This could be increased by changing the complexity of $F(\omega)$ in (8.28). For example, $F(\omega)$ could be squared, or the passband zeros could be brought closer together, and ϵ could be allowed to set the loss at the ends of the passband.

This example shows that we are limited only by our imagination in creating functions $F(\omega)$ that yield desired (and nonstandard) filter response.

At this point, we have developed procedures to handle prototype function design for the classical Butterworth and Chebyshev all-pole filters, along with the inverted Chebyshev function. In addition, the last example showed how a function can be created to design a custom filter function. The procedures required to obtain a

8.2. GENERAL METHODS

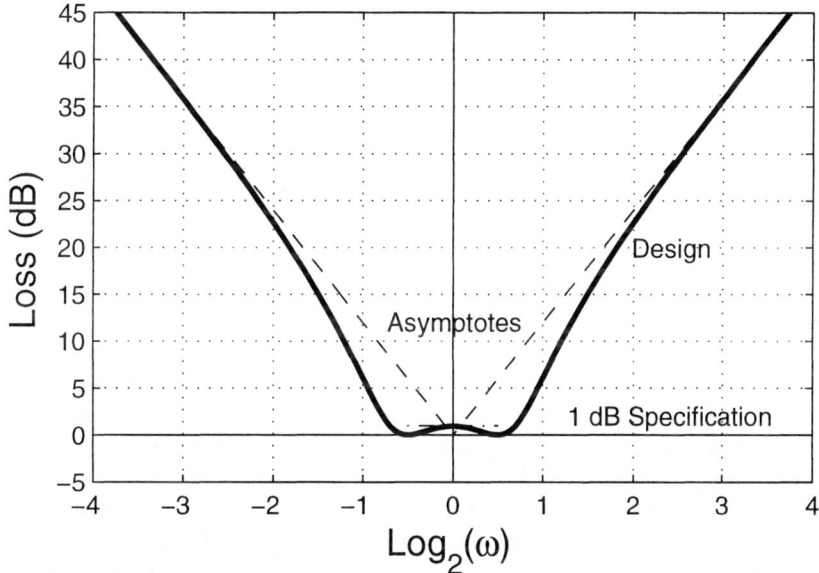

Figure 8.16. Frequency response for the octave band example.

network function for these different approaches are all straightforward, as witnessed by the computer code supplied for each example, so an unlimited number of active filter designs can be achieved through the procedures presented up to this point.

The approximants developed thus far are for constant response in a specified band of frequency. Quite often, in modern applications, one must fit exact frequency response shapes for signal-processing purposes, such as pulse shaping, equalizing, and other applications, so it is useful to now consider alternate methods for approximating nonconstant frequency response requirements.

8.2 General Methods

As concluded in the previous section, other methods are sometimes required in order to obtain network functions for unusual filter requirements. A method involving the use of the Fourier series approximation to a periodic function provides a brute-force solution to almost any requirement that can be imagined. This Fourier method, which is developed first, leads to a simple iterative structure. The method also introduces the Hilbert transform relationship between the network magnitude function and its required minimum phase function. The chapter concludes by presenting an algorithm that finds a network function from the associated minimum phase to fit magnitude (attenuation) specifications that are described by a series of straight-line connected points on a magnitude (in decibels) frequency response curve.

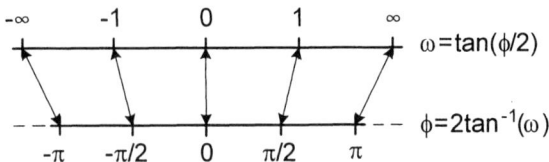

Figure 8.17. Mapping of frequency, ω, into a periodic variable, ϕ.

8.2.1 A Fourier Series Solution

Fourier series have the advantage that by simply adding more terms, convergence to a desired (objective) function is obtained. Normally, the requirement for a large number of terms is undesirable, but in the case to be presented here, each pole is obtained from a two-transistor unit-gain amplifier, and so, from an integrated circuit point of view, the iterated circuit thus obtained is sometimes a practical realization. Consequently, this section presents the theory for an inefficient Fourier series approach that leads to a more efficient pole-placement method, which concludes the chapter. Theoretically, however, we gain more understanding about the network function and the properties of its sinusoidal steady-state magnitude and phase function dependencies.

In order to obtain a Fourier series, we need a periodic function. The transformation given by (8.30) provides a desirable transform for mapping the frequency $-\infty < \omega < \infty$ onto the variable ϕ over a principal period $-\pi < \phi < \pi$:

$$\omega = \tan(\phi/2) \tag{8.30}$$

The mapping of ω onto ϕ, or vice versa, is shown in Figure 8.17. Equation (8.31) provides an equivalent relation between ϕ and ω through an all-pass-type network function:

$$e^{-j\phi} = \frac{1 - j\omega}{1 + j\omega} \tag{8.31}$$

Thus, any given real-part function, $R(\omega)$, is a periodic function on the variable ϕ with a period of 2π. $R(\phi)$ can be written in terms of its Fourier cosine series, as in (8.32):

$$R(\phi) = \sum_{k=0}^{\infty} A_k \cos(k\phi) \equiv \mathcal{R}e\left[\sum_{k=0}^{\infty} A_k e^{-jk\phi}\right] \tag{8.32}$$

Substitution of (8.31), for $e^{-j\phi}$, converts (8.32) back into a function of ω as desired:

$$R(\omega) = \mathcal{R}e\left[\sum_{k=0}^{\infty} A_k \left(\frac{1-j\omega}{1+j\omega}\right)^k\right] \tag{8.33}$$

The corresponding complex system function is found by associating the imaginary part of the exponential function of $-j\phi$ with the real part, $R(\omega)$, and by extending

8.2. GENERAL METHODS

$j\omega$ to complex frequency s to obtain $H(s)$:

$$H(s) = \sum_{k=0}^{\infty} A_k \left(\frac{1-s}{1+s}\right)^k \tag{8.34}$$

The network function, $H(s)$, provides an approximation to $R(\omega)$ when $s = j\omega$.

There are several issues that we now need to consider in order to obtain useful solutions by this method. First, we need a simple circuit that iterates easily to construct the sum in (8.34) in order to build the network function, $H(s)$; second, we need to obtain a suitable real-part function from a specified network transmission, or loss function; third, we need to compute the desired Fourier coefficients A_k in (8.32); and fourth, we need to know what issues are involved and how to deal with them when the Fourier series is truncated to a finite sum.

Before doing a lot of math, it is helpful to know there is a candidate circuit to realize the desired network function of (8.34). A basic circuit is shown in Figure 8.18 for the kth stage of a cascade architecture. The circuit uses complementary NPN-PNP general-purpose transistors to provide a dc signal path with small dc offset in the provision of an all-pass transfer function, as will be shown next. (See Problem 8.18 for an opamp version of the all-pass cascade stage.) Conceptually, the desired network is obtained by a cascade of the unit-gain stages, as shown in Figure 8.19. The output of each stage is used, with appropriate weights for the Fourier coefficients, to form the sum required for a truncated version of (8.34). The circuit shown implements the constant term, A_0, plus N-stages of the all-pass term to provide an N-cosine approximation to $R(\omega)$. The fact that some Fourier coefficients, A_k, are negative is allowed by the proposed architecture. In addition, intermodulation between the weights is reduced since each stage provides a low impedance (voltage source) output off the emitter of each output transistor.

Now we can sketch an analysis of how the proposed circuit in Figure 8.18 provides the desired transfer function. To simplify the analysis, it is assumed that the base current is negligible. In addition, for incremental signal response, it is further assumed that the (dynamic) node voltage on the base also appears on the emitter node with negligible loss (unit-gain). A simple analysis is as follows. Let $Z = R + 1/Cs$:

KVL:

$$R_2(\alpha \mathbf{I}_e + \mathbf{I}_c) + Z\mathbf{I}_c + \mathbf{V}_k = 0$$
$$\text{and} \quad \mathbf{V}_{k+1} = R\mathbf{I}_c + \mathbf{V}_k$$

KCL:

$$\mathbf{I}_e + \mathbf{I}_c = G_1 \mathbf{V}_k$$
$$\therefore \mathbf{I}_e = G_1 \mathbf{V}_k - \mathbf{I}_c$$

Substitution of $\mathbf{I}_e$ into KVL yields

$$\alpha R_2(G_1 \mathbf{V}_k - \mathbf{I}_c) + R_2 \mathbf{I}_c + Z\mathbf{I}_c = -\mathbf{V}_k$$

340 CHAPTER 8. THE APPROXIMATION PROBLEM

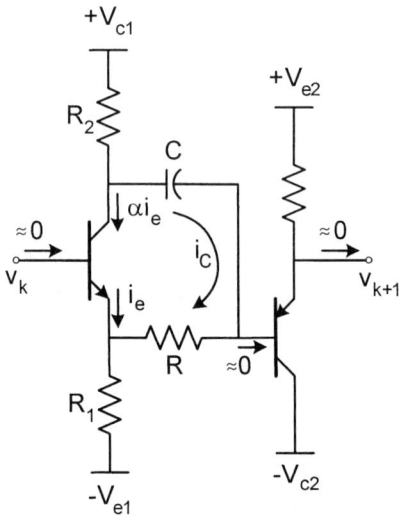

Figure 8.18. Example unit-gain all-pass transistor cascade circuit.

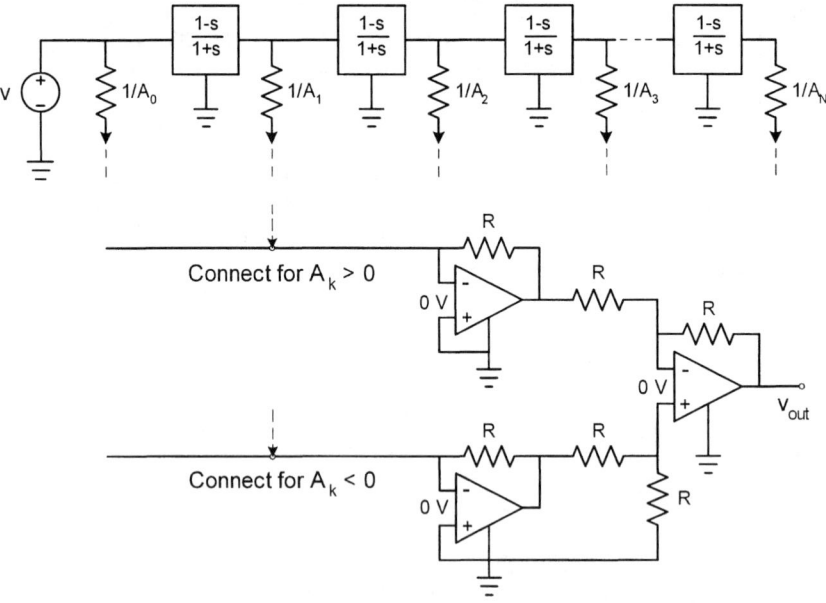

Figure 8.19. N-stage implementation of the Fourier network function, (8.34).

8.2. GENERAL METHODS

$$\therefore [R_2(1-\alpha) + Z]\mathbf{I}_c = -[1 + \alpha R_2 G_1]\mathbf{V}_k$$

$$\text{so} \quad \frac{\mathbf{I}_c}{\mathbf{V}_k} = \frac{-(1 + \alpha R_2 G_1)}{[Z + (1-\alpha)R_2]} \quad \text{and then}$$

$$\mathbf{V}_{k+1} = \mathbf{V}_k + R\mathbf{I}_c = \mathbf{V}_k - \frac{R(1 + \alpha R_2 G_1)}{Z + (1-\alpha)R_2}\mathbf{V}_k$$

Let $\rho = (1 + \alpha R_2 G_1)$ be a constant, and note that $(1-\alpha)R_2 << R$ to obtain the following:

$$\frac{\mathbf{V}_{k+1}}{\mathbf{V}_k} \approx 1 - \frac{\rho R}{Z} = \frac{R + 1/Cs - \rho R}{R + 1/Cs}$$

$$\approx \frac{1 - (\rho-1)RCs}{1 + RCs} \tag{8.35}$$

Setting $\rho = 2$ yields the desired result:

$$\frac{\mathbf{V}_{k+1}}{\mathbf{V}_k} = \frac{1 - RCs}{1 + RCs} = \left(\frac{1-s}{1+s}\right)$$

$$\text{where} \quad \alpha R_2 = R_1$$

$$\text{and} \quad RC = 1 \tag{8.36}$$

The choice of R_2 turns out to be critical in obtaining all-pass response over the widest possible bandwidth. A suitable value can be found by simulation using six or more stages in cascade and adjusting R_2 to obtain a flat (constant) response over the widest range possible. This method of finding R_2 was used for the example to follow this discussion.

The circuit concept, Figure 8.19, for the implementation of an N-term approximation to $H(s)$, (8.34), has some pros and cons. Conceptually, the iterated design is very straightforward for N identical stages where all element values are the same. Even with 50 stages, having 100 transistors is not a serious problem for modern IC technology. Neither is the requirement for three opamps. However, the idea of 50 stages is not appealing as a solution for most problems, but larger numbers (>100) are quite often used for digital FIR filters. A disadvantage is the requirement for $N+1$ summing resistors. These probably have to be external to the chip for general applications, so a large number of stages causes a small packaging and tuning (adjustment) problem in producing the circuit. Finally, as we will see, the Fourier coefficients quite often tend to have a wide range of values, which also contributes to implementation difficulty.

Next, we need to obtain a suitable real-part function from a specified network transmission (or loss) magnitude function. Normally, only a magnitude function is specified with little or no information about phase. The magnitude and loss functions are related as follows:

$$|T(j\omega)| = e^{-\alpha(\omega)} \quad \text{or}$$

$$\alpha(\omega) = -\log_e |T(j\omega)| \quad \text{Nep} \tag{8.37}$$

Both $\alpha(\omega)$ and $|T(j\omega)|$ are even functions of ω. The complex transmission function has both real and imaginary parts, which determine the phase lag angle, β, which in turn determines the network propagation constant, $\gamma = \alpha + j\beta$:

$$T(j\omega) = e^{-(\alpha+j\beta)} = R(\omega) + jX(\omega) \qquad (8.38)$$
$$\text{where} \quad R(\omega) = e^{-\alpha}\cos(\beta) \text{ and}$$
$$X(\omega) = -e^{-\alpha}\sin(\beta) \qquad (8.39)$$

R and X are the real and imaginary parts of $T(j\omega)$, respectively. It is important to realize that whenever a loss function is specified, there is a unique associated phase function implied that makes $\alpha(\omega)$ a realizable, stable (lhp poles), network function. This makes it very risky to try to specify both magnitude and phase since it is easier to come up with incompatible (unrealizable) or contradictory conditions than it is to find realizable conditions. It should also be noted that magnitude-phase relationships are not unique since it is possible to cascade an all-pass network (unit-gain) with an existing network to obtain a different phase response combined with the original magnitude response. This procedure is referred to as equalization. Thus, there is a minimum phase associated with any given realizable magnitude function of $s = j\omega$. The relationship of imaginary part to real part for any function having only lhp critical frequencies is provided by the Hilbert transform (see [1], pp. 296–308):

$$\beta(\omega) = \frac{-1}{\pi} \int_{-\infty}^{\infty} d\xi \frac{\alpha(\xi)}{(\omega - \xi)} \qquad (8.40)$$

Equation (8.40) shows that the relation between phase, β in radians, and loss, α in Nepers, is a convolution of $\alpha(\omega)$ with $-1/\pi\omega$. Appendix 8A provides a detailed discussion of how this convolution relationship is obtained. Once $\beta(\omega)$ is determined from a specified or implied loss function, $\alpha(\omega)$, then (8.39) is applied to obtain a suitable $R(\omega) = \mathcal{R}e[T(j\omega)]$ so as to calculate required Fourier coefficients A_k. Further discussion on obtaining $\beta(\omega)$ from a specific $\alpha(\omega)$ will be deferred for a short time while we consider the calculation of the Fourier coefficients, the third item on our list.

As shown in Figure 8.20, $R(\omega)$ becomes an even and periodic function of the variable $\phi = 2\tan^{-1}(\omega)$. In terms of the Fourier complex exponential series, we get (8.41) as an approximation to $R(\phi)$. Note that the period of ϕ is $P = 2\pi$, so $\omega_0 = 2\pi/P = 1$ for the expansion.

$$R(\phi) \;(=)\; \sum_{n=-\infty}^{\infty} r_n e^{+jn\cdot 1 \cdot \phi}$$

$$\text{where} \quad r_k = \frac{1}{P}\int_{-P/2}^{P/2} d\phi\, e^{-jk\cdot 1 \cdot \phi} R(\phi) \quad k = 0, \pm 1, \pm 2, \ldots$$

$$= \frac{1}{2\pi}\int_{-\pi}^{\pi} d\phi\, R(\phi)[\cos(k\phi) - j\sin(k\phi)]$$

$$\equiv \frac{1}{2\pi}\int_{-\pi}^{\pi} d\phi\, R(\phi)\cos(k\phi)$$

8.2. GENERAL METHODS

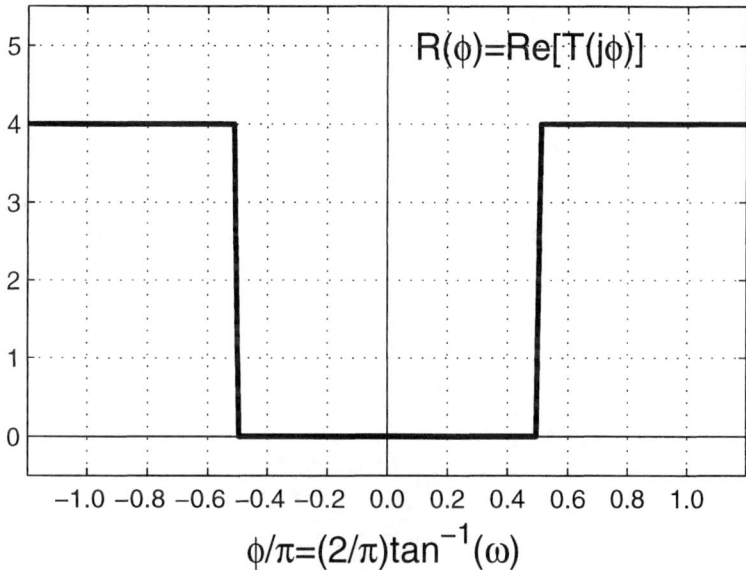

Figure 8.20. The $\mathcal{R}e[T(j\phi)]$ is even and periodic on the mapping, $\phi = 2\tan^{-1}(\omega)$.

$$\equiv \frac{1}{\pi} \int_0^\pi d\phi R(\phi) \cos(k\phi) \tag{8.41}$$

The r_k are all real due to the property that the integral over symmetrical limits of the odd function, $R(\phi)\sin(k\phi)$, is identically zero. The expansion for $R(\phi)$ is then converted from a sum of complex sinusoids to the desired cosine series. Note that since $\cos(-k\phi) = \cos(+k\phi)$, then $r_{-k} = r_{+k}$.

$$\begin{aligned}
R(\phi) \quad (=) \quad & \sum_{k=-\infty}^{\infty} r_k e^{jk\phi} = \ldots + r_{-3}e^{-j3\phi} + r_{-2}e^{-j2\phi} + r_{-1}e^{-j\phi} \\
& + r_0 + r_1 e^{j\phi} + r_2 e^{j2\phi} + r_3 e^{j3\phi} + \ldots \\
= \quad & r_0 + r_1(e^{j\phi} + e^{-j\phi}) + r_2(e^{j2\phi} + e^{-j2\phi}) + r_3(e^{j3\phi} + e^{-j3\phi}) + \ldots \\
\equiv \quad & r_0 + 2r_1 \cos(\phi) + 2r_2 \cos(2\phi) + 2r_3 \cos(3\phi) + \ldots
\end{aligned} \tag{8.42}$$

Comparison of (8.42) with (8.32) yields the following values for the Fourier cosine series:

$$\begin{aligned}
A_0 &= r_0 \\
A_k &= 2r_k \text{ for } k = 1, 2, 3, \ldots
\end{aligned} \tag{8.43}$$

Now that we know how to determine the coefficients for the infinite series, we need to look at how to deal with the effects of truncating the series (taking only N cosine terms) to approximate the infinite series. This is the fourth item on our list.

An example will be used to help understand the effect of using a truncated Fourier series to approximate a periodic function. Truncation of an infinite series is necessary in practical problems since it is not possible to implement an infinite number of stages in a real circuit. A truncated series, where all terms for $|k| > N$ are dropped, is usually referred to as a partial sum.

Example 5 - Take the ideal LP filter function, $F(\phi)$, shown as $R(\phi)$ in Figure 8.20, to be an objective function and construct a partial sum approximant for the first 10 harmonics:

$$F(\phi) = 4[1 - u(\phi + \pi/2) + u(\phi - \pi/2)]$$

$$(=) \sum_{k=-\infty}^{\infty} f_k e^{jk\phi}$$

$$\text{where} \quad f_k = \frac{1}{2\pi}\int_{-\pi}^{\pi} d\phi\, e^{-jk\phi} F(\phi)$$

$$= -2\frac{\sin(k\pi/2)}{(k\pi/2)} + 4\delta_{k,0}$$

$$= \begin{cases} 0 \text{ for } k \text{ even and } \neq 0 \\ \frac{-4}{k\pi}(-1)^{(k-1)/2} \text{ for } k \text{ odd} \\ 2 \text{ for } k = 0 \end{cases} \qquad (8.44)$$

$\delta_{j,k}$ is the Kronecker delta function, which equals 1 when $j = k$ and 0 for all $j \neq k$.

Solution - The partial sum for the first N harmonics is given by F_P:

$$F_P = \sum_{k=-N}^{N} f_k e^{jk\phi} \qquad (8.45)$$

Figure 8.21 shows the partial sum for $N = 10$ compared to the infinite sum objective, $F(\phi)$. The result shows a periodic-like error, in this example, nine full periods, across the fundamental period of 2π. As terms are added to the partial sum, the period of the error decreases, but at any discontinuity, the percentage of overshoot remains the same. This overshoot, associated with a truncated series, is referred to as Gibb's phenomenon.

Lanczos [2] observed that the ripple in the partial sum has a period of either the last term kept, $2\pi/N$, or the first term neglected, $2\pi/(N+1)$. In this example, 10 harmonics were kept, but $f_{10} = 0$ anyway, so the 9 periods of ripple agree with Lanczos' claim. Lanczos proposed that a partial sum could be smoothed by averaging the sum over the ripple period. The smoothing lowers periodic error by filtering out undesired ripple. Letting F_S be the smoothed version of F_P and averaging over the period, $2\pi/(N+1)$, of the first term neglected, we get the following useful result:

$$F_S = \frac{N+1}{2\pi} \int_{\phi-\pi/(N+1)}^{\phi+\pi/(N+1)} dp\, F_P(p) \qquad (8.46)$$

8.2. GENERAL METHODS

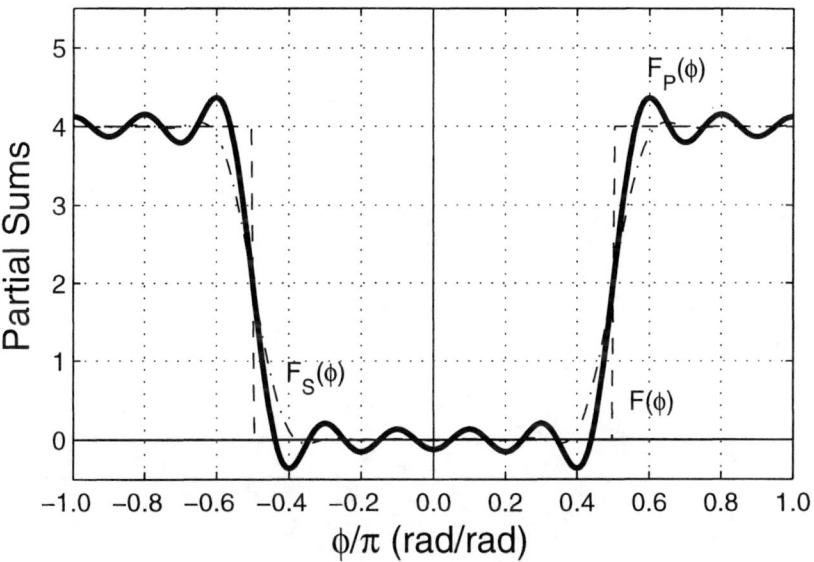

Figure 8.21. A truncated series has a periodic error component.

Substitution for F_P, along with interchange of summation and integration, yields a partial sum with weighted coefficients:

$$\begin{aligned}
F_S &= \frac{(N+1)}{2\pi} \int_{\phi-\pi/(N+1)}^{\phi+\pi/(N+1)} dp \sum_{k=-N}^{N} f_k e^{jkp} \\
&= \frac{(N+1)}{2\pi} \sum_{k=-N}^{N} f_k \int_{\phi-\pi/(N+1)}^{\phi+\pi/(N+1)} dp\, e^{jkp} \\
&= \frac{(N+1)}{2\pi} \sum_{k=-N}^{N} f_k \frac{e^{jk\phi}}{jk}[e^{jk\pi/(N+1)} - e^{-jk\pi/(N+1)}] \\
&= \frac{(N+1)}{2\pi} \sum_{k=-N}^{N} \frac{2\sin(k\pi/(N+1))}{k} f_k e^{jk\phi} \\
&= \sum_{k=-N}^{N} \sigma(N,k) f_k e^{jk\phi} \\
\text{where} \quad \sigma(N,k) &= \frac{\sin(k\pi/(N+1))}{k\pi/(N+1)}
\end{aligned} \qquad (8.47)$$

The $\sigma(N,k)$ values are referred to as Lanczos weighting factors, and F_S is a weighted partial sum. Note that $\sigma(N,K) = 0$ for $k = N+1$, the first term to be neglected, in forming the partial sum.

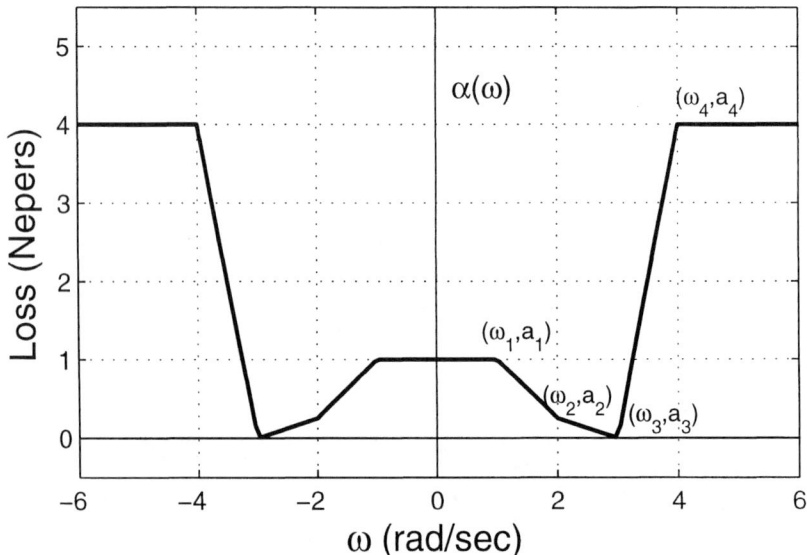

Figure 8.22. A piecewise, straight-line, connected description of the loss function, $\alpha(\omega)$, for $nk = 4$ breakpoints.

The response for F_S is also shown in Figure 8.21 in comparison to both the partial sum, F_P, and the infinite series limit, $F(\phi)$. Clearly, the smoothed function has less ripple than does the partial sum. The overshoot is also greatly reduced, but in some regions the error is larger. Smoothing generally widens transition bands at discontinuous points, but otherwise, a fit with less ripple is usually obtained. In any event, the infinite Fourier series is the starting point for obtaining an approximation to a desired periodic function.

We now turn our attention back to evaluating the Hilbert transform, (8.40), to obtain the phase, $\beta(\omega)$, from a specific loss function, $\alpha(\omega)$. A specific form of $\alpha(\omega)$ to be considered is shown in Figure 8.22. It is an even function of ω with zero slope as $|\omega|$ goes to either ∞ or zero. The variation is described by straight-line connected segments that form nk breakpoints over the range $0 < \omega_1 < \omega < \omega_{nk} < \infty$. The behavior for $\omega < 0$ is implied by the evenness requirement for a loss function. Functionally, this $\alpha(\omega)$ can be described using a constant plus a sum of weighted unit ramp functions, as in (8.48):

$$\alpha(\omega) = a_{nk} + \sum_{k=1}^{nk} c_k[(\omega - \omega_k)u(\omega - \omega_k) + (\omega + \omega_k)u(\omega + \omega_k)] \qquad (8.48)$$

The weighting coefficients c_k come from the second derivative of α with respect to ω, as shown in Figure 8.23. As the figure shows, the second derivative for this specific $\alpha(\omega)$ yields an even function of ω that consists of a set of impulse functions with weights, c_k. Figure 8.23(b) provides the basis for writing $\alpha(\omega)$ as a sum of

8.2. GENERAL METHODS

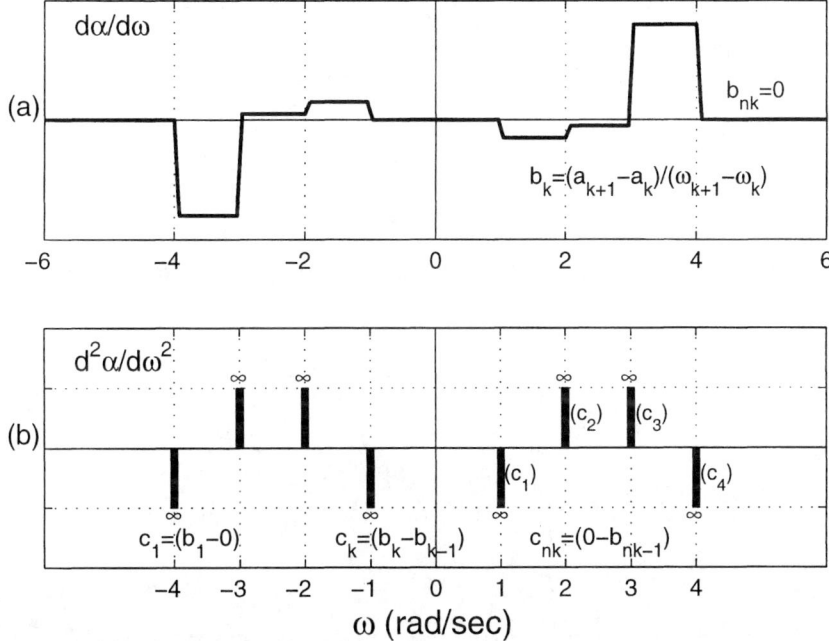

Figure 8.23. The derivatives of α: (a) the first derivative is a pulse train with odd ω-dependence, and (b) the second derivative is an impulse-train with even ω-dependence.

ramp functions plus the additive constant, a_{nk}. Note that for this even function of ω, which starts and ends with zero slope, the weights, c_k, are required to sum to zero. This fact is easy to verify inductively as follows:

$$\begin{aligned}
\sum_{k=1}^{nk} c_k &= c_1 + c_2 + c_3 + \ldots + c_{nk} \\
&= (b_1 - 0) + (b_2 - b_1) + (b_3 - b_2) + \ldots + (0 - b_{nk-1}) \\
&\equiv 0
\end{aligned} \qquad (8.49)$$

Now, in order to find $\beta(\omega)$ associated with the given form of $\alpha(\omega)$, we need to apply the Hilbert transform to each term in the sum that defines $\alpha(\omega)$.

First, we find that the phase associated with the constant, a_{nk}, is zero:

$$\begin{aligned}
\beta_{nk} &= \frac{-1}{\pi} \int_{-\infty}^{\infty} d\xi \frac{\alpha(\xi)}{(\omega - \xi)} = \frac{1}{\pi} \int_{-\infty}^{\infty} d\xi \frac{a_{nk}}{(\xi - \omega)} \\
&= \frac{a_{nk}}{\pi} [\log_e |\xi - \omega|]_{-\infty}^{\infty} = \frac{a_{nk}}{\pi} \lim_{E \to \infty} [\log_e |E - \omega| - \log_e |-E - \omega|] \\
&= \frac{a_{nk}}{\pi} \lim_{E \to \infty} \log_e \left|\frac{E - \omega}{E + \omega}\right| \to 0
\end{aligned} \qquad (8.50)$$

We conclude that adding a flat loss (versus ω) for any loss function does not affect the network phase function.

Next, we find the phase, p_{k+}, associated with the unit ramp at $\omega = +\omega_k$:

$$\text{let} \quad \alpha_{ramp} = \alpha_{k+} = (\omega - \omega_k)u(\omega - \omega_k)$$

$$\text{then} \quad p_{k+} = \frac{+1}{\pi}\int_{-\infty}^{\infty} d\xi \frac{(\xi - \omega_k)u(\xi - \omega_k)}{-(\omega - \xi)}$$

$$\equiv \frac{1}{\pi}\int_{-\infty}^{\infty} d\xi \left[1 + \frac{(\omega - \omega_k)}{(\xi - \omega)}\right] u(\xi - \omega_k)$$

Integration by parts yields

$$p_{k+} = \frac{1}{\pi}\{[(\xi + (\omega - \omega_k)\ln|\xi - \omega|)u(\xi - \omega_k)]_{-\infty}^{\infty}$$
$$- \int_{-\infty}^{\infty} d\xi[\xi + (\omega - \omega_k)\ln|\xi - \omega|]\delta(\xi - \omega_k)\}$$

$$= \frac{1}{\pi}\{\lim_{E\to\infty}[E + (\omega - \omega_k)\ln|E - \omega|]$$
$$- [(\omega_k + (\omega - \omega_k)\ln|\omega_k - \omega|)u(\xi - \omega_k)]_{-\infty}^{\infty}\}$$

for $E >> \omega$, $\ln|E - \omega| \to \ln|E|$, and then

$$p_{k+} = \frac{-1}{\pi}\{\omega_k + (\omega - \omega_k)\ln|\omega - \omega_k| - \lim_{E\to\infty}[E + (\omega - \omega_k)\ln|E|]\} \quad (8.51)$$

The phase associated with the ramp at $\omega = -\omega_k$ is then obtained from p_{k+} by just changing $+\omega_k$ to $-\omega_k$. Thus, for $\alpha_{k-} = (\omega + \omega_k)u(\omega + \omega_k)$, we get

$$p_{k-} = [p_{k+}]_{\omega_k \to -\omega_k} = \frac{-1}{\pi}\{-\omega_k + (\omega + \omega_k)\ln|\omega + \omega_k|$$
$$- \lim_{E\to\infty}[E + (\omega + \omega_k)\ln|E|]\} \quad (8.52)$$

Now, for the kth term in $\alpha(\omega)$, the last two results are combined to obtain the phase associated with the two ramps combined to form $\alpha_k(\omega)$:

$$\alpha_k(\omega) = c_k[(\omega - \omega_k)u(\omega - w_k) + (\omega + \omega_k)u(\omega + \omega_k)]$$
$$\beta_k(\omega) = c_k[p_{k+} + p_{k-}]$$
$$= \frac{-c_k}{\pi}\{(\omega - \omega_k)\ln|\omega - \omega_k| + (\omega + \omega_k)\ln|\omega + \omega_k|$$
$$- \lim_{E\to\infty}[2E + 2\omega\ln|E|]\} \quad (8.53)$$

Note that combining the two ramps yields a limit in the associated phase function, $\beta_k(\omega)$, that has no k dependence. When the β_k are summed over all k to form $\beta(\omega)$,

8.2. GENERAL METHODS

the limit term gets eliminated by the zero sum of the weights, c_k. Thus, for

$$\alpha(\omega) = \sum_{k=1}^{nk} \alpha_k(\omega)$$

$$\beta_k(\omega) = \sum_{k=1}^{nk} \beta_k(\omega)$$

$$\equiv \frac{-1}{\pi} \sum_{k=1}^{nk} c_k[(\omega - \omega_k)\ln|\omega - \omega_k| + (\omega + \omega_k)\ln|\omega + \omega_k|]$$

$$+ \frac{1}{\pi}\left(\sum_{k=1}^{nk} c_k\right) \cdot \lim_{E \to \infty}[2E + 2\omega \ln|E|]$$

Substitution of (8.49) yields

$$\beta(\omega) = \frac{-1}{\pi} \sum_{k=1}^{nk} c_k[(\omega - \omega_k)\ln|\omega - \omega_k| + (\omega + \omega_k)\ln|\omega + \omega_k|] \qquad (8.54)$$

due to $\Sigma c_k = 0$. The zero sum of the weights, c_k, also constrains $\beta(\omega)$ to go to zero as $\omega \to \pm\infty$, in spite of the sum looking like a logarithmic infinity. Thus, for $|\omega| \gg |\omega_{nk}|$, $\beta(\omega)$ goes to zero as follows:

$$\lim_{|\omega| \to \infty} \beta(\omega) \to \frac{-1}{\pi} \sum_{k=1}^{nk} c_k(\omega \ln|\omega| + \omega \ln|\omega|)$$

$$= \frac{-1}{\pi}\left(\sum_{k=1}^{nk} c_k\right)(2\omega \ln|\omega|) = 0 \qquad (8.55)$$

Now that we know how to determine a phase function from a specified loss function, we are ready to work examples to demonstrate the Fourier series network approximation method.

Example 6 - Use the Fourier series approximation method to fit the requirement for a loss function described by the breakpoints, $\omega_k = [0.5\ 1.5]$ rad/sec and $a_k = [0\ 2]$ Nep. Demonstrate convergence for both the partial and smoothed sums for the number of cosine terms, nt, equal to 4, 8, and 16. Design a circuit using the smoothed sum for $nt = 4$, and compare the circuit response to the theoretical smoothed response for the truncated series.

Solution - The following Matlab code was written to obtain the truncated response for any specified piecewise straight-line connected loss function (in Nepers) that has nk breakpoints for $0 < \omega < \infty$.

350   CHAPTER 8. THE APPROXIMATION PROBLEM

```
% Example 6 - find Fourier coefficients for given loss
% for specified number of cosine terms

wk=[0.5 1.5];    % frequency brkpts, rad/sec
ak=[0 2];   % attenuation (loss), Nepers
nk=length(wk); % should be same as length of ak
if (length(ak) ~= nk)
   error('length(ak) .neq. length(wk)')
end
nw=512; % no. of frequencies for objective function
dp=pi/nw;
x=linspace(0,1-1e-5,nw)';
phi=pi*x;
w=tan(phi/2);    % objective optimizing frequency set 0<=w<6.4e4
nt=4; % number of cosine terms in series approx.
% calculate attenuation spec, alfa, over frequency set
alfa=ak(1)*(w<=wk(1))+ak(nk)*(w>wk(nk));
for k=1:nk-1
   a1=ak(k);
   a2=ak(k+1);
   w1=wk(k);
   w2=wk(k+1);
   alfa=alfa+(a1+(a2-a1)*(w-w1)/(w2-w1)).*((w>w1)-(w>w2));
end
% get Hurwitz minimum phase
p_1=0;
for k=1:(nk-1)
bk(k)=(ak(k+1)-ak(k))./(wk(k+1)-wk(k));   % First derivative
ck(k)=bk(k)-p_1;                          % Second derivative
p_1=bk(k);
end
bk(nk)=0;
ck(nk)=-bk(nk-1);
beta=zeros(1,nw)';
for k=1:nk
  uk=wk(k);
  beta=beta+ck(k)*((w-uk).*log(abs(w-uk)+eps)...
     +(w+uk).*log(abs(w+uk+eps)));
end
beta=-beta/pi;
Rw=exp(-alfa).*cos(beta); % Desired real part of |T|
u=exp(-j*phi);
nt1=nt+1;
for k=1:nt1
   Ik=integral(phi,Rw.*cos((k-1)*phi)); % includes DC term
   rk(k)=Ik(nw)/pi; % Fourier coefficients/exponential series
```

8.2. GENERAL METHODS

```
end
apk=rk(1)*ones(size(u));  % the DC term
apL=apk;
rL(1)=rk(1);  % the weighted coefficients
for k=1:nt
   nk=k+1;
   Lwt=sin(pi*k/nt1)/(pi*k/nt1);  %Lanczo weight factor
   rL(nk)=rk(nk)*Lwt;  %weighted coefficients
   apk=apk+2*rk(nk)*u.^k;
   apL=apL+2*rL(nk)*u.^k;
end
apprk=-log(abs(apk));
apprL=-log(abs(apL));
```

This code is about the longest we have used in the whole book, but it does several things. The breakpoints, wk and ak, used to define $\alpha(\omega)$ can be of any length or even written as functions. However, wk and ak must be the same length and are checked by the program. Next, the computing space, phi for ϕ, is defined in order to estimate the Fourier exponential series coefficients, r_k, over one full period of $\alpha(\phi)$. The variable ϕ is set up on a linear space using 512 points for series that have 100 cosine terms or fewer. This number should be increased for better resolution to estimate r_k values for longer series. The loss function, α (alfa in the code), is calculated on the ω-variable space where the breakpoints are defined. Note that the range for ω works out to $0 \leq \omega < 6.4 \cdot 10^4$. Next, the Hilbert phase is evaluated using the expansion as developed in (8.54) from the set of impulse coefficients, c_k. With $\beta(\omega)$ thus determined, it is then possible to determine the desired real part using (8.39) to get $R(\omega)$. Finally, the Fourier coefficients, r_k, as in (8.43), are estimated for the desired number of cosine terms (circuit stages) to implement the partial series form for $H(s)$.

Approximation responses to the specified loss function are shown in Figure 8.24 for series having 4, 8, and 16 cosine terms. Clearly, the solutions exhibit convergence to the desired response. Note that the truncated sum exhibits ripple error as discussed earlier and that the weighted sum approaches the objective function with greatly reduced ripple as expected.

The graphs in Figure 8.24 are plotted over one half-period of the variable space for ϕ/π. This amounts to spanning the range zero to ∞ on the ω-variable and, so, illustrates one problem with the method (i.e., loss functions should always be frequency-scaled so their principal variation occurs around $\omega = 1$). Frequency-scaling of the prototype function can affect accuracy in the estimates of the Fourier coefficients whenever the variation in $\alpha(\omega)$ occurs for large values of ω. Better results are obtained when principal variation occurs around $\omega = 1$.

The second part of the example requires a circuit design and simulation to test the ability of the circuit in Figure 8.19 to yield the response for a partial sum of the network function, (8.34), for four cosine terms.[4] Operation of the previous code for

---

[4]The reason for the small number of terms is that the student version of μCAP will not accommodate a more complex circuit.

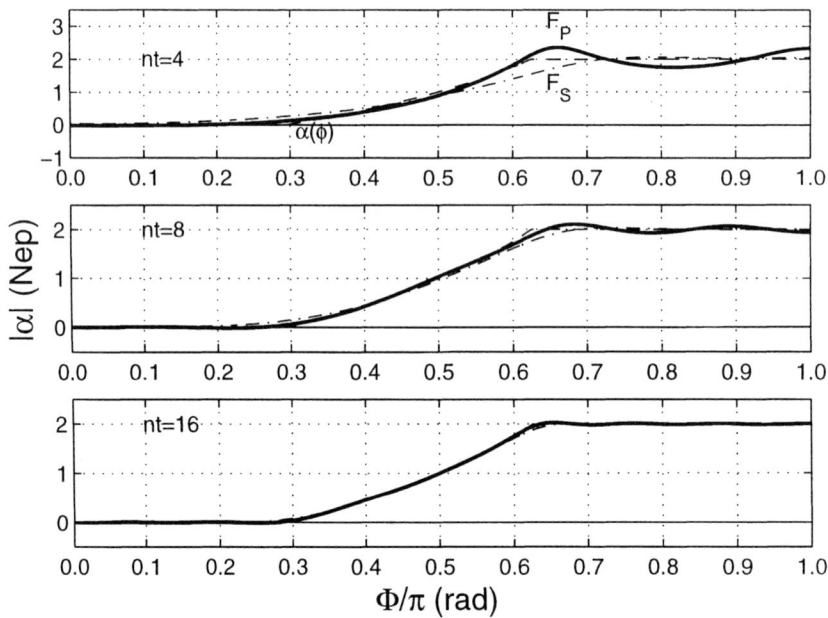

Figure 8.24. Approximations of the example loss, α, versus the number of terms.

a single run of four cosine terms yields the following values for the weighted Fourier coefficients, $rL_k = \sigma(4,k)r_k$:

$$\begin{aligned} r_0 &= 0.35827 = rL_0 \\ \sigma_1 r_1 &= 0.20388 = rL_1 \\ \sigma_2 r_2 &= 0.10535 = rL_2 \\ \sigma_3 r_3 &= 0.0063521 = rL_3 \\ \sigma_4 r_4 &= -0.0090665 = rL_4 \end{aligned}$$

The above coefficients are all multiplied by the factor 2 for $k > 0$, per (8.43). The μCAP circuit is shown in Figure 8.25, where the coefficients are only entered to three significant figures for practical reasons. Note that the negative coefficient, $2\sigma_4 r_4$, goes to the bottom opamp input, whereas the positive coefficients all go to the top summing opamp input. Overall, the circuit is straightforward to implement. Note that in each all-pass stage, R_2 is 5% larger than R_1. This factor was determined empirically so as to make the frequency response to the last stage, $Q8$, flat over as wide a frequency range as possible. With $R_2 = 1.05 R_1$, the all-pass response was flat out to nearly 10 MHz. In order to test the example response at a practical frequency, $\omega = 1$ was scaled to $f = 1$ kHz. Thus, $k_\omega = 2\pi 10^3$ for the example, along with $k_R = 2$ kΩ.

8.2. GENERAL METHODS

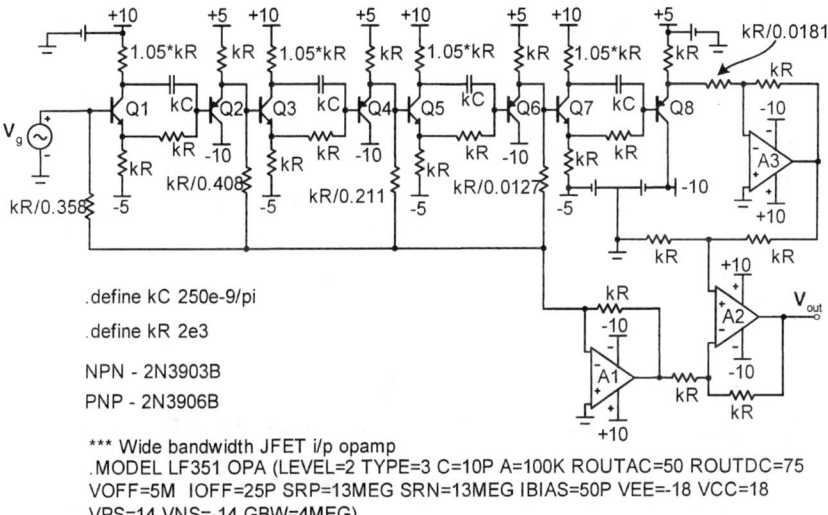

Figure 8.25. Circuit used to simulate the Fourier approximation method, $nt = 4$.

Figure 8.25 shows the circuit used in μCAP to simulate the performance of a smoothed partial sum approximation to the specified loss function. The circuit is easy to implement for any number of stages and follows the design laid out in connection with Figures 8.18 and 8.19.

The frequency response for the circuit in Figure 8.25 is shown in Figure 8.26, where it is compared to both the desired $\alpha(\omega)$ (in decibel gain) and the theoretical partial weighted sum response for a four cosine sum. The μCAP circuit response, shown as a dash-dot line, follows the theoretical sum response extremely well through the passband and transition band but yields deteriorated performance for frequencies above 500 kHz. Note that the response drops off rapidly about three decades above the $\omega = 1$ frequency of 1 kHz. Thus, the cascaded all-pass seems restricted to applications below 1 MHz for this transistor-opamp implementation. The fact that the circuit response deviates slightly from the theoretical response for the sum is due to approximations made in arriving at the nominal all-pass response for the circuit of Figure 8.18. The validity for making the approximations decreases for high-frequency values. Nevertheless, the procedure does have merit for low-frequency applications and where a sufficient number of stages can be implemented to obtain a desired network response.

8.2.2 Finding Polynomial Ratio Network Functions

The previous section has led us through a brute-force approach to solving the approximation problem. The method has several disadvantages, namely, the requirement for many stages and a restricted frequency range for the simple implementation that was developed. There are advantages to the method: all stages are identical,

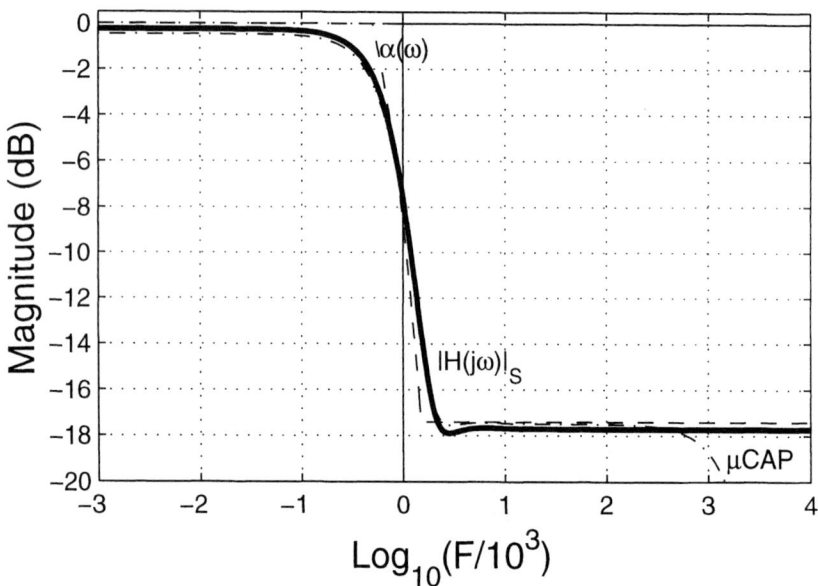

Figure 8.26. Circuit response compared to theory for a partial sum approximation.

with resistor values used to do all the tuning (implement the Fourier coefficients), and the desired filter response shape can be of any form. This last fact is very important since one is usually interested in the approximation problem only in those cases where standard classical filter solutions are inadequate for the application in question.

Guillemin [1] points out in his synthesis text that the Fourier method, which uses N cascaded all-pass stages, is an inefficient approach because the poles of the resulting network function have no freedom. Inspection of the $H(s)$ in (8.34) for an N-term series shows that the N poles of $H(s)$ are all constrained to lie at $s = -1$ (i.e., an Nth-order pole is obtained). With the synthesis to an arbitrary loss function achieved by means of the Fourier series coefficients, A_k, the resulting $H(s)$ tries to meet the required performance by placing zeros in the s-plane so as to obtain a desired response. So far, then, we have seen how standard LP filters are obtained from all-pole functions with all zeros fixed at $s = \infty$, and now, for the Fourier method with all poles fixed at $s = -1$, we have a more general filter shape attainable from an all-zeros approach. Thus, it would seem intuitive to expect that letting zeros and poles work together to fit a desired network response should lead to a more efficient use of both the poles and the zeros. Lower-order network functions should be expected for this case.

The problem is how to proceed to achieve this goal. With today's computers and math tools to assist, it is possible to take a ratio of polynomials, with all

8.2. GENERAL METHODS

coefficients initially unknown, to represent a desired network function, then let computer optimization routines iterate the coefficient values (or pole-zero locations) to find a best fit to the specified response. Procedures to do this have been around for years. In fact, the whole field of optimization theory evolved over the last 30 years to solve those kinds of problem formulations. The problem is that most of them do not work unless you have a good guess about how many poles and zeros to use and about what the initial values should be for starting the iteration. If the order of polynomials is too small (not enough poles and zeros), a proposed network function cannot possibly achieve the goal, and if the starting point is not close to what is called the global minimum (the answer), then the optimization often ends at what are called local minima. Thus having a given order set of polynomials not reach a good answer does not necessarily mean that the order is too low, as it could just be that the starting values were too close to a local minimum. Consequently, if we want to resort to the use of optimization tools to assist in the solution of the approximation problem, it is very helpful to develop a procedure to tell us how to pick an appropriate polynomial order and also a good starting point for the initial set of poles and zeros (polynomials).

Fortunately, there is a way to do this, and it involves the minimum phase associated with a given loss function [3]. We already know about the Hilbert phase function, $\beta(\omega)$, of (8.54) for the typical loss function, $\alpha(\omega)$, of (8.48). The other thing we need to know is how a ratio of polynomials yields the phase (lag) function and how that is related to both the order and the coefficients of the polynomials. This latter fact can be deduced by looking at how the network function phase response is related to polynomials for a given network function. When we know how to choose a starting function, then we can turn it over to Matlab to tweak the polynomial parameters to obtain an optimum solution for the polynomials thus chosen.

Example 7 - Using the following network function, evaluate the loss function, α, and phase lag function, β. Use frequency mapping, $\phi = 2\tan^{-1}(\omega)$, to display the infinite ω-range over $0 < x < 1$ for $x = \phi/\pi$, as was done in the previous section. Relate the network polynomial parameters to the phase response.

$$H(s) = \frac{P(s)}{Q(s)} = \frac{Q(s/3)}{Q(s)}$$
$$\text{where} \quad Q(s) = s^5 + 3s^4 + 5s^3 + 5s^2 + 3s + 1 \quad (8.56)$$

Solution - Let $s = j\omega$ for the sinusoidal steady-state response function. Then α and β are found as follows:

$$H(j\omega) = e^{-(\alpha+j\beta)} = \frac{|P(j\omega)|e^{j\beta_p}}{|Q(j\omega)|e^{j\beta_q}}$$
$$\alpha(\omega) = -\ln(|P(j\omega)|/|Q(j\omega)|) \text{ Nep}$$
$$\beta(\omega) = -(\angle P(j\omega) - \angle Q(j\omega)) \text{ rad} \quad (8.57)$$

(The loss function in decibels is 8.6859 times the loss in Nepers). For $s = j\omega$, each

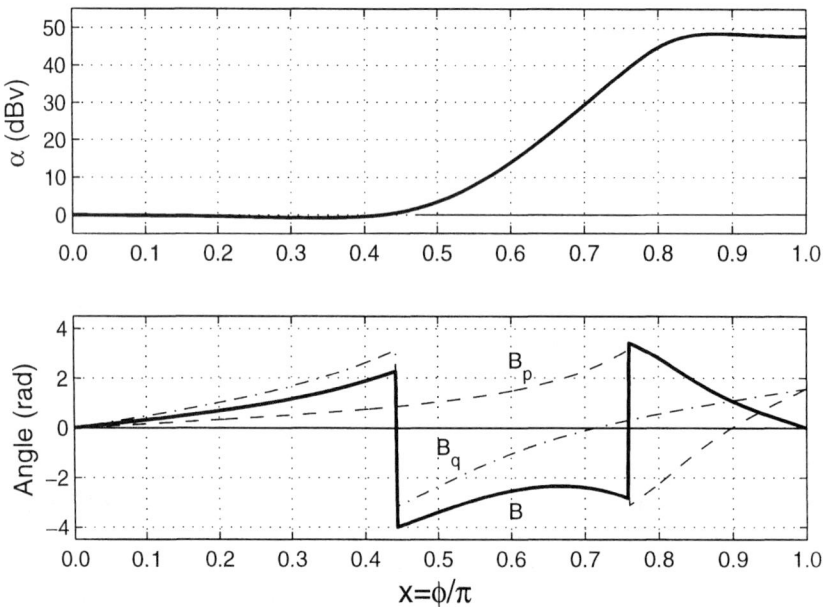

Figure 8.27. Example 7 loss and phase-lag functions of $x = \phi/\pi = 2\tan^{-1}(\omega)/\pi$.

polynomial is just a complex number, so we get the following for the angle of $P(j\omega)$:

$$\beta_p = \angle P(j\omega) = \tan^{-1}(\mathcal{I}m[P(j\omega)]/\mathcal{R}e[P(j\omega)])$$

Plots for the loss and phase lag associated with this example function are shown in Figure 8.27. The loss function exhibits LP behavior, while the resulting phase lag, β, exhibits discontinuous behavior while starting and ending at zero.

We can deduce that the phase discontinuities in β_q, β_p, and $\beta = \beta_q - \beta_p$ are due to ambiguities associated with computing difficulty in the evaluation of the inverse tangent of the real and imaginary parts of a complex number. Plots of the computed polynomial angles are shown in Figure 8.28 compared to the signs (polarities) of the polynomial real and imaginary parts. The plots are displayed against the transformed ω-variable, $x = \phi/\pi$, so that the full range of variation can readily be shown on one graph.

Observe that as the angle for $Q(j\omega)$ reaches $\pi/2$ rad (denoted by a circle on the graph) the real part of $Q(j\omega)$ goes through zero and changes sign from $+$ to $-$. This means the angle enters the second quadrant. As frequency increases, the angle approaches π rad, which means the imaginary part is going to zero. At the point where the imaginary part of $Q(j\omega)$ changes from $+$ to $-$, the angle jumps from $+\pi$ to $-\pi$. In other words, the computer calculation of third-quadrant angles starts at $-\pi$ rather than at $+\pi$. Continuing with increasing frequency, the real part of $Q(j\omega)$ becomes $+$ again when the angle enters the fourth quadrant at $-\pi/2$ rad, and so forth. In the limit for large ω, the fifth-order polynomial reaches an angle

8.2. GENERAL METHODS

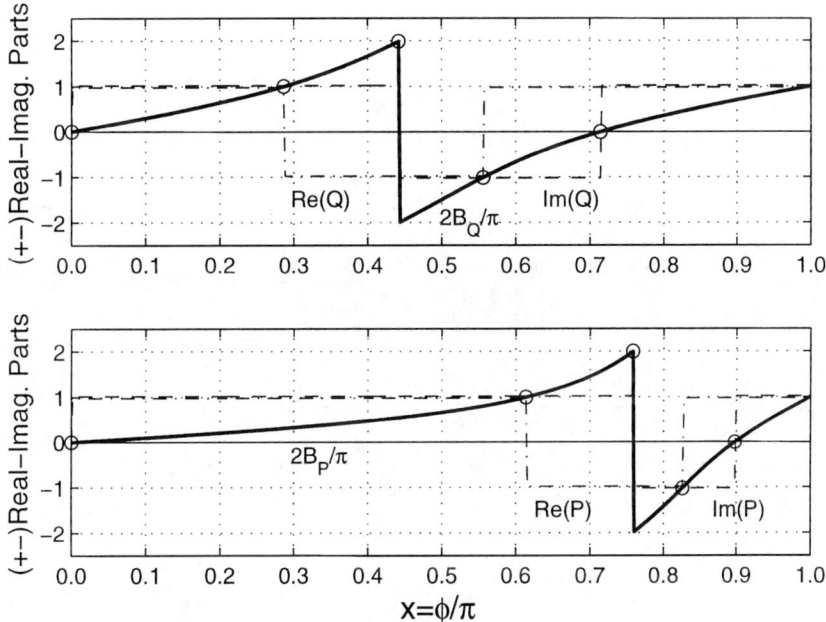

Figure 8.28. The computed angle for polynomials, $P(j\omega)$ and $Q(j\omega)$, compared to the polarities of the polynomial real and imaginary parts.

approaching $+\pi/2$ rad. Thus, as ω goes from zero to ∞, the angle β_q completes 1.25 revolutions to complete a total angular variation of zero to $5\pi/2$ rad. The polynomial angle increases monotonically for increasing ω-values.

The "cut" in the computed angle is a mathematical convention to compute a principal value of angle ranging between $\pm\pi$ (4 quadrants) since the computation of a single ratio knows nothing about a progression, or continuum, of values. It is up to the user (or programmer) to add or subtract multiples of 2π to the answer in order to remove any unwanted discontinuities. Note that the variation for $\angle P(j\omega)$ is the same as for $Q(j\omega)$ but displaced to ω-values that are three times those for Q since this example uses $P(j\omega) = Q(j\omega/3)$.

Before we obtain corrected plots of the polynomial phase functions, it will be useful first to observe how the polynomial parameters are tied to the ω-values where the real and imaginary parts go to zero or change signs versus ω. This will help us see how the network phase lag function can lead back to the polynomials.

For the denominator polynomial, we have the following relationships in the calculation of its angle dependence upon ω:

$$\beta_q = \angle Q(j\omega) = \tan^{-1}\left(\frac{\mathcal{I}m[Q(j\omega)]}{\mathcal{R}e[Q(j\omega)]}\right)$$

$$Q(j\omega) = \left[s^5 + 3s^4 + 5s^3 + 5s^2 + 3s + 1\right]_{s=j\omega}$$

$$= (3\omega^4 - 5\omega^2 + 1) + j\omega(\omega^4 - 5\omega^2 + 3) \text{ so}$$

$$\tan(\beta_q) = \frac{\omega(\omega^4 - 5\omega^2 + 3)}{(3\omega^4 - 5\omega^2 + 1)}$$

$$= \frac{\omega(-\omega^2 + 0.6972)(-\omega^2 + 4.3028)}{3(-\omega^2 + 0.2324)(-\omega^2 + 1.4343)} \tag{8.58}$$

This last form shows that as ω increases, β_q starts at zero, goes to $\pi/2$ as $\omega^2 \to 0.2324$ where the tangent goes to ∞, then goes to π as the numerator goes to zero as $\omega^2 \to 0.6972$. These first two roots of (8.58) take β_q halfway (2 quadrants) around the complex number plane. As ω continues to increase, the angle enters the third quadrant and goes to $3\pi/2$ as $\omega^2 \to 1.4343$, where the tangent again goes to ∞. As $\omega^2 \to 4.3028$, the tangent goes to zero, and the angle goes to 2π. Finally, as $\omega \to \infty$, the tangent goes through the first quadrant a second time, ending at $\beta_q = 2\pi + \pi/2 = 5\pi/2$ rad. This angle agrees with the angle obtained from the dominant term, s^5 in $Q(s)$ as s becomes large and equal to $j\omega$.

To compare the finite quadrant frequency values just described with the circled sign change points in Figure 8.28, we need to transform the ω-values to $x = \phi/\pi$ values as follows:

$$\omega_{quad} = [0.2324\ 0.6972\ 1.4343\ 4.3028]^{1/2}$$

$$= [0.4821\ 0.8350\ 1.1976\ 2.0743]$$

$$x = \frac{\phi}{\pi} = \frac{2}{\pi}\tan^{-1}(\omega_{quad})$$

$$= [0.2860\ 0.4429\ 0.5571\ 0.7140]$$

8.2. GENERAL METHODS

An eyeball comparison of these numbers with the circled points on Figure 8.28 for β_q shows that these numbers exactly define the locations for the $\pi/2$ angle locations of the polynomial, $Q(j\omega)$.

All polynomials that have their roots restricted to lie in the lhp exhibit this same character of a monotonically increasing angle versus increasing frequency. Such polynomials are referred to as Hurwitz polynomials and can be written in what is called the Hurwitz form, as follows, for the network function, $H(s)$ (see [1], p. 16, for more details):

$$H(s) = \frac{P(s)}{Q(s)}$$

$$\text{where } P(s) = \sum_{k=0}^{n} p_k s^k = (p_0 + p_2 s^2 + p_4 s^4 + \ldots) + (p_1 s + p_3 s^3 + p_5 s^5 + \ldots)$$

$$= Ev[P(s)] + Od[P(s)] \quad \text{and}$$

$$Q(s) = \sum_{k=0}^{n} q_k s^k = (q_0 + q_2 s^2 + q_4 s^4 + \ldots) + (q_1 s + q_3 s^3 + q_5 s^5 + \ldots)$$

$$= Ev[Q(s)] + Od[Q(s)] \tag{8.59}$$

The network function under consideration is one where both polynomials have the same order. Whether n is an even or odd number determines which polynomial part, even or odd, has the highest-order term. The polynomial angle, β_q for $Q(j\omega)$, can thus be written in terms of the Hurwitz form for the polynomial:

$$\tan(\beta_q) = \left[\frac{Od[Q(s)]/j}{Ev[Q(s)]} \right]_{s=j\omega}$$

$$\equiv \left[\frac{K_q (s/j)(s^2 + \omega_{q2}^2)(s^2 + \omega_{q4}^2)\ldots}{(s^2 + \omega_{q1}^2)(s^2 + \omega_{q3}^2)\ldots} \right]_{s=j\omega} \tag{8.60}$$

There are $n-1$ frequencies, ω_{qk}, for the nth-order polynomial. The monotonic increasing value of β_q for increasing ω-values is a consequence of the angle passing through each quadrant in sequence as ω is increased. This requires the following relation for the frequencies, ω_{qk}, in (8.60):

$$0 < \omega_{q1} < \omega_{q2} < \omega_{q3} < \ldots < \omega_{qn-1} \tag{8.61}$$

A similar form can be written for the angle, β_p, for the polynomial, $P(j\omega)$.

The ω_{qk} are known as the Hurwitz roots for $Q(s)$. The fact that a polynomial with all lhp roots can always be written as in (8.59) can be deduced from the following consideration.

Starting with a polynomial in factored form and multiplying all of the factors yields a polynomial with $n+1$ terms. LHP factors only have two forms $(s+a)$ with $a > 0$ for a root on the negative real axis, and $(s^2 + 2as + a^2 + b^2)$ with a and b real and greater than zero for complex conjugate roots at $-a \pm jb$. Multiplication of these type factors yields a polynomial with positive real coefficients with no missing terms

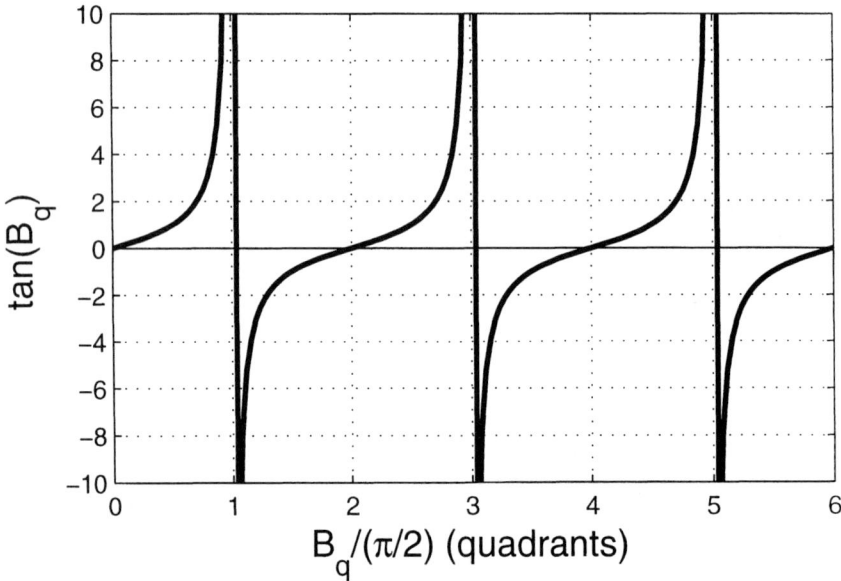

Figure 8.29. Showing the relation between the tangent function and its argument.

since all a and b being positive real means that no terms cancel in the multiplication procedure. The remarkable result is that the roots for the even and odd parts are ordered such that they interlace, as in (8.61), to yield the monotonically increasing phase function. Each property is a consequence of the other and is tied to the property of the tangent function shown in Figure 8.29. Equation (8.60) shows the functional relation of $\tan(\beta_q)$ to the Hurwitz form of the polynomial, and Figure 8.29 shows that in order for β_q to increase uniformly, the function must successfully pass through $n-1$ first-order poles and zeros.

So far, we have seen how a polynomial phase response can lead us to the Hurwitz roots of the polynomial. This amounts to identifying those $n-1$ frequencies that yield phase shifts that are integer multiples of $\pi/2$ rad. Once this special set of frequencies is identified, it is then possible to determine the constant multiplier on the odd part of the polynomial by using an estimate for the initial slope of the tangent function, (8.60). Thus, choose a $\Delta\omega > 0$ such that $\Delta\omega^2 \ll \omega_{qk}^2$ for all $k < n-1$ so that the following result is obtained:

$$\tan(\beta_q) = \left[\frac{K_q(s/j)(s^2+\omega_{q2}^2)(s^2+\omega_{q4}^2)\cdots}{(s^2+\omega_{q1}^2)(s^2+\omega_{q3}^2)\cdots}\right]_{s=j\Delta\omega}$$

$$\approx \frac{K_q(\Delta\omega)(\omega_{q2}^2\cdot\omega_{q4}^2\cdots)}{(\omega_{q1}^2\cdot\omega_{q3}^2\cdots)} \quad \text{or}$$

8.2. GENERAL METHODS

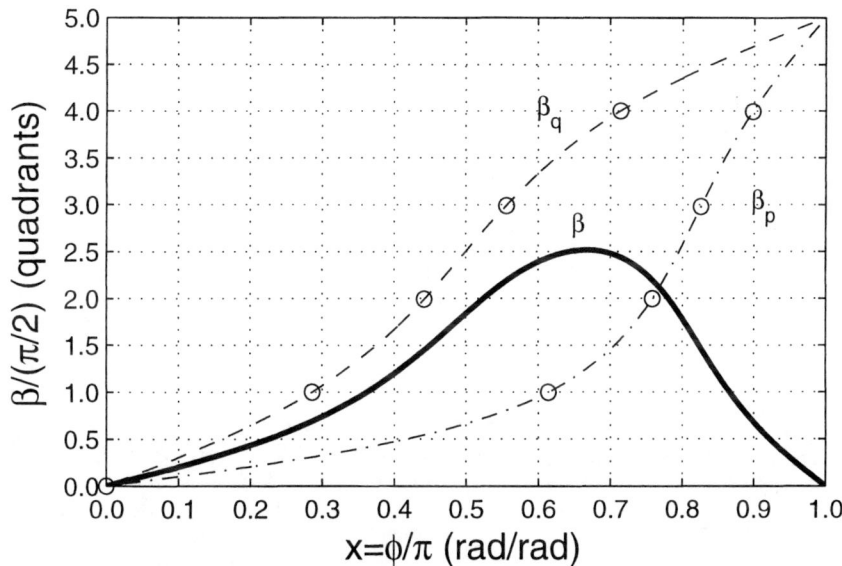

Figure 8.30. Corrected phase calculation for monotonic response.

$$K_q \approx \frac{\tan(\beta_q(\Delta\omega))}{(\Delta\omega)} \frac{(\omega_{q1}^2 \cdot \omega_{q3}^2 \ldots)}{(\omega_{q2}^2 \cdot \omega_{q4}^2 \ldots)} \tag{8.62}$$

Thus, the phase response for a polynomial with all lhp roots can lead us to the polynomial to within an unknown multiplicative constant. Using the Hurwitz form for each polynomial, the network function now looks as follows:

$$H(s) = \frac{P(s)}{Q(s)} = k_H \frac{(s^2 + \omega_{p1}^2)(s^2 + \omega_{p3}^2)(\ldots) + K_p s(s^2 + \omega_{p2}^2)(\ldots)}{(s^2 + \omega_{q1}^2)(s^2 + \omega_{q3}^2)(\ldots) + K_q s(s^2 + \omega_{q2}^2)(\ldots)} \tag{8.63}$$

The constant multiplier, k_H, accounts for the multiplicative ambiguity in both polynomials. k_H can be determined by specifying, or measuring, the loss at a specified frequency, such as at $s = 0$ or some other convenient point. Thus, (8.63) tells us how the polynomial phase information can be used to determine a formulation of the corresponding network function to within a constant multiplication factor.

The polynomial phase response, corrected for the numerical cuts due to the inverse tangent calculation, is shown in Figure 8.30, along with the network phase-lag function, β, obtained from the difference of the two polynomial phase functions. The monotonic response is obtained as shown in the following example Matlab code:

```
Q=[1 3 5 5 3 1];
nw=501;
x=linspace(0,1-1e-4,nw)';
```

362   CHAPTER 8. THE APPROXIMATION PROBLEM

```
w=tan(pi*x/2);
s=j*w;
% x is a uniformly increasing (monotonic) variable
num=polyval(Q,s/3);
den=polyval(Q,s);
alpha=-20*log10(abs(num./den));
Bp=angle(num);
Bq=angle(den);
goff=0.;
poff=0.;
for k=2:nw
  bqk=Bq(k);
  bpk=Bp(k);
  if (bqk+qoff<Bq(k-1))qoff=qoff+2*pi; end
  if (bpk+poff<Bp(k-1))poff=poff+2*pi; end
  Bq(k)=bqk+qoff;
  Bp(k)=bpk+poff;
end
BHs=Bq-Bp;
```

Note that the unit of phase shift in Figure 8.30 is normalized by $\pi/2$ rads and is thus in terms of quadrants. An nth-order polynomial always has a total variation of n quadrants, or $n\pi/2$ rad. The circled points identify the frequencies where each phase function enters the next quadrant. Thus, these points would be used to obtain the Hurwitz frequencies for each polynomial.

At this point, we have completed the rather lengthy Example 7, but now it is appropriate to interpret further the phase response and its relation to a given loss function and the polynomials required to yield a desired response.

8.2.3   Network Function Phase Versus Loss Function Phase

It is interesting to observe that the phase-lag function evaluated from a network function ratio of polynomials agrees with the Hilbert phase associated with the loss function as given by (8.54). Results from the previous example can be used to check this observation as follows.

Use a subset of points from the loss curve shown in Figure 8.27 to create a straight-line connected approximation to the example polynomial loss function; then, use the subset to evaluate the associated phase function, (8.54), and compare the result to the polynomial-derived phase lag (shown in Figure 8.30). The added code (to that used for Figure 8.30) required to make this calculation is given below, with the results shown in Figure 8.31:

```
xk=[0.4:0.05:0.85];
wk=tan(pi*xk/2);
dbk=-20*log10(abs(polyval(Q,j*wk/3)./polyal(Q,j*wk)));
ak=dbk/8.6859; % loss in Nepers
nk=length(wk); % should be same as length of ak
```

8.2. GENERAL METHODS

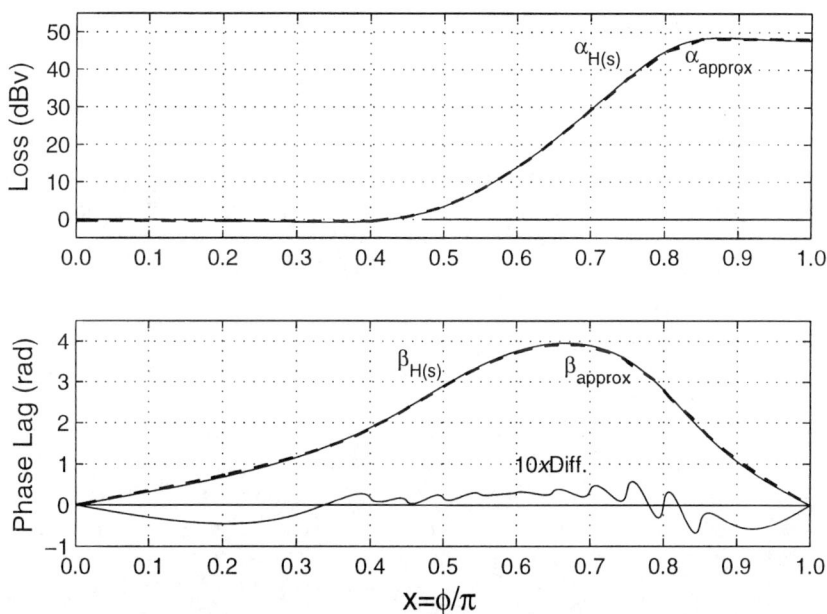

Figure 8.31. Phase from either an approximation or a polynomial ratio is equal.

```
% calculate straight-line approximation over frequency set
alfa=ak(1)*(w<wk(1))+ak(nk)*(w>wk(nk));
for k=1:nk-1
  a1=ak(k);
  a2=ak(k+1);
  w1=wk(k);
  w2=wk(k+1);
  alfa=alfa+(a1+(a2-a1)*(w-w1)/(w2-w1)).*((w>w1)-(w>w2));
end
alfdb=alfa*8.6859; % approximation to alpha
p_1=0;
for k=1:(nk-1)
  bk(k)=(ak(k+1)-ak(k))./(wk(k+1)-wk(k)); % First derivative
  ck(k)=bk(k)-p_1;                        % Second derivative
  p_1=bk(k);
end
bk(nk)=0;
ck(nk)=-bk(nk-1);
beta=zeros(size(w));
for k=1:nk
  uk=wk(k);
  beta=beta+ck(k)*((w-uk).*log(abs(w-uk)+eps)...
```

```
       +(w+uk).*log(abs(w+uk+eps)));
end
Bhilb=-beta/pi;   % Hilbert phase associated with approximation
```

On the loss graph, the smooth curve is computed from the given network function (as before), and the dashed curve is based on a 10-point straight-line (in ω-space) approximation to the network function. On the phase graph, the Hilbert phase, (8.54) derived from the approximation to the loss curve, is labeled β_{approx} and plotted as a dashed line. Ten times the phase difference between the two curves is also shown on the graph.

The remarkable result is that the two phase curves are virtually identical! We can conclude that the phase-lag function obtained from a ratio of polynomials has the same dependence upon its loss function as does the Hilbert phase lag associated with a specified loss function. The commonality is that both are minimum phase functions. Polynomials with all lhp roots yield minimum phase functions.

From the point of view of design, we have arrived at the following problem: if we are given a specified loss function, α, represented as a connected straight-line function (e.g., Figure 8.22) for which we can calculate the associated phase-lag function, β, (8.54), how do we work backwards from the phase function to find a suitable ratio of polynomials to obtain the required response?

8.2.4 Obtaining Polynomials from a Phase Response

When a network function is represented in the standard polynomial form [e.g., (8.59)], the relationship between a specified loss function, α, and the polynomial parameters is very obscure. No one has ever found straightforward procedures to relate the polynomial coefficients, p_k and q_k, to specific values on either the loss response or the phase response. However, when the network function is expressed as a ratio of polynomials written in their Hurwitz form, we have seen how the Hurwitz roots can be specifically identified at the points where the polynomial phase angle is equal to integer multiples of $\pi/2$ rad. Thus, it is helpful to find suitable polynomials for the required network function by forming required phase curves for each polynomial and then identifying the Hurwitz parameters required to produce that phase response. The next problem is how to create target phase curves for each polynomial, $P(s)$ and $Q(s)$, to provide the phase lag, β, associated with a given specified loss function α.

From Bob Dylan's music we may have heard, "The answer is blowin' in the wind." We do know that the phase lag, β, is the difference between the denominator phase, β_q, and the numerator phase, β_p (i.e., $\beta = \beta_q - \beta_p$), as shown in Figure 8.30 for the last example. We also know that polynomial phase functions are always positive (for $\omega > 0$) and increase monotonically with increasing frequency. Polynomial phase curves never decrease (for minimum phase polynomials, i.e., those with lhp roots only) for the class of polynomials in which we are interested. Therefore, we can conclude that all increases in β come from the denominator phase, β_q, and all decreases come from the numerator phase, β_p. Once we have computed the phase lag, β, associated with a specified loss, α, then as a first step to determin-

8.2. GENERAL METHODS

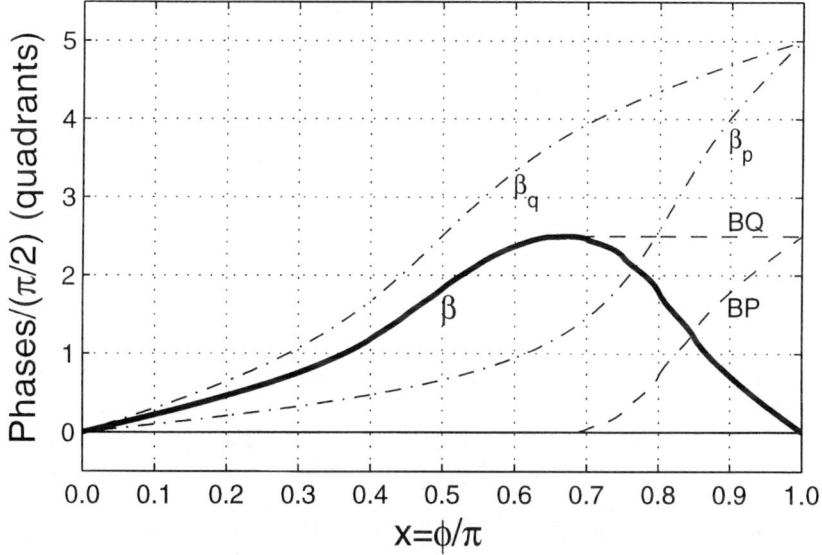

Figure 8.32. Plot of phase-lag variation curves, BQ and BP.

ing suitable polynomials, we can separate β into two curves: one derived from its positive variation only, and the other derived from its negative variation. These would be first-cut phase functions to represent phase requirements for the selection of polynomials to obtain the required loss response.

The following Matlab code shows how the separation is done. The result is plotted in Figure 8.32 for the ongoing $H(s)$ example, (8.56). The positive variation of β is assigned to the curve labeled BQ, and the magnitude of the negative variation is assigned to the curve labeled BP. β_q and β_p are the polynomial phase curves and the phase lag, β, is the difference, $\beta_q - \beta_p$, as plotted previously in Figures 8.30 and 8.31.

```
% separation of beta by its variation
BP=zeros(1,nw)'; % initializing BP
BQ=zeros(1,nw)'; % and BQ
for k=2:nw
  dbet=beta(k)-beta(k-1);
  if dbet>0     % increment BQ
    BQ(k)=BQ(k-1)+dbet;
    BP(k)=BP(k-1);
  else
    BP(k)=BP(k-1)-dbet;  % increment BP
    BQ(k)=BQ(k-1);
  end
end
```

First of all, we note that the separated curves, BQ and BP, do not go to an integer number of quadrants in the limit as $\omega \to \infty$ or $x = \phi/\pi \to 1$. In addition, the polynomial curves, β_q and β_p, go to a limit of five quadrants ($5\pi/2$ rad), and so, each has more variation than the curves obtained from the variation of their difference. Consequently, there is some function, Ψ, that is added to both BQ and BP in order to form, or yield, the fifth-order polynomial angle functions, β_q and β_p:

$$\begin{aligned} \beta_q &= BQ + \Psi \\ \beta_p &= BP + \Psi \\ \text{where} \quad \beta &= \beta_q = \beta_p \equiv BQ - BP \end{aligned} \quad (8.64)$$

Clearly, BQ and BP represent the smallest angle functions that can be found to yield the required phase lag, but a Ψ-function has to be found to convert these minimum functions into angle functions obtainable from polynomials. Since BQ and BP go to 2.5 quadrants in the limit as $x \to 1$, and since an nth-order polynomial angle goes to $n\pi/2$ rad (n quadrants) in the limit, it appears that the lowest order that could meet this example loss function is $n = 3$. Note that the data for the example were generated using fifth-order polynomials. Apparently, the five poles and zeros are not used as efficiently as they could be to obtain the resulting loss function. The total variation from the phase lag for a given loss function thus provides us with a minimum polynomial order that is feasible to obtain the specified loss. For this example, $n \geq 3$ would be feasible.

The problem now is how to find the mysterious Ψ-function that separates the minimum variation curves from the polynomial phase curves. Apparently it could be anything since it has no effect upon the resulting phase-lag function. It does, however, greatly affect the polynomial parameters and, hence, would affect the resulting loss function. There do not appear to be any useful physical conditions that can be imposed to determine these functions. Some examples are shown in Figure 8.33 for different candidate functions compared to the actual Ψ-function that generates the fifth-order polynomial angle curves for this example. The solid-line curve, Ψ, represents the difference between β_q and BQ (or between β_p and BP).[5] The dashed curve, Ψ_1, is proportional to the sum, BQ+BP, derived from the variation of β. It wraps around Ψ and is asymptotic to Ψ as $x \to 0$ or $x \to 1$. The dash-dot curve, Ψ_2, is proportional to the integral of the phase lag, β. Generally, this curve lies above the actual Ψ-function. The dotted curve, Ψ_3, is proportional to x^2. Generally, this curve lies below the actual Ψ-function. Neither Ψ_2 nor Ψ_3 is asymptotic to Ψ in any region of x.

The result of many trials, in connection with an optimization procedure, has led to the conclusion that Ψ_1 consistently leads to solutions with the least error for a wide variety of loss function shapes. It is felt, intuitively, that the choice of Ψ needs to be tied to problem requirements in some manner. The proportionality to x^2 does not work well for BP filter shapes. The use of the integral of β seems to provide starting functions too far away from the global minimum and, so, leads to answers with larger error when compared to Ψ_1. Thus, Ψ_1, proportional to the absolute

---

[5]This is the actual Ψ-function present in the example network function.

8.2. GENERAL METHODS

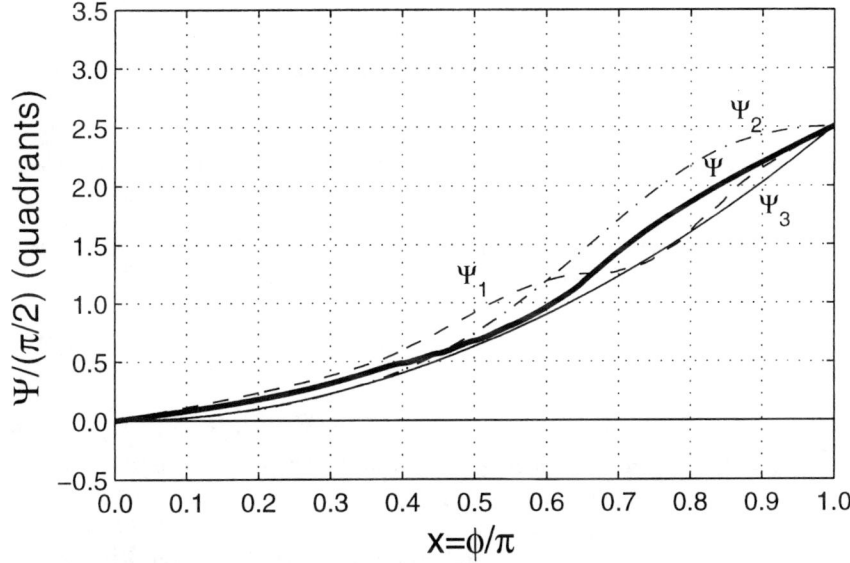

Figure 8.33. Comparison of different candidate Ψ-functions.

variation of β, has been chosen as the basis for choosing a Ψ-function in order to create polynomial angle functions from BQ and BP.

We can now construct a systematic procedure to obtain a starting set of polynomials, $P(s)$ and $Q(s)$, to provide a required phase-lag function, $\beta(\omega)$, associated with a specified loss function, $\alpha(\omega)$.

1. Start with a piecewise, straight-line connected representation for a specified loss function. This requires the vectors of length nk: for frequency, wk; for loss, ak. wk is in rad/sec in the range of $\omega = 1$, and ak is in Nep (dB/8.6859).

2. Calculate the phase lag, β, associated with the specified loss. This is determined from the second derivative of α to get the impulse values, ck, and the use of (8.54).

3. Calculate the minimum phase curves, BQ and BP, from the variation of β. From the limit of BQ (and BP) determine the minimum order, n_{min}, to be the next integer $\pi/2$ rad after the limit of BQ. Decide what order to use to solve the problem. N_{order} equals $n_{min} + 0, +1, +2, \ldots$, depending upon the perceived difficulty of the problem. $n_{min} + 1$ works for most LP functions, while $n_{min} + 2$ works for simple BP functions. A complicated requirement (multiple passbands or odd-shaped passbands) may require $n_{min} + 4$.

4. Calculate a Ψ-function proportional to BQ+BP to add to BQ and BP to obtain the desired-order polynomial angle functions, β_q and β_p. The limit of the angle function is $N_{order}\pi/2$ rad, so the limit of $\Psi + BQ = N_{order}\pi/2$ rad.

5. From β_q and β_p thus obtained, find the Hurwitz parameters for each polynomial. The $n-1$ Hurwitz roots are obtained at the points where the angle curve yields integer values of $\pi/2$ rad (shown as circles on Figure 8.30). The odd-part multiplier is obtained from the slope of the phase curve at some $\Delta\omega > 0$ through the use of (8.62). [For K_p just change q to p in (8.62)]. These Hurwitz parameters define polynomials P and Q to within an arbitrary constant multiplier.

6. Determine the required constant multiplier, k_H, for the desired network function, $H(s)$. This is done by choosing $\omega = 0$ and evaluating (8.63) using the specified loss (in Nep) at $\omega = 0$ and solving for k_H, as in (8.65):

$$k_H = |H(0)| \cdot \frac{(\omega_{q1}^2 \cdot \omega_{q3}^2 \cdots)}{(\omega_{p1}^2 \cdot \omega_{p3}^2 \cdots)} \qquad (8.65)$$

Completion of Steps 1 to 6 determines an initial approximation to a network function required to meet a given loss function.

Example 8 - Use a straight-line approximation to the loss function, α, for the network function in Example 7, and construct polynomial approximants, following Steps 1 to 6 above, to obtain approximations to the straight-line loss function for nmin = $+0, 1, 2$. Compare the resulting loss, from the polynomials thus obtained, to the specified straight-line approximation.

Solution - The 10-point approximation, used previously to obtain the separated phase function, is a sufficient approximation for the loss function. This yields the minimum phase functions, BQ and BP, shown in Figure 8.32. This result shows that the phase variation for the specified loss amounts to 2.5 quadrants, or minimum order nmin = 3. Thus, the Ψ-function, determined from the sum of BQ and BP, must be scaled to take the resulting polynomials to third, fourth, and fifth order, respectively.

Procedures used to get from the minimum phase curves (Step 3) to the initial polynomial selection (Step 6) will be discussed briefly. The discussion begins with Step 4, selection of the Ψ-function, which is added to BQ and BP to take the phase variation to an integer number of $\pi/2$ rad in order to provide a polynomial limit to the phase curve.

The following code shows how polynomial phase curves are formed using the specified variable, nplus, to obtain a desired polynomial order, Norder = nmin + nplus.

```
nplus=2;
nmin=max(ceil(2*[max(BP) max(BQ)]/pi));
Norder=nmin+nplus   % desired order for poly's P & Q
SEP=BQ+BP;
xm=(pi*Norder/2-max([max(BP) max(BQ)]))/SEP(nw);
Psi=xm*SEP;
bp=BP+Psi;
```

8.2. GENERAL METHODS

```
bq=BQ+Psi;
```

The code is self-explanatory, as nmin is rounded up from the peak value of either BQ or BP in quadrants. The separation function Ψ (Psi) is then formed from the sum of BQ+BP and scaled to provide the desired angle limit, Norder times $\pi/2$, for the polynomial angle curves, bp and bq. This completes Step 4.

Step 5 obtains the Hurwitz roots for each polynomial from the points where the phase curves are equal to integer values of $\pi/2$ rad. This procedure is carried out in two steps due to the fact that the frequencies are represented with a set of discrete values. First, the discrete frequency that provides the phase value closest to $k\pi/2$ rad is chosen for the Hurwitz root, ω_{hk}. Second, these closest values are interpolated to obtain first-order estimates to the actual frequency that provides the phase $k\pi/2$ rad. The first selection code is as follows:

```
for k=1:Norder-1
  [mq(k), ibq(k)]=min(abs((2*bq/pi)-k*ones(nw,1)));
  [mp(k), ibp(k)]=min(abs((2*bp/pi)-k*ones(nw,1)));
end
whq=w(ibq); % 1st order Est. Hurwitz roots for Q(s)
whp=w(ibp); % 1st order Est. Hurwitz roots for P(s)
```

The interpolation is explained using the graph shown in Figure 8.34. ω_2 is assigned to the discrete frequency for which the phase exceeds $k\pi/2$ rad, and ω_1 is assigned to the preceding frequency (in the set of frequency values) for which the phase is less than $k\pi/2$ rad. Forming a triangle ratio then yields a refined estimate for the kth Hurwitz root:

$$\omega_k = \omega_1 + \frac{(\omega_2 - \omega_1)}{(p_2 - p_1)} \left(\frac{k\pi}{2} - p_1 \right) \tag{8.66}$$

The interpolation is applied to bp for whp(k) and to bq for whq(k). This interpolation completes the calculations for estimating the Hurwitz roots for forming a set of polynomials to provide the required phase-lag function. In order to complete Step 5, the odd-part multipliers must be determined. The following code lists both the interpolation and the odd-part multiplier estimation:

```
% The Hurwitz roots need to be "fine tuned," i.e.,
% by linear interpolation on the set of frequencies
% just selected to lie closest to the integer pi/2 values
for k=1:Norder-1
  pik=pi*k/2;
  Iqk=ibq(k);
  Ipk=ibp(k);
  if ((bp(Ipk)-pik)<0)
    w2=w(Ipk+1);
    w1=w(Ipk);
    p2=bp(Ipk+1);
    p1=bp(Ipk);
  else
```

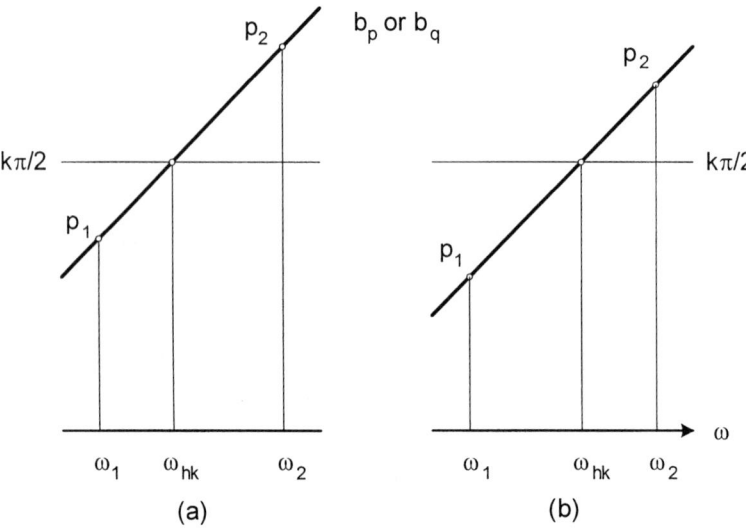

Figure 8.34. Interpolation geometry for Hurwitz frequency estimates for: (a) selected frequency with phase less than $k\pi/2$ rad, and (b) selected frequency with phase greater than $k\pi/2$ rad.

```
    w2=w(Ipk);
    w1=w(Ipk-1);
    p2=bp(Ipk);
    p1=bp(Ipk-1);
  end
  whp(k)=w1+(w2-w1)*(pik-p1)/(p2-p1);
  if ((bq(Iqk)-pik)<0)
    w2=w(Iqk+1);
    w1=w(Iqk);
    p2=bq(Iqk+1);
    p1=bq(Iqk);
  else
    w2=w(Iqk);
    w1=w(Iqk-1);
    p2=bq(Iqk);
    p1=bq(Iqk-1);
end
whq(k)=w1+(w2-w1)*(pik-p1)/(p2-p1);
end
Kp=(bp(2)-bp(1))/(w(2)-w(1)); % Kp and Kq are initially set
Kq=(bq(2)-bq(1))/(w(2)-w(1)); % to match slopes at the origin
for k=2:2:Norder-1
```

8.2. GENERAL METHODS

```
      Kp=Kp/whp(k)^2;
      Kq=Kq/whq(k)^2;
   end
   for k=1:2:Norder-1
      Kp=Kp*whp(k)^2;
      Kq=Kq*whq(k)^2;
   end
```

The multipliers, Kp and Kq, are evaluated near $\omega = 0$ using the first two frequency values in the frequency set [normally $\omega(1) = 0$, and $\omega(2)$ is a very small value]. The for-loops in the above code just implement (8.62) for both polynomials.

Having the Hurwitz roots and odd-part multipliers in hand, all that remains to complete Step 6 is to find k_H, the constant multiplier for $H(s)$, and to cast the polynomials from their Hurwitz form back to standard polynomial form. These procedures are implemented with the following code:

```
kH=exp(ak(1));
for l=1:2:length(whq)
   kH=kH*(whq(1)^2)/(whp(1)^2);
end


%find std. polynomial forms for P and Q
if rem(Norder,2)==1 % Is Norder odd?
   Peven=[0 1];
   Qeven=[0 1]; % Even part has a leading zero
   Podd=Kp*[1 0];
   Qodd=Kq*[1 0];
else
   Peven=[1];   % Norder is even
   Qeven=[1];
   Podd=Kp*[0 1 0];
   Qodd=Kq*[0 1 0]; % Odd part has a leading zero
end
for k=2:2:Norder-1
   Podd=conv(Podd,[1 0 whp(k)^2]);
   Qodd=conv(Qodd,[1 0 whq(k)^2]);
end
for k=1:2:Norder-1
   Peven=conv(Peven,[1 0 whp(k)^2]);
   Qeven=conv(Qeven,[1 0 whq(k)^2]);
end
Pest=Peven+Podd;
Qest=Qeven+Qodd;

alfest=-20*log10(abs(kH*polyval(Pest,s)./polyval(Qest,s)));
```

The calculation for k_H just follows the evaluation of (8.65). ak(1) is the specified loss, in Nepers, from the first breakpoint of the loss curve. To convert the polynomials from Hurwitz to standard form, we note that when the order is odd, the polynomial even part needs a leading zero in its structure; when the order is even, the odd part needs a leading zero in its structure so that odd and even parts can be added (vectorially) in the computation process. Consider the two following forms for n odd and even, respectively:

n odd:

$$\begin{aligned} Q &= q_5 s^5 + q_4 s^4 + q_3 s^3 + q_2 s^2 + q_1 s + q_0 = Od[Q] + Ev[Q] \\ &= [q_5 s^5 + 0 s^4 + q_3 s^3 + 0 s^2 + q_1 s + 0] \\ &\quad + [0 s^5 + q_4 s^4 + 0 s^3 + q_2 s^2 + 0 s + q_0] \end{aligned} \quad (8.67)$$

$Ev[Q]$ has a leading zero.

n even:

$$\begin{aligned} Q &= q_4 s^4 + q_3 s^3 + q_2 s^2 + q_1 s + q_0 = Ev[Q] + Od[Q] \\ &= [q_4 s^4 + 0 s^3 + q_2 s^2 + 0 s + q_0] \\ &\quad + [0 s^4 + q_3 s^3 + 0 s^2 + q_1 s + 0] \end{aligned} \quad (8.68)$$

$Od[Q]$ has a leading zero.

Thus, the first part of the code prepares the leading-zero requirement in both even and odd order cases for subsequent multiplication by the Hurwitz factors, $[s^2 + 0 + \omega_{hk}^2]$, for each part of the polynomial (even-numbered roots to the odd part and odd-numbered roots to the even part, respectively).

The polynomials thus obtained, $Pest$ and $Qest$, are each to within an unknown multiplier so that the desired network function is obtained by multiplication by k_H [i.e., $H(s) = k_H Pest/Qest$].

The above description completes the initial selection of polynomials to fit a specified loss function. The graphs in Figure 8.35 illustrate the performance of this initial fit for nplus = $0, 1, 2$, respectively (i.e., Norder = $3, 4, 5$). Recall that the original specification originates from a fifth-order network function. Inspection of these results shows that a third-order solution fits the specification reasonably well almost all the way to the stopband. The fourth-order solution appears to provide a closer fit through both the pass- and transition bands, while the fifth order has much more deviation than either the third- or fourth-order curves through both the transition band and stopband. This result seems to indicate that it is harder to fit too many poles and zeros to a given loss function.

This concludes Example 8. We have rather carefully gone through and explained all the steps to obtain an initial set of polynomials from a piecewise straight-line connected loss function of frequency. It now remains only to optimize, or tweak, the initial choice so that a function can be found to give less error between its loss function and the specified function.

8.2. GENERAL METHODS

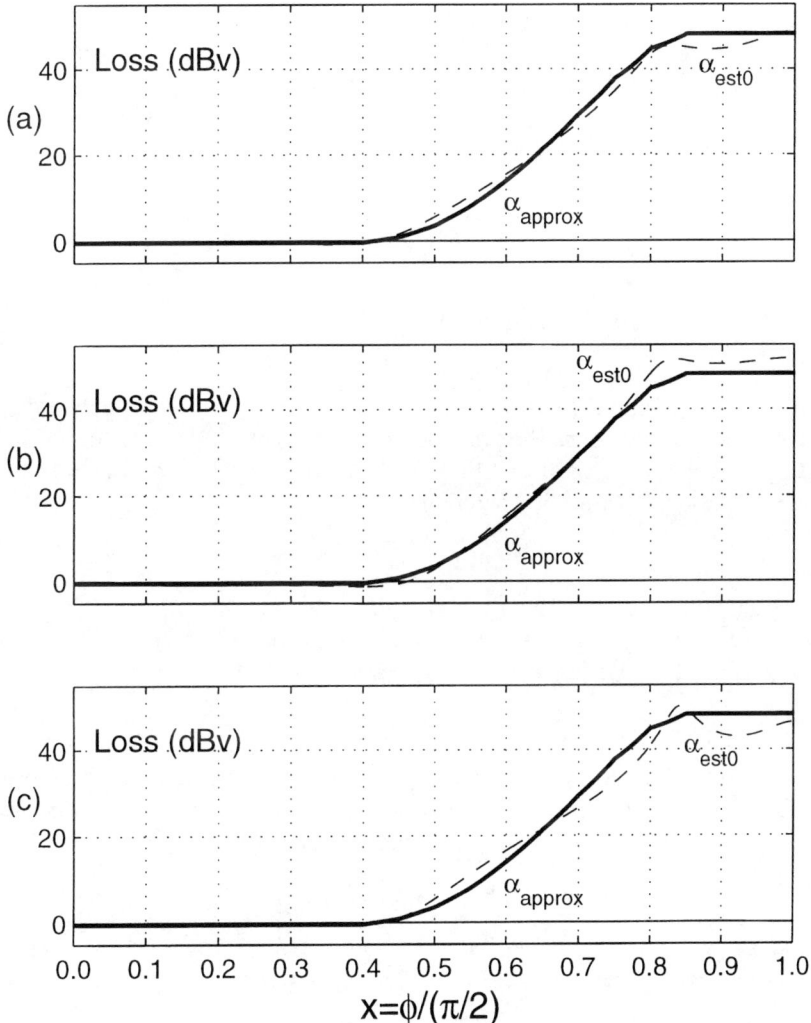

Figure 8.35. Comparison of loss estimates for phase derived polynomials for (a) nplus = 0, Norder = 3; (b) nplus = 1, Norder = 4; and (c) nplus = 2, Norder = 5.

8.2.5 Optimizing Polynomial Parameters

The last example has carefully shown us how to find an initial set of polynomials to fit a specified loss function versus frequency. The results obtained for the example show that the initial function does not fit the specified function uniformly across the entire frequency range. This is due to not knowing the optimum separation function, Ψ, for the given loss function so that the polynomial Hurwitz roots can be placed precisely in the best, or optimum, locations to minimize the total error in the fit to the objective loss function. This section explains how to use Matlab to modify the initial choice of polynomial parameters easily in order to obtain a better fit to a specified objective loss function, $\alpha(\omega)$.

At this point, the network function has the following form:

$$H(j\omega) = e^{-(\alpha+j\beta)} \quad (8.69)$$

$$= k_H \left[\frac{(s^2 + \omega_{p1}^2)(s^2 + \omega_{p3}^2)(\ldots) + K_p s(s^2 + \omega_{p2}^2)(\ldots)}{(s^2 + \omega_{q1}^2)(s^2 + \omega_{q3}^2)(\ldots) + K_q s(s^2 + \omega_{q2}^2)(\ldots)} \right]_{s=j\omega}$$

For nth-order polynomials there are $n-1$ Hurwitz roots and one odd-part multiplier. The parameter k_H multiplies the polynomial ratio to control the loss value at $\omega = 0$ (dc). Consequently, the total number of parameters used to represent $H(s)$ is as follows for a ratio of nth-order polynomials:

$$n_{par} = 2(n-1) + 3 = 2n + 1 \quad (8.70)$$

The result in (8.70) is one less than would be used for a ratio of two nth-order polynomials in a normal power series form with $n+1$ coefficients for each polynomial. Let $\mathbf{P}$ represent a vector of these parameters as in (8.71):

$$\mathbf{P} = [\omega_{p1}^2 \ \omega_{p2}^2 \ldots \omega_{pn-1}^2 \ \omega_{q1}^2 \ \omega_{q2}^2 \ldots \omega_{qn-1}^2 \ K_p \ K_q \ k_H] \quad (8.71)$$

Now we can represent the loss, α in (8.69), as a function of both the given parameter vector, $\mathbf{P}$, and the frequency, ω. Further, let this loss be represented as α_{est} as the estimate is made to fit, or match, the specified loss function α_{spec}:

$$\alpha_{est} = \alpha(\omega, \mathbf{P}) = -\ln|H(j\omega, \mathbf{P})| \quad \text{Nep} \quad (8.72)$$

The total error in the fit is then based on the squared error summed over all frequencies. The total error is thus only a function of the parameter vector, $\mathbf{P}$:

$$E(\mathbf{P}) = 8.6859^2 \sum_{\omega} |\alpha_{est}(\omega, \mathbf{P}) - \alpha_{spec}|^2 \quad \text{dBv}^2 \quad (8.73)$$

Both αs are in Nepers for the above sum. Decibels with respect to voltage (dBv) could alternately be used in the above sum, or especially when plotting error versus frequency, since most people are more familiar with visualizing ratio values in decibels.

Matlab uses a program called `fminsearch`[6] to optimize parameter selection for functions of a set of parameters, like $E(\mathbf{P})$ described above. All the user has to do

---
[6]Prior to Version 7, the Matlab optimizing function was referred to as `fmins`.

8.2. GENERAL METHODS

is provide the parameter vector, the name of the .m file that calculates the desired objective function, along with other data, such as the set of ω-values, required to evaluate the objective function for the problem. Basically fminsearch seeks to adjust the vector, **P**, in a systematic manner so as to find a global minimum in the objective function, $E(\mathbf{P})$. This procedure involves many iterations to estimate a gradient and to make changes along a path of steepest descent to arrive at a minimum. The program will stop after a default number of iterations. The number of allowed iterations and other variables related to obtaining iterative solutions can be set through the proper use of an options-setting function called optimset. optimset will not be discussed here as it is best addressed through help files in Matlab. Our only concern here is the number of allowed iterations, and that will be set to 5000. Code used to tweak the parameters for Example 8 is as follows:

```
% Section 1
nw=501; % no. of freq. for obj. function should be modest
dp=pi/nw;
x=linspace(0,1-1e-4,nw); % sets range, 0<w<6.4e3
w=tan(pi*x/2);   % objective optimizing frequency set

% Section 2
Qspec=[1 3 5 5 3 1];
xk=[0.4:0.05:0.85];
wk=tan(pi*xk/2);
dbk=-20*log10(abs(polyval(Qspec,j*wk/3)./polyval(Qspec,j*wk)));
ak=dbk/8.6859; % loss in Nepers
nk=length(wk);
nplus=2;

% Section 3 calc atten spec over optimizing frequencies
alfa=ak(1)*(w<=wk(1))+ak(nk)*(w>wk(nk)); % Nepers
for k=1:nk-1
   a1=ak(k);
   a2=ak(k+1);
   w1=wk(k);
   w2=wk(k+1);
   alfa=alfa+(a1+(a2-a1)*(w-w1)/(w2-w1)).*((w>w1)-(w>w2));
end
alpha=8.6859*alfa; % dBv - This is the specified loss function

% Section 4 get initial start from min phase in Hurwitz form
jw=j*w;
[whp,whq,Kp0,Kq0,kH0,Norder]=hurw2(w,wk,ak,nplus);
[p0,q0]=pq(Norder,whp,whq,Kp0,Kq0);  %initial polynomials
aest=-20*log10(abs(kH0*polyval(p0,jw)./polyval(q0,jw))); % dBv
esterr=abs(aest-alpha);   % initial error, dBv
```

```
% Section 5 now optimize solution
w2=w.^2;
x0=[whp.^2 whq.^2 Kp0 Kq0 kH0];   % initial parameter vector
e0=obj2(x0,alpha,w2);   % initial objective error
%y=foptions;
%y(14)=5000;
%xopt=fmins('obj2',x0,y,[],alpha,w2);  % Matlab code before Version 7
options=optimset('Maxiter',5000,'Maxfunevals',5000);
xopt=fminsearch('obj2',x0,options,alpha,w2);  % Matlab code after Version 7
eopt=obj2(xopt,alpha,w2);   % final objective error

% Section 6
n0=length(x0);
nhr=(n0-3)/2;
xm=[sort(xopt(1:nhr)) sort(xopt(nhr+1:2*nhr))];
Kp=xopt(n0-2);
Kq=xopt(n0-1);
kH=xopt(n0);
[P,Q]=pq(Norder,sqrt(xm(1:nhr)),sqrt(xm(nhr+1:2*nhr)),Kp,Kq);
% the network function is kH*P(s)/Q(s)
aopt=-20*log10(abs(kH*polyval(P,jw)./polyval(Q,jw)));
opterr=abs(aopt-alpha);
```

The above code is nearly self-explanatory. Section 1 sets up the optimizing frequency space determined from a linspace on the variable $x = \phi/\pi$. In Section 2 the specified loss points are defined (wk, ak) and put into Neper units for the subsequent phase-lag calculation. The variable *nplus* sets the order above *nmin*, determined from the associated phase functions.

Section 3 computes the loss function over all frequencies to use for calculating error, as described above. So far, these are all steps that have been illustrated via previous discussion and examples. Section 4 then obtains the initial polynomials from two subroutines, hurw2 and pq. These two routines cover the steps and code used in Example 8 to obtain the initial parameter choice for the polynomials to fit the specified loss function.[7] The input to hurw2 consists of the vectors w for the set of frequency values; wk and ak for the sets of breakpoints used to represent the piecewise straight-line connected loss function; and nplus.

hurw2 returns a vector defining the numerator, $P(s)$, $(n-1)$ Hurwitz roots; the denominator, $Q(s)$, $(n-1)$ Hurwitz roots; the numerator odd-part multiplier, Kp; the denominator odd-part multiplier, Kq; the network function multiplier, kH; and the polynomial order, Norder. pq takes the Hurwitz parameters for each polynomial and returns polynomials $P(s)$ and $Q(s)$ in standard coefficient form for use with the routine polyval. The initial loss estimate, α_{est}, is then computed for this initial choice of polynomial parameters along with error magnitude versus frequency.

---

[7] Complete listings for all Matlab files used to solve the approximation problem are given in Appendix 8A.

8.2. GENERAL METHODS

Section 5 of the program turns the problem over to Matlab to tweak the choice of parameter values to minimize the error versus frequency. The initial parameter vector, $x0$, is formed from the initial choices, assuming the Hurwitz form of (8.69), and then fminsearch returns an optimized $xopt$ vector when it either reaches a minimum or uses up the allowed number of iterations. The input to fminsearch consists of the objective error function file name, obj2 (the .m extension is defaulted); the initial vector, x0; the revised options statement; the specified loss function alpha (in dBv); and the frequency vector squared, w2.

Section 6 takes the optimized vector, xopt, from fminsearch and converts back to standard polynomial forms, P and Q, and then computes a final optimized estimate to the specified loss function. The estimated error is also calculated in this section. Thus, the goal to find a network function to fit a specified loss function is completed as we have obtained an optimized choice for the polynomials.

Figure 8.36(a–c) shows optimized results in the form of error magnitudes obtained from the starting functions for nplus $= 0, 1, 2$, respectively. The minimum-order solution (nplus $= 0$) improves the error from an initial peak of 3.6 dB in the stopband to a peak of 2.2 dB in the transition band. Note that in an attempt to reduce large out-of-band error, the passband error increased for this case. Nevertheless, a passband ripple of 1 dB is acceptable for some applications, so this third-order solution could be acceptable. Clearly, this solution does not appear to have enough flexibility to adjust the parameters to provide arbitrarily small error in its fit to the specified loss function.

Figure 8.35(b), for nplus $= 1$, shows an excellent result for the fourth-order fit to the specified loss function. The peak absolute error is 0.616 dB over the whole frequency band. Clearly, the fourth-order solution is a good solution to the specified loss function. Recall that the specified loss was generated from a ratio of fifth-order polynomials.

In Figure 8.36(c), for nplus $= 2$, the extra pole for the fifth-order solution does not appear to offer much advantage over the fourth-order solution. Its peak error over the whole band is 0.613 dB, which is virtually the same as the fourth-order solution; however, the passband error is considerably smaller for this case. The error is distributed differently with the largest error occurring in the transition band. There is very little to warrant using the fifth-order solution over the fourth-order solution.

This section has illustrated the use of Matlab's fminsearch to optimize polynomial parameters to find a network function that best fits a desired loss function. We have seen that the method, which is based upon the associated phase-lag function of a given loss function, provides a bare bones estimate of the minimum order needed to meet the requirement. Sometimes the minimum order provides suitable answers. Experience shows that it most often does not. Increasing the order provides smaller error in the fit until further increase offers no additional advantage. Thus, the procedure converges to a limit by adding poles to the approximating function. Generally, the limit is affected by the number of breakpoints in the specified function, as these provide cusp-like error behavior versus frequency. Cusps provide nonanalytic behavior and are difficult to approximate with continuous functions, such as the polynomial ratio being used. The effect is reduced by using more closely spaced points to describe the specified loss function rather than just a few widely separated points.

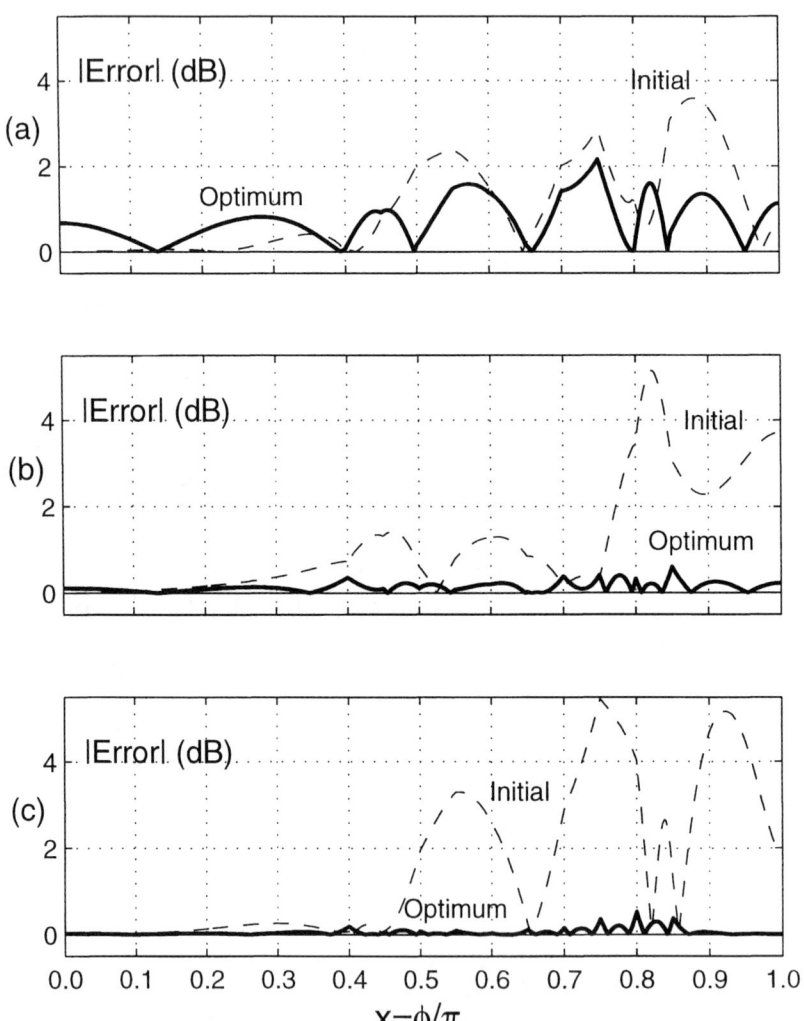

Figure 8.36. Optimized response for Example 8 specification for: (a) nplus = 0, Norder = 3; (b) nplus = 1, Norder = 4; and (c) nplus = 2, Norder = 5.

8.2. GENERAL METHODS

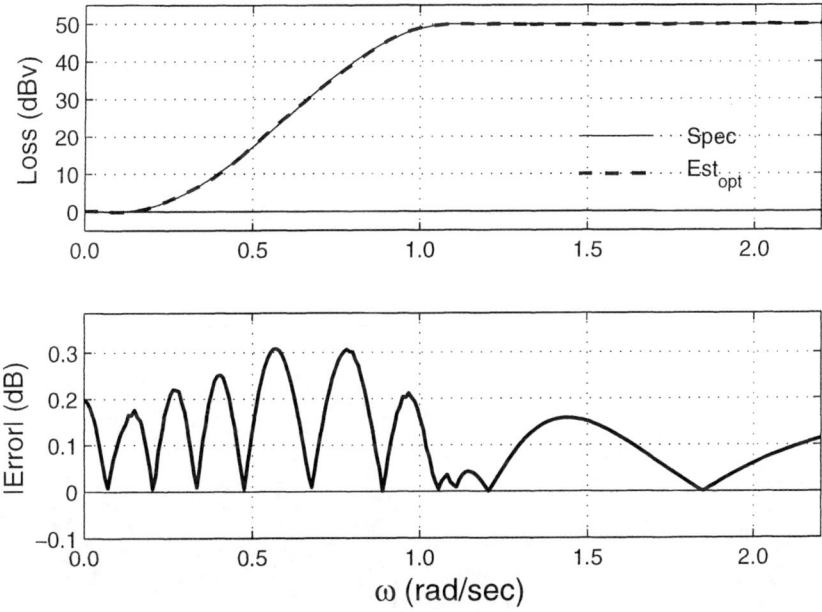

Figure 8.37. Response of solution found for Example 9.

Example 9 - Use the procedures just developed to find a network function to provide the following prototype LP loss function versus frequency. Specify the function, plot the responses (specified and chosen network function loss) and the error, in decibels, versus frequency, and make a pole-zero plot for the result. Use nplus = 1 for the design.

```
wk=linspace(0.1,1.1,51);
dbk=25*(1-cos(pi*(wk-0.1)));
```

Solution - The program apprx8 is used by typing in the loss function description and setting nplus = 1 for the desired solution. (See Appendix 8A for a full listing of the program apprx8.)

The responses for both the specification and optimized network functions are shown in Figure 8.37. The optimized answer appears to coincide exactly with the specified cosine function on the function loss curve, but the difference (error) between the two curves yields a maximum absolute error of 0.31 dB when plotted on a separate curve, as shown. The optimized answer has its error spread smoothly across the full frequency band so as to provide a minimum squared error sum for the fit.

The pole-zero diagram is shown in Figure 8.38 for the following polynomials (and roots) of the resulting optimized network function:

380   CHAPTER 8. THE APPROXIMATION PROBLEM

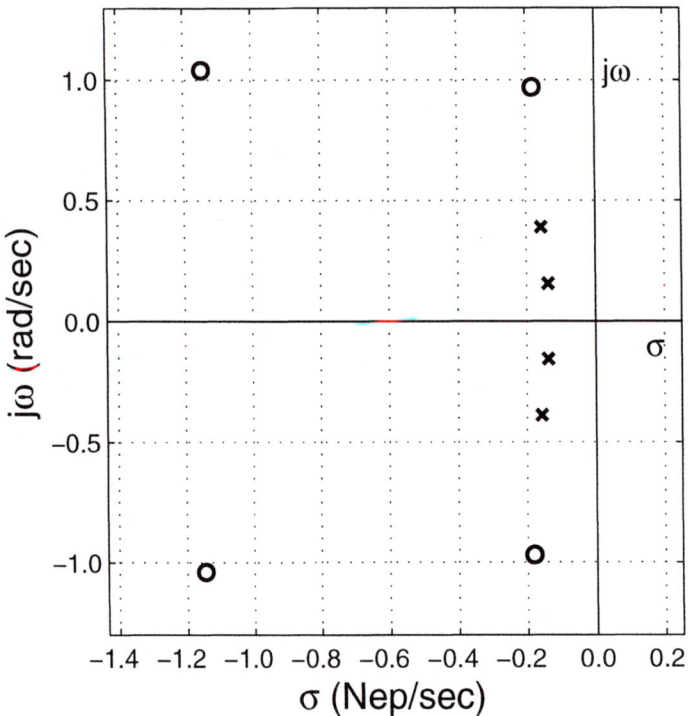

Figure 8.38. Pole-zero pattern for LP filter requirement of Example 9.

8.2. GENERAL METHODS

$$H(s) = k_H \frac{P(s)}{Q(s)}$$

where
$$k_H = 1/311.0 = 3.215 \cdot 10^{-3}$$
$$P(s) = s^4 + 2.6547s^3 + 4.1990s^2 + 3.0975s + 2.3283$$
$$Q(s) = s^4 + 0.5909s^3 + 0.3066s^2 + 0.0623s + 7.657 \cdot 10^{-3}$$
$$\text{Zeros} = roots[P(s)]$$
$$= -0.1815 \pm j0.9692; \quad -1.1458 \pm j1.0401$$
$$\text{Poles} = roots[Q(s)]$$
$$= -0.1378 \pm j0.1564; \quad -0.1577 \pm j0.3891$$

The resulting pole-zero distribution is interesting in that the poles are arranged in a convex pattern (bending away from the $j\omega$-axis) for this LP function, whereas conventional all-pole solutions, like Butterworth and Chebyshev, are arranged in a concave pattern (bending toward the $j\omega$-axis). The zeros adjust the approach to the stopband, where one complex pair is aligned with the poles and near the $j\omega$-axis, while the other complex pair is on a line with the beginning of the stopband but removed from the poles.

The significance of this example is that a solution was found to fit a specified loss function of frequency. Traditional procedures and tabulated designs provide a fit (to within a specified tolerance) to a constant passband and to a minimum loss in the stopband. The procedure shown here converges to a specified loss function over the full frequency band. An example of where fully specified loss functions are desirable is in the area of signal-processing matched filters for optimum receiver design where specified chip detectors are required.

Example 10 - Use `apprx8` to find a solution for the following prototype BP loss function. Obtain solutions for `nplus` $= 1, 2, 3$, and show the convergence obtained as the number of poles is increased. Specify the function, and show a pole-zero plot for `nplus` $= 3$.

```
wk=linspace(0.5,1.5,51);
dbk=25*(1-cos(2*pi*(wk-1.0)));
```

Solution - The input to `apprx8` is modified to find a solution for the above 51-point specification and run three times for the different values of `nplus`. It is noted that nmin, from the separated phase-lag curves, is found to equal 5 for this function.

Frequency response curves obtained for the three values of `nplus` are shown in Figures 8.39, 8.40, and 8.41. The maximum error for each solution is 2.27, 1.40, and 0.228 dB, respectively, for `nplus` $= 1, 2, 3$. Clearly, the procedure converges to the desired function as the number of poles is increased for the problem.

The resulting pole-zero pattern for `nplus` $= 3$ is shown in Figure 8.42. The optimized network function is as follows:

$$H(s) = k_H \frac{P(s)}{Q(s)}$$

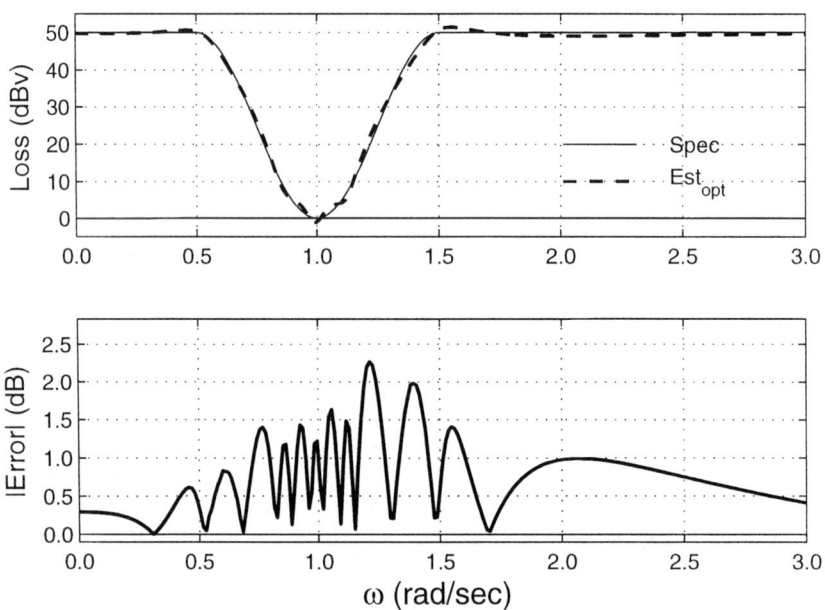

Figure 8.39. Example 10 response and error curves for `nplus = 1`, `Norder = 6`.

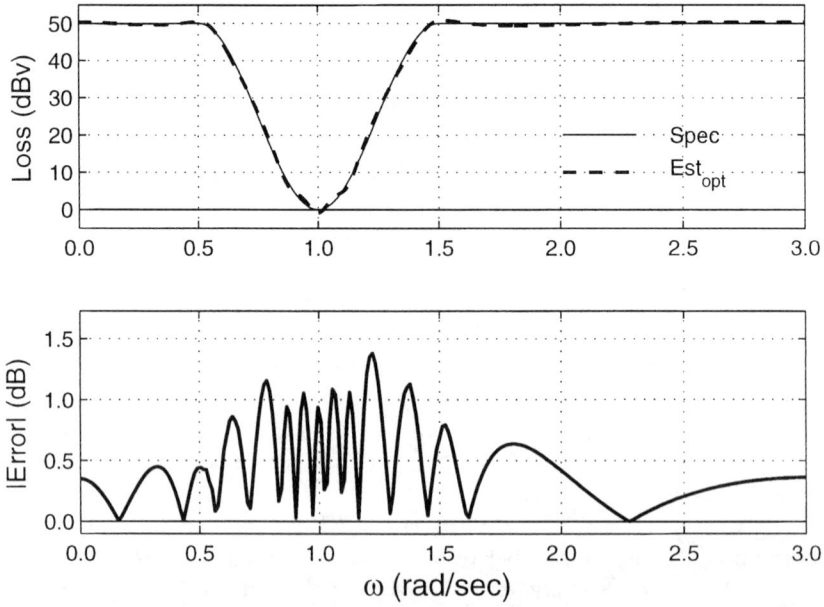

Figure 8.40. Example 10 response and error curves for `nplus = 2`, `Norder = 7`.

8.2. GENERAL METHODS

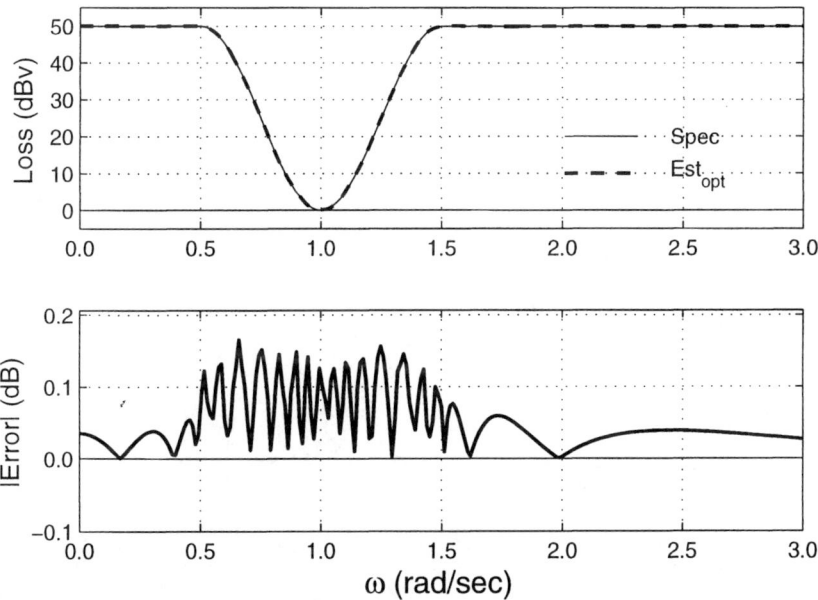

Figure 8.41. Example 10 response and error curves for nplus = 3, Norder = 8.

where k_H = $3.1668 \cdot 10^{-3}$
P = [1.0 2.4636 7.1052 10.0505 13.6923 11.1807 7.6352 2.7630 1.0088]
Q = [1.0 0.6349 4.2798 1.9741 6.5878 1.9960 4.3201 0.6559 1.0160]

Note that the resulting polynomials are very close to being symmetrical about the unit circle in the s-plane (i.e., the coefficients are symmetrical from end to end toward the center of the polynomial). In addition, the pole-zero pattern for this BP prototype resembles a translated version of the LP function plotted in Figure 8.38 with some resulting distortion of the LP pole-zero pattern. Thus, the BP cosine loss function result obtained here bears a strong resemblance to a frequency translation of the LP cosine function used in Example 9. This is a reassuring result for the procedure and shows that a specific value of nplus is required in order to obtain a best fit to a specified BP loss function.

A further significance of this example is that the specified loss function describes a BP filter that has arithmetic symmetry with respect to the center frequency at $\omega = 1$. This is shown in Figure 8.43, where the response is plotted versus frequency for $0 < \omega < 2$. The graph shows equidistant ω-values (from the center, $\omega = 1$) for each loss value going from 0 to 50 dBv. The standard LP-to-BP transform procedures (discussed in Chapter 2) cannot ever provide true arithmetic symmetry since that procedure always yields a geometric symmetry (by construction). Thus, the approximation procedure developed in this chapter is very versatile and relatively unconstrained in the type of response for which a solution can be found. The only

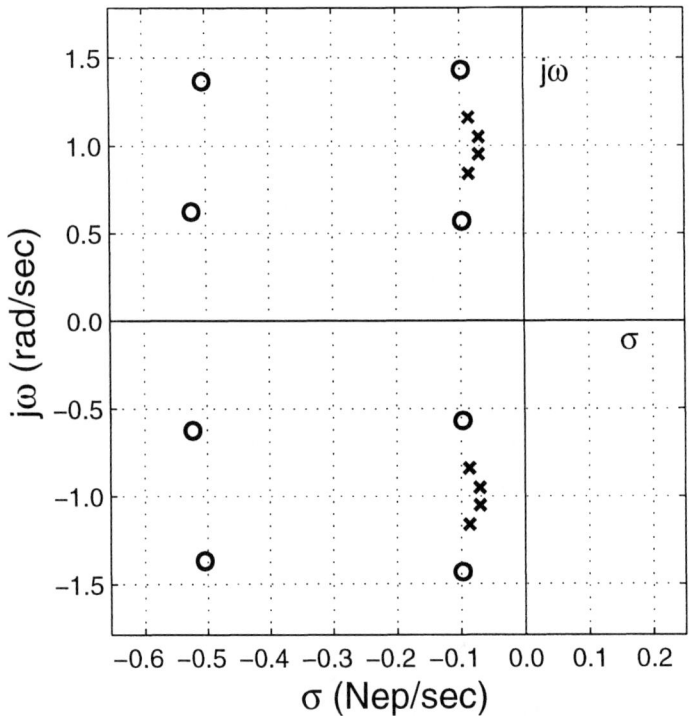

Figure 8.42. Pole-zero pattern for Example 10, nplus = 3, Norder = 8.

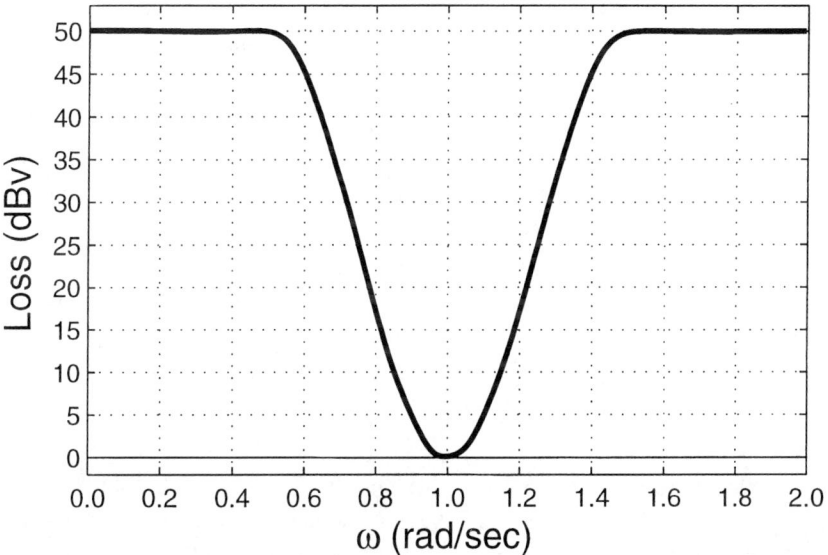

Figure 8.43. Expanded plot versus ω for the BP loss function of Example 10.

constraint is that the loss must approach a constant value for both large ($\omega \to \infty$) and small ($\omega \to 0$) frequencies.

8.3 Problems

In the following problems, loss specifications are loosely meant to be in terms of relative loss unless otherwise specified. Flat loss is therefore ignored in most problem definitions.

8.1 Determine the order of a Butterworth polynomial so that the following requirements will be met. Also specify an appropriate frequency scale factor so the prototype design will meet the requirements. The requirements are that the passband should have less than 2-dB loss for all frequencies less than 3 kHz, and the stopband should have more than 30-dB loss for all frequencies greater than 9 kHz.

8.2 Find values for ϵ and n so that a Chebyshev filter will provide less than 1 dB of ripple in the passband and more than 70 dB of loss for all frequencies greater than 2.5 times the passband cutoff frequency.

8.3 A customer wants an active Chebyshev LP filter to provide relative losses of less than 1 dB for all frequencies less than 3 kHz and more than 66 dB for all frequencies greater than 9 kHz. Explain why you might try to get the stopband specification lowered by 1 or 2 dB.

8.4 An LP filter has the following requirements: the passband loss should be less than 1 dB for all frequencies less than 3 kHz, and the stopband loss should be greater than 60 dB for all frequencies greater than 5 kHz.

(a) What minimum-order Butterworth function is required to meet the requirement? Find the poles, and specify a suitable prototype for the filter design. Use a computer to confirm the design, and plot the network function frequency response.

(b) Repeat part a using a Chebyshev function. Specify the maximum ϵ parameter value and the minimum order to meet the specified requirements.

(c) Invent your own $F^2(\omega)$ function to meet the requirement. Specify the resulting prototype network function, and check the response with a Matlab calculation and plot. Hopefully you are able to use fewer poles through the inclusion of finite zeros in your network function.

8.5 A customer wants a five-pole Chebyshev filter to give at least 40 dB of attenuation at $\omega = 1.75$, while having the least possible ripple in the passband. The customer is firm on the five-pole requirement. What value of passband ripple (in dB) would you recommend using for the prototype? Explain your answer.

8.6 Consider the definition of Chebyshev polynomials, $V_n(x) = \cos(n\phi)$, with $\cos(\phi) = x$, for $x > 1$. What value of x yields $V_{10}(x) = 100$? *Hint*: Let the real variable $u = jn\phi$, therefore, $j\phi = u/n$, and go hyperbolic, or you can apply other polynomial properties.

8.7 The following LP prototype function has been selected to meet specified filter performance specifications for some application.

$$|T(j\omega)|^2 = \frac{1}{1 + 25F^2(\omega)}$$
$$\text{where} \quad F(\omega) = \frac{(0.5 - \omega^2)(0.1 - \omega^2)}{(1 - 0.5\omega^2)(1 - 0.1\omega^2)}$$

(a) Find the implied $T(s) \cdot T(-s)$ required for the prototype filter and sketch $|T|^2$ in dB versus ω. How many poles and zeros will the final network function have? You are not required to factor the polynomials.

(b) What are the apparent minimum stopband and maximum passband attenuations for this filter function?

8.8 The following functions are used to define a filter prototype requirement:

$$|T(j\omega)|^2 = e^{-\omega^2}$$
$$= \frac{1}{1 + F^2(\omega)}$$
$$\text{where} \quad F^2(\omega) = \sum_{k=1}^{5} a_k \omega^{2k}$$

8.3. PROBLEMS

(a) Choose a_k for the given function, $F^2(\omega)$, so as to make a best fit to the specified network function over the range $-3 < \omega < +3$ rad/sec. Use any method you know or can invent to find these coefficients. What property does $F^2(\omega)$ have to satisfy for all ω values? This requirement will limit or constrain choices for a_k.

(b) Use the result of part a, choosing all lhp poles, and find the network function, $T(s)$. Specify $T(s)$ as a ratio of polynomials, and list the poles and zeros. Use Matlab to check the response of $T(s)$ against the specified function in both magnitude and dB plots versus ω. Plot an error curve versus ω, and show what happens for frequencies greater than $+3$ rad/sec. In what sense is the five-pole approximation adequate to fit this $T(s)$ requirement?

8.9 An LP filter is required to have a maximum passband loss of 0.5 dB and a minimum stopband loss of 50 dB. The passband cutoff frequency is required to be 3 kHz, while the stopband is required to start at 6 kHz.

(a) Specify the order, n, required in order to meet the LP filter specifications with a Butterworth filter. Find the network function, and use Matlab to check the result. What scale factor would be required in order to meet the frequency specification?

(b) Specify parameters ϵ and n to meet the LP filter specifications with a Chebyshev filter. Find the network function, and use Matlab to check the result.

(c) Obtain a state variable design for the Chebyshev filter of part b, and scale the resulting design to meet the required frequency specifications. Check your result using μCAP.

8.10 The following requirements are given to determine an LP filter design:

(1) *Passband loss*: ≤ 0.5 dB for $0 < f < 3$ kHz
(2) *Stopband loss*: ≥ 60 dB for 10 kHz $< f$

(a) Determine Chebyshev polynomial parameters to meet the given requirements.

(b) Specify the prototype network function in both polynomial and factored (quadratic) forms, and determine the number of opamps required for a state variable solution versus a cascade of biquad circuits.

(c) Specify a suitable frequency scale factor for scaling the prototype to the required application.

8.11 The following filter magnitude function is given for all $\omega > 0$:

$$|T(j\omega)|^2 = 10^{-4} + 2(\omega - 0.5)u(\omega - 0.5) - 2(\omega - 1)u(\omega - 1)$$
$$-(\omega - 2)u(\omega - 2) + (\omega - 2.999)u(\omega - 2.999)$$

Use any method to obtain a filter function, $T(s)$, whose squared magnitude for $s = j\omega$ fits the given function as closely as possible for all $\omega > 0$.

Try to fit with small error in the passband and asymptotically in the stopband. You can allow up to 3-dB error at the breakpoints (corners) of the function, but the bidder who gets as close as possible gets the contract. Specify your $T(s)$ as a ratio of polynomials, and list the poles and zeros for your result. Use the computer to check your answer against the specified function in both straight magnitude and decibel plots versus ω.

8.12 Use the given fifth-order inverted Chebyshev network function to answer the following questions:

$$|T(j\omega)|^2 = \frac{1}{1 + h^2/V_5^2(1/\omega)}$$

(a) Sketch the behavior of $|T(j\omega)|^2$ versus ω, and specify the parameter h so that the filter will have a minimum stopband attenuation of 60 dB.

(b) Sketch the behavior of $|T(j\omega)|^2$ versus ω, and determine the passband frequency where the attenuation reaches 0.5 dB when $h = 100$.

8.13 Specify n and ϵ for a Chebyshev filter so as to obtain a 1 dB passband for all $f < 16$ kHz and a minimum attenuation of 80 dB for all $f > 40$ kHz.

8.14 The following requirements are given for an LP filter design. The maximum passband attenuation should be less than 0.5 dB for all frequencies less than 1 kHz; the minimum stopband attenuation should be greater than 45 dB for all frequencies greater than 2 kHz (i.e., a 45 dB to 0.5 dB shape factor of 2 is required for the prototype design).

(a) Design a prototype state variable Butterworth LP filter to meet the given requirements.

(b) Scale the prototype found in part a so that the circuit resistors will have values greater than 1 kΩ and the response function will meet frequency requirements. Use μCAP to check the design performance for all frequencies out to 10 MHz.

8.15 Repeat Problem 8.14 using Chebyshev polynomials.

8.16 Use the inverted Chebyshev network function to find a suitable LP filter transfer function so that passband loss will be less than 1 dB for frequencies less than 1 kHz, and stopband loss will be greater than or equal to 36 dB for all frequencies greater than 2.5 kHz. Specify the prototype network function, its roots, and the frequency scale factor required to meet the specifications. Use Matlab to check your results.

8.17 An LP filter requires maximum passband loss of 0.5 dB for frequencies less than 100 Hz and minimum stopband loss of 30 dB for frequencies greater than 150 Hz.

(a) Find the number of poles and their location in the s-plane for a Butterworth filter to realize the given requirements.

8.3. PROBLEMS

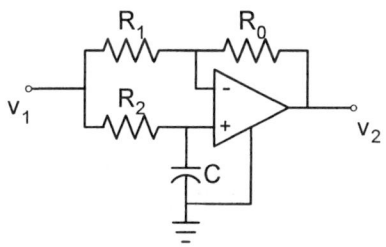

Figure 8.44. Circuit for Problem 8.18.

(b) Confirm the prototype design using Matlab to check the transfer function versus frequency.

(c) Design an active filter to obtain the specified response and check the results using μCAP. Be sure to redistribute gain so that clipping will occur at the output stage before other stages begin to saturate or cut off.

8.18 Analyze the circuit in Figure 8.44 and find the transfer function, $\mathbf{V}_2/\mathbf{V}_1$, in terms of parameters $R_{0,1,2}$ and C. Specify prototype values for the parameters so as to obtain the all-pass function $(-s+1)/(s+1)$, thus obtaining a useful iterative stage for the Fourier series solution to the approximation problem.

8.19 The following wk-ak breakpoint sets describe desired frequency response magnitude functions in Nepers versus frequency, ω, in accordance with terminology developed in the text. For each specified response, find a set of truncated and weighted Fourier coefficients to obtain a candidate solution to the approximation problem. Increase the number of Fourier coefficients used to solve the problem until passband errors of less than 1 dB are obtained for the answer. Plot your result against the transformed frequency variable, $x = \phi/\pi$, compared to the specified function in dBv relative loss. Also, create and plot the error between the two magnitude curves (in decibels).

(a) wk = [0.1 0.5 0.8 1.2 1.5 2.0] rad/sec
         ak = [1.0 0.5 0.0 0.0 6.0 5.0] Nep

(b) wk = linspace(0.5, 1.5, 51)
         ak = $3.5(1 + \cos(6\pi(wk - 0.5)))$

(c) wk = [0.75 1.5]
         ak = [0 6]

(d) wk = [0.70 0.85 1.15 1.30]
         ak = [6 0 0 7]

(e) wk = [0.70 0.85 1.15 1.30]
         ak = [7 0 0 6]

8.20 Repeat Problem 8.19 by converting the a_k values to db_k values and using `apprx8` to find a polynomial ratio to fit the specified magnitude functions. Increase the polynomial order until the passband error in your answer is less than 1 dB. Specify the polynomials and the roots that you obtain for each answer.

8.21 For the circuit in Figure 8.25, verify by circuit analysis why the opamp configuration provides for either $A_k < 0$ or $A_k > 0$, as stated in the text.

8.22 In Example 4, for the design of an octave BP filter, repeat the design by moving the zeros in $F(\omega)$ of (8.28) closer together and allow ϵ to set the maximum passband loss as discussed in the example. Use Matlab to check your answer and iterate to get a best response.

8.23 Repeat Problem 8.22 by squaring the $F(\omega)$ given in (8.28) as discussed in the text.

References

[1] Guillemin, E. A., *Synthesis of Passive Networks*, New York, NY: Wiley, 1957.

[2] Lanczos, C., *Applied Analysis*, Upper Saddle River, NJ: Prentice Hall, 1956.

[3] Irons, F. H., and Gilbert, M. J., "A New Formulation of the Approximation Problem," *IEEE Trans on Circuits and Systems*, Vol. CAS-24, No. 5, May 1977.

Appendix 8A Approximation Problem Details

This appendix presents a plausibility derivation for the Hilbert transform used in this chapter and a complete code listing for the approximation problem, `apprx8`.

8A.1 Derivation of the Hilbert Transform

The Hilbert transform, (8.40), was used without derivation to find the phase lag, β, from a given loss, α, in the following network function representation:

$$\begin{aligned} H(j\omega) &= e^{-(\alpha+j\beta)} = |H|\angle H \\ \text{where} \quad \alpha(\omega) &= -log_e|H(j\omega)| \text{ Nep} \\ \beta(\omega) &= -\angle H(j\omega) \text{ rad} \end{aligned} \quad (8A.1)$$

$\gamma = \alpha + j\beta$ is the network propagation constant, and it is a complex number function of frequency ω with real part α and imaginary part β. Generally, in filter design, the loss, $\alpha(\omega)$, is specified, and the phase lag, $\beta(\omega)$, is unknown or unspecified. In order to find a ratio of polynomials to best fit the specified loss function, it was found useful to use the phase-lag function in order to develop an initial set of

APPENDIX 8A. APPROXIMATION PROBLEM DETAILS

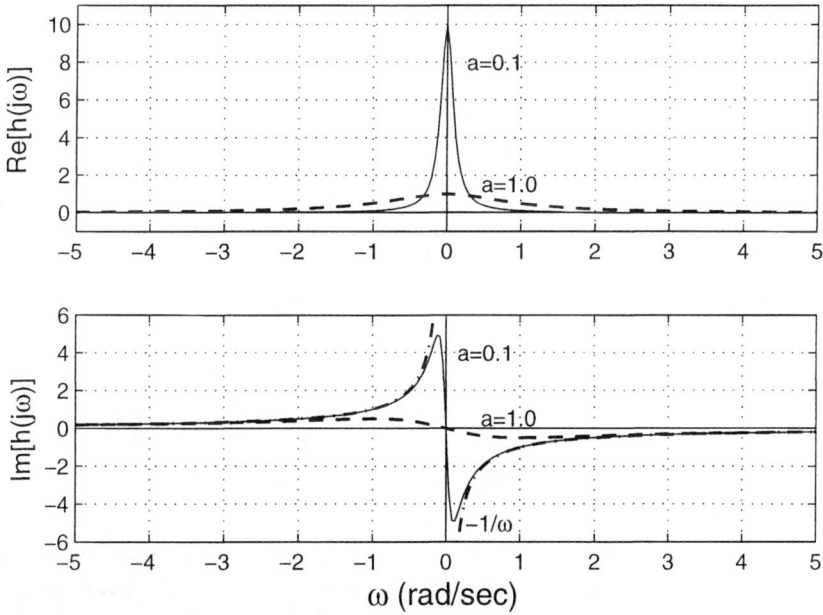

Figure 8A.1. Real and imaginary parts for a one-pole network function.

polynomials derived from the specified function, $\alpha(\omega)$. This required the use of the Hilbert transform to obtain an estimate of β from α.

A straightforward way to obtain this transform is to consider the imaginary part associated with the real part of a single pole as it is allowed to coincide with the origin. The result obtained from this limiting procedure yields an impulse for the real part associated (or paired) with $-j/\omega$ for the imaginary part. Thus, any given real function of ω can be broken into a series of differential areas versus frequency, and each associated imaginary part is then summed to yield a function for the imaginary part associated with the given real part. The procedure goes as follows. Let $h(j\omega)$ represent a network function with a single pole located at $s = -a$. The real and imaginary parts of $h(j\omega)$ are given in (8A.2) and plotted in Figure 8A.1 for $a = 1$ and 0.1.

$$\begin{aligned} h(j\omega) &= \frac{1}{(a + j\omega)} = \frac{(a - j\omega)}{(a^2 + \omega^2)} \\ &= \frac{a}{a^2 + \omega^2} - \frac{j\omega}{a^2 + \omega^2} \end{aligned} \quad (8A.2)$$

Inspection of (8A.2) shows that the imaginary part of $h(j\omega)$ appears clearly to approach $-1/\omega$ as $a \to 0$. The complete real-part behavior is not obvious until it is determined that the area under its curve is a constant. It is clear in (8A.2) that as $\omega \to 0$, the value of the real part approaches $1/a$. Thus, as $a \to 0$, the real part tends toward ∞ as $\omega \to 0$.

The area under the real part is found as follows:

$$\begin{aligned} A &= \int_{-\infty}^{\infty} d\omega \frac{a}{a^2 + \omega^2} \\ &= \lim_{\Omega \to \infty} \left[\tan^{-1}(\omega/a) \right]_{-\Omega}^{\Omega} = \frac{\pi}{2} - \frac{-\pi}{2} = \pi \end{aligned} \qquad (8A.3)$$

The area, equal to π, is found to be independent of the value of the parameter, a. As $a \to 0$, the real part of $h(j\omega)$ goes to a narrow pulse of infinite amplitude with area equal to π. Thus, it is an impulse of value π. The result is summarized in (8A.4):

$$\begin{aligned} h(j\omega) &= \lim_{a \to 0} \frac{1}{(a + j\omega)} \\ &= \pi \delta(\omega) - \frac{j}{\omega} \end{aligned} \qquad (8A.4)$$

This last result forms the basis for the Hilbert transform used to find the imaginary part associated with a given real function of ω for the network propagation constant, $\gamma = \alpha + j\beta$.

Equation (8A.4) tells us that a unit impulse located at $\omega = 0$ yields an associated imaginary part, $-1/(\pi\omega)$. We then deduce the following results through the use of properties of proportionality and frequency translation:

$$\begin{aligned} \mathcal{R}eal\,Part &\to \mathcal{I}mag\,Part \\ 1\delta(\omega) &\to \frac{-j}{\pi\omega} \\ k\delta(\omega) &\to \frac{-jk}{\pi\omega} \\ k\delta(\omega - \xi) &\to \frac{-jk}{\pi(\omega - \xi)} \end{aligned} \qquad (8A.5)$$

Letting the proportionality, k, be the differential area, $\alpha(\xi)d\xi$, at $\omega = \xi$, for a given real function, $\alpha(\omega)$, we get the following associated differential imaginary part:

$$\begin{aligned} d\alpha &\to jd\beta \text{ so that} \\ d\alpha = \alpha(\xi)d\xi\, \delta(\omega - \xi) &\to jd\beta = \frac{-j\alpha(\xi)d\xi}{\pi(\omega - \xi)} \text{ for all } \xi \end{aligned} \qquad (8A.6)$$

An interpretation of (8A.6) is shown in Figure 8A.2.
Integration over all ξ then yields both α and β.

$$\alpha(\omega) = \int_{-\infty}^{\infty} d\xi \alpha(\xi) \delta(\omega - \xi) \qquad (8A.7)$$

$$\beta(\omega) = -\int_{-\infty}^{\infty} \frac{d\xi \alpha(\xi)}{\pi(\omega - \xi)} \qquad (8A.8)$$

APPENDIX 8A. APPROXIMATION PROBLEM DETAILS

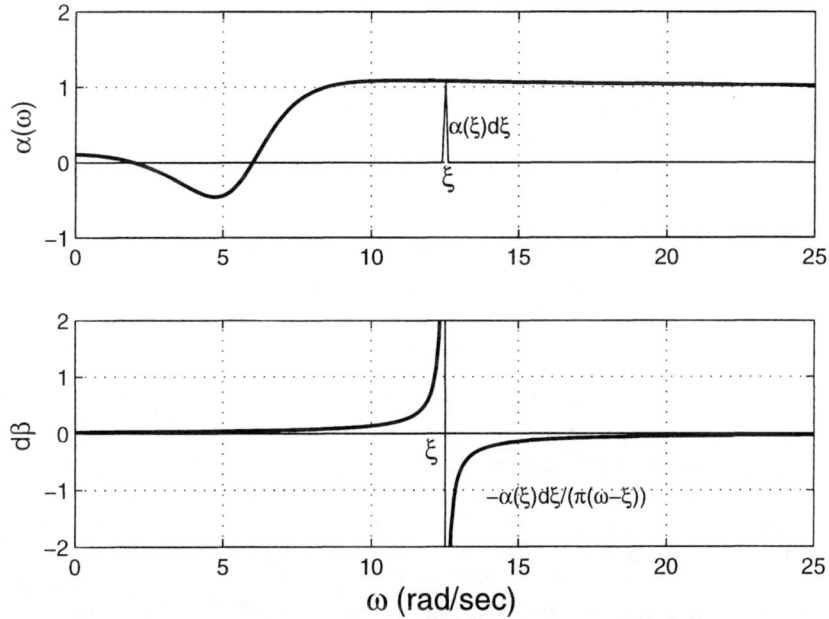

Figure 8A.2. The differential area of α generates a differential component of β.

The integral relationship for $\alpha(\omega)$, (8A.7), is sometimes referred to as a scanning operation, or integral. The result for β in (8A.8) agrees with (8.40). It shows how the associated phase, β, is related to a specified loss, α, as a function of frequency.

8A.2 Program Description

The program `apprx8` was used for Examples 9 and 10 in this chapter. This appendix provides a complete listing for the working program, along with all of its function subroutines. The key concepts and steps for this program have already been explained in this chapter and will not be repeated here. Procedures not previously explained will be discussed briefly in this appendix.

The complete program uses the following routines:

  `appx_mgr.m` - routine that sets up parameters for a specific problem and calls `apprx8` to find an answer.

  `apprx8` - the main function used to fit a ratio of polynomials to a specified loss function. It takes a set of loss values versus ω and a specified excess order above the variable `nmin` for the solution it is to find. It returns the polynomial coefficients, plus a constant multiplier, for the proposed solution.

hurw2 - the first subfunction of `apprx8`. It takes the input to `apprx8`, plus a set of objective frequency values, and returns a first-choice network function description to match the specified network function. The return network function is described in terms of the polynomial Hurwitz parameters plus a constant multiplier. This is the longest routine in the program, and its operation (phase separation and interpolation of the Hurwitz roots) has been described in this chapter.

pg - routine that converts the polynomial parameters from Hurwitz form to standard polynomial coefficients. The requirement for augmentation by adding a zero was discussed in this chapter for the difference in form based upon even- versus odd-order polynomials. The code is self-explanatory.

obj2 - routine that evaluates the network function, using Hurwitz polynomial parameters, over the set of objective frequency values and determines the squared error sum for the given vector of parameters in their fit to the specified loss function (in dBv). This function is called by the Matlab function `fminsearch`, which optimizes the parameters initially chosen by `hurw2`. The evaluation is carried out as follows.

The program is given the computed loss function, alpha (in dBv), versus the set of ω-values and the current parameter vector **x**. **x** consists of the network function parameters in Hurwitz form and has the following form:

$$\mathbf{x} = [\omega_{p_1}^2 \ \omega_{p_2}^2 \ \ldots \ \omega_{p_{n-1}}^2 \ \omega_{q_1}^2 \ \omega_{q_2}^2 \ \ldots \ \omega_{q_{n-1}}^2 \ Kp \ Kq \ kH]$$

where $\omega_{p_k}^2$ and $\omega_{q_k}^2$ are the squared Hurwitz roots for $P(s)$ and $Q(s)$, respectively. Kp and Kq are the odd-part multipliers for each polynomial, and kH is the constant multiplier for the ratio of polynomials. $H(j\omega)$ has the following form:

$$H(j\omega) = kH \frac{(\omega_{p_1}^2 - \omega^2)(\omega_{p_3}^2 - \omega^2)\ldots + jKp\omega(\omega_{p_2}^2 - \omega^2)(\omega_{p_4}^2 - \omega^2)\ldots}{(\omega_{q_1}^2 - \omega^2)(\omega_{q_3}^2 - \omega^2)\ldots + jKq\omega(\omega_{q_2}^2 - \omega^2)(\omega_{q_4}^2 - \omega^2)\ldots}$$

The number of Hurwitz roots, $nhr = n - 1$, equals the polynomial order minus 1. The routine `obj2` calculates $H(j\omega)$ in the following form for each value of ω:

$$H(j\omega) = kH \frac{(a + jb)}{(c + jd)}$$

with the result:

$$a_{est} = -20 log_{10} |H| \ (\text{dBv})$$

and

$$E = \sum_\omega |a_{est} - \alpha|^2$$

E is the measure of fit returned to `fminsearch`.

8A.3 Code Listing

This section provides a complete listing for all the programs used to run the polynomial solution to the approximation problem as developed in this chapter. The program is designed to run in Matlab and uses the Matlab tool fminsearch.

appx_mgr.m - This program is used to run the routine apprx8. It basically sets up the filter specifications and plots comparative results to show the fit to the specification.

```
% program to run apprx8.m
% appx_mgr.m  Feb 25, 2003
% modify Aug 22,2003

npts=51;

% cosine LP Example 9 in Chapter 8
%nplus=1;
%wk=linspace(0.1,1.1,npts);
%dbk=25*(1-cos(pi*(wk-0.1)));   % loss in dBv

% cosine BP Example 10 in Chapter 8
nplus=3;
wk=linspace(0.5,1.5,npts);
dbk=25*(1-cos(2*pi*(wk-1.0)));   % loss in dBv

[P,Q,kH]=apprx8(wk,dbk,nplus);
Norder=length(P)-1;

w=linspace(0,max(wk)*2,251);   % response frequencies
jw=j*w;
% calculate attenuation spec over response frequencies
nwk=length(wk);
alpha=dbk(1)*(w<=wk(1))+dbk(nwk)*(w>wk(nwk));  % dBv
for k=1:nwk-1
   a1=dbk(k);
   a2=dbk(k+1);
   w1=wk(k);
   w2=wk(k+1);
   alpha=alpha+(a1+(a2-a1)*(w-w1)/(w2-w1)).*((w>w1)-(w>w2));
end

aopt=-20*log10(abs(kH*polyval(P,jw)./polyval(Q,jw)));
eopt=abs(aopt-alpha);

figure(1)
hold off
```

```
subplot(211),plot(w,alpha,'k',w,aopt,'k--'),grid
wu=2*max(wk);
axis([0 wu -5 55])
legend('Spec','Est_{opt}')
ylabel('Loss (dBv)','Fontsize',14)
subplot(212),plot(w,eopt),grid
em=max(eopt);
axis([0 wu -0.1 1.25*em])
ylabel('|Error| (dB)','Fontsize',14)
xlabel('\omega (rad/sec)','Fontsize',14)

figure(2)
hold off
zros=roots(P);
pols=roots(Q);
subplot(211)
plot(real(zros),imag(zros),'o',...
    real(pols),imag(pols),'x'),grid
sl=max(abs(real([pols;zros])));
wu=max(abs(imag([pols;zros])));
axis([-1.25*sl 0.25 1.25*[-wu wu]])
axis('square')
ylabel('j\omega (rad/sec)','Fontsize',14)
xlabel('\sigma (nep/sec)','Fontsize',14)
```

apprx8 - The following is a complete listing of all the routines (except fminsearch) that were developed and used to solve the approximation problem as described in this chapter:

```
function [P,Q,kH]=apprx8(wk,dbk,np)
%  function [P,Q,kH]=apprx8(wk,dbk,np)
% General purpose routine to find network function ratio
% to fit a piece-wise straight-line connected loss function
% Input:
% wk = set of frequency points normalized around w=1 rad/sec
% dbk = set of loss values at wk frequencies (dBv)
%  np = desired excess above minimum order
% P and Q are final polynomial coefficient vectors
% The network function is kH*P(s)/Q(s)
%  Feb 24, 2003

ak=dbk/8.6859;   % ak=set of loss points in Nepers
nplus=np;
nk=length(wk);
if (length(ak) ~= nk)
    error('length(ak) .neq. length(wk)')
end
```

APPENDIX 8A. APPROXIMATION PROBLEM DETAILS

```
nw=501; % no. of frequencies for objective function
% nw can be increased for complex loss functions
x=linspace(0,1-1e-4,nw); % sets range, 0<=w<6.4e3, phi=pi*x
w=tan(pi*x/2);  % objective optimizing frequency set

% calculate attenuation spec over optimizing frequencies
alfa=ak(1)*(w<=wk(1))+ak(nk)*(w>wk(nk)); % Nepers
for k=1:nk-1
   a1=ak(k);
   a2=ak(k+1);
   w1=wk(k);
   w2=wk(k+1);
   alfa=alfa+(a1+(a2-a1)*(w-w1)/(w2-w1)).*((w>w1)-(w>w2));
end
alpha=8.6859*alfa; % dBv

% get initial start from min phase in Hurwitz form
jw=j*w;
[whp,whq,Kp0,Kq0,kH0,Norder]=hurw2(w,wk,ak,nplus);

% now optimize solution
w2=w.^2;
x0=[whp.^2 whq.^2 Kp0 Kq0 kH0];   % initial parameter vector
%y=foptions;
%y(14)=5000;
%xopt=fmins('obj2',x0,y,[],alpha,w2); % Matlab code before Version 7
options=optimset('Maxiter',5000,'Maxfunevals',5000);
xopt=fminsearch('obj2',x0,options,alpha,w2); % Matlab code after Version 7
n0=length(x0);
nhr=(n0-3)/2;
xm=[sort(xopt(1:nhr)) sort(xopt(nhr+1:2*nhr))];
Kp=xopt(n0-2);
Kq=xopt(n0-1);
kH=xopt(n0);
[P,Q]=pq(Norder,sqrt(xm(1:nhr)),sqrt(xm(nhr+1:2*nhr)),Kp,Kq);
return

function [whp, whq, Kp, Kq, kH, Norder] = hurw2(w,wk,ak,nplus)
% This subroutine returns the Hurwitz roots, the constant multipliers
% Kp and Kq and a constant kH to normalize polynomials to fit the
% desired loss.  Interpolation is used to refine the estimates for
% the Hurwitz roots.  Jan 2, 2003

nw=length(w);
p_1=0;
nk=length(wk);
```

```
for k=1:(nk-1)
p1(k)=(ak(k+1)-ak(k))./(wk(k+1)-wk(k));   % First derivative
p2(k)=p1(k)-p_1;                           % Second derivative
p_1=p1(k);
end
p1(nk)=0;
p2(nk)=-p1(nk-1);
beta=zeros(1,length(w));
for k=1:nk
  uk=wk(k);
  beta=beta+p2(k)*((w-uk).*log(abs(w-uk)+eps)...
     +(w+uk).*log(abs(w+uk+eps)));
end
beta=-beta/pi;

BP=zeros(1,nw);                 % initializing BP
BQ=zeros(1,nw);                 % and BQ
for k=2:nw
  BETA=beta(k)-beta(k-1);       % Is beta increasing or decreasing?
 if BETA>0                      % If it is increasing, increment BQ
  BQ(k)=BETA+BQ(k-1);           % by the amount of the increase
  BP(k)=BP(k-1);
    else
  BP(k)=BP(k-1)-BETA;           % If it is decreasing, increment BP
  BQ(k)=BQ(k-1);                % by the amount of the decrease
    end
end
nmin=max(ceil(2*[max(BP) max(BQ)]/pi));   %minimum order
Norder=nmin+nplus        %this is desired order for polynomials P & Q

SEP=(BQ+BP);
% x=2*atan(w)/pi; % Jan 22, 2003
% SEP=integral(x,beta);
xm=(pi*Norder/2-max([max(BP) max(BQ)]))/SEP(nw);
SEP=xm*SEP;
bp=BP+SEP;
bq=BQ+SEP;
tq=2*bq(nw)/pi;
tp=2*bp(nw)/pi;
if (tp < Norder-1 | tq < Norder-1)
   error('Objective frequency range is too small')
end
for k=1:Norder-1
  [mq(k),ibq(k)]=min(abs((2*bq/pi)-k*ones(1,nw)));
  [mp(k),ibp(k)]=min(abs((2*bp/pi)-k*ones(1,nw)));
end
```

APPENDIX 8A. APPROXIMATION PROBLEM DETAILS

```
% whq=w(ibq); % 1st order estimated Hurwitz roots for Q(s)
% whp=w(ibp); % 1st order estimated Hurwitz roots for P(s)
% The Hurwitz roots need to be "fine tuned."
% Better estimates are found by linear interpolation
% on the discrete frequency set of values
for k=1:Norder-1
    pik=pi*k/2;
    Iqk=ibq(k);
    Ipk=ibp(k);
    if ((bp(Ipk)-pik)<0)
        w2=w(Ipk+1);
        w1=w(Ipk);
        p2=bp(Ipk+1);
        p1=bp(Ipk);
    else
        w2=w(Ipk);
        w1=w(Ipk-1);
        p2=bp(Ipk);
        p1=bp(Ipk-1);
    end
    whp(k)=w1+(w2-w1)*(pik-p1)/(p2-p1);
    if ((bq(Iqk)-pik)<0)
        w2=w(Iqk+1);
        w1=w(Iqk);
        p2=bq(Iqk+1);
        p1=bq(Iqk);
    else
        w2=w(Iqk);
        w1=w(Iqk-1);
        p2=bq(Iqk);
        p1=bq(Iqk-1);
    end
    whq(k)=w1+(w2-w1)*(pik-p1)/(p2-p1);
end
Kp=(bp(2)-bp(1))/(w(2)-w(1)); % Kp and Kq are initially set
Kq=(bq(2)-bq(1))/(w(2)-w(1)); % to match slopes at the origin
Pk=Kp;
Qk=Kq;
for k=2:2:Norder-1
      Pk=Pk/whp(k)^2;
      Qk=Qk/whq(k)^2;
end
for k=1:2:Norder-1
      Pk=Pk*whp(k)^2;
      Qk=Qk*whq(k)^2;
end
```

```
Kp=Pk;
Kq=Qk;
kH=exp(-ak(1));
for l=1:2:length(whq)
  kH=kH*(whq(l)^2)/(whp(l)^2);
end

function [P,Q] = pq(Norder,whp,whq,Kp,Kq)
% This routine converts the polynomial data from
% Hurwitz form to standard polynomial coefficients
if rem(Norder,2)==1 % Is Norder odd?
        Peven=[0 1];
        Qeven=[0 1];
        Podd=Kp*[1 0];
        Qodd=Kq*[1 0];
   else
        Peven=[1];   % Norder is even
        Qeven=[1];
        Podd=Kp*[0 1 0];
        Qodd=Kq*[0 1 0];
end
for k=2:2:Norder-1
  Podd=conv(Podd,[1 0 whp(k)^2]);
  Qodd=conv(Qodd,[1 0 whq(k)^2]);
end
for k=1:2:Norder-1
  Peven=conv(Peven,[1 0 whp(k)^2]);
  Qeven=conv(Qeven,[1 0 whq(k)^2]);
end
  P=Peven+Podd;
  Q=Qeven+Qodd;
return

function e = obj2(x,alpha,w2)
% e = obj2(x,alpha,w2)
% fits to a given attenuation function (alpha) vs w (jw = j*w)
% using the Hurwitz variables to find the network function polynomials
% format x = [whp2(1:nhr) whq2(1:nhr) Kp Kq kH]
% whp2, whq2 are squared Hurwitz roots for P(S) & Q(S) respectively
% Kp, Kq are the odd part multipliers and
% kH is the constant multiplier for the network function,
% i.e., H(S)=kH*P(S)/Q(S)
nx = length(x);
nhr=(nx-3)/2;
   xs=[sort(x(1:nhr)) sort(x(nhr+1:nx-3))];
     % The Hurwitz roots must be sorted to be
```

APPENDIX 8A. APPROXIMATION PROBLEM DETAILS

```
    % monotonically increasing values
norder = nhr + 1;
w=sqrt(w2);

nw = length(w2);
a = ones(1,nw);
c = ones(1,nw);
b = w*x(nx-2);
d = w*x(nx-1);

for k = 1:2:nhr
    a = a.*(xs(k)-w2);
    c = c.*(xs(nhr+k)-w2);
end
for k = 2:2:nhr
    b = b.*(xs(k)-w2);
    d = d.*(xs(nhr+k)-w2);
end
h = x(nx)*(a+j*b)./(c+j*d);
aest = -20*log10(abs(h));  % dBv
    e = sum((aest-alpha).^2);
    %e = sum(abs(aest-alpha));  % Jan 2, 2003 (not as good)
```

About the Author

Fred H. Irons received his BSEE from The Ohio State University in 1956, his MSEE from Massachusetts Institute of Technology (MIT) in 1959, and his Ph.D. from Lehigh University in 1971. Currently, he is a professor emeritus at the University of Maine, where his on-and-off teaching career spanned the years from 1967 to 2000. In 1990 he was appointed to be the first Castle Professor of Electrical and Computer Engineering, a position that he held until his retirement in 2000. His research specialty, from 1990 to 2000, was in the areas of quantization techniques and ultra high-frequency (UHF) devices. He also served on the subcommittee for analog-to-digital converters (ADCs) of the IEEE Standards Committee, TC-10, Waveform Measurements and Analysis, from 1993 to 2000. From 1977 to 1990, he worked in the Advanced Techniques Group at MIT Lincoln Laboratory, where he was involved in the design and evaluation of spread spectrum guidance and control systems and the development of phase-plane compensation techniques for the characterization of high-speed ADCs. From 1962 to 1967, he was chief engineer for Guillemin Networks, Inc., which was involved with the pioneering development of active filters for many practical applications. You may consult the University of Maine Web site at http://www.eece.maine.edu/~irons to obtain more details about Dr. Irons' publications and other career details.

Index

active filter definition, 11
all-pass stage, opamp, 389
all-pole filters, 40
all-pole function, 167
all-zeros approach, 354
analog computer component, 154
angle, discontinuity, 358
angle, monotonic increase, 359
angle, principal value, 358
approximation problem
    general, 337
    traditional, 313
appx_mgr.m, 395
arithmetic symmetry, 383
associated phase, 342
associated phase, straight-line loss, 349
asymptotes, Butterworth polynomials, 322
asymptotes, Chebyshev polynomials, 327
asymptotic representation, 170
attenuation function, 3

bandpass filter, octave, 334
bandwidth, diffamp, 95
bare bones, 377
basis, Hilbert transform, 392
Bell, 3
biasing, 58
bipolar current flow, 260
bipolar operation, 58
biquad circuit, 191
biquad, state variable, 198
biquadratic ratio, 193
Bode characterization, 169
branches, series, 223
branches, shunt, 223
breakpoints, 170
buffer, noninverting multiplying, 155
buffer, voltage follower, 154
bypass capacitors, 64

capacitance, bottom plate, 265

capacitance, top plate, 265
capacitor, CMOS, 267
capacitor, floating, 265
cascade circuit, Fourier series, 340
cascade circuits, concept, 170
cascade stage, all-pass, 340
CCCS, 81
charge per cycle, 262
circuit
    lead-lag, 277
    subtract and integrate, 276
    sum and integrate, 275
clock, two-phase, 273
CMOS switch, 260
common mode, 97
    input, 94
compensation, 190
compensation network function, 173
complex pole Q, 29
complementary follower output, 113
conductance unit, Siemens, 65
current controlled opamp, 124
current mirror, 82
current sink, 82
current, diffamp output, 88
curve, tangent function, 360
cutoff frequency, f_α, 130
cutoff frequency, f_β, 130

Darlington transistor, 133
dB, 3
dead zone, 114
definition, active filter, 9
delay factor, τ, 314
delta function, 344
describing function, 158
difference equation, 288
differential amplifier stage, 87
differential mode, 97
    input, 95

diode offset bias, 115
discrimination ratio, 261
distortion, s-plane, 287
double-pole response, 67
dual, double terminated, 47
dynamic circuit, 143
dynamic transconductance, 101

Early
    effect, 143
    voltage, 102
effective bias current, 118
element value ratios, 159
equiripple loss, 324
equivalence, BP parallel LC, 37

feedback
    frequency effects, 57
    negative, 65
feedforward circuit, 237
filter
    bandpass, 7
    bandstop, 8
    causal, 314
    design factors, 11
    highpass, 7
    ideal, 313
    lowpass, 6
    terminology, 2
filter Q, 28
first derivative of loss, 347
first-cut phase, 365
flat loss, 10
flip-flop circuit, 272
flow diagram, signal, 158
flow graph, 154
forcing function, 158
Fourier approximations, graphs, 352
Fourier coefficients, Matlab code, 349
Fourier series, 338
Fourier transform, inverse, 313
fractional discharge, 264
frequency response, octave BP, 337
frequency scaling, 21
frequency transform, periodic, 338
frequency transformation, 24

gain
    closed-loop, 65
    frequency-dependent, 57
    open-loop, 59
gain bandwidth product, 66
gain distribution, 159
gain equalization, 161
geometrical mean, 32
Gibb's phenomenon, 344
group delay, 318

high-gain circuits, 81
Hilbert transform, 342
    derivation, 390
Hurwitz form, 359
Hurwitz roots, 359
hybrid models, 93
hybrid-π model, 143

ideal transformer, 41
idle-power ratio, 110
image, s-plane, 318
impedance scaling, 20
impedance transforms, 40
insertion loss, 5
integrator, inverting, 157, 274
intermediate variable, ϕ, 325
intermodulation, 157
interpolation geometry, 370
invariant bandwidth response, 74
isolation, 154

ladder analysis, 188
ladder network
    structure, 223
    terminology, 223
latch-up, 65
leapfrog, 41
    filters, 223
linear region, 64
listing, **apprx8**, 395
loop parameter matrices, 15
loop scaled equivalent, 47
loss, 3
    flat, 6
    insertion, 5
    parasitic, 9

mapping
    LP to BP, 28
    LP to BS, 39
    periodic transform, 338
matched filters, 381

INDEX

Miller effect, 108
minimum phase functions, 366
mirror gain factor, α_M, 82
mirror, EFA, 83
modeling dynamic response, 143
multiplier
    V_{be}, 129
    inverting summing, 156
    noninverting summing, 155

network function, 3
    general form, 333
    LP to BP, 31
network propagation constant, 390
node parameter matrices, 16

off-chip power consumption, 111
offset voltage, 64
opamp, 58
    design considerations, 81
    model, dependent source, 153
    model, simple, 153
    three-stage, 119
operational amplifier, dual, 72
operational circuit, 158
operator, design method, 153
optimized results, graphs, 377
optimizing polynomials, 374

parallel subnetworks, 188
parameter matrices, 15
parameter variance, 219
parasitic capacitance, switches, 264
parasitic response, 69
partial fraction expansion, 35
partial sum, 344
periodic error component, 345
phase lead, 318
phasor representation, 3
pole Q, 184
polynomial parameter vector, 374
polynomials
    Butterworth, 319
    Chebyshev, 324
    Hurwitz, 359
    initial choice, 374
    inverted Chebyshev, 330
positive real coefficients, 359
power ratio, 5
precompensation, 294

principal period, 338
procedure, systematic, 367
propagation constant, 342
prototype, 19

quadratic factor, Bode form, 183
quadratic factor, single amplifier, 183

recipe, initial network function, 367
recursion, Chebyshev, 324
recursion, squared Chebyshev, 325
renegotiate, 330
resistance, equivalent, 267
resonant branches
    Type I, 236
    Type II, 240
response
    closed-loop, 66
    inverted Chebyshev, 331

Sallen and Key circuit, 183
scaling operations, 19
scaling, internal loop, 41
scanning integral, 393
second derivative of loss, 347
second-cut gain distribution, 168
sensitivity
    component, 9
    definition, 217
    network, 217
    network function, 219
    performance, 231
    terminated network, 222
separated curves, 366
separation function, Ψ, 366
shape factor, 313
signal mode, definitions, 97
signal processor, 154
single-pole response, 66
small-signal response, 93
source transformation, 229
specmanship, 111
starting function, 355
state variables, 166
static circuit, 143
static transfer function, 106
statistical records, 218
straight-line connected segments, 346
switch
    bbm, 264

DPDT, 264
inverting DPDT, 265
mbb, 264
SPDT, 264
switched capacitor, 259
j-axis zeros, 291
all-pole filter, 285
scale factors, 298
state variable circuit, 297
switching, symmetrical, 265

table, Chebyshev polynomials, 326
temperature compensation, 121
temperature compensation, three-stage, 123
temperature stability, 9
template
LP to BP, 30
LP to BS, 40
temptation, 222
threshold voltage, 261
topology, all-pole bandpass, 239
transconductance amplifier, 200
transform
LP to BP, 27
LP to BS, 38
LP to HP, 25
LP to LP, 24
transistor modeling, Matlab, 135
transition band, 6
transmission gate, 259
transmission zeros, 164
triple-pole response, 70
truncated series, 344
twin-T circuit, 186

unbalanced mode input, 95
unit ramp functions, 346
unstable network function, 71

VCVS, 81

weighted coefficients, 345
weighted partial sum, 345
weighting factors, Lanczos, 345
Widlar current source, 85
Widlar mirror, 84

zero-mean process, 218

Recent Titles in the Artech House Microwave Library

Active Filters for Integrated-Circuit Applications, Fred H. Irons

Advanced Techniques in RF Power Amplifier Design, Steve C. Cripps

Automated Smith Chart, Version 4.0: Software and User's Manual, Leonard M. Schwab

Behavioral Modeling of Nonlinear RF and Microwave Devices, Thomas R. Turlington

Broadband Microwave Amplifiers, Bal S. Virdee, Avtar S. Virdee, and Ben Y. Banyamin

Computer-Aided Analysis of Nonlinear Microwave Circuits, Paulo J. C. Rodrigues

Design of FET Frequency Multipliers and Harmonic Oscillators, Edmar Camargo

Design of Linear RF Outphasing Power Amplifiers, Xuejun Zhang, Lawrence E. Larson, and Peter M. Asbeck

Design of RF and Microwave Amplifiers and Oscillators, Pieter L. D. Abrie

Distortion in RF Power Amplifiers, Joel Vuolevi and Timo Rahkonen

EMPLAN: Electromagnetic Analysis of Printed Structures in Planarly Layered Media, Software and User's Manual, Noyan Kinayman and M. I. Aksun

FAST: Fast Amplifier Synthesis Tool—Software and User's Guide, Dale D. Henkes

Feedforward Linear Power Amplifiers, Nick Pothecary

Generalized Filter Design by Computer Optimization, Djuradj Budimir

High-Linearity RF Amplifier Design, Peter B. Kenington

High-Speed Circuit Board Signal Integrity, Stephen C. Thierauf

Intermodulation Distortion in Microwave and Wireless Circuits, José Carlos Pedro and Nuno Borges Carvalho

Lumped Elements for RF and Microwave Circuits, Inder Bahl

Microwave Circuit Modeling Using Electromagnetic Field Simulation, Daniel G. Swanson, Jr. and Wolfgang J. R. Hoefer

Microwave Component Mechanics, Harri Eskelinen and Pekka Eskelinen

Microwave Engineers' Handbook, Two Volumes, Theodore Saad, editor

Microwave Filters, Impedance-Matching Networks, and Coupling Structures, George L. Matthaei, Leo Young, and E.M.T. Jones

Microwave Materials and Fabrication Techniques, Third Edition, Thomas S. Laverghetta

Microwave Mixers, Second Edition, Stephen A. Maas

Microwave Radio Transmission Design Guide, Trevor Manning

Microwaves and Wireless Simplified, Thomas S. Laverghetta

Modern Microwave Circuits, Noyan Kinayman and M. I. Aksun

Neural Networks for RF and Microwave Design, Q. J. Zhang and K. C. Gupta

Nonlinear Microwave and RF Circuits, Second Edition, Stephen A. Maas

QMATCH: Lumped-Element Impedance Matching, Software and User's Guide, Pieter L. D. Abrie

Practical Analog and Digital Filter Design, Les Thede

Practical RF Circuit Design for Modern Wireless Systems, Volume I: Passive Circuits and Systems, Les Besser and Rowan Gilmore

Practical RF Circuit Design for Modern Wireless Systems, Volume II: Active Circuits and Systems, Rowan Gilmore and Les Besser

Production Testing of RF and System-on-a-Chip Devices for Wireless Communications, Keith B. Schaub and Joe Kelly

Radio Frequency Integrated Circuit Design, John Rogers and Calvin Plett

RF Design Guide: Systems, Circuits, and Equations, Peter Vizmuller

RF Measurements of Die and Packages, Scott A. Wartenberg

The RF and Microwave Circuit Design Handbook, Stephen A. Maas

RF and Microwave Coupled-Line Circuits, Rajesh Mongia, Inder Bahl, and Prakash Bhartia

RF and Microwave Oscillator Design, Michal Odyniec, editor

RF Power Amplifiers for Wireless Communications, Steve C. Cripps

RF Systems, Components, and Circuits Handbook, Ferril Losee

Stability Analysis of Nonlinear Microwave Circuits, Almudena Suárez and Raymond Quéré

TRAVIS 2.0: Transmission Line Visualization Software and User's Guide, Version 2.0, Robert G. Kaires and Barton T. Hickman

Understanding Microwave Heating Cavities, Tse V. Chow Ting Chan and Howard C. Reader

For further information on these and other Artech House titles, including previously considered out-of-print books now available through our In-Print-Forever® (IPF®) program, contact:

Artech House
685 Canton Street
Norwood, MA 02062
Phone: 781-769-9750
Fax: 781-769-6334
e-mail: artech@artechhouse.com

Artech House
46 Gillingham Street
London SW1V 1AH UK
Phone: +44 (0)20 7596-8750
Fax: +44 (0)20 7630 0166
e-mail: artech-uk@artechhouse.com

Find us on the World Wide Web at: www.artechhouse.com